Lecture Notes in Mathematics

Edited by A. Dold and B. Eckmann

950

Complex Analysis

Proceedings of the Summer School
Held at the International Centre for Theoretical Physics Trieste
July 5–30, 1980

Edited by J. Eells

Springer-Verlag
Berlin Heidelberg New York 1982

Editor
James Eells
University of Warwick, Mathematics Institute
Coventry CV4 7AL, England

AMS Subject Classifications (1980): 32-XX, 53-XX, 58-XX

ISBN 3-540-11596-X Springer-Verlag Berlin Heidelberg New York
ISBN 0-387-11596-X Springer-Verlag New York Heidelberg Berlin

2141/3140-543210

FOREWORD

A summer seminar in complex analysis took place in Trieste (Italy) during July 1980 at the International Centre for Theoretical Physics of the International Atomic Energy Agency (Vienna, Austria) and the United Nations Educational, Scientific and Cultural Organization (Paris, France), on the invitation of its Director, Professor Abdus Salam. The course was organised by the late Professor Aldo Andreotti and myself. The main lectures included the following:

M. G. Eastwood	Twistor Theory (The Penrose Transform)
L. Lemaire and J. C. Wood	Introduction to Analysis on Complex Manifolds
M. Nacinovich	Complex Analysis and Complexes of Differential Operators
M. S. Narasimhan	Deformations of Complex Structures and Holomorphic Vector Bundles
W. Stoll	Introduction to Value Distribution Theory of Meromorphic Maps
D. Sundararaman	Compact Hausdorff Transversally Holomorphic Foliations
G. Trautman	Holomorphic Vector Bundles and Yang Mills Fields

In addition to these main lecture courses there were mini-courses (3 to 6 lectures), as well as research lectures. The present volume is a collection of revised versions of some of these. They are largely self-contained; however, an accessible reference during the summer was Complex Analysis and Its Applications, edited by A. Andreotti, J. Eells, and F. Gherardelli, ICTP (1976, Volumes 1 - 3).

I would like to thank all the lecturers (named and unnamed) for their efforts during the course. I know that in turn they join me in expressing our appreciation of the truly exceptional administrative and secretarial staff of the Centre.

James Eells

CONTENTS

TWISTOR THEORY

(THE PENROSE TRANSFORM)

M.G. Eastwood
Mathematical Institute
24-29 St. Giles'
Oxford, OX1 3LB
U.K.

§0 Introduction

Where does complex mathematics intervene in our real world? [5]

Answer: Twistor Theory! [19]

Twistors were introduced by Penrose [11, 13] in order to provide an alternative description of Minkowski-space which emphasizes the light rays rather than the points of space-time. Minkowski-space constructions must be replaced by corresponding constructions in twistor-space. The twistor programme [17] has met with much success:

(1) The description of massless free fields (the Penrose transform)

(2) The description of self-dual Einstein manifolds

(3) The description of self-dual Yang-Mills fields

(4) The description of elementary particles (rather tentative).

Brief Comments: In all cases the twistor description is simpler!

(1) This is the subject of these notes

(2) The idea here is that complex deformations of appropriate portions of twistor-space give rise to curved space-times. It turns out [16] that the curved space-times which arise in this way are conformally half-flat and that, locally, all such spaces arise in this way. By preserving a little more structure when deforming the twistor space a metric can be constructed and this will automatically satisfy the Einstein vacuum equations. This non-linear graviton construction has been used by Hitchin [8] to produce asymptotically locally Euclidean space-times.

(3) This was first described by Ward in his D.Phil. thesis and in [22]. It has led to a complete solution of the equations on S^4 by Atiyah, Hitchin, Manin and Drinfeld [1]. Earlier in this conference Trautman [21] showed how this may be achieved for the gauge group SU(2) for bundles with second Chern class (topological quantum number) 1. This line of investigation is related to both (1) and (2). (1) and (3) have a special case in common namely the description of a free electromagnetic field. This is a massless field of helicity ± 1 or a Yang-Mills field for the gauge group U(1). Also (3) can be used to modify (1) so as to incorporate a background to which the massless fields are minimally coupled. (3) is related to (2) since the vector-bundles on twistor-space generating the Yang-Mills fields may be regarded as arising by deformation of the trivial bundle.

(4) The hope here is that the success of (1) may be extended to massive particles and interactions thereof. This is extremely tentative but see [15, 9, 3]. Scattering of massless particles may be described to some extent by twistor diagrams [14] and Ginsberg has succeeded in making this approach much more precise [6 (with Huggett), 7, and his D.Phil. thesis]. This success allows an investigation [4] of the relationship between using twistors and dual twistors in the description of massless fields.

The details omitted from the notes may be found in [23, 2]. Much related material together with a good introduction to twistor theory can be found in [10]. Another good introduction is [14].

§1 Spinors

Real <u>Minkowski-space</u> M^I is a real affine space of dimension 4 equipped with a metric $\| \ \|$ of signature (+,-,-,-). This metric partitions the space according to causality:

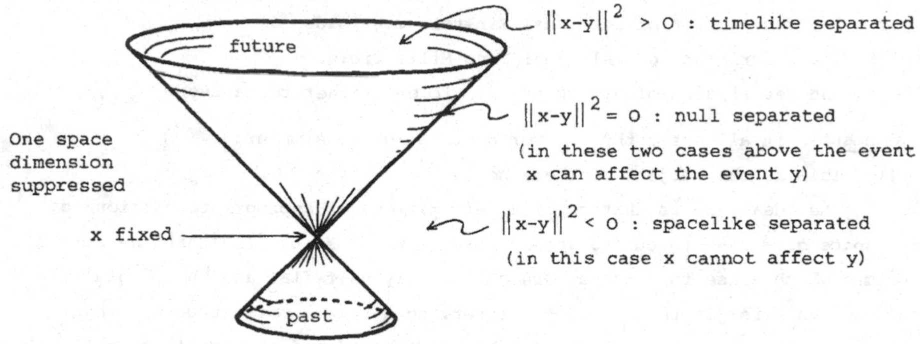

If we choose an origin for M^I then we can find coordinates (x^0, x^1, x^2, x^3) such that the metric takes the form

$$\|x^a\|^2 = (x^0)^2 - (x^1)^2 - (x^2)^2 - (x^3)^2 .$$

Suppose we now embed $\mathbb{R}^4 \hookrightarrow \mathbb{C}^4$ in the usual way and extend the Minkowski metric to a holomorphic metric

$$\|z^a\|^2 = (z^0)^2 - (z^1)^2 - (z^2)^2 - (z^3)^2 .$$

The resulting space is called <u>complexified</u> Minkowski space $\mathbb{M}^I$. Note that the metric no longer has well-defined signature (we could equally well embed $\mathbb{R}^4$ with its usual Euclidean metric). If we make a complex change of coordinates

$$z^{AA'} = \begin{bmatrix} z^{00'} & z^{01'} \\ z^{10'} & z^{11'} \end{bmatrix} = \frac{1}{\sqrt{2}} \begin{bmatrix} z^0+z^1 & z^2+iz^3 \\ z^2-iz^3 & z^0-z^1 \end{bmatrix}$$

we see that $\|z^a\|^2 = 2 \det z^{AA'}$ and that z^a is real if and only if $z^{AA'}$ is Hermitian. Clearly $z^{AA'} = \zeta^A \xi^{A'} \Rightarrow \|z^a\| = 0$ and it is easy to check that the converse if true. What we have done therefore is to write

$$\mathbb{M}^I = \mathbb{C}^4 = \mathbb{C}^A \otimes \mathbb{C}^{A'}$$

so that the null vectors (those with vanishing norm) are the simple vectors. Moreover, if we introduce complex conjugation

$$\begin{array}{lcl} \mathbb{C}^A & \longleftrightarrow & \mathbb{C}^{A'} \\ \zeta^A & \longmapsto & \bar{\zeta}^{A'} \\ \bar{\xi}^A & \longleftarrow & \xi^{A'} \end{array} \qquad \text{by } \bar{\zeta}^{0'} = \overline{\zeta^0} \text{ etc.}$$

then the real vectors are those $z^{AA'}$ such that $z^{AA'} = \bar{z}^{AA'}$. This process alows us to take "square roots" of null vectors in real Minkowski-space i.e. if v^a is null and future-pointing (i.e. $v^0 > 0$) then

$$v^{AA'} = \zeta^A \bar{\zeta}^{A'} \text{ for some } \zeta^A \in \mathbb{C}^A .$$

This square root, however, is defined only up to phase $\zeta^A \mapsto e^{i\theta}\zeta^A$. Finally we introduce skew forms ε_{AB}, $\varepsilon_{A'B'}$ on $\mathbb{C}^A$ and $\mathbb{C}^{A'}$ by

$$\varepsilon_{AB} = \begin{bmatrix} 0 & 1 \\ -1 & 0 \end{bmatrix} = \varepsilon_{A'B'}$$

and use then to <u>raise and lower indices</u>:

$$\zeta_A = \zeta^B \varepsilon_{BA}, \ \zeta^A = \varepsilon^{AB}\zeta_B, \ \xi_{A'} = \xi^{B'}\varepsilon_{B'A'}, \ \xi^{A'} = \varepsilon^{A'B'}\xi_{B'}$$

where we are using the Einstein symmation convention. So, for example: $\zeta_1 = \zeta^0\varepsilon_{01} + \zeta^1\varepsilon_{11} = \zeta^0$, $\zeta_0 = \ldots = -\zeta^1$. Using these ε's we may recover the metric $\|z^a\|^2 = z^{AA'}z_{AA'}$, or in other words $\varepsilon_{AB}\varepsilon_{A'B'}$ <u>is</u> the metric $\| \ \|$. Thus, ε may be regarded as the "square-root" of the original Minkowski metric. A <u>spinor</u> is an element of a tensor product

$$\mathbb{C}^A \otimes \mathbb{C}^B \otimes \ldots \otimes \mathbb{C}^D \otimes \mathbb{C}^{A'} \otimes \ldots \otimes \mathbb{C}_E \otimes \ldots \otimes \mathbb{C}_{E'} \otimes \ldots$$

just as a <u>tensor</u> is an element of a tensor product

$$\mathbb{C}^a \otimes \mathbb{C}^b \otimes \ldots \otimes \mathbb{C}^d \otimes \mathbb{C}_e \otimes \ldots$$

For more on spinors see [12].

What we have just described (rather vaguely) is a <u>spin structure</u> on M^I or $\mathbb{M}^I$. This procedure may be carried out locally on a general Lorentzian manifold (since each tangent space may be regarded as a Minkowski-space with origin).

It is often possible to patch these local constructions together to form a global spin structure - there is a topological obstruction to doing so. We will only be concerned with the flat (Minkowski-space) case.

§2 Twistor Geometry

In §1 we observed how it was natural to regard Minkowski-space after choice of origin as a subspace of the space 2×2 complex matrices $\mathbb{C}^{AA'}$. Such a matrix may be thought of as a linear transformation

$$\mathbb{C}_{A'} \longrightarrow \mathbb{C}^{A}$$

$$\pi_{A'} \longmapsto iz^{AA'}\pi_{A'} \quad \text{(the factor of i is conventional)}$$

and such a linear transformation is determined by its graph in $\mathbb{C}^A \oplus \mathbb{C}_{A'}$. This 4-dimensional complex vector-space is called twistor-space $\mathbb{T}$. Thus, a point $z^{AA'}$ in complexified Minkowski space $\mathbb{M}^I$ gives rise to a plane in $\mathbb{T}$ (i.e. a 2-dimensional complex subspace) given by

$$\{(\omega^A, \pi_{A'}) \in \mathbb{T} \text{ s.t. } \omega^A = iz^{AA'}\pi_{A'}\} .$$

"Most" planes in $\mathbb{T}$ arise in this way. More precisely, if we let $\mathbb{M}$ denote the Graussmannian of all planes in $\mathbb{T}$ then $\mathbb{M}^I$ is an open and dense subset of $\mathbb{M}$. From this point of view it is clear that thereis nothing special about the particular planes which comprise $\mathbb{M}^I$ and so $\mathbb{M}$ is a natural compactification of $\mathbb{M}^I$. It is convenient to factor out scale from $\mathbb{T}$ to form projective twistor-space $\mathbb{PT}$ or just $\mathbb{P}$. Then a plane in $\mathbb{T}$ corresponds to a line (a linearly embedded Riemann sphere) in $\mathbb{P}$. It is easy to check that two points $w^{AA'}$, $z^{AA'}$ in $\mathbb{M}^I$ are null separated if and only if the planes $\{\omega^A = iw^{AA'}\pi_{A'}\}$ and $\{\omega^A = iz^{AA'}\pi_{A'}\}$ intersect in a line or, equivalently, the lines in $\mathbb{P}$ intersect. We use this property as a definition of null separation on $\mathbb{M}$. Hence we obtain a metric defined only up to scale on $\mathbb{M}$ i.e. a conformal structure. This is the basic twistor correspondence:

$$\{\text{points in } \mathbb{M}\} \longleftrightarrow \{\text{lines in } \mathbb{P}\}$$

$$z \longleftrightarrow L_z$$

$$z \text{ is null separated from } w \Longleftrightarrow L_z \cap L_w \neq \emptyset .$$

If we fix a point Z in $\mathbb{P}$ and consider the set of all lines passing through Z then we obtain a $\mathbb{P}_2(\mathbb{C})$ embedded in $\mathbb{M}$ such that any two points on it are null separated. Such a subspace is called totally null. The totally null 2-planes (embedded 2-dimensional complex manifolds) are easily shown to fall into two distinct families. One family is the one just obtained from points in $\mathbb{P}$. Its members are called α-planes. Thus we have a correspondence:

$\{\alpha\text{-planes in } \mathbb{M}\} \longleftrightarrow \{\text{points in } \mathbb{P}\}$

$\hat{Z} \longleftrightarrow Z$

The other family, β-planes, is obtained by using dual twistor-space $\mathbb{T}^*$ (or projectively $\mathbb{P}^*$) instead of twisor-space. This amounts to interchanging primed and unprimed spin-spaces in the definition of $\mathbb{T}$. Equivalently a projective dual twistor $W \in \mathbb{P}^*$ may be regarded as a plane (a linearly embedded $\mathbb{P}_2(\mathbb{C})$) in $\mathbb{P}$ and then the β-plane corresponds to the set of lines in that plane:

$\{\beta\text{-planes in } \mathbb{M}\} \longleftrightarrow \{\text{planes in } \mathbb{P}\}$

$\hat{W} \longleftrightarrow W .$

Thus, we have the following picture of the twistor correspondence

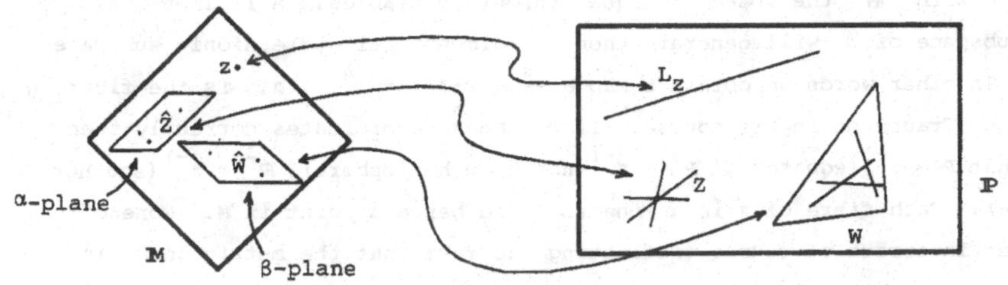

Finally we ask if there is a good way of determining which lines in $\mathbb{P}$ correspond to points in real Minkowski-space. If $z^{AA'}$ is Hermitian and $(\omega^A,\pi_{A'})$ is a point on L_z then

$$\omega^A\bar{\pi}_A + \bar{\omega}^{A'}\pi_{A'} = iz^{AA'}\pi_{A'}\bar{\pi}_A + \overline{iz^{AA'}\pi_{A'}}\,\pi_{A'}$$
$$= iz^{AA'}\pi_{A'}\bar{\pi}_A - iz^{AA'}\bar{\pi}_A\pi_{A'} = 0 .$$

So let us introduce the Hermitian form

$$\Phi(Z) = \omega^A\bar{\pi}_A + \bar{\omega}^{A'}\pi_{A'}$$

on $\mathbb{T}$. This form divides $\mathbb{P}$ into three pieces

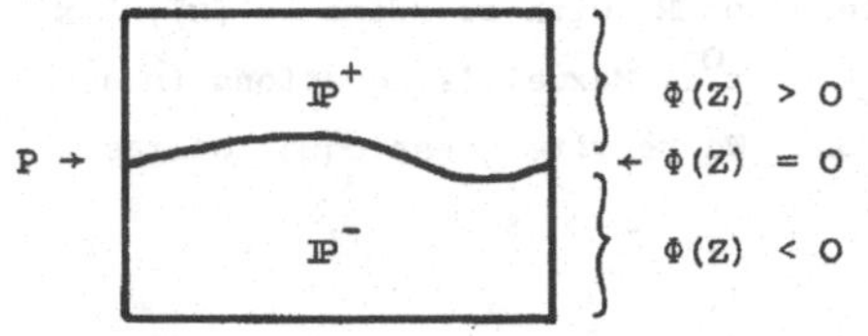

and we have shown that $z \in M^I \Rightarrow L_z \subset P$. Conversely it is not hard to show that "most" lines in $\mathbb{P}$ which lie entirely in P arise in this way i.e. if we let M denote the space of all lines in $\mathbb{P}$ then $\mathbb{M}^I$ is an open dense subset of $\mathbb{M}$. What's missing is a particular line I given by $\pi_{A'} = 0$ together with all lines that intersect it. M is called compactified Minkowski-space. Thus M is obtained by adding to M^I a single point together with its null cone. P is then the space of light rays in M, and for $z \in M$ the points of L_z represent the celestial sphere of light rays through z. It may be verified that M is diffeomorphic to $S^3 \times S^1$ and that P is diffeomorphic to $S^3 \times S^2$. Since P is embedded as a real hypersurface in a complex manifold it inherits a CR-structure. This fact, that the space of light rays in Minkowski-space carries a structure related to the complex, was one of the original motivations for twistor theory [11]. The Levi form has signature (+,-). More details can be found in, for example, [20].

It is worth remarking as to how the above geometry fits in with twistor theory for S^4 as described by Trautman [21]. We may choose coordinates so as to identify $\mathbb{T}$ with $\mathbb{H}^2$ the space of 2 quaternionic variables. A 1-dimensional complex subspace of $\mathbb{T}$ will generate then a 1-dimensional quaternionic subspace of $\mathbb{H}^2$. In other words we obtain a map $\mathbb{P} \xrightarrow{\pi} \mathbb{P}_1(\mathbb{H}) = S^4$. This is the fibering described by Trautmann in his course. If we choose coordinates correctly then we find that $P = \pi^{-1}$(equator), $\mathbb{P}^+ = \pi^{-1}$(northern hemisphere), $\mathbb{P}^- = \pi^{-1}$(southern hemisphere). Each fibre of π is a line in $\mathbb{P}$ and hence a point in $\mathbb{M}$. None of these lines intersect of course (reflecting the fact that the metric on S^4 is Riemannian). In this way S^4 is embedded in $\mathbb{M}$. Thus, both M and S^4 are realized as real slices of the same complex manifold $\mathbb{M}$.

§3 Massless Fields

The prototype massless fields equations are Maxwell's equations of electromagnetism

$$\left.\begin{aligned} &\text{Div } D = \rho \\ &\text{Curl } H = J + \frac{\partial D}{\partial t} \\ &\text{Curl } E + \frac{\partial B}{\partial t} = 0 \\ &\text{Div } B = 0 \end{aligned}\right\} \quad \text{or in a vacuum} \quad \left\{\begin{aligned} &\text{Div } E = 0 \\ &\text{Curl } B = \frac{\partial E}{\partial t} \\ &\text{Curl } E + \frac{\partial B}{\partial t} = 0 \\ &\text{Div } B = 0 \end{aligned}\right.$$

E and B are usually regarded as vector-fields on $\mathbb{R}^3$ with coordinates (x^1,x^2,x^3) depending on an additional parameter time t or x^0. Maxwell's equations in a vacuum describe the propagation of a photon. We rewrite these equations as follows. Form the skew-symmetric matrix:

$$[F_{ab}] = \begin{bmatrix} 0 & E_1 & E_2 & E_3 \\ -E_1 & 0 & B_3 & -B_2 \\ -E_2 & -B_3 & 0 & B_1 \\ -E_3 & B_2 & -B_1 & 0 \end{bmatrix}$$

Then $F = F_{ab}\, dx^a \wedge dx^b$ and it is easy to check that

(i) $dF = 0 \Leftrightarrow \text{Curl } E + \frac{\partial B}{\partial t} = 0$ & $\text{Div } B = 0$

(ii) $d{*}F = 0 \Leftrightarrow \text{Curl } B = \frac{\partial E}{\partial t}$ & $\text{Div } E = 0$.

Here $*$ is the Hodge $*$-operator, $* : \Lambda^2 M^I \to \Lambda^2 M^I$. Actually $*$ depends only on the metric up to scale (when acting on 2-forms) and so makes sense as an operator $* : \Lambda^2 M \to \Lambda^2 M$. It is easy to verify that $*^2 = -1$ (as opposed to $*^2 = 1$ in the Euclidean case). Thus we obtain $\Lambda^2 M = \Lambda^2_+ M \oplus \Lambda^2_- M$ where $\Lambda^2_\pm M = \{F \in \Lambda^2 M \text{ s.t. } *F = \pm iF\}$ Here $\Lambda^2 M$ denotes the space of two forms on M with complex coefficients. If we use this decomposition to write $F = F_+ + F_-$ then Maxwell's equations become $dF_+ = 0$ and $dF_- = 0$. These equations may be solved separately. F_+ may be interpreted as the self-dual (i.e. $*F_+ = iF_+$) curvature of a U(1)-bundle i.e. a self-dual Yang-Mills field for the gauge group U(1) (cf. [21]). If we write all this out in terms of spinors then the decomposition $F = F_+ + F_-$ becomes

$$F_{ab} = \varepsilon_{AB}\phi_{A'B'} + \varepsilon_{A'B'}\psi_{AB}$$

where $\phi_{A'B'}$ and ψ_{AB} are symmetric i.e. $\phi_{0'1'} = \phi_{1'0'}$ etc. and the equations $dF_+ = 0$ and $dF_- = 0$ become

$$\nabla^{AA'}\phi_{A'B'} = 0 \qquad \nabla^{AA'}\psi_{AB} = 0 \qquad \text{where } \nabla_{AA'} = \frac{\partial}{\partial z^{AA'}} . \text{ Okay on } M^I \text{ or } \mathbb{M}^I .$$

For the details of this translation see [23]. These equations are the field equations for a massless free particle of helicity ±1 (i.e. a photon). More generally a massless free field of helicity n/2 is a solution of

$$\nabla^{AA'}\phi_{\overbrace{A'B'\ldots D'}^{n}} = 0 \qquad \text{if } n > 0$$

$$\Box\phi = 0 \text{ if } n = 0 \qquad (\Box = \nabla^{AA'}\nabla_{AA'} \text{, the wave operator})$$

$$\nabla^{AA'}\phi_{\underbrace{AB\ldots D}_{}}{}^{\!\!-n} = 0 \quad \text{if } n < 0 .$$

The spinor fields in these equations are symmetric in the indices. If the equations are interpreted on $\mathbb{M}^I$ rather than M^I when the fields are supposed to be holomorphic also. Just as for Maxwell fields it is possible, by the introduction of "conformal weights", to make sense of these equations globally on M or $\mathbb{M}$ but we will not do this here (see [2]). For helicity ± 1 the equations describe a a photon, for helicity $\pm\frac{1}{2}$ the equations describe a neutrino (if neutrinos are indeed massless), and for helicity 2 a graviton i.e. they are the linearized (weak field limit of) Einstein's vacuum equations.

It is possible to solve the massless free field equations by Fourier analysis. This breaks the solutions into two components $\phi = \phi^+ + \phi^-$ where ϕ^+ is of <u>positive frequency</u> and ϕ^- is of <u>negative frequency</u>. Roughly speaking a field is of positive (resp. negative) frequency if it is the boundary value on M^I of a field defined on

$$\mathbb{M}^+ = \{z^a = x^a - iy^a \text{ s.t. } y^0 > 0 \text{ and } (y^0)^2 - (y^1)^2 - (y^2)^2 - (y^3)^2 > 0\}$$

(resp. $\mathbb{M}^- = \ldots\ldots\ldots\ldots\ldots\ldots < \ldots\ldots\ldots\ldots\ldots\ldots\ldots\ldots\ldots\ldots\ldots\ldots$).
Thus we see that it is of particular importance to study fields defined on $\mathbb{M}^\pm$. For the rest of this course we shall look only at holomorphic fields on $\mathbb{M}^+$ although the methods apply much more generally [2].

§4 <u>The Penrose Transform</u>

Penrose introduced contour integral formulae into the theory of twistors in [13] in order to solve the massless field equations. Later he realized that the functions appearing in the integrand of these formulae should be interpreted as representing a cocycle in Čech cohomology [18]. The aim of these notes is to at least sketch a proof of

<u>Theorem</u> (Penrose): There is a canonical isomorphism $\mathcal{P} : H^1(\mathbb{P}^+, \mathcal{O}(-n-2)) \to$ {holomorphic massless fields on $\mathbb{M}^+$ of helicity $n/2$}. □

Actually there are three different cases ($n > 0$, $n = 0$, $n < 0$) as one might expect by looking at the fields equations and, although we will give a method which works for all three cases, we will concentrate just on $n > 0$. It is remarkable that the different cases are incorporated so uniformly in the twistor description. It is also apparent that in the twistor description the field equations have completely disappeared - they have been absorbed into the geometry. This is surely a significant simplification.

<u>Sketch of Proof</u>: A simple calculation shows that

$$\mathbb{P}^+ = \bigcup_{z \in \mathbb{M}^+} L_z$$

and that all lines in $\mathbb{P}^+$ arise from points in $\mathbb{M}^+$. The idea behind the construction of $\mathcal{P}$ is to take a cohomology class in $H^1(\mathbb{P}^+, \mathcal{O}(-n-2))$ and to restrict it to L_z for $z \in \mathbb{M}^+$. As z varies this restriction will generate the field. Let us first recall the definition of $\mathcal{O}(k)$ on $\mathbb{P}$. For U an open set in $\mathbb{P}$ let $\tilde{U}$ denote the open subset of $\mathbb{T} - \{0\}$ swept out by the corresponding lines.

$\mathcal{O}(k)$ is defined by the assignment

$$U \mapsto \{\text{homogeneous holomorphic functions of degree } k \text{ on } \tilde{U}\}.$$

It is easy to check that $\mathcal{O}(k)$ is isomorphic to the sheaf of holomorphic sections of the line bundle H^k where H is the hyperplane section bundle which has as fibre over a point in $\mathbb{P}$ the dual as a complex vector space of the line in $\mathbb{T}$ which the point represents. Comparing coefficients in Laurent series expansions allows us to compute

$$H^p(\mathbb{P}_m(\mathbb{C}),\ \mathcal{O}(k)) \text{ for all } k,m,p.$$

In particular for the Riemann sphere S with homogeneous coordinates $\pi_{A'}$, we obtain

$$* \quad H^1(S, \mathcal{O}(-n-2)) = \mathbb{C}_{\underbrace{(A'B'\ldots D')}_{n}}, \text{ the space of symmetric primed spinors with } n \text{ indices.}$$

The twistor correspondence may be described by the following diagram

$$\begin{array}{ccc} & \mathbb{M}^+ \times S \ni (z^{AA'}, \pi_{A'}) & \\ {}^{\nu}\swarrow & \searrow^{\mu} \qquad \downarrow & \\ \mathbb{M}^+ & & \mathbb{P}^+ \ni (iz^{AA'}\pi_{A'}, \pi_{A'}) \end{array}$$

so that $L_z = \mu(\nu^{-1}(z))$.

The Penrose transform is best studied by two steps:

1 $\quad H^1(\mathbb{P}^+, \mathcal{O}(-n-2)) \longrightarrow H^1(\mathbb{M}^+ \times S,\ \mu^{-1}\mathcal{O}(-n-2))$

2 $\quad H^1(\mathbb{M}^+ \times S,\ \mu^{-1}\mathcal{O}(-n-2)) \longrightarrow$ Massless fields.

Here $\mu^{-1}\mathcal{O}(k)$ is the sheaf of holomorphic functions on $\mathbb{M}^+ \times S$ twisted with respect to S and constant on the fibres of μ. The map in 1 is easily constructed. Showing that it is an isomorphism involves using a Dolbeault resolution of $\mathcal{O}(-n-2)$ and an integral on the fibres of μ using the fact that all the fibres are convex. This argument works for any twisting or, more generally, for coefficients $\mathcal{O}(V)$ where V is an arbitrary holomorphic vector-bundle. It is in step 2 that the field equations occur. To compute $H^1(\mathbb{M}^+ \times S,\ \mu^{-1}\mathcal{O}(-n-2))$ we use the holomorphic deRham complex relative to the fibration μ:

$$0 \to \mu^{-1}\mathcal{O}(-n-2) \to \mathcal{O}(-n-2) \xrightarrow{d_\mu} \Omega^1_\mu(-n-2) \xrightarrow{d_\mu} \Omega^2_\mu(-n-2) \to 0.$$

A computation in local coordinates on S shows that this sequence may be rewritten more explicitly as

$$0 \longrightarrow \mu^{-1}\mathcal{O}(-n-2) \longrightarrow \mathcal{O}(-n-2) \xrightarrow{\pi_{A'}\nabla^{AA'}} \mathcal{O}^{A}(-n-1) \xrightarrow{\pi_{B'}\nabla^{B'}_{A}} \mathcal{O}(-n) \longrightarrow 0.$$

It follows as for ✱ that

$$H^1(\mathbb{M}^+ \times S, \mathcal{O}(-n-2)) = \{\text{holomorphic fields on } \mathbb{M}^+ \text{ with values in } \mathbb{C}_{\overbrace{(A'B'\ldots D')}^{n}}\}.$$

Now the field equations arise from the effect of $\pi_{A'}$, $\nabla^{AA'}$ on cohomology. To show that 2 is an isomorphism involves diagram chasing using the whole relative deRham sequence. The step 2 is better formulated using direct image sheaves and a spectral sequence (see [2]).

References

1. M.F. Atiyah, N.J. Hitchin, V.G. Dinfeld, Yu.I. Manin: Construction of instantons, Phys. Lett. 65A, 185-187 (1978).

2. M.G. Eastwood, R. Penrose, R.O. Wells, Jr.: Cohomology and massless fields, Comm. Math. Phys., 78, 305-351 (1981).

3. M.G. Eastwood: On the twistor description of massive fields, Proc. Roy. Soc. Lond. A 374, 431-445 (1981).

4. M.G. Eastwood and M.L. Ginsberg: Duality in twistor theory, Duke Math. J. 48 (1981).

5. J. Eells: Complex analysis and geometry, lecture notes, this seminar.

6. M.L. Ginsberg, S.A. Huggett: Sheaf cohomology and twistor diagrams, in [10], 287-292.

7. M.L. Ginsberg: A cohomological scalar product construction, in [10], 293-300.

8. N.J. Hitchin: Polygons and gravitons, Math. Proc. Camb. Phil. Soc. 85, 465-476 (1979).

9. L.P. Hughston: Twistors and particles, Springer lecture notes in physics 97 (1979).

10. L.P. Hughston, R.S. Ward (eds.): Advances in twistor theory, Pitman research notes in math. 37 (1979).

11. R. Penrose: Twistor algebra, J. Math. Phys. 8, 345-366 (1967).

12. R. Penrose: The structure of space-time, in Batelle rencontres, eds. C.M. DeWitt, J.A. Wheeler, 121-235, Benjamin (1967).

13. R. Penrose: Twistor quantization and curved space-time, Int. J. Th. Phys. 1, 61-99 (1968).

14. R. Penrose: Twistor theory, it aims and achievements, in Quantum gravity, an Oxford symposium, eds. C.J. Isham, R. Penrose, D.W. Sciama, 268-407, Oxford Clarendon Press (1975).

15. R. Penrose: Twistors and particles - an outline, in Quantum theory and the structure of space-time, eds. L. Castell, M. Drieschner, C.F. von Weizsäcker, 129-145, Carl Hanser Verlag (1975).

16. R. Penrose: Non-linear gravitons and curved twistor theory, Gen. Rel. Grav. 7, 31-52 (1976).

17. R. Penrose: The twistor programme, Reps. on Math. Phys. 12, 65-76 (1977).

18. R. Penrose: On the twistor description of massless fields, in Complex manifold technqiues in theoretical physics, eds. D.E. Lerner, P.D. Sommers, 55-91, Pitman research notes in math. 32 (1979).

19. R. Penrose: Is nature complex?, in the Encyclopedia of ignorance, eds. R. Duncan, M. Weston-Smith, Pergamon (1977).

20. R. Penrose, R.S. Ward: Twistors for flat and curved space-time, in Einstein centennial volume, ed. A.P. Held, Plenum (1980).

21. G. Trautman: Holomorphic vector-bundles and Yang-Mills fields, lecture notes, this seminar.

22. R.S. Ward: On self-dual gauge fields, Phys. Lett. 61A, 81-82 (1977).

23. R.O. Wells, Jr.: Complex manifolds and mathematical physics, Bull. Amer. Math. Soc. 1, (new series), 296-336 (1979).

AN INTRODUCTION TO ANALYSIS ON COMPLEX MANIFOLDS

Luc Lemaire
Chercheur qualifié au Fonds National Belge de la Recherche Scientifique, Université Libre de Bruxelles, Belgium.

John C. Wood
University of Leeds, England. Supported by the Sonderforschungsbereich "Theoretische Mathematik", Universität Bonn during some of the preparation of these notes.

We give an account of some of the analysis on complex manifolds leading in particular to Dolbeault cohomology and the Hodge decomposition Theorem. On the way we introduce some basic notions such as sheaves, connections in fibre bundles and Kähler manifolds. We thank Frl. A. Thiedemann for the difficult job of typing our illegible manuscript.

§1 ELEMENTARY SEVERAL COMPLEX VARIABLES

A C^1 function of one complex variable is holomorphic if and only if it satisfies the Cauchy Riemann equations. We shall define holomorphicity of a function of several complex variables similarly and show that it is equivalent to the existence of a power series expansion. We shall then discuss some elementary properties of a holomorphic function, including the phenomenon of domains of holomorphy. The exposition of this and other sections owes much to the treatise of Griffiths and Harris [G-H].

A. Definitions and basic theorems

We write $z = (z^1,\ldots,z^m)$ for a point of $\mathbb{C}^m$ and set $z^j = x^j + iy^j$ where x^j, y^j are real numbers and $i = \sqrt{-1}$. We set $|z| = \max_j |z^j|$, the "1^∞ norm" of z.

Let U be an open subset of $\mathbb{C}^m$, let $C^o(U)$ denote the set of continuous complex-valued functions f on U and $C^r(U)$ $(r = 1,2,\ldots.)$ those $f \in C^o(U)$ such that all derivatives of order $\leq r$ with respect to the real variables x^j, y^j exist and are continuous on U. We define first order partial differential operators $C^1(U) \to C^o(U)$:

$$(1.1)\qquad \frac{\partial}{\partial z^j} = \frac{1}{2}\left(\frac{\partial}{\partial x^j} - i\,\frac{\partial}{\partial y^j}\right),\qquad \frac{\partial}{\partial \bar z^j} = \frac{1}{2}\left(\frac{\partial}{\partial x^j} + i\,\frac{\partial}{\partial y^j}\right)\qquad (j = 1,\ldots,m).$$

A simple computation shows that the total differential $df = \sum \frac{\partial f}{\partial x^j} dx^j + \sum \frac{\partial f}{\partial y^j} dy^j$ of a function $f \in C^1(U)$ may be written

$$(1.2)\qquad df = \sum \frac{\partial f}{\partial z^j} dz^j + \sum \frac{\partial f}{\partial \bar z^j} d\bar z^j$$

where $dz^j = dx^j + idy^j$, $d\bar z^j = dx^j - idy^j$.

<u>Example</u>: By direct computation we find

$$(1.3)\qquad \frac{\partial}{\partial z^j}(z^j)^k = k(z^j)^{k-1}\quad,\quad \frac{\partial}{\partial \bar z^j}(z^j)^k = 0\quad .$$

If $m = 1$, writing $z = x + iy$ and $f(z) = u(x,y) + iv(x,y)$, the equation $\frac{\partial f}{\partial \bar{z}} = 0$ is equivalent to the Cauchy-Riemann equations:

$$(1.4) \qquad \mathrm{Re}\left(\frac{\partial f}{\partial \bar{z}}\right) = \frac{1}{2}\left(\frac{\partial u}{\partial x} - \frac{\partial v}{\partial y}\right) = 0 \quad , \quad \mathrm{Im}\left(\frac{\partial f}{\partial \bar{z}}\right) = \frac{1}{2}\left(\frac{\partial u}{\partial y} + \frac{\partial v}{\partial x}\right) = 0 \quad .$$

This suggests the

(1.5) <u>Definition</u>: A function $f \in C^1(U)$ is said to be <u>holomorphic</u> if $\frac{\partial f}{\partial \bar{z}^j} = 0 \quad \forall j = 1,\ldots,m.$

(1.6) <u>Remark</u>: So a C^1 function is holomorphic if and only if it is holomorphic in each variable separately. In fact this is true without any differentiability or continuity hypotheses on f (a theorem of Hartogs see [Hö]).

We now note an important formula:

(1.7) <u>The general Cauchy Integral Formula in One Complex Variable</u>. Let Δ be a disc in $\mathbb{C}$ with boundary $\partial\Delta$ traversed anticlockwise and let $f \in C^0(\bar{\Delta}) \cap C^1(\Delta)$. Then for any $z \in \Delta$,

$$f(z) = \frac{1}{2\pi i}\int_{\partial\Delta} \frac{f(w)\,dw}{w-z} + \frac{1}{2\pi i}\int_{\Delta} \frac{\partial f(w)}{\partial \bar{w}}\,\frac{dw \wedge d\bar{w}}{w-z}$$

the right-most integral being convergent.

<u>Proof</u>. We use Stokes' theorem for a differential form on $\mathbb{R}^2$ [Sp]. Consider the 1-form

$$\eta = \frac{1}{2\pi i}\,\frac{f(w)}{w-z}\,dw \quad .$$

Since $\frac{\partial}{\partial \bar{w}}\left(\frac{1}{w-z}\right) = 0,$

$$d\eta = -\frac{1}{2\pi i}\,\frac{\partial f}{\partial \bar{w}}\,\frac{1}{w-z}\,dw \wedge d\bar{w} \qquad (w \neq z) \qquad .$$

Let $\Delta_\varepsilon = \Delta(z,\varepsilon)$ be the open disc with centre z and radius $\varepsilon > 0$. The form η is C^1 on $\Delta \setminus \Delta_\varepsilon$ so we may apply Stokes' theorem to obtain

$$-\frac{1}{2\pi i}\int_{\partial\Delta_\varepsilon} \frac{f(w)\,dw}{w-z} + \frac{1}{2\pi i}\int_{\partial\Delta} \frac{f(w)\,dw}{w-z} = -\frac{1}{2\pi i}\int_{\Delta\setminus\Delta_\varepsilon} \frac{\partial f}{\partial \bar{w}}\,\frac{dw \wedge d\bar{w}}{w-z} \qquad .$$

Parametrizing $\partial\Delta_\varepsilon$ by $w = z + \varepsilon e^{i\theta}$,

$$\frac{1}{2\pi i}\int_{\partial\Delta_\varepsilon}\frac{f(w)dw}{w-z} = \frac{1}{2\pi}\int_0^{2\pi} f(z+\varepsilon e^{i\theta})d\theta \to f(z) \quad \text{as} \quad \varepsilon \to 0 \quad .$$

As for the area integral on the right-hand side, let (r,θ) be polar coordinates centred on z, then $dw\wedge d\bar{w} = -2i\, dx \wedge dy = -2ir dr \wedge d\theta$, hence

$$\left|\frac{\partial f}{\partial \bar{w}}\frac{dw\wedge d\bar{w}}{w-z}\right| = 2\left|\frac{\partial f}{\partial \bar{w}}\, dr\wedge d\theta\right| \le M|dr\wedge d\theta|$$

where M is an upper bound on $2\left|\frac{\partial f}{\partial \bar{w}}\right|$. Hence $\frac{\partial f}{\partial \bar{w}}\frac{dw\wedge d\bar{w}}{w-z}$ is absolutely integrable over Δ and

$$\int_{\Delta_\varepsilon}\frac{\partial f}{\partial \bar{w}}\frac{dw\wedge d\bar{w}}{w-z} \to 0 \quad \text{as} \quad \varepsilon \to 0 \quad .$$

Letting $\varepsilon \to 0$ yields the result.

Note that, if f is holomorphic, only the line integral appears. To generalize to higher dimensions we replace the disc Δ not by a ball but by a polydisc: an <u>(open) polydisc</u> (sometimes called a <u>polycylinder</u>) in $\mathbb{C}^m$ is a subset $\Delta = \Delta(w,\underline{r}) = \Delta(w^1,\ldots,w^m, r^1,\ldots,r^m) = \{z \in \mathbb{C}^m: |z^j-w^j| < r^j, 1 \le j \le m\}$; here $w = (w^1,\ldots,w^m) \in \mathbb{C}^m$ is called the <u>centre</u> and $\underline{r} = (r^1,\ldots,r^m) \in \mathbb{R}^m$ $(r^j > 0)$ the <u>polyradius</u>. Note that Δ is the cartesian product of the discs $D_j = \{z \in \mathbb{C}: |z-w^j| < r^j\}$. The product of the boundaries $\partial D_1 \times \ldots \times \partial D_m = \{z \in \mathbb{C}^m: |z^j-w^j| = r^j\}$ is called the <u>distinguished boundary</u> of the polydisc $\Delta = D_1 \times \ldots \times D_m$ and is denotes by $\partial_0\Delta$. The closure $\bar{\Delta} = \bar{\Delta}(w,\underline{r})$ of Δ is called the closed polydisc with centre w and polyradius $\underline{r}$.

(1.8) <u>Theorem (Cauchy's Integral Formula for a Polydisc)</u>. Let f be continuous on a closed polydisc $\bar{\Delta}(w,\underline{r})$ and holomorphic on the open polydisc $\Delta = \Delta(w,\underline{r})$, then for $z \in \Delta$,

$$f(z) = \left(\frac{1}{2\pi i}\right)^m \int_{\partial_0\Delta}\frac{f(\rho)d\rho^1\ldots d\rho^m}{(\rho^1-z^1)\ldots(\rho^m-z^m)} \quad .$$

<u>Proof</u>. Since f is holomorphic in each variable separately we may repeatedly apply Cauchy's Integral Formula in one Complex Variable viz:

$$f(z) = \frac{1}{2\pi i}\int_{|\rho^1-w^1|=r^1}\frac{d\rho^1}{\rho^1-z^1}\left\{\frac{1}{2\pi i}\int_{|\rho^2-w^2|=r^2}\frac{d\rho^2}{\rho^2-z^2}\, f(\rho^1,\rho^2,z^3,\ldots,z^m)\right\}$$

$$= \ldots$$

$$= \left(\frac{1}{2\pi i}\right)^m \int_{|\rho^1-w^1|=r^1} \frac{d\rho^1}{\rho^1-z^1} \int_{|\rho^2-w^2|=r^2} \frac{d\rho^2}{\rho^2-z^2} \cdots \int_{|\rho^m-w^m|=r^m} \frac{d\rho^m}{\rho^m-z^m} f(\rho^1,\ldots,\rho^m).$$

Since f is continuous on $\bar{\Delta}$, Fubini's theorem tells us that this iterated integral is the same as the stated multiple integral.

Remark: For other integral representation formulae on more general domains see [Fu] [Ha].

We can immediately deduce an important convergence theorem:

(1.9) Theorem: Let (f_k) be a sequence of holomorphic functions on an open set U of $\mathbb{C}^m$. Suppose that (f_k) is pointwise convergent to some function f on U, the convergence being uniform on compact subsets of U. Then the limit function f is holomorphic.

Proof. Firstly, by uniform convergence, f is continuous on U. Let $w \in U$ and choose $\underline{r} > 0$ (i.e. $r^j > 0$ $\forall j$) such that $\bar{\Delta}(w,\underline{r}) \subseteq U$. Then, for any point $z \in \Delta = \Delta(w,\underline{r})$, we have Cauchy's Integral Formula for f_k:

$$f_k(z) = \left(\frac{1}{2\pi i}\right)^m \int_{\partial_0 \Delta} \frac{f_k(\rho)\,d\rho^1 \ldots d\rho^m}{(\rho^1-z^1)\ldots(\rho^m-z^m)} .$$

Since $f_k \to f$ uniformly on the domain of integration, we may take the limit to get

$$f(z) = \left(\frac{1}{2\pi i}\right)^m \int_{\partial_0 \Delta} \frac{f(\rho)\,d\rho^1 \ldots d\rho^m}{(\rho^1-z^1)\ldots(\rho^m-z^m)} .$$

Differentation under the integral sign (cf. [Na]) shows that the first partial derivatives $\frac{\partial f}{\partial z^j}$ and $\frac{\partial f}{\partial \bar{z}^j}$ exist and are continuous; further $\frac{\partial f}{\partial \bar{z}^j} = 0$ $(j = 1,\ldots,m)$, thus f is holomorphic on U.

B. Power Series

A power series "in z-w", "centred on w" or "about w" is a multiple series

$$(1.10)\quad s(z) = \sum_{\nu_1,\ldots,\nu_m=0}^{\infty} a_{\nu_1,\ldots,\nu_m} (z^1-w^1)^{\nu_1} \ldots (z^m-w^m)^{\nu_m}$$

where $a_{\nu_1,\ldots,\nu_m} \in \mathbb{C}$, $w = (w^1,\ldots,w^m)$, $z = (z^1,\ldots,z^m)$. It is convenient to use multi-index notation $\nu = (\nu_1,\ldots,\nu_m)$, $|\nu| = \nu_1 + \ldots + \nu_m$, $(z-w)^\nu = (z^1-w^1)^{\nu_1} \ldots (z^m-w^m)^{\nu_m}$, then (1.10) can be written

$$(1.11) \quad s(z) = \sum_{|\nu|=0}^{\infty} a_\nu (z-w)^\nu \quad .$$

As in the case of one variable, it is easily seen that if $s(z)$ converges for some $z_* \in \mathbb{C}^m$ with $z_*^j \neq w^j$ $\forall j = 1,\ldots,m$, then, writing $R^j = |z_*^j - w^j|$, $j = 1,\ldots,m$, (i) $s(z)$ converges absolutely for all z in the open polydisc $\Delta = \Delta(w,\underline{R})$; (ii) the convergence is uniform on any compact subset of Δ; (iii) hence by (1.9) the sum $s(z)$ is holomorphic on Δ; (iv) derivatives of $s(z)$ of all orders with respect to the real or complex variables may be found by differentiating the series term by term, the resulting series converging on Δ; (v) for $z \in \Delta$, the sum may be rearranged, or arranged in any fashion as a single series, for example, we may arrange (1.11) as a convergent power series in $z^1 - w^1$:

$$(1.12) \quad s(z) = \sum_{k=0}^{\infty} g_k(z^2,\ldots,z^m)(z^1-w^1)^k$$

where $g_k(z^2,\ldots,z^m) = \sum_{\nu_2,\ldots,\nu_m=0}^{\infty} a_{k,\nu_2,\ldots,\nu_m} (z^2-w^2)^{\nu_2} \ldots (z^m-w^m)^{\nu_m}$

is holomorphic for $|z^j-w^j| < R^j$, $j = 2,\ldots,m$.

As in the one variable case we can expand a holomorphic function locally in a power series as follows:

(1.13) Taylor's Theorem: Let f be holomorphic on an open subset U of $\mathbb{C}^m$. Then each point $w \in U$ has an open neighborhood N such that, on N, f is the sum of a convergent power series:

$$(1.14) \quad f(z) = \sum_{|\nu|=0}^{\infty} a_\nu (z-w)^\nu \quad .$$

In fact, (1.14) converges on any polydisc $\Delta(w,\underline{R})$ contained in U, the coefficients a_ν being uniquely determined by

$$(1.15) \quad a_\nu = \frac{1}{\nu!} \frac{\partial^\nu f}{\partial z^\nu}(w) \quad .$$

(Here we use the multi-index notation $\nu! = \nu_1! \dots \nu_m!$ and $\frac{\partial^\nu}{\partial z^\nu} = \frac{\partial^{\nu_1}}{\partial z^{\nu_1}} \dots \frac{\partial^{\nu_m}}{\partial z^{\nu_m}}$). The series (1.14) is called the Taylor series of f about w.

Proof. Let $\Delta(w,\underline{R})$ be a polydisc contained in U, let $z \in \Delta(w,\underline{R})$. Choose a polydisc $\Delta(w,\underline{r})$ containing z with $\underline{r} < \underline{R}$ (i.e. $r^j < R^j$, $\forall j = 1,\dots,m$). It can easily be shown that the series

$$(1.16)\qquad \sum_{\nu_1,\dots,\nu_m=0}^{\infty} \frac{(z^1-w^1)^{\nu_1}\dots(z^m-w^m)^{\nu_m}}{(\rho^1-w^1)^{\nu_1+1}\dots(\rho^m-w^m)^{\nu_m+1}}$$

is absolutely and uniformly convergent in ρ for $\rho \in \partial_0\Delta(w,\underline{r})$ to the sum $1/\{(\rho^1-z^1)\dots(\rho^m-z^m)\}$. Substituting this expression into Cauchy's Integral Formula (1.8) and interchanging the summation and integration gives the series (1.14) with coefficients

$$(1.17)\qquad a_\nu = \left(\frac{1}{2\pi i}\right)^m \int_{\partial_0\Delta(w,\underline{r})} \frac{f(\rho)\,d\rho^1\dots d\rho^m}{(\rho^1-w^1)^{\nu_1+1}\dots(\rho^m-w^m)^{\nu_m+1}} .$$

The alternative formula (1.15) for the coefficients follows by differentiating (1.14) term by term and putting $z = w$.

(1.18) Remarks: (i) It follows that any holomorphic function $f \in C^1(U)$ is C^∞. (ii) A function $f \in C^\infty(U)$ admitting a power series expansion (1.14) about any point of U is often called (complex) analytic. We have thus shown that f is holomorphic if and only if it is complex analytic.

C. Properties of holomorphic functions

Many familiar properties in one complex variable generalize to several complex variables. We give some important examples:

(1.19) Theorem (Sums, Products, Quotients of Holomorphic Functions): If f,g are holomorphic functions on an open subset U of $\mathbb{C}^m$,

(i) $f + g$ and $f \cdot g$ are holomorphic on U;
(ii) if f is nowhere zero on U, then $1/f$ is holomorphic on U.

Proof. (i) By direct computation we find

$$(1.20)\quad \frac{\partial}{\partial \bar{z}^j}(f+g) = \frac{\partial f}{\partial \bar{z}^j} + \frac{\partial g}{\partial \bar{z}^j}$$

$$(j = 1,\dots,m)$$

$$(1.21)\quad \frac{\partial}{\partial \bar{z}^j}(f\cdot g) = \frac{\partial f}{\partial \bar{z}^j}\cdot g + f\cdot\frac{\partial g}{\partial \bar{z}^j}$$

(ii) Use (1.21) on $f\cdot\frac{1}{f} = 1$.

Remark: The set of holomorphic functions on U is denoted by $\mathcal{O}(U)$ after Oka. We have shown that $\mathcal{O}(U)$ is a ring. Its importance will be shown in §3.

For the next result, say that a C^∞ mapping $g: U \to \mathbb{C}^r$ ($U \subseteq \mathbb{C}^m$ open) is holomorphic if each component $g^\alpha: U \to \mathbb{C}$ is holomorphic.

(1.22) Theorem (Composition of holomorphic functions): Let $U \subseteq \mathbb{C}^m$, $U' \subseteq \mathbb{C}^n$ be open and let $f: U \to U'$, $z \to f(z)$, and $g: U' \to \mathbb{C}^r$, $w \to g(w)$ be holomorphic. Then $g\circ f: U \to \mathbb{C}^r$ is holomorphic.

Proof. By direct computation, we can establish the "complex form of the chain rule" valid for any C^1 functions $f: U \to U'$, $g: U' \to \mathbb{C}^r$:

$$(1.23)\quad \frac{\partial}{\partial \bar{z}^j}(g^\alpha\circ f) = \sum_{k=1}^{n}\left(\frac{\partial g^\alpha}{\partial w^k}\cdot\frac{\partial f^k}{\partial \bar{z}^j} + \frac{\partial g^\alpha}{\partial \bar{w}^k}\cdot\frac{\partial \bar{f}^k}{\partial \bar{z}^j}\right) \qquad (\alpha = 1,\dots,r)$$

(Here $\bar{f}: U \to U'$ is defined by $\bar{f}(z) = \overline{f(z)}$.) If g,f are holomorphic, the right-hand side is zero, hence $g^\alpha\circ f$ is holomorphic for $\alpha = 1,\dots,r$ as required.

(1.24) Remarks: (i) For any C^1 functions f,g, we have "the other" chain rule

$$(1.25)\quad \frac{\partial}{\partial z^j}(g^\alpha\circ f) = \sum_{k=1}^{n}\left(\frac{\partial g^\alpha}{\partial w^k}\,\frac{\partial f^k}{\partial z^j} + \frac{\partial g^\alpha}{\partial \bar{w}^k}\,\frac{\partial \bar{f}^k}{\partial z^j}\right)$$

Note that $\dfrac{\partial \bar{f}^k}{\partial \bar{z}^j} = \overline{\left(\dfrac{\partial f^k}{\partial z^j}\right)}$ and $\dfrac{\partial \bar{f}^k}{\partial z^j} = \overline{\left(\dfrac{\partial f^k}{\partial \bar{z}^j}\right)}$.

(ii) Equations (1.20), (1.21) and similar formulae for $\dfrac{\partial}{\partial z^j}$, (1.23) and (1.25) illustrate how the operators $\dfrac{\partial}{\partial z^j}, \dfrac{\partial}{\partial \bar{z}^j}$ may be manipulated in much the same way as $\frac{\partial}{\partial x^j}$, $\frac{\partial}{\partial y^j}$.

(1.26) Identity Theorem: If $f,g\colon D \to \mathbb{C}$ are holomorphic functions on a domain (= connected open set) D of $\mathbb{C}^m$ and $f(z) = g(z)$ for all points z in a non-empty open subset U of D then $f(z) = g(z)$ for all $z \in D$.

Proof. Let E be the interior of the set of points z for which $f(z) = g(z)$. Then E is a non-empty open subset of D. To show that $E = D$ it suffices to show that E is relatively closed in D: Let $w \in D \cap \bar{E}$, then for sufficiently small $\underline{r} > 0$, the polydisc $\Delta(w,\underline{r}) \subseteq D$. Since $w \in \bar{E}$, we may choose $w' \in \Delta(w,\frac{\underline{r}}{2})$ with $w' \in E$, then $\Delta(w',\frac{\underline{r}}{2}) \subseteq D$. The function $f-g$ is holomorphic on this polydisc and therefore has a power series expansion convergent on this polydisc. But since $w' \in E$, $f-g$ is identically zero on some neighborhood of w' and so by (1.15) the power series is identically zero on $\Delta(w',\frac{\underline{r}}{2})$. In particular $f=g$ around w and so $w \in E$ showing that E is relatively closed as desired.

(1.27) Maximum Modulus Theorem: Let f be holomorphic on a domain D of $\mathbb{C}^m$. If there exists $w \in D$ with $|f(z)| \le |f(w)|$ for all $z \in D$, then f is constant on D.

Proof. It can easily be seen from Cauchy's Integral Formula that on any closed polydisc $\bar{\Delta} = \bar{\Delta}(w,\underline{r})$ in D, we have a mean value property:

$$(1.28)\qquad f(w) = \frac{1}{\mathrm{vol}(\bar{\Delta})} \int_{\bar{\Delta}} f(\rho)\, V(\rho)$$

where $\mathrm{vol}(\bar{\Delta})$ denotes the volume of $\bar{\Delta}$ and $V(\rho)$ the volume element in $\mathbb{C}^m = \mathbb{R}^{2m}$. Hence

$$0 \le \int_{\bar{\Delta}} \{|f(w)|-|f(\rho)|\}\, V(\rho) = \mathrm{vol}(\bar{\Delta})\,|f(w)| - \int_{\bar{\Delta}} |f(\rho)|\, V(\rho) \le 0 \quad .$$

It follows that $|f(w)| - |f(\rho)| = 0$ for all $\rho \in \bar{\Delta}$, so f has constant modulus on the polydisc. But then, for all z on the polydisc,

$$0 = \frac{\partial}{\partial z^j}\{f(z)\bar{f}(z)\} = \frac{\partial f}{\partial z^j}\,\bar{f}(z) \qquad \text{whence} \quad f(z) = 0$$

or all the first order partial derivatives of f vanish at z. It easily follows that f is constant on the polydisc, and so, by the identity theorem, is constant on D.

D. Domains of holomorphy

We here give some properties peculiar to the case of more than one complex variable.

(1.29) Definition: Let f be a holomorphic function on an open set U of $\mathbb{C}^m$ $(m \geq 1)$. A holomorphic (or analytic) extension of f is a holomorphic function F defined on a larger domain $V \supsetneq U$ such that $F|U=f$.

(1.30) Theorem: Let $0 < \underline{r}' < \underline{r} \leq \infty$. Any holomorphic function on a neighborhood U of $\Delta(0,\underline{r}) - \Delta(0,\underline{r}')$ in $\mathbb{C}^m$ $(m \geq 2)$ extends uniquely to a holomorphic function on $\Delta(0,\underline{r})$.

Proof. For simplicity of notation we take m = 2. We extend f along each slice z^1 = constant by Cauchy's Integral Formula, i.e. define

$$F(z^1,z^2) = \frac{1}{2\pi i} \int_{|\rho^2|=r^2} \frac{f(z^1,\rho^2)}{z^2-\rho^2}\, d\rho^2 .$$

F is defined throughout $\Delta(0,\underline{r})$, is holomorphic in z^2 and, since $\frac{\partial f}{\partial \bar{z}^1} = 0$, is holomorphic in z^1 also. On the open set $(r')^1 < z^1 < r^1$, by Cauchy's Integral Formula, F = f. Hence, by the identity theorem F = f on the whole of U and so F is an extension. Uniqueness is immediate from the Identity Theorem.

In the same spirit we have

(1.31) Riemann Removable Singularity Theorem: Any holomorphic function defined on a deleted neighborhood $U \setminus \{w\}$ of a point w in $\mathbb{C}^m$ $(m \geq 2)$ can be extended uniquely to a holomorphic function on U.

These results are clearly false for m = 1, for example $\frac{1}{z}$ on $\mathbb{C} \setminus \{0\}$ cannot be extended to $\mathbb{C}$. (In contrast, $\frac{1}{z^1}$ on $\mathbb{C}^2 \setminus \{0\}$ does not satisfy the hypotheses of (1.31).) To study this phenomenon further we need some definitions:

(1.32) Definition: Let D be a domain of $\mathbb{C}^m$ $(m \geq 1)$. We say that D is a domain of holomorphy if given any non-empty connected open subsets

U, U_1 of $\mathbb{C}^m$ with $U \not\subseteq D$ and $U_1 \subseteq U \cap D$ there exists a holomorphic function f on D such that we can find no holomorphic function g on U with $g = f$ on U_1.

(1.33) Remarks: (i) Roughly, D is a domain of holomorphy if for any given part of the boundary, there is a holomorphic function f on D which does not extend across it.

(ii) Clearly any domain of $\mathbb{C}$ is a domain of holomorphy (consider $1/(z-w)$ for $w \in \partial D$). In contrast, theorems (1.30), (1.31) exhibit domains of $\mathbb{C}^m$ which are not domains of holomorphy for $m \geq 2$.

Domains of holomorphy may be studied using sheaf theory (see §3), see also references [G-R], [Gr-R], [Hö], [Na].

§2 ANALYSIS ON MANIFOLDS

We first recall (without details) the basic elements of analysis on real manifolds. We then examine the new structures introduced on passing to the complex case.

A. Manifolds

Let M be a topological space which we suppose Hausdorff or T_2 (which means that any two points of M are contained in disjoint open neighborhoods), and (for convenience) connected. At times, we shall also suppose that M is paracompact, i.e. any covering $\{V_\alpha\}$ of M by open sets admits a locally finite refinement $\{U_j\}$ ($\forall j\ \exists\alpha: U_j \subset V_\alpha$, $\{U_j\}$ is a covering of M and each point of M has a neighborhood intersecting only a finite number of U_j's). This would for instance be satisfied if M had a countable basis for its topology.

(2.1) A real chart on M is an open set $U \subset M$ and a homeomorphism φ from U to an open set of $\mathbb{R}^m$. If (U,φ) and (V,ψ) are two charts and $U \cap V \neq \emptyset$, we can consider the transition functions $\varphi \circ \psi^{-1}$ from $\psi(U \cap V)$ to $\varphi(U \cap V)$. We shall say that the charts are compatible if $\varphi \circ \psi^{-1}$ is a C^∞ diffeomorphism, i.e. a bijective map which is C^∞ together with its inverse.

A covering of M by a family of compatible charts will be called a differentiable atlas. The union of all charts compatible with those of a given atlas is a maximal atlas, and we shall say that is defines a differentiable structure on M.

(2.2) Definition: A differentiable manifold is a Hausdorff topological space equipped with a differentiable structure.

Its dimension is m, the dimension of the Euclidean spaces used for the charts.

We could also have defined C^k manifolds $(0 \leq k < \infty)$ by requesting the transition maps and their inverses to be C^k differentiable.

In the same way:

(2.3) Definition: A complex manifold (of complex dimension m) is a differentiable manifold together with a maximal atlas of charts $U \to \mathbb{C}^m$ such that the transition functions are holomorphic.

The maximal atlas is called the <u>complex structure</u> of the manifold. If $(x^1,\dots,x^m)$ are coordinates in $\mathbb{R}^m$ (or $(z^1,\dots,z^m)$ in $\mathbb{C}^m$), their composition with φ defines a local system of coordinates in U. We shall denote a point x of U by $(x^1,\dots,x^m)$, meaning $(x^1(\varphi(x)),\dots$ $\dots, x^m(\varphi(x)))$.

A real function on M is called C^∞ if, in each chart, $f\circ\varphi^{-1}$ is C^∞ from $\varphi(U)\subset\mathbb{R}^m$ to $\mathbb{R}$. Similarly, a map between manifolds is C^∞ if its representations in charts are, and is holomorphic if its representations are.

B. Examples of complex manifolds

(2.4) The Riemann surfaces are complex manifolds of dimension one.

(2.5) Let $\Lambda = \mathbb{Z}^k$ be a lattice affinely embedded in $\mathbb{C}^m$. The quotient $\mathbb{C}^m/\Lambda$ is a complex manifold. If $k = 2m$, it is compact and is called a complex torus.

(2.6) An essential example in complex theory is the <u>complex projective space</u> $\mathbb{P}^m = P^m(\mathbb{C})$ defined as the set of complex lines through the origin in $\mathbb{C}^{m+1}$. Such a line ℓ is determined by any $Z(\neq 0)\in\ell$, so that $P^m(\mathbb{C}) = \left\{Z\in\mathbb{C}^{m+1}\setminus\{0\}\right\}/Z\sim\lambda Z$. Denote by Z^j the coordinates of Z.

A holomorphic atlas (defining the complex structure of $P^m(\mathbb{C})$) can be obtained as follows: on the subset $U_j = \{Z:\ Z^j \neq 0\}$, the map

$$\varphi_j:\ (Z^0,\dots,Z^m)\to\left(\frac{Z^0}{Z^j},\dots,\widehat{\frac{Z^j}{Z^j}},\dots,\frac{Z^m}{Z^j}\right) = (z^1_{(j)},\dots,z^m_{(j)})\in\mathbb{C}^m$$

is bijective ($\widehat{\cdot}$ means that the element $\cdot$ is deleted).
In $\varphi_j(U_j\cap U_k)\subset\mathbb{C}^m$, with $j<k$, we observe that the transition maps

$$\varphi_k\circ\varphi_j^{-1}(z^1_{(j)},\dots,z^m_{(j)}) = \left(\frac{z^1_{(j)}}{z^k_{(j)}},\dots,\frac{z^j_{(j)}}{z^k_{(j)}},\frac{1}{z^k_{(j)}},\frac{z^{j+1}_{(j)}}{z^k_{(j)}},\dots,\widehat{\frac{z^k_{(j)}}{z^k_{(j)}}},\dots,\frac{z^m_{(j)}}{z^k_{(j)}}\right)$$

are holomorphic.
The numbers $Z^0,\dots,Z^m$ are called homogeneous coordinates.

Note that $P^m(\mathbb{C})$ can also be seen as the quotient of the unit sphere in $\mathbb{C}^{m+1}$ by the action of $U(1)$ defined by $U(1)\times S^{2m+1}\to S^{2m+1}$: $(e^{i\alpha},Z)\to e^{i\alpha}Z$. We deduce that $P^m(\mathbb{C})$ is compact.

$P^m(\mathbb{C})$ can also be pictured as the completion of $\mathbb{C}^m$ by a hyperplane $H = P^{m-1}(\mathbb{C})$ at infinity. In coordinates, the inclusion $\mathbb{C}^m \to P^m(\mathbb{C})$ is $(z^1,\dots,z^m) \to (1,z^1,\dots,z^m)$ and H has equation $z^0 = 0$. H can be seen as the set of complex directions going to infinity in $\mathbb{C}^m$.

In particular, $P^1(\mathbb{C}) = \mathbb{C} \cup \{\infty\}$ is the real 2-sphere, or Riemann sphere.

C. Partition of unity

The existence of partitions of unity is an important tool in the study of manifolds.

(2.7) Definition: Let $\{U_j\}$ be a locally finite covering of M. A partition of unity subordinate to $\{U_j\}$ is a family of smooth (= C^∞ differentiable) functions $\theta_j: M \to \mathbb{R}$ such that the support of θ_j is included in U_j and $\sum_j \theta_j = 1$ (At each point, this sum is well defined and finite as it contains only a finite number of non-zero terms.)

Theorem: Any open covering of a paracompact manifold M by open sets admits a locally finite refinement relative to which there exists a partition of unity.

For proofs, we refer for instance to ([D] vol. 3, [KN] vol. 1, [N]).

In the course of the proof, one can also establish the existence, for two disjoint closed sets A and B in M, of a smooth function with value 0 on A and 1 on B.

Note that this theorem has no equivalent in the holomorphic case - indeed θ_j is zero outside U_j and cannot be holomorphic because of the identity theorem of section 1.C. This is one motivation for the study of holomorphic objects by means of real ones (see in section 3.D the proof of the Dolbeault theorem).

D. Vector bundles

(2.8) Definition: A real (resp. complex) vector bundle is a triple (E,M,Π) such that E and M are real manifolds, $\Pi: E \to M$ a C^∞ map, $\Pi^{-1}(x) = \mathbb{R}^k$ (resp. $\mathbb{C}^k$) $\forall x \in M$, and $\forall x_0 \in M$, there is a neighborhood U of x_0 in M and a diffeomorphism $\varphi_U: \Pi^{-1}(U) \to U \times \mathbb{R}^k$

(resp. $U \times \mathbb{C}^k$) which for all $x \in U$ induces a linear isomorphism from $\Pi^{-1}(x)$ to $\mathbb{R}^k$ (resp. $\mathbb{C}^k$).

Π is called the <u>projection</u>, φ_U a <u>trivialisation</u> over U, $E_x = \Pi^{-1}(x)$ the <u>fibre</u> at x and k the <u>rank</u> of the bundle. If $k = 1$, E is called a <u>line bundle</u>.

If φ_U and φ_V are two trivialisations with $U \cap V \neq \emptyset$, the map $g_{UV}: U \cap V \to GL(k)$ defined by

$$g_{UV}(x) = (\varphi_U \circ \varphi_V^{-1})\Big|_{x \times \mathbb{R}^k \ (\text{or } x \times \mathbb{C}^k)}$$

is C^∞ and is called a <u>transition function</u>.

These satisfy the "cocycle conditions" (see §3) $g_{UV} \circ g_{VU} = I$ and $g_{UV} \circ g_{VW} \circ g_{WU} = I$.

Conversely, given M, an open covering and a family of functions satisfying these conditions, there is a unique vector bundle of which they are transition functions.

In fact, $E \to M$ is defined (as a set) as the union $\bigcup_\alpha U_\alpha \times \mathbb{R}^k$ (or $\times \mathbb{C}^k$) with points $(x,\lambda) \in U_\beta \times \mathbb{R}^k$ (or ...) and $(x, g_{\alpha\beta}(x).\lambda) \in U_\alpha \times \mathbb{R}^k$ identified. The manifold structure is induced by the inclusions $U_\alpha \times \mathbb{R}^k \to E$.

<u>Definition</u>: A smooth <u>section</u> of E is a smooth map $\sigma: M \to E$ such that $\Pi \circ \sigma = \mathrm{Id}$.

A <u>frame</u> for E over $U \subset M$ is a collection $(\sigma_1,\ldots,\sigma_k)$ of sections of M such that at each point x, $(\sigma_1(x),\ldots,\sigma_k(x))$ is a basis of E_x.

Giving a frame for E over U amounts to specifying a trivialisation. Indeed, given $(\sigma_1,\ldots,\sigma_k)$, a trivialisation is obtained by setting

$$\varphi_U(\lambda) = (x,(\lambda^1,\ldots,\lambda^k)) \quad \text{where} \quad \lambda = \lambda^j\, \sigma_j(x)$$

(We use the summation convention that $\lambda^j \sigma_j = \sum_{j=1}^{k} \lambda^j \sigma_j$); and given a trivialisation φ_U, the sections $\sigma_j(x) = \varphi_U^{-1}(x, e_j)$ form a frame (e_i is a basis of $\mathbb{R}^k$ or $\mathbb{C}^k$).

A bundle is called <u>trivial</u> if it admits a global trivialisation (over all M).

In general, operations on vector spaces induce similar operations on bundles, by applying them to each fibre. For bundles E and F, defined by transition functions $g_{\alpha\beta}$ and $h_{\alpha\beta}$ on the same covering $\{V_\alpha\}$ of M , we can then construct: (in the complex case, $\mathbb{C}^k$ will simply replace $\mathbb{R}^k$)

(2.9) The <u>dual bundle</u> $E^* \to M$, with fibre $E^*_x = (E_x)^*$ and transition functions $j_{\alpha\beta} = {}^t g_{\alpha\beta}^{-1}$.

(2.10) The <u>direct sum</u>, or Whitney sum, $E \oplus F$, with fibre $E_x \oplus F_x$ and transition functions

$$\begin{pmatrix} g_{\alpha\beta} & 0 \\ 0 & h_{\alpha\beta} \end{pmatrix} \in GL(\mathbb{R}^k \oplus \mathbb{R}^\ell)$$

and the <u>tensor product</u> $E \otimes F$, with fibre at x $E_x \otimes F_x$ and transition functions $j_{\alpha\beta} = g_{\alpha\beta}(x) \otimes h_{\alpha\beta}(x) \in GL(\mathbb{R}^k \otimes \mathbb{R}^\ell)$.

(2.11) Denote by $\Lambda^p(V)$ the space of antisymmetric p times contravariant tensors of a vector space V. Recall that a p-tensor v is antisymmetric if $s.v = \varepsilon(s)v$ for all permutations s of p elements, where $\varepsilon(s) = \pm 1$ is the signature of the permutation and s acts on v by $(s.v)^{j_1,\dots,j_p} = v^{s^{-1}j_1,\dots,s^{-1}j_p}$.

$\Lambda^p(V)$ is the image of the space of p-tensors by the projection

$$A: v \to \frac{1}{p!} \sum_s \varepsilon(s) s.v$$

$$\text{Set} \quad \Lambda(V) = \sum_{p=0}^{\infty} \Lambda^p(V), \text{ where } \Lambda^o(V) = \mathbb{R} .$$

With the product $\wedge$ defined for $\lambda \in \Lambda^p(V)$ and $\mu \in \Lambda^q(V)$ by $\lambda \wedge \mu = \frac{(p+q)!}{p!q!} A(\lambda \otimes \mu)$
$\Lambda(V)$ becomes an exterior algebra. $\lambda \wedge \mu$ belongs to $\Lambda^{p+q}(V)$ and $\lambda \wedge \mu = (-1)^{pq} \mu \wedge \lambda$.

If (e_j) is a basis of V, a basis of $\Lambda^p(V)$ is given by $(e_{j_1} \wedge \dots \wedge e_{j_p})$ where $1 \le j_1 < j_2 < \dots < j_p \le m = \dim V$. This implies that $\Lambda^p(V) = 0$ for $p > m$ and that $\dim \Lambda^p V = \binom{m}{p}$ and $\dim \Lambda(V) = 2^m$.

The coefficient $\frac{(p+q)!}{p!q!}$ in the definition of Λ is not essential and does not always appear in the texts to which we refer. It is used in [Sp] and helps to simplify the coefficients of other formulae.

If E is a vector bundle, we denote by $\Lambda^p(E^*)$ the bundle whose fibre at x is $\Lambda^p(E_x^*)$.

(2.12) If $f: M \to N$ is a differentiable map and E a vector bundle over N, the pull-back bundle $f^{-1}E \to M$ has fibre $(f^{-1}E)_x = E_{f(x)}$ and transition functions equal to the composition with f of those of E.

E. Real tangent and cotangent bundles

(2.13) A vector X tangent to M at a point x can be defined in two different ways:

i) it is the equivalence class of functions (or paths) $F: \mathbb{R} \to M$ having at 0 same value x and same first derivatives.

ii) it is a derivation of the space $C(M)$ (or $C^\infty(M)$) of smooth real functions on M, i.e. a map $X: C(M) \to \mathbb{R}$ such that

$$\begin{aligned} X(f+g) &= Xf + Xg & f,g \in C(M) \\ X(k\cdot f) &= k\cdot Xf & k \in \mathbb{R} \\ X(f\cdot g) &= Xf\cdot g(x) + f(x)\cdot Xg \end{aligned}$$

The equivalence of these notions can be observed in a coordinate chart. On one hand, the equivalence class of paths containing F defines a derivation by

$$Xf = \frac{d}{dt}(f\circ F) = \frac{\partial f}{\partial x^j}\frac{dx^j}{dt} \quad \text{or}$$
$$X = \frac{dx^j}{dt}\frac{\partial}{\partial x^j}$$

On the other hand, applying a derivation X to the Taylor series with remainder $f(x) = f(x_o) + (x^j - x_o^j)\left(\frac{\partial f}{\partial x^j}\right)_{x_o} + (x^j - x_o^j)(x^k - x_o^k)\alpha_{jk}(x)$ and using its properties, we get $Xf = X(x^j)\frac{\partial f}{\partial x^j}$ or $X = X(x^j)\frac{\partial}{\partial x^j}$.

A natural basis for the real tangent space (space of tangent vectors) in the chart $(x^1,\dots,x^m)$ is therefore $\left(\frac{\partial}{\partial x^1},\dots,\frac{\partial}{\partial x^m}\right)$.

(2.14) Let T_xM denote the <u>tangent space</u> at the point $x \in M$. If $f\colon M \to N$ is a smooth map, it induces a map $f_{*x}\colon T_xM \to T_{f(x)}N$, using naturally either definition of tangent vectors. In charts (x^j) around x and (v^α) around $f(x)$, we have

$$f_{*x}\colon T_xM \to T_{f(x)}N$$
$$X^j \frac{\partial}{\partial x^j} \to \frac{\partial f^\alpha}{\partial x^j} X^j \frac{\partial}{\partial v^\alpha} .$$

In some texts, f_* is denoted by df (and note that if $N = \mathbb{R}$, it coincides with definition (2.15) below).

This justifies the following definition: let TM be the disjoint union of the tangent spaces T_xM, $x \in M$.

TM is a vector bundle with fibre at x T_xM, projection mapping T_xM on x and with the transition function for two charts U_α and U_β with coordinates (x^j) and (y^k) defined by

$$g_{\alpha\beta}\colon U_\alpha \cap U_\beta \to GL(m,\mathbb{R})$$
$$x \to \frac{\partial x^j}{\partial y^k} .$$

For two vector fields X and Y, their Lie bracket $[X,Y]$ defined by $[X,Y]f = XYf - YXf$ is again a vector field.

(2.15) The dual to TM is called the <u>cotangent bundle</u> T^*M. Its sections are called <u>one-forms</u>.

For $f \in C(M)$ and $X \in C(TM)$ (i.e. X a vector field), we define the one-form df by $df(X) = Xf$. The dual basis to $\left(\frac{\partial}{\partial x^j}\right)$ is seen to be (dx^k), since $dx^k\left(\frac{\partial}{\partial x^j}\right) = \frac{\partial x^k}{\partial x^j} = \delta^k_j$.

If $f\colon M \to N$, a one-form φ on N induces a one-form $f^*\varphi$ on M by $(f^*\varphi)X = \varphi(f_*X)$. In a chart, if $\varphi = \varphi_a\, dx^a$ then $f^*\varphi = \frac{\partial f^a}{\partial x^j} \varphi_a\, dx^j$.

(2.16) A tensor field of type (r,s) on M is a section of the space $\otimes^r TM \otimes \otimes^s T^*M$.

F. p-forms and the Rham cohomology

(2.17) A real p-form on M is a section of $\Lambda^p T^*M$. Thus, it acts on p vector fields in an antisymmetric manner. Locally, it can be expressed as

$$\sum \lambda_{j_1,\dots,j_p} dx^{j_1} \wedge \dots \wedge dx^{j_p} \text{ , with } j_1 < j_2 < \dots < j_p .$$

We shall denote by $A^p(M)$ or $A^p(M,\mathbb{R})$ the space of real p-forms.

For $\lambda \in A^p(M)$, the exterior differential of λ, $d\lambda$, is defined by its action on $p+1$ vectors as

$$(2.18)\quad d\lambda(X_1,\dots,X_{p+1}) = \sum_{j=1}^{p+1} (-1)^{j-1} X_j\ \lambda(X_1,\dots,\hat{X}_j,\dots,X_{p+1})$$

$$+ \sum_{j<k} (-1)^{j+k}\ \lambda([X_i,X_j]X_1,\dots,\hat{X}_i,\dots,\hat{X}_j,\dots,X_{p+1}) .$$

One can check that $d\lambda$ is a $p+1$ form (in particular, it depends only on the values of the X_j's at the point in consideration).

In local coordinates, with the above expression of λ,

$$d\lambda = \frac{\partial\lambda_{j_1,\dots,j_p}}{\partial x^k} dx^k \wedge dx^{j_1} \wedge \dots \wedge dx^{j_p} .$$

A basic property of d is that for all p-forms λ, $dd\lambda = d^2\lambda = 0$, or $d^2 = 0$.

Note that d is an antiderivation in the sense that $d(\lambda \wedge \mu) = d\lambda \wedge \mu + (-1)^p \lambda \wedge d\mu$. d is in fact determined by this condition together with df = differential of f and $d.df = 0$.

If $f\colon M \to N$ is a smooth map, f induces $f^*\colon A^p(N) \to A^p(M)$ by $(f^*\lambda)\ (X_1,\dots,X_p) = \lambda(f_*X_1,\dots,f_*X_p)$. One can then show that $d\circ f^* = f^*\circ d$.

(2.19) Definition: A p-form λ is closed, or is a cocycle, if $d\lambda = 0$ and is exact, or is a coboundary, if $\lambda = d\mu$ for a certain $p-1$-form μ.

Let $Z^p(M,\mathbb{R})$ denote the space of closed forms and $B^p(M,\mathbb{R}) = dA^{p-1}(M,\mathbb{R})$ the space of exact forms. Since $d^2 = 0$, $B^p(M,\mathbb{R}) \subset Z^p(M,\mathbb{R})$.

(2.20) <u>Definition</u>: The pth <u>de Rham cohomology group</u> of M is the quotient

$$H^p_{DR}(M,\mathbb{R}) = \frac{Z^p(M,\mathbb{R})}{B^p(M,\mathbb{R})} .$$

For the spaces of complex valued p-forms $\Lambda^p(M,\mathbb{C}) = \Lambda^p(M,\mathbb{R}) \otimes \mathbb{C}$, the same construction leads to $H^p_{DR}(M,\mathbb{C}) = H^p_{DR}(M,\mathbb{R}) \otimes \mathbb{C}$.

(2.21) A basic tool in the study of these groups (see section 3.D) is the

<u>Poincaré lemma</u>: Any closed form on a ball of $\mathbb{R}^m$ is exact.

For a proof of this classical (and not too difficult) result, we refer e.g. to [G], [N], [Wa].

We shall later prove a complex analogue of this result.

G. <u>Integration of forms</u>

An m-dimensional manifold M is called <u>orientable</u> if it admits a nowhere zero m-form, or equivalently if it admits an atlas for which all transition functions have positive Jacobian determinant.

If a specific m-form is chosen, M is said to be oriented.

Let M be a paracompact oriented manifold. (The condition of orientability is not necessary - see e.g. [E] - but it simplifies matters and we shall see in (2.26) that is is satisfied for complex manifolds).

Let λ be a m-form on M.

Choose a locally finite covering of M by coordinate charts $\{U_j,\varphi_j\}$ with transition functions respecting orientation and a smooth associated partition of unity (η_j).

The form $\eta_j\cdot\lambda$ has support in U_j and can be transported to $\mathbb{R}^m$ by φ_j^{-1}. $(\varphi_j^{-1})^*\eta_j\cdot\lambda$ is a multiple of $dx^1\wedge\ldots\wedge dx^m$.

(2.22) <u>Definition</u>:

$$\int_M \lambda = \sum_j \int_{\mathbb{R}^m} (\varphi_j^{-1})^*(\eta_j\cdot\lambda)$$

One checks - using the change of variables formula - that this expression is independent of the choice of the partition of unity.

If M is not compact, we take the usual precaution in supposing this summation absolutely convergent to a finite quantity.

Let M be a manifold with boundary (i.e. a space with an atlas such that some charts have as image an open set of a closed half Euclidean space $\mathbb{R}^{m-1} \times [0,\infty)$, the boundary ∂M being the inverse image of the pieces of hyperplane $\mathbb{R}^{m-1} \times \{0\}$). We have the important

(2.23) Stokes theorem: If M is an m-manifold with boundary ∂M and λ an (m-1)-form on M with compact support, then

$$\int_{\partial M} \lambda = \int_M d\lambda .$$

This result is quite easily proved in local charts (see e.g. [G], [N]).

H. Complex manifolds

With more details, we now turn to the case of complex manifolds.

Let M be a complex m-dimensional manifold (so that its real dimension is $2m$).

In a chart U, denote by (z^j) a system of coordinates, where $z^j = x^j + iy^j$.

The tangent space to M at a point z in U is then generated over $\mathbb{R}$ by $\left(\frac{\partial}{\partial x^j}, \frac{\partial}{\partial y^j}\right)$.

In that space, multiplication by i induces an operator J, mapping $\frac{\partial}{\partial x^j}$ to $\frac{\partial}{\partial y^j}$ and $\frac{\partial}{\partial y^j}$ to $-\frac{\partial}{\partial x^j}$. We have $J^2 = -I$ (I = identity).

Consider now the complexified tangent space $T_z^{\mathbb{C}}M = T_zM \otimes \mathbb{C}$. This space is generated over $\mathbb{C}$ by $\left(\frac{\partial}{\partial x^j}, \frac{\partial}{\partial y^j}\right)$ or, using the notations of (1.1) by $\left(\frac{\partial}{\partial z^j}, \frac{\partial}{\partial \bar{z}^j}\right)$.

J extends by linearity to an operator still called J on $T_z^{\mathbb{C}}M$, and again $J^2 = -I$. J has therefore two eigenvalues i and $-i$ and we denote the associated eigenspaces by $T_z'M$ and $T_z''M$. Then $T_z^{\mathbb{C}}M = T_z'M \oplus T_z''M$.

We observe that $J\frac{\partial}{\partial z^j} = \frac{1}{2}J\left(\frac{\partial}{\partial x^j} - i\frac{\partial}{\partial y^j}\right) = \frac{1}{2}\left(\frac{\partial}{\partial y^j} + i\frac{\partial}{\partial x^j}\right) =$ $\frac{1}{2}i\left(\frac{\partial}{\partial x^j} - i\frac{\partial}{\partial y^j}\right) = i\frac{\partial}{\partial z^j}$ and similarly that $J\frac{\partial}{\partial \bar{z}^j} = -i\frac{\partial}{\partial \bar{z}^j}$.

Over $\mathbb{C}$, $T_z'M$ is hence generated by $\left(\frac{\partial}{\partial z^j}\right)$ and $T_z''M$ by $\left(\frac{\partial}{\partial \bar{z}^j}\right)$.

In order to define J, we restricted attention to a chart. We shall now see that J is unchanged by holomorphic changes of charts, so that it is globally defined on M.

Indeed, let U and V be two charts, with coordinates (z^j) and (w^a) and call $f = \varphi_V \circ \varphi_U^{-1}$ the change of coordinates.

In the basis $\left(\frac{\partial}{\partial z^j}, \frac{\partial}{\partial \bar{z}^j}\right)$ and $\left(\frac{\partial}{\partial w^a}, \frac{\partial}{\partial \bar{w}^a}\right)$, the matrix expression of the transition function f_* induced by f on $T^{\mathbb{C}}M$ is

$$\begin{pmatrix} \frac{\partial f^a}{\partial z^j} & \frac{\partial f^a}{\partial \bar{z}^j} \\ \frac{\partial \overline{f^a}}{\partial z^j} & \frac{\partial \overline{f^a}}{\partial \bar{z}^j} \end{pmatrix} = \begin{pmatrix} \frac{\partial f^a}{\partial z^j} & \frac{\partial f^a}{\partial \bar{z}^j} \\ \overline{\frac{\partial f^a}{\partial \bar{z}^j}} & \overline{\frac{\partial f^a}{\partial z^j}} \end{pmatrix}.$$

f is holomorphic iff the matrix reduces to

$$\begin{pmatrix} \frac{\partial f^a}{\partial z^j} & 0 \\ 0 & \overline{\frac{\partial f^a}{\partial z^j}} \end{pmatrix}.$$

Since $J = \begin{pmatrix} iI & 0 \\ 0 & -iI \end{pmatrix}$, we have clearly $J \circ f_* = f_* \circ J$, insuring that J is globally defined on M.

Moreover, we observe that $f_*(T_z'M) \subset T_z'M$ iff f is holomorphic.

(2.24) <u>Definition</u>: J is called the induced <u>almost complex structure</u> of M. (See section 2.L below for a study of this structure.)

(2.25) <u>Definition</u>: $T'M$ and $T''M$ are called respectively the <u>holomorphic</u> and <u>antiholomorphic tangent bundles</u>.

As real bundles, $T'M$ and $T''M$ are both isomorphic to TM. Indeed,

$T'M = \{X - iJX \mid X \in TM\}$ and $T''M = \{X + iJX \mid X \in TM\}$. This relates real and holomorphic geometry.

Denoting, as above, complex conjugation by $\bar{\ }$, we see that $\overline{T'M} = T''M$ and that $(T'M, i)$ is complex-isomorphic to (TM, J) and $(\overline{T''M, i})$.

Other consequences of the above calculation are:

(2.26) <u>Proposition</u>: Any complex manifold is orientable.

Indeed, the Jacobian determinant of a change of charts is given by

$$\det \begin{pmatrix} \frac{\partial f^a}{\partial z^j} & 0 \\ 0 & \overline{\frac{\partial f^a}{\partial z^j}} \end{pmatrix} = \left| \det\left(\frac{\partial f^a}{\partial z^j} \right) \right|^2 > 0 \quad .$$

(2.27) <u>Definition</u>: The natural orientation on $\mathbb{C}^m$ is given by the 2m-form

$$\begin{aligned} \eta &= (\tfrac{i}{2})^m \, dz^1 \wedge d\overline{z^1} \wedge \ldots \wedge dz^m \wedge d\overline{z^m} \\ &= dx^1 \wedge dy^1 \wedge \ldots \wedge dx^m \wedge dy^m \quad . \end{aligned}$$

(2.28) <u>Proposition</u>: Let $f: M \to N$ be a C^∞ map. f is holomorphic iff $f_* T'_z M \subset T'_{f(z)} N \quad \forall z \in M$.

I. <u>(p,q)-forms and Dolbeault cohomology</u>

The decomposition $T^{\mathbb{C}}M = T'M + T''M$ induces a dual decomposition $T^{*\mathbb{C}}M = T^{*\prime}M + T^{*\prime\prime}M$, where $T^{*\prime}M$ is the dual of $T'M$ and $T^{*\prime\prime}M$ that of $T''M$. Bases of $T^{*\prime}_z M$ and $T^{*\prime\prime}_z M$ are given by (dz^j) and $(d\bar{z}^j)$, since

$$dz^j \frac{\partial}{\partial z^k} = d\bar{z}^j \frac{\partial}{\partial \bar{z}^k} = \delta^j_k \quad \text{and}$$

$$dz^j \frac{\partial}{\partial \bar{z}^k} = d\bar{z}^j \frac{\partial}{\partial z^k} = 0 \ .$$

Extending this decomposition to exterior products, we obtain $\Lambda^r T^{*\mathbb{C}}_z M = \bigoplus_{p+q=r} \left(\Lambda^p T^{*\prime}_z M \otimes \Lambda^q T^{*\prime\prime}_z M \right)$ and, going to sections,

(2.29) $A^r(M) = \bigoplus_{p+q=r} A^{p,q}(M)$

where $A^{p,q}(M) = \{\varphi \in A^r(M) \,|\, \varphi(z) \in \Lambda^p T_z{}'M \otimes \Lambda^q T_z{}''M\}$.

In local coordinates, a (p,q)-form can be written

(2.30) $\varphi(z) = \sum_{\substack{\#J=p\\ \#K=q}} \varphi_{JK}\, dz^J \wedge d\bar{z}^K$

where for $J = (j_1,\ldots,j_p)$, $dz^J = dz^{j_1} \wedge \ldots \wedge dz^{j_p}$ and for $K = (k_1,\ldots,k_q)$, $d\bar{z}^K = d\bar{z}^{k_1} \wedge \ldots \wedge d\bar{z}^{k_q}$.

We denote by $\Pi^{p,q}$ the projection $A^*(M) \to A^{p,q}(M)$, so that for $\varphi \in A^*M$, $\varphi = \sum \Pi^{p,q}\varphi = \sum \varphi^{(p,q)}$. If $\varphi \in A^{p,q}(M)$, then we see in local coordinates that $d\varphi(z) \in (\Lambda^p T_z^*{}'M \otimes \Lambda^q T_z^*{}''M) \wedge T_z^{*\mathbb{C}}M$ i.e. $d\varphi \in A^{p+1,q}(M) \oplus A^{p,q+1}(M)$.

(2.31) <u>Definition</u>: The operators

$$\partial: A^{p,q}(M) \to A^{p+1,q}(M)$$
$$\bar{\partial}: A^{p,q}(M) \to A^{p,q+1}(M)$$

are defined by

$$\partial = \Pi^{p+1,q} \circ d$$
$$\bar{\partial} = \Pi^{p,q+1} \circ d \quad .$$

(2.32) Therefore, $d = \partial + \bar{\partial}$.

In local coordinates, with φ given by (2.31):

$$\partial\varphi(z) = \sum_{J,K,j} \frac{\partial\varphi_{JK}}{\partial z^j}(z)\, dz^j \wedge dz^J \wedge d\bar{z}^K$$
$$\bar{\partial}\varphi(z) = \sum_{J,K,k} \frac{\partial\varphi_{JK}}{\partial \bar{z}^k}(z)\, d\bar{z}^k \wedge dz^J \wedge d\bar{z}^K \quad .$$

For a holomorphic map $f: M \to N$, we have seen in (2.28) that the decomposition into types is preserved for vector fields, and this implies that

$$f^*(A^{p,q}(N)) \subset A^{p,q}(M)$$

and hence $\bar\partial\circ f^* = f^*\circ\bar\partial$.

From (2.32) and $d^2 = 0$, we observe that for $\varphi \in A^{p,q}(M)$,

$$(\partial+\bar\partial)(\partial+\bar\partial)\varphi = \partial^2\varphi + \bar\partial^2\varphi + (\bar\partial\partial+\partial\bar\partial)\varphi = 0 \quad .$$

This sum contains terms in $A^{p+2,q}$, $A^{p,q+2}$ and $A^{p+1,q+1}$ which must all be zero. Hence:

(2.33) $\partial^2 = 0 \qquad \bar\partial^2 = 0 \qquad \partial\bar\partial = -\bar\partial\partial$.

The relation $\bar\partial^2 = 0$ allows to define a cohomology as in the real case (2.20).

Calling $Z^{p,q}_{\bar\partial}$ the space of $\bar\partial$-closed (p,q)-forms (i.e. such that $\bar\partial\varphi = 0$) and $B^{p,q}_{\bar\partial}$ the space $\bar\partial A^{p,q-1}(M)$, we have $B^{p,q}_{\bar\partial} \subset Z^{p,q}_{\bar\partial}$.

(2.34) <u>Definition</u>: The <u>Dolbeault cohomology groups</u> are

$$H^{p,q}_{\bar\partial}(M) = \frac{Z^{p,q}_{\bar\partial}(M)}{B^{p,q}_{\bar\partial}(M)} \quad .$$

Again, these groups will be extensively studied in further sections, and a basic tool will be an analogue of (2.21):

(2.35) <u>The $\bar\partial$-Poincaré lemma</u>: For $\Delta = \Delta(\underline{r})$ a polydisk in $\mathbb{C}^m$, $H^{p,q}_{\bar\partial}(\Delta)=0$ for $q \geq 1$.

<u>Proof</u>: We must show that if $\bar\partial\varphi = 0$, there exists a form ψ such that $\varphi = \bar\partial\psi$. We shall proceed in three steps.

i) <u>$\bar\partial$-Poincaré lemma in one variable</u>: given $g(z) \in C^\infty(\bar\Delta)$, where Δ is an open ball in $\mathbb{C}$, the function

$$f(z) = \frac{1}{2\pi i}\int_\Delta \frac{g(w)}{w-z}\, dw \wedge d\bar w$$

is defined and C^∞ in Δ and satisfies $\dfrac{\partial f}{\partial \bar z} = g$.

<u>Proof of i)</u>: For $z_0 \in \Delta$, choose $\varepsilon > 0$ such that the disk $\Delta(z_0,2\varepsilon)\subset\Delta$. By multiplication with a smooth function with value 1 on $\Delta(z_0,\varepsilon)$ and 0 outside $\Delta(z_0,2\varepsilon)$, and 1 minus that function, we can write

$$g(z) = g_1(z) + g_2(z)$$

where $g_1(z)$ is zero outside $\Delta(z_o,2\varepsilon)$ and $g_2(z)$ is zero inside $\Delta(z_o,\varepsilon)$.

Since for $z \in \Delta(z_o,\varepsilon)$, $g_2(z) = 0$, the integral

$$f_2(z) = \frac{1}{2\pi i} \int_\Delta g_2(w) \frac{dw \wedge d\overline{w}}{w-z}$$

is well defined and C^∞ in that domain, and

$$\frac{\partial f_2(z)}{\partial \overline{z}} = \frac{1}{2\pi i} \int_\Delta \frac{\partial}{\partial \overline{z}}\left(\frac{g_2(w)}{w-z}\right) dw \wedge d\overline{w} = 0 .$$

Since $g_1(z)$ has compact support in Δ, we see that

$$f_1(z) = \frac{1}{2\pi i} \int_\Delta g_1(w) \frac{dw \wedge d\overline{w}}{w-z} = \frac{1}{2\pi i} \int_{\mathbb{C}} g_1(w) \frac{dw \wedge d\overline{w}}{w-z} .$$

Setting $v = w - z$ and changing to polar coordinates, we see using (2.27) that

$$f_1(z) = \frac{1}{2\pi i} \int_{\mathbb{C}} g_1(v+z) \frac{dv \wedge d\overline{v}}{v} = -\frac{1}{\pi} \int_{\mathbb{C}} g_1(z+re^{i\theta})e^{-i\theta} dr \wedge d\theta$$

which is defined and C^∞ in z. Then

$$\frac{\partial f_1(z)}{\partial \overline{z}} = -\frac{1}{\pi} \int_{\mathbb{C}} \frac{\partial g_1}{\partial \overline{z}}(z+re^{i\theta})e^{-i\theta}\, dr \wedge d\theta$$

$$= \frac{1}{2\pi i} \int_\Delta \frac{\partial g_1}{\partial \overline{w}}(w) \frac{dw \wedge d\overline{w}}{w-z} .$$

But $g_1 = 0$ on $\partial\Delta$, so by Cauchy's formula (see §1), for $f = f_1 + f_2$, we get

$$\frac{\partial f}{\partial \overline{z}}(z) = \frac{\partial f_1(z)}{\partial \overline{z}} = g_1(z) = g(z) .$$

ii) $\overline{\partial}$-Poincaré lemma in a sub-polydisk.

First observe that if $\varphi = \sum_{\substack{\# J=p \\ \# K=q}} \varphi_{JK}\, dz^J \wedge d\overline{z}^K$ is $\overline{\partial}$ closed, then the forms $\varphi_J = \sum_{\# K=q} \varphi_{JK}\, d\overline{z}^K$ are also $\overline{\partial}$ closed. Indeed,

$$\overline{\partial}\varphi = \Sigma \frac{\partial \varphi_{JK}}{\partial \overline{z}^k} d\overline{z}^k \wedge dz^J \wedge d\overline{z}^K = 0$$

so that, separating types, all coefficients of the different dz^J's must be zero.

Also, if $\varphi_J = \bar\partial\eta_J$, then $\varphi = \pm\bar\partial(\sum_J dz^J \wedge \eta_J)$.

Hence, it suffices to prove the result for $(0,q)$-forms.

We shall prove in ii) that if φ is a $\bar\partial$-closed $(0,q)$-form on $\Delta(\underline{r})$ (where $\underline{r} = (r^1,\ldots,r^m)$ is the set of radii in the poly-disk) then for $\underline{s} < \underline{r}$ (i.e. $s^j < r^j$), there exists $\psi \in A^{0,q-1}(\Delta(\underline{s}))$ with $\bar\partial\psi = \varphi$ in $\Delta(\underline{s})$.

We'll say that $\varphi = \sum \varphi_K\, d\bar z^K \equiv 0$ modulo $(d\bar z^1,\ldots,d\bar z^n)$ iff $\varphi_K = 0$ for $K \not\subset \{1,\ldots,n\}$, or equivalently if $\{1,\ldots,n\}$ are the only values of k appearing in the expression. We shall show that if $\bar\partial\varphi = 0$ and $\varphi \equiv 0$ modulo $(d\bar z^1,\ldots,d\bar z^n)$, then there exists $\eta \in A^{0,q-1}(\Delta(s'))$ such that $\varphi - \bar\partial\eta \equiv 0$ modulo $(d\bar z^1,\ldots,d\bar z^{n-1})$. By recurrence (applying the result here to $\varphi-\bar\partial\eta$) we see that $\varphi - \bar\partial\mu \equiv 0$ in $\Delta(s)$.

So assume $\varphi \equiv 0$ modulo $(d\bar z^1,\ldots,d\bar z^n)$ and set

$$\varphi_1 = \sum_{K:n\in K} \varphi_K\, d\bar z^{K-(n)}$$

$$\varphi_2 = \sum_{K:n\notin K} \varphi_K\, d\bar z^{K}$$

so that $\varphi = \varphi_1 \wedge d\bar z^n + \varphi_2$ with $\varphi_2 \equiv 0$ modulo $(d\bar z^1,\ldots,d\bar z^{n-1})$.

If $l > n$, $\bar\partial\varphi_2$ contains no term with the factor $d\bar z^n \wedge d\bar z^l$. Since $0 = \bar\partial\varphi = \bar\partial\varphi_1 \wedge d\bar z^n + \bar\partial\varphi_2$, it follows that

$$\frac{\partial}{\partial \bar z^l}\varphi_K = 0 \quad \text{for } l > n \text{ and } K \text{ such that } n \in K. \tag{2.37}$$

Now set $\eta = \sum_{K:n\in K} \eta_K\, d\bar z^{K-(n)}$, where

$$\eta_K(z) = \frac{1}{2\pi i}\int_{|w^n|\le s^n} \varphi_K(z^1,\ldots,w^n,\ldots,z^m)\,\frac{dw^n\wedge d\bar w^n}{w^n-z^n} .$$

By i), $\dfrac{\partial\eta_K}{\partial \bar z^n}(z) = \varphi_K(z)$ and by (2.37), for $l > n$

$$\frac{\partial}{\partial \bar z^l}\eta_K(z) = \frac{1}{2\pi i}\int_{|w^n|\le s^n} \frac{\partial\varphi_K}{\partial \bar z^l}(z^1,\ldots,w^n,\ldots,z^m)\frac{dw^n\wedge d\bar w^n}{w^n-z^n} = 0.$$

Thus $\varphi-\bar\partial\eta = 0$ modulo $(d\bar z^1,\dots,d\bar z^{n-1})$ in $\Delta(s)$.

iii) $\bar\partial$-Poincaré lemma in $\Delta(r)$.

To obtain the full result, we shall approach $\underline{r}$ by an increasing sequence $\underline{r}_\nu$.

By ii), we can find $\psi'_\nu \in A^{o,q-1}(\Delta(\underline{r}_{\nu+1}))$ such that $\bar\partial\psi'_\nu = \varphi$ on $\Delta(\underline{r}_\nu)$. Multiplying this by a smooth function with value 1 in $\Delta(\underline{r}_\nu)$ and 0 outside $\Delta(\underline{r}_{\nu+1})$, we obtain $\psi_\nu \in A^{o,q-1}(\Delta(\underline{r}))$ such that $\bar\partial\psi_\nu = \varphi$ in $\Delta(\underline{r}_\nu)$.

We shall see how to choose the sequence (ψ_ν) so that it converges to a solution of the problem.

Suppose first $q \geq 2$. Construct as above ψ_ν and $\alpha \in A^{o,q-1}(\Delta(\underline{r}))$ with $\bar\partial\alpha = \varphi$ in $\Delta(\underline{r}_{\nu+1})$. Then in $\Delta(\underline{r}_\nu)$, $\bar\partial(\psi_\nu-\alpha) = 0$ and there exists $\beta \in A^{o,q-2}(\Delta(\underline{r}))$ such that $\bar\partial\beta = \psi_\nu-\alpha$ in $\Delta(\underline{r}_{\nu-1})$. Set $\psi_{\nu+1} = \alpha + \bar\partial\beta$. Then $\bar\partial\psi_{\nu+1} = \bar\partial\alpha = \varphi$ in $\Delta(\underline{r}_{\nu+1})$ and $\psi_{\nu+1} = \psi_\nu$ in $\Delta(\underline{r}_{\nu-1})$. Hence, the sequence (ψ_ν) converges uniformly on compact subsets to $\psi \in A^{o,q-1}(\Delta(\underline{r}))$ with $\bar\partial\psi = \varphi$.

Suppose now $q = 1$. As above, take $\psi_\nu \in C(\Delta(\underline{r}))$ such that $\bar\partial\psi_\nu = \varphi$ in $\Delta(\underline{r}_\nu)$ and $\alpha \in C(\Delta(\underline{r}))$ such that $\bar\partial\alpha = \varphi$ in $\Delta(\underline{r}_{\nu+1})$. $\psi_\nu-\alpha$ is then holomorphic in $\Delta(\underline{r}_\nu)$ and has a power series expansion around the origin in $\mathbb{C}^m$.

Truncate this series to obtain a polynomial β with $\sup_{\Delta(\underline{r}_{\nu-1})} |(\psi_\nu-\alpha)-\beta| < \frac{1}{2^\nu}$ and set $\psi_{\nu+1} = \alpha+\beta$.

Then $\bar\partial\psi_{\nu+1} = \bar\partial\alpha = \varphi$ in $\Delta(\underline{r}_{\nu+1})$, $\psi_{\nu+1} - \psi_\nu$ is holomorphic in $\Delta(\underline{r}_\nu)$ and $\sup_{\Delta(\underline{r}_{\nu-1})} |\psi_{\nu+1}-\psi_\nu| < \frac{1}{2^\nu}$, so that on a compact C (included say in $\Delta(\underline{r}_\mu)$), the sequence ψ_ν converges uniformly. On C, the sequence of holomorphic maps $(\psi_\nu-\psi_\mu)$ (where μ is fixed) converges uniformly, so that by a theorem of section 1.A, $\psi-\psi_\mu$ is holomorphic. Hence, in $\Delta(\underline{r}_\mu)$, $\bar\partial\psi = \bar\partial\psi_\mu = \varphi$.

This concludes the proof of lemma (2.36).

(2.38) Remark: A variation of this proof using annuli and Laurent series shows that

$$H^{p,q}_{\bar\partial}((\Delta-\{0\})^k \times \Delta^l) = 0 \quad \text{for } q \geq 1.$$

J. Metrics

(2.39) Definition: A Riemannian structure on a real manifold M is a twice covariant tensor field h whose value at each point $x \in M$ satisfies

$$h_x(X_x,Y_x) = h_x(Y_x,X_x) \qquad \forall X_x,Y_x \in T_xM$$

$$h_x(X_x,X_x) \geq 0 \quad \text{and} \quad h_x(X_x,X_x) = 0 \quad \text{iff} \quad X_x = 0.$$

A manifold together with such a structure is called a Riemannian manifold.

Recall that h induces an Euclidean structure on each tangent space, and that the length of a curve $F(t)$ with respect to h is

$$l(F) = \int_{t_0}^{t_1} \left(h\left(\frac{dF}{dt}, \frac{dF}{dt}\right)\right)^{\frac{1}{2}} dt.$$

This leads to the definition of a distance function on M, as the infimum of lengths of curves between two points.

Because of this relation with a "length element", the Riemannian structure is often denoted by ds^2.

(2.40) Proposition: Any paracompact manifold admits a Riemannian structure (and in fact, infinitely many).

This is proven by defining in each chart of a locally finite atlas an Euclidean metric, then glueing them together by means of a partition of unity.

(2.41) Definition: Let M be a complex manifold. A Riemannian metric h on M is said to be Hermitian iff it satisfies

$$h_z(JX_z,JY_z) = h_z(X_z,Y_z) \quad \forall z \in M,\ X_z,Y_z \in T_zM .$$

M,h is then called a Hermitian manifold.

(2.42) Proposition: Any paracompact complex manifold admits a Hermitian structure.

Indeed, let $g(X,Y)$ be a Riemannian structure. Then $h(X,Y) = g(X,Y) + g(JX,JY)$ is Hermitian.

h is $\mathbb{R}$-bilinear on TM and extends to a $\mathbb{C}$-bilinear tensor on

$T^{\mathbb{C}}M$, still denoted by h.

We observe that for all $V,W \in T_z'M$:

$$\begin{aligned} h_z(V,W) &= h_z(JV,JW) \\ &= h_z(iV,iW) \\ &= -h_z(V,W) \end{aligned}$$

so that it must be zero. Similarly, for all $V,W \in T_z''M$, $h_z(V,W) = 0$.

However, for $V \in T_z'M$, $W \in T_z''M$, this calculation yields no restriction since

$$h_z(V,W) = h_z(JV,JW) = h_z(iV,-iW) = h_z(V,W) \quad .$$

(2.43) Definition: The fundamental two-form associated with J is defined on TM by

$$F(X,Y) = h(JX,Y).$$

It can be extended to $T^{\mathbb{C}}M$ by bilinearity.

Remark: In [KN], F is defined with opposite sign.

Remark: We shall not always mention explicitly whether h and F are considered on TM or $T^{\mathbb{C}}M$.

F is a form, since

$$F(X,Y) = h(JX,Y) = h(J^2X,JY) = -h(JY,X) = -F(Y,X) \ .$$

By the vanishing of h on $T'M \times T'M$ and $T''M \times T''M$, it is a (1,1)-form. By definition, it is a real form (i.e. real on TM).

Consider two vectors $V = X - iJX \in T'M$ and $W = Y + iJY \in T''M$, where $X,Y \in TM$. We have

$$\begin{aligned} h(X-iJX,Y+iJY) &= h(X,Y) + h(JX,JY) + i[h(X,JY)-h(JX,Y)] \\ &= 2h(X,Y) - 2i\,F(X,Y) \quad . \end{aligned} \tag{2.44}$$

Hence, on TM, the Riemannian structure is $\frac{1}{2}\operatorname{Re} h$ and the fundamental form $-\frac{1}{2}$ Im h, where h acts on $T^{\mathbb{C}}M$ as indicated.

Using the isomorphism $T''M = \overline{T'M}$, we define on T'M a Hermitian product H by

$$H(V,W) = h(V,\overline{W}) \quad \text{for} \quad V,W \in C(T'M) \ .$$

We have

$$H(V,W) = \overline{H(W,V)}$$

$$H(aV,bW) = a\overline{b}H(V,W) \quad \forall\, a,b \in \mathbb{C}$$

$$H(V,V) > 0 \quad \text{if} \quad V \neq 0 \quad ,$$

i.e. H is positive definite.

Note that the word Hermitian is used both for h (bilinear on $T^{\mathbb{C}}M$) and H (linear on the left and conjugate linear on the right on T'M).

If a real (1,1)-form F is given, we can find a tensor h with $h(JX,Y) = F(X,Y)$, but it may not be positive definite.

(2.45) If it is, h is a Hermitian structure and F is called positive.

In local coordinates, a real metric will be defined by $h\left(\frac{\partial}{\partial x^j}, \frac{\partial}{\partial x^k}\right) = h_{jk}$ and $ds^2 = h_{jk}\, dx^j \otimes dx^k$, which means that $h\left(X^1 \frac{\partial}{\partial x^1}, Y^n \frac{\partial}{\partial x^n}\right) = h_{jk}X^jY^k$.

Similarly, in the complex case, calling j and $\overline{j}$ the indices of the coordinates z^j and $\overline{z}^j$ $(j = 1,\dots,m,\ \overline{j} = \overline{1},\dots,\overline{m})$ and $J = j$ or $\overline{j} \in \{1,\dots,m,\overline{1},\dots,\overline{m}\}$, we set

$$h_{JK} = h\left(\frac{\partial}{\partial z^J}, \frac{\partial}{\partial z^K}\right) \ .$$

The conditions above imply that

$$h_{jk} = h_{\overline{j}\overline{k}} = 0 \quad ,$$

(2.46) $ds^2 = h_{j\overline{k}}(dz^j \otimes d\overline{z}^k + d\overline{z}^k \otimes dz^j)$ and

(2.47) $F = i\, h_{j\overline{k}}\, dz^j \wedge d\overline{z}^k$.

Remark: Often we shall describe ds^2 by its action on $T'M$ only by $ds^2 = h_{j\bar{k}}\, dz^j\, d\bar{z}^k \equiv H_{jk}\, dz^j\, d\bar{z}^k$. To describe it on TM we must add the complex conjugate term. We mention also that some authors use the notation $ds^2 = 2h_{j\bar{k}}\, dz^j\, d\bar{z}^k$. This will introduce some coefficients 2 and $\frac{1}{2}$ in different formulae.

(2.48) Another way of expressing locally the Hermitian structure is by defining a coframe for H as an m-tuple $(\varphi^1,\ldots,\varphi^m)$ of forms of type (1,0), orthonormal for the dual structure of H.

This is always possible locally by a diagonalisation process and we have

$$ds^2 = \sum \varphi^j \otimes \overline{\varphi^j} \quad (+ \sum \overline{\varphi^j} \otimes \varphi^j)$$
$$F = i \sum \varphi^j \wedge \overline{\varphi^j} .$$

We note finally that a Riemannian structure induces a volume element on an orientable manifold.

(2.49) Definition: The canonical volume form on M,h is defined in local coordinates by

$$V_h = \det(h_{jk})^{\frac{1}{2}}\, dx^1 \wedge \ldots \wedge dx^m \quad .$$

The change of variable formula implies that V_h is globally defined.

Naturally, this applies to the case of a Hermitian structure, and in that case

(2.50) $V_h = \frac{1}{m!} F^m$,

where F is the fundamental form and $F^m = F \wedge \ldots \wedge F$.

K. Examples

(2.51) $\mathbb{R}^m$, $\mathbb{C}^m$

$\mathbb{R}^m$ carries the standard metric $ds^2 = \sum_j dx^j\, dx^j$. On $\mathbb{C}^m$, we'll consider the associated structure $ds^2 = \frac{1}{2} \sum_j dz^j\, d\bar{z}^j$, so that $F = \frac{i}{2} \sum_i dz^j \wedge d\bar{z}^j$. These induce similar structures on quotients by lattices.

(2.52) Submanifolds

Definition: Let M and N be manifolds and $f: M \to N$ a map such that f_{*x} is injective at every point $x \in M$. M is called an immersed submanifold of N. If M and N are complex and f holomorphic, M is a complex submanifold.

More precisely, a submanifold is an equivalence class of such maps with same image, but we shall consider here a fixed f.

If N carries a Riemannian structure h, the induced Riemannian structure on M is defined by

$$g(X,Y) = h(f_*X, f_*Y) \quad \forall X,Y \in TM.$$

For instance, the canonical metric on the sphere S^m is induced by the standard mapping of S^m on $\{(x^1,\ldots,x^{m+1}) \in \mathbb{R}^{m+1} \mid \sum (x^j)^2 = 1\}$.

(2.53) If, moreover, f is a holomorphic map from the complex manifold M to the Hermitian manifold N, then the induced metric is Hermitian.

In this case, we can find in a neighbourhood (in N) of each point of $f(M)$ a local frame field $(\varphi^1,\ldots,\varphi^n)$ such that $f^*\varphi^{m+1} = \ldots = f^*\varphi^n = 0$ and $(f^*\varphi^1,\ldots,f^*\varphi^m)$ is a local frame field on M (here $m = \dim M < n = \dim N$).

Hence, for the fundamental forms on M and N, we have

$$\begin{aligned} F_M &= i \sum_{j=1}^{m} f^*\varphi^j \wedge f^*\overline{\varphi^j} \\ &= i \sum_{j=1}^{n} f^*\varphi^j \wedge f^*\overline{\varphi^j} \\ &= f^*(i \sum_{j=1}^{n} \varphi^j \wedge \overline{\varphi^j}) \\ &= f^*F_N \quad , \end{aligned}$$

so that

(2.54) Proposition: The fundamental form of the induced metric on M is the pull-back of the fundamental form on N.

(2.55) Submersions

Let M,g be a Riemannian manifold and $f: M \to N$ a surjective submersion, i.e. a surjective map such that f_* is surjective at every point.

The vector space T_xM can then be split as the orthogonal direct sum, $T_xM = \ker f_{*x} \oplus H_x$, which defines the horizontal space H_x.

$$f_{*x}|_{H_x} \text{ is an isomorphism } H_x \to T_{f(x)}N \quad .$$

For any $x,y \in M$ such that $f(x) = f(y)$, suppose that the isomorphism

$$\left(f_{*y}|_{H_y}\right)^{-1} \circ f_{*x}|_{H_x} : H_x \to H_y$$

is an isometry.

We can then define the induced metric h on N as the unique Riemannian structure such that $\forall x \in M$, $f_{*x}|_{H_x} : H_x \to T_{f(x)}N$ is an isometry, and this does not depend on the choice of x. $F:M,g \to N,h$ is called a Riemannian submersion.

(2.56) $\mathbb{P}^m(\mathbb{C})$

Let us go back to the charts U_j with coordinates $z^k_{(j)}$ defined in (2.6). Consider in U_j the function $f_{(j)} = 1 + \sum_k z^k_{(j)}\bar{z}^k_{(j)}$. By the form of the transition functions, we have $f_{(j)} = f_{(k)}\underline{z^k_{(j)}}\bar{z}^k_{(j)}$ on $U_j \cap U_k$ (no summation on j and k). Since $\bar{\partial} z^k_{(j)} = \partial \bar{z}^k_{(j)} = 0$, we have

$$\partial\bar{\partial} \log f_{(j)} = \partial\bar{\partial} \log f_{(k)} \quad \text{on} \quad U_j \cap U_k \quad ,$$

so that the (1,1)-form defined by

$$F = -\frac{i}{2}\, \partial\bar{\partial} \log f_{(j)} \quad \text{on} \quad U_j$$

is globally defined on $\mathbb{P}^m(\mathbb{C})$.

F will be the fundamental form of a Hermitian structure $h(X,Y) = -F(JX,Y)$.

To check that h is positive definite, one can calculate it in local coordinates, to obtain in U_o:

(2.57) $$ds^2 = \frac{1}{2}\,\frac{(1+\sum z^j\bar{z}^j)(\sum dz^k d\bar{z}^k)-(\sum \bar{z}^j dz^j)(\sum z^k d\bar{z}^k)}{(1+\sum z^j\bar{z}^j)^2} \;.$$

This can be shown to be positive, but this fact will also be deduced from the following construction.

We have seen in (2.6) that $\mathbb{P}^m(\mathbb{C}) = S^{2m+1}/U(1)$, so that it is the quotient of the sphere by an isometry group (group of maps preserving the metric). This defines a submersion $S^{2m+1} \to \mathbb{P}^m(\mathbb{C})$ satisfying the condition of (2.55) for the canonical metric on S^{2m+1}. A (long) calculation shows that the induced metric on $\mathbb{P}^m(\mathbb{C})$ is (2.57). It is called the Fubini-Study metric.

L. Almost complex structures

The following definition is motivated by (2.24).

(2.58) Definition: An almost complex structure on a real manifold M is a field of endomorphisms J on TM such that $J^2 = J\circ J = -I$. If such a J exists, (M,J) is called an almost complex manifold.

Some properties of complex manifolds are shared by almost complex ones: e.g. they are even-dimensional and orientable, and J induces the splitting $T^{\mathbb{C}}M = T'M \oplus T''M$. However, we don't in general have complex coordinates with holomorphic changes of charts, so that various calculations made above don't apply.

For example, on an almost complex manifold, one has $d\,A^{p,q} \subset A^{p-1,q+2} \oplus A^{p,q+1} \oplus A^{p+1,q} \oplus A^{p+2,q-1}$ and in general we do not have $d = \partial + \bar{\partial}$ as in the complex case. In fact, we mention without proof the following characterizations.

(2.59) Theorem: Let M,J be an almost complex manifold. Then the following conditions are equivalent:

(i) If $Z,W \in C(T'M)$, then $[Z,W] \in C(T'M)$,

(ii) $d\,A^{p,q} \subset A^{p+1,q} \oplus A^{p,q+1}$

(iii) The torsion tensor field defined by

$$N(X,Y) = 2([JX,JY]-[X,Y]-J[X,JY]-J[JX,Y])$$

is identically zero.

When these conditions are satisfied, the almost complex structure is called <u>integrable</u>.

For a proof of this rather easy result, see [K-N vol. 2].

(2.60) <u>Theorem</u>: An almost complex structure J is integrable iff it is induced by a complex structure.

Different proofs of this deep theorem of Newlander and Nirenberg can be found in [N-N.],[Hö], [Ma] and (in the real analytic case) in [K-N. vol. 2] and [N].

§3 SHEAVES AND COHOMOLOGY

In §2 we saw how partitions of unity could be used to convert locally defined C^∞ objects into globally defined objects (e.g. (2.7), (2.22), (2.40)). As holomorphic partitions of unity don't exist, it is not possible to do the same thing with holomorphic objects. Instead, obstructions to globalising lie in sheaf cohomology groups (e.g. (3.15), (3.26)). We give the basic elements of this theory and show its use in important problems in complex analysis.

A. Sheaves

Let M be a complex manifold. For each open set U of M we may consider several sets of functions, e.g. (i) $C^\infty(U)$ = set of C^∞ functions on U, (ii) $\mathcal{O}(U)$ = the set of holomorphic functions on U, (iii) $\mathcal{O}^*(U)$ = the set of non-zero holomorphic functions on U. Each of these sets is a group, the first two under addition, the last one under multiplication. In each case for any pair $V \subseteq U$ of open sets of M we have group homomorphisms, for example $r_{uv}: C^\infty(U) \to C^\infty(V)$, given by restriction: $r_{uv}(f) = f|V$; we also have the property that functions defined on each set of an open cover $\{U_\alpha\}$ of an open set $U \subseteq M$ which agree on the intersections may be "glued" together to give a unique function on U. A sheaf is simply an abstraction of these properties. Now, for the definitions: throughout this chapter let X be a topological space.

(3.1) <u>Definition</u>: A <u>sheaf (of abelian groups)</u> S <u>on</u> X is an association to each open set U of X of an abelian group $S(U)$ (whose elements are called <u>sections</u> of S over U), and to each pair $V \subseteq U$ of open sets of a homomorphism $r_{uv}: S(U) \to S(V)$ called a <u>restriction map</u> such that

(i) for each open set U, r_{uu} = identity; for each triple $W \subseteq V \subseteq U$ of open sets, $r_{uw} = r_{vw} \circ r_{uv}$. So for $f \in S(U)$ we may write $f|V$ for $r_{uv}(f)$ without loss of information.

(ii) ("completeness axiom") for each open set U and each cover $\{U_\alpha\}$ of U, if $s_\alpha \in S(U_\alpha)$ is a collection of sections with $s_\alpha|U_\alpha \cap U_\beta = s_\beta|U_\alpha \cap U_\beta$, then there exists a <u>unique</u> section $s \in S(U)$ such that $s|U_\alpha = s_\alpha$ for all α.

(3.2) <u>Remarks</u>: (i) For the moment "sections of S over U" are just elements of an abstract group, the reason for the terminology will appear later. (ii) If condition (ii) is omitted, S is called a presheaf or stack. A sheaf is thus a "complete presheaf". (iii) Sheaves of rings, modules etc. can be defined similarly. (iv) The notation $\Gamma(U,S)$ is often used for $S(U)$.

(3.3) Examples: (i) On a C^∞ manifold M we may define sheaves C^∞, C^*, A^p, Z^p, $\mathbb{Z}$, $\mathbb{Q}$, $\mathbb{R}$, $\mathbb{C}$ by:

$C^\infty(U)$ = additive group of C^∞ functions on U,
$C^*(U)$ = multiplicative group of non-zero C^∞ functions on U,
$A^p(U)$ = additive group of C^∞ p-forms on U,
$Z^p(U)$ = additive group of C^∞ closed p-forms on U,
$\mathbb{Z}(U)$, $\mathbb{Q}(U)$, $\mathbb{R}(U)$, $\mathbb{C}(U)$ = additive groups of locally constant $\mathbb{Z}$-, $\mathbb{Q}$-, $\mathbb{R}$-, $\mathbb{C}$-valued functions on U.

(ii) On a complex manifold M, we may further define sheaves $\mathcal{O}$, $\mathcal{O}^*$, Ω^p, $A^{p,q}$, $Z^{p,q}_{\bar\partial}$ by

$\mathcal{O}(U)$ = additive group of holomorphic functions on U,
$\mathcal{O}^*(U)$ = multiplicative group of non-zero holomorphic functions on U,
$\Omega^p(U)$ = additive group of holomorphic p-forms on U,
$A^{p,q}(U)$ = C^∞ forms of type (p,q) on U,
$Z^{p,q}(U)$ = $\bar\partial$-closed C^∞ forms of type (p,q) on U.

In all the above sheaves, the restriction maps are the obvious ones.

An important construction is the following:

(3.4) Let S be a sheaf on a topological space X and let $x \in X$. Define an equivalence relation $\sim_x$ on the disjoint union of the sets $\{S(U): U$ an open neighbourhood of $x\}$ as follows: if $f \in S(U)$, $g \in S(V)$ write $f \sim_x g$ if $f|W = g|W$ for some $W \subseteq U \cap V$. The set of equivalence classes $\{S(U)\}/\sim_x$ is called the direct limit of the groups $S(U)$, written $\varinjlim_{U \ni x} S(U)$. It is clearly a group.

(3.5) Definition: Let S be a sheaf on a topological space X. The stalk S_x of S at $x \in X$ is the direct limit $S_x = \varinjlim_{U \ni x} S(U)$.

Denoting the equivalence class of $f \in S(U)$ by f_x, we call f_x the germ of f at x.

Sheaves can be seen from another viewpoint as follows:

(3.6) Definition: A sheaf space on X is a topological space $\underline{S}$ together with a continuous map $\pi: \underline{S} \to X$ called the projection and an abelian group structure in each stalk $S_x = \pi^{-1}(x)$ such that

(i) π is a local homeomorphism

(ii) the group operations in $\underline{S}$ are continuous (i.e. setting $E = \{(u,v) \in S \times S: \pi(u) = \pi(v)\}$, the mapping $E \to S$ given by $(u,v) \to u-v$ is continuous).

Given a sheaf S we define its associated sheaf space as follows. Let $\underline{S}$ = the disjoint union of the stalks S_x, $x \in X$. For U open and $f \in S(U)$ set $V(f,U) = \{f_x: x \in U\}$; then the sets $V(f,U)$ define a base for a topology on $\underline{S}$. Giving $\underline{S}$ this topology, we see that the group operations are continuous. We define the projection $\pi: \underline{S} \to X$ by $\pi(S_x) = x$. Then $\underline{S}$ is a sheaf space.

Conversely we may recover a sheaf if we know its associated sheaf space as follows:

(3.7) Theorem: Let the sheaf S have associated sheaf space $\underline{S}$. Then $S(U)$ = sections of $\underline{S}$ over U. (By a section we here mean a continuous map $f: U \to \underline{S}$ such that $\pi \circ f$ = identity.) The proof of this theorem may be found, for example, in [Wa], [T]. The completeness axiom is used in an essential way.

(3.8) Remarks: (i) The notation $\Gamma(U,\underline{S})$ is used to denote the sections of $\underline{S}$ over U. Under the isomorphism $S(U) \cong \Gamma(U,\underline{S})$ the restriction maps $r_{uv}: S(U) \to S(V)$ become genuine restriction maps $\Gamma(U,\underline{S}) \to \Gamma(V,\underline{S})$ so justifying , as promised, the terminology used in definition (3.1).

(ii) In a sheaf space $\underline{S}$, the stalks at different points need not be isomorphic, the topology on each stalk is the discrete topology, and the local homeomorphism nature of π means that there exist "thin" open sets in $\underline{S}$, lying over open sets of X. (By "thin" open set we mean an open set whose intersection with each stalk is empty or is a single point.) These act as "guides" for moving germs from one place to another and explain the use of sheaves to "globalize" local data.

(3.9) Examples: (i) The skyscraper sheaf is the sheaf S on (say) $\mathbb{C}$

defined by $S(U) = \mathbb{Z}$ if $0 \in U$, $S(U) = 0$ otherwise, with restriction maps r_{UV} = identity if $V \ni 0$, $r_{UV} = 0$ otherwise. Its stalks are $S_z = \mathbb{Z}$ if $z = 0$, $S_z = 0$ otherwise. The topology of the associated sheaf space is not Hausdorff. Indeed, the open neighbourhoods of a point $\lambda \in S_0$ are of the form $\lambda \cup \{0 \in S_z : z \in U \setminus 0\}$ where U is an open subset of $\mathbb{C}$.

(ii) The sheaf space associated to the sheaf $\mathcal{O}$ of holomorphic functions on a complex manifold M is given by $\underline{\mathcal{O}}$ = the disjoint union of the additive goups $\mathcal{O}_x$ of germs of holomorphic functions at x $(x \in M)$. The topology on $\underline{\mathcal{O}}$ is given by the base $V(f,U) = \{f_x : x \in U\}$ where U is open and f is holomorphic on U, and the projection $\pi: \underline{\mathcal{O}} \to M$ is given by $\pi(\mathcal{O}_x) = x$. Note that $\mathcal{O}_x$ can be thought of as the set of Taylor series on x; germs of C^∞ functions have no such description.

Now we discuss homomorphisms of sheaves.

(3.10) <u>Definition</u>: Let S, S' be sheaves on X. A <u>sheaf homomorphism</u> $\alpha: S \to S'$ is a collection of homomorphisms $\alpha_U: S(U) \to S'(U)$ which commute with the restriction maps, i.e. for all open sets $V \subseteq U$, $\alpha_V \circ r_{UV} = r'_{UV} \circ \alpha_U$.

Defining a <u>homomorphism of sheaf spaces</u> $\underline{S}, \underline{S}'$ to be a continuous map $\alpha: \underline{S} \to \underline{S}'$ which maps S_x to S'_x homomorphically for each x, we see that a sheaf homomorphism $\alpha: S \to S'$ induces a homomorphism of the associated sheaf spaces $\alpha: \underline{S} \to \underline{S}'$ defining $\alpha_x: S_x \to S'_x$ by $\alpha_x(f_x) = (\alpha_U(f))_x$ for $f \in S(U)$, $U \ni x$.

If each α_x is an inclusion map we say that S is a <u>subsheaf</u> of S'. For example, over a complex manifold, $\mathcal{O}$ is a subsheaf of C^∞.

(3.11) <u>Definition</u>: A sequence of sheaves and sheaf homomorphisms

$$\cdots \to S_n \overset{\alpha_n}{\to} S_{n+1} \overset{\alpha_{n+1}}{\to} S_{n+2} \to \cdots$$

is said to be <u>exact</u> if it is exact on each stalk, i.e.

$$\ker(\alpha_{n+1})_x = \operatorname{im}(\alpha_n)_x \quad \text{for all } n \text{ and } x \in X.$$

(3.12) <u>Remarks</u>: (i) If $\cdots \to S_n \overset{\alpha_n}{\to} S_{n+1} \overset{\alpha_{n+1}}{\to} S_{n+2} \to \cdots$ is exact, it does <u>not</u> follow that the sequence of groups $\cdots \to S_n(U) \overset{(\alpha_n)_U}{\to} S_{n+1}(U) \to \cdots$

is exact, for example, if α_n is surjective, inverse images of the points of a section in $S_{n+1}(U)$ may not necessarily form a section in $S_n(U)$ (see example (3.13)). (ii) An exact sequence of the form

$$0 \to S' \to S \to S'' \to 0 \quad ,$$

where 0 denotes the "zero sheaf" defined by $0(U)$ = trivial group, is called a <u>short exact sequence</u>.

(3.13) On any complex manifold M, we have a short exact sequence

$$0 \to \mathbb{Z} \xrightarrow{i} \mathcal{O} \xrightarrow{\exp} \mathcal{O}^* \to 0$$

where $i_u: \mathbb{Z}(U) \to \mathcal{O}(U)$ is inclusion and $\exp_u: \mathcal{O}(U) \to \mathcal{O}^*(U)$ is defined by $\exp(f) = e^{2\pi i f}$. This important sequence is called the <u>exponential sheaf sequence</u>.

We illustrate (3.12)(i) as follows: if $M = \mathbb{C}\setminus\{0\}$ and $U = M$ then the function $z \in \mathcal{O}^*(U)$ cannot be expressed in the form $\exp(f)$ for any $f \in \mathcal{O}(U)$, so $\exp_u: \mathcal{O}(U) \to \mathcal{O}^*(U)$ is not surjective. (Note that if U is contractible in $\mathbb{C}\setminus\{0\}$ then $\exp_u$ is surjective).

Connected with the idea of a short exact sequence is the idea of kernel and cokernel:

(3.14) <u>Definition</u>: The <u>kernel</u> $\ker\phi$ of a sheaf homomorphism $\phi: R \to S$ is the subsheaf of R given by $(\ker\phi)(U) = \ker\phi_u$. The sheaf space associated to $\ker\phi$ is the disjoint union of the stalks $(\ker\phi)_x = \ker(\phi_x)$ topologized as a subspace of R. The definition of $\operatorname{coker}\phi$ is not so easy; it is tempting to set $(\operatorname{coker}\phi)(U) = \operatorname{coker}\phi_u = S(U)/\phi_u(R(U))$, unfortunately this doesn't in general satisfy the completeness axiom (ii) of definition (3.1). For example, for the exponential sheaf sequence on $M = \mathbb{C}\setminus\{0\}$, the section $z \in \mathcal{O}^*(U)$ is in the zero coset of $\mathcal{O}^*(U)/\exp_u\mathcal{O}(U)$ if U is contractible, but is not in the zero coset if $U = M$ contradicting the completeness axiom (3.1)(ii). However, it becomes easy if we use sheaf spaces.

(3.15) <u>Definition</u>: The <u>cokernel</u> $T = \operatorname{coker}\phi$ of a sheaf homomorphism $\phi: R \to S$ is the sheaf with associated sheaf space $\underline{T}$ = disjoint union $\bigcup_{x\in X} T_x$ where $T_x = \operatorname{coker}(\phi_x) = S_x/\phi_x(R_x)$ topologized as a quotient space of S.

If $\phi: R \to S$ is an inclusion map, coker ϕ is called the quotient sheaf S/R.

(3.16) Remark: It can be checked that the description of T is as follows: For U open in X, an element of $T(U)$ is an open cover $\{U_\beta\}$ of U together with sections $\sigma_\beta \in S(U_\beta)$ such that, for all β, γ, we have $\sigma_\beta | U_\beta \cap U_\gamma - \sigma_\gamma | U_\beta \cap U_\gamma \in \phi_{U_\beta \cap U_\gamma}(R(U_\beta \cap U_\gamma))$, with two collections (U_β, σ_β), $(U'_\gamma, \sigma'_\gamma)$ identified if, for all $x \in U$, and U_β, U'_γ we have $(\sigma_\beta)_x - (\sigma'_\gamma)_x \in \phi_x(R_x)$.

B. Čech Cohomology of Sheaves

We first motivate the definition of Čech cohomology of sheaves by studying an important problem. Let S be a Riemann surface. For $p \in S$, a principal part of p is the polar part $\sum_{k=1}^{n} a_k z^{-k}$ of a Laurent series where z is a local complex coordinate centred on p.

(3.17) The Mittag-Leffler Question. Given a discrete set $\{p_\alpha\}$ of points of S and a principal part at p_α for each α, does there exist a meromorphic function f on S which is holomorphic on $S \setminus \{p_\alpha\}$ and has the specified principal part at each p_α?

To attack this, note that we can solve it locally, viz we can choose an open cover $\mathcal{U} = \{U_\alpha\}$ of S such that each U_α contains at most one p_α and a meromorphic function f_α on U_α with the specified principal part at p_α. Now consider the functions $f_{\alpha\beta} = f_\alpha - f_\beta$. Clearly $f_{\alpha\beta} \in \mathcal{O}(U_\alpha \cap U_\beta)$ for all α, β. On $U_\alpha \cap U_\beta \cap U_\gamma$ we have

(3.18) $f_{\alpha\beta} + f_{\beta\gamma} + f_{\gamma\alpha} = 0.$

To solve the problem, it is sufficient to find $\{g_\alpha \in \mathcal{O}(U_\alpha)\}$ such that on $U_\alpha \cap U_\beta$,

(3.19) $g_\beta - g_\alpha = f_{\alpha\beta}$,

for then $f = f_\alpha + g_\alpha$ is a globally defined meromorphic function with the specified principal parts. The Čech theory is defined in such a way that (3.18) says that $\{f_{\alpha\beta}\}$ is a "1-cocycle" and (3.19) asks whether this cocycle is the "coboundary" of some "0-cochain" $\{g_\alpha\}$. The problem thus has a solution if and only if the 1-cocycle $\{f_{\alpha\beta}\}$ is a coboundary, i.e. the obstruction of solving the problem lies in the "first Čech cohomology group" 1-cocycles/1-coboundaries.

Now for the general definitions. Throughout this section, let X be a paracompact topological space.

(3.20) <u>Definition</u>: Let S be a sheaf on X. Let $\mathcal{U} = \{U_\alpha\}$ be a locally finite open cover of X (We assume $U_\alpha \neq U_\beta$ for $\alpha \neq \beta$.). A <u>p-cochain</u> is an assignment to each ordered $(p+1)$-tuple of distinct sets $U_{\alpha_0}, \ldots, U_{\alpha_p}$ of $\mathcal{U}$ of an element $\sigma_{\alpha_0 \ldots \alpha_p}$ of $S(U_{\alpha_0} \cap \ldots \cap U_{\alpha_p})$.

Thus, denoting the set of p-cochains by $C^p(\mathcal{U},S)$,

(3.21) $$C^p(\mathcal{U},S) = \prod_{\alpha_0;\ldots,\alpha_p \text{ distinct}} S(U_{\alpha_0} \cap \ldots \cap U_{\alpha_p}) \; .$$

Here, $\prod$ denotes Cartesian product of the abelian groups. The <u>coboundary operator</u> $\delta = \delta_p : C^p(\mathcal{U},S) \to C^{p+1}(\mathcal{U},S)$ is defined by

$$(\delta\sigma)_{\alpha_0,\ldots,\alpha_{p+1}} = \sum_{k=0}^{p+1} (-1)^k \, \sigma_{\alpha_0 \ldots \widehat{\alpha_k} \ldots \alpha_{p+1}} \big|_{U_{\alpha_0} \cap \ldots \cap U_{\alpha_{p+1}}} \; .$$

In particular, if $\sigma = \{\sigma_\alpha\} \in C^0(\mathcal{U},S)$, $\tau = \{\tau_{\alpha\beta}\} \in C^1(\mathcal{U},S)$, omitting the restriction signs (c.f. 3.18, 3.19),

$$(\delta\sigma)_{\alpha\beta} = \sigma_\beta - \sigma_\alpha$$
$$(\delta\tau)_{\alpha\beta\gamma} = \tau_{\alpha\beta} + \tau_{\beta\gamma} - \tau_{\alpha\gamma} \; .$$

Consider now the sequence of abelian groups

$$\ldots \to C^{p-1}(\mathcal{U},S) \xrightarrow{\delta_{p-1}} C^p(\mathcal{U},S) \xrightarrow{\delta_p} C^{p+1}(\mathcal{U},S) \to \ldots$$

A direct calculation shows that this is a "cochain complex" i.e. $\delta_p \circ \delta_{p-1} = 0 \;\forall p$. We define the set of p-cocycles $Z^p(\mathcal{U},S) = \ker \delta_p = \{\sigma \in C^p(\mathcal{U},S) : \delta\sigma = 0\}$ and the set of p-coboundaries $B^p(\mathcal{U},S) = \operatorname{Im} \delta_{p-1} = \{\sigma \in C^p(\mathcal{U},S) : \sigma = \delta\tau \text{ for some } \tau \in C^{p-1}(\mathcal{U},S)\}$. Since $\delta^2 = 0$, $B^p(\mathcal{U},S)$ is a subgroup of $Z^p(\mathcal{U},S)$ and we set $\check{H}^p(\mathcal{U},S) = Z^p(\mathcal{U},S)/B^p(\mathcal{U},S)$.

To get the Čech cohomology groups, we must remove the dependence of the definition on the choice of cover $\mathcal{U} = \{U_\alpha\}$. Given two locally finite open covers $\mathcal{U} = \{U_\alpha\}_{\alpha\in I}$, $\mathcal{U}' = \{U'_\beta\}_{\beta\in I'}$ of X, say that $\mathcal{U}'$ is a refinement of $\mathcal{U}$, written $\mathcal{U}' < \mathcal{U}$, if for every $\beta \in I'$ there exists $\alpha \in I$ such that $U'_\beta \subseteq U_\alpha$. If $\mathcal{U}' < \mathcal{U}$ we may choose a map $\phi: I' \to I$ such that $U'_\beta \subseteq U_{\phi(\beta)}$ for all β, then we have a map $\rho_\phi: C^p(\mathcal{U},S) \to C^p(\mathcal{U}',S)$ given by $(\rho_\phi\sigma)_{\beta_0,\ldots,\beta_p} = \sigma_{\phi\beta_0,\ldots,\phi\beta_p}| |U'_{\beta_0} \cap \ldots \cap U'_{\beta_p}$. Clearly $\delta \circ \rho\phi = \rho\phi \circ \delta$ and so $\rho\phi$ induces a homomorphism $\rho: \check{H}^p(\mathcal{U},S) \to \check{H}^p(\mathcal{U}',S)$. This is independent of the choice of ϕ as may be seen by showing that the chain maps $\rho\phi, \rho\psi$ associated to two maps $\phi,\psi: I' \to I$ are chain homotopic (see [T]).

(3.22) Definition: Let S be a sheaf on a paracompact topological space X. The p'th Čech cohomology group "of S" (or "of X with coefficients in S") is the direct limit taken over all locally finite open covers $\mathcal{U}$ of X ordered by refinement:

$$\check{H}^p(X,S) = \lim_{\overrightarrow{\mathcal{U}}} \check{H}^p(\mathcal{U},S) \quad .$$

(3.23) It is easily seen that $\check{H}^0(X,S) = S(X)$, the (continuous) sections of S over X.

(3.24) Theorem: Let K be a locally finite simplicial complex with underlying C^0 manifold X . Then there is a canonical isomorphism: $\check{H}^p(X,\mathbb{Z}) \to H^p(K,\mathbb{Z})$, i.e. the sheaf cohomology with coefficients in $\mathbb{Z}$ and the simplicial cohomology $H^p(K,\mathbb{Z})$ are canonically isomorphic.

Proof: For each vertex v_α of K define an open set $U_\alpha = St(v_\alpha)$ called the star of v_α as the interior of the union of all (closed) simplices having v_α as vertex. Then $\mathcal{U} = \{U_\alpha\}$ is a locally finite open cover of X. It can easily be seen that $\bigcap_{i=0}^{p} St(v_{\alpha_i})$ is a non-empty connected set if $v_{\alpha_0},\ldots,v_{\alpha_p}$ are vertices of a p-simplex, otherwise, it is empty. Thus a p-cochain $\sigma \in C^p(\mathcal{U},\mathbb{Z})$ associates to every $(\alpha_0,\ldots,\alpha_p)$ an element $\sigma_{\alpha_0,\ldots,\alpha_p} \in \mathbb{Z}(\bigcap_{i=0}^{p} St(v_\alpha)) =$

$$= \begin{cases} \mathbb{Z} & \text{if } v_{\alpha_0},\ldots,v_{\alpha_p} \text{ span a p-simplex} \\ 0 & \text{otherwise} \end{cases}$$

Hence, given σ, we may associate to each p-simplex $<v_{\alpha_0},\dots,v_{\alpha_p}>$ an integer $\sigma_{\alpha_0,\dots,\alpha_p}$, thus we have a mapping

$$S: C^p(\mathcal{U},\mathbb{Z}) \to C^p(K,\mathbb{Z})$$

defined by

$$(S\sigma)(<v_{\alpha_0},\dots,v_{\alpha_p}>) = \sigma_{\alpha_0,\dots,\alpha_p} .$$

It can easily be seen that S is an isomorphism of groups, also S commutes with the coboundary operators for

$$\delta(S\sigma)(<v_{\alpha_0},\dots,v_{\alpha_{p+1}}>) = \sum_{i=0}^{p+1} (-1)^i \quad (S\sigma)(<v_{\alpha_0},\dots,\widehat{v_{\alpha_i}},\dots,v_{\alpha_{p+1}}>)$$

$$= S(\delta\sigma)$$

hence S induces an isomorphism from $\check{H}^p(\mathcal{U},\mathbb{Z})$ to $H^p(K,\mathbb{Z})$ and taking the direct limit over finer and finer open covers gives the result.

Although we took the direct limit over arbitrarily fine covers in the definition of Čech cohomology, to do any calculations we often need to go back to a specific cover $\mathcal{U}$ and work with $\check{H}^p(\mathcal{U},S)$. The best sort of open cover to work with is given by:

(3.25) Definition: A locally finite open cover $\mathcal{U} = \{U_\alpha\}_{\alpha\in I}$ is called an acyclic cover or a Leray cover for the sheaf S if $\check{H}^q(U_{\alpha_1} \cap \dots \cap U_{\alpha_p},S) = 0$ for all $q \geq 1$, $p \geq 0$, $\alpha_0,\dots,\alpha_p \in I$.

(3.26) Theorem (Leray): If $\mathcal{U}$ is an acyclic locally finite open cover for S,

$$\check{H}^*(\mathcal{U},S) \cong \check{H}^*(X,S) .$$

Proof: A spectral sequence argument, c.f. [Sw].

C. The Long Exact Cohomology Sequence and its Applications

Let $0 \to R \overset{\phi}{\to} S \overset{\psi}{\to} T \to 0$ be a short exact sequence of sheaves on a paracompact topological space X. We shall construct a long exact sequence of sheaf cohomology groups.

Firstly, given a homomorphism $\phi: R \to S$ of sheaves, there are induced homomorphisms $C^p(\mathcal{U},R) \to C^p(\mathcal{U},S)$ given by composition with ϕ. These clearly commute with cohomology operators and refinement maps and hence induce homomorphisms $\phi^*: \check{H}^p(X,R) \to \check{H}^p(X,S)$.

Secondly, given the above short exact sequence of sheaves, we define the <u>coboundary map</u> $\delta^*: \check{H}^p(X,T) \to \check{H}^{p+1}(X,R)$ as follows: Represent an element of $\check{H}^p(X,T)$ by a p-cocycle $\sigma \in C^p(\mathcal{U},T)$ where $\mathcal{U}$ is a locally finite open cover of X. Since ψ is surjective, by passing to a refinement $\mathcal{U}'$ of $\mathcal{U}$ we may find $\tau \in C^p(\mathcal{U}',S)$ with $\psi(\tau)=\rho\sigma$. Consider $\delta\tau$. Since $\psi\delta\tau = \delta\psi\tau = \delta\rho\sigma = 0$, by the exactness of the sheaf sequence we may choose a refinement $\mathcal{U}''$ of $\mathcal{U}'$ and $\mu \in C^{p+1}(\mathcal{U}'',R)$ such that $\phi(\mu) = \rho\delta\tau$. Then $\phi(\delta\mu) = \delta(\phi\mu) = \delta(\rho\delta\tau) = \rho\delta^2\tau = 0$, since ϕ is injective, this implies $\delta\mu = 0$ so $\mu \in Z^{p+1}(\mathcal{U}'',R)$. It is easily seen that the cohomology class $[\mu] \in \check{H}^{p+1}(X,R)$ of μ is independent of the various choices made. We define $\delta^*: \check{H}^p(X,T) \to \check{H}^{p+1}(X,R)$ by $\delta^*[\sigma] = [\mu]$. (C.f. diagram after proof of (3.28)).

(3.27) <u>Remark</u>: To understand this and the next theorem we may avoid the problem of passing to refinements if we assume that we may choose a cover $\mathcal{U}$ so fine that $0 \to R(U) \xrightarrow{\phi_U} S(U) \xrightarrow{\psi_U} T(U) \to 0$ is exact for all intersections $U = U_{\alpha_0} \cap \ldots \cap U_{\alpha_p}$ of sets of $\mathcal{U}$. This may always be done for the sheaves given in example (3.3). For such a cover, the sequence $0 \to C^p(\mathcal{U},R) \xrightarrow{\phi} C^p(\mathcal{U},S) \xrightarrow{\psi} C^p(\mathcal{U},T) \to 0$ is exact.

(3.28) <u>Theorem</u>: Given a short exact sequence $0 \to R \xrightarrow{\phi} S \xrightarrow{\psi} T \to 0$, the sequence

$$0 \to \check{H}^0(X,R) \xrightarrow{\phi^*} \check{H}^0(X,S) \xrightarrow{\psi^*} \check{H}^0(X,T) \xrightarrow{\delta^*} \check{H}^1(X,R) \xrightarrow{\phi^*} \check{H}^1(X,S) \xrightarrow{\psi^*}$$
$$\check{H}^1(X,T) \xrightarrow{\delta^*} \ldots \to \check{H}^p(X,R) \xrightarrow{\phi^*} \check{H}^p(X,S) \xrightarrow{\psi^*} \check{H}^p(X,T) \xrightarrow{\delta^*} \check{H}^{p+1}(X,R)$$
$$\to \ldots$$

is exact.

<u>Proof</u>: We prove exactness at $\check{H}^p(X,T)$, exactness elsewhere being proved (see diagram)
similarly. We will assume that a cover $\mathcal{U}$ can be chosen as in Remark (3.27). (If this is not the case the initial cover $\mathcal{U}$ may need to be refined at each stage of the following argument.) Let $\sigma \in C^p(\mathcal{U},T)$

be a cocycle representing a cohomology class $[\sigma]$ in $H^p(X,T)$. Suppose $\delta^*[\sigma] = 0$. We must show that $[\sigma] = \psi^*[\nu]$ for some cocycle $\nu \in C^p(\mathcal{U},S)$. Now as before we can find $\tau \in C^p(\mathcal{U},S)$ with $\psi\tau = \sigma$, however τ may not be a cocycle. However, we can find $\mu \in C^{p+1}(\mathcal{U},R)$ with $\phi(\mu) = \delta\tau$ and then $\delta^*[\sigma] = [\mu]$. Since $\delta^*[\sigma] = 0$ this means $\mu = \delta\xi$ for some $\xi \in C^p(\mathcal{U},R)$, but then $\delta(\phi\xi) = \phi\delta\xi = \phi\mu = \delta\tau$, hence $\nu = \tau - \phi\xi$ is a cocycle. Clearly $\psi(\nu) = \psi(\tau) = \sigma$ so we have found a cocycle $\nu \in C^p(\mathcal{U},S)$ such that $[\sigma] = \psi^*[\nu]$. Hence $\ker \delta^* \subseteq \operatorname{Im} \phi^*$. The opposite inclusion is proved similarly.

Diagram for proof of (3.28)

$$\begin{array}{ccccc}
\overset{\xi}{\hat{C^p(\mathcal{U},R)}} & \xrightarrow{\phi} & \overset{\tau}{\hat{C^p(\mathcal{U},S)}} & \xrightarrow{\psi} & \overset{\sigma}{\hat{C^p(\mathcal{U},T)}} \\
\delta\downarrow & & \delta\downarrow & & \delta\downarrow \\
\overset{\mu}{\hat{C^{p+1}(\mathcal{U},R)}} & \xrightarrow{\phi} & \overset{\delta\tau}{\hat{C^{p+1}(\mathcal{U},S)}} & \xrightarrow{\psi} & C^{p+1}(\mathcal{U},T) \\
\delta\downarrow & & \delta\downarrow & & \delta\downarrow \\
C^{p+2}(\mathcal{U},R) & \xrightarrow{\phi} & C^{p+2}(\mathcal{U},S) & \xrightarrow{\psi} & C^{p+2}(\mathcal{U},T)
\end{array}$$

We now show how the long exact sequence of Theorem (3.28) can help in our study of the Mittag-Leffler question and its higher dimensional generalization, the Cousin 1-problem. Firstly we must define the concept of meromorphic functions on a complex manifold M. Roughly, this is locally the quotient of two holomorphic functions. To say this properly, first note that for each $x \in M$, $\mathcal{O}_x$, the set of germs of holomorphic functions at x, is an integral domain under addition and multiplication. Hence we may form the field of fractions of $\mathcal{O}_x$, namely $\mathcal{M}_x = \{f_x/g_x : f_x, g_x \in \mathcal{O}_x,\ f_x, g_x$ relatively prime, $g_x \neq 0\}$, note that $\mathcal{M}_x$ is an abelian group under addition. Let $\mathcal{M}$ = disjoint union $\bigcup_{x\in M} \mathcal{M}_x$ with topology given by the base of open sets

$$V(f,g,U) = \{f_x/g_x : x \in U,\ f,g \in \mathcal{O}(U),\ f_x, g_x \text{ relatively prime } \forall x \in U\}.$$

That this is a base follows from the

(3.29) Lemma: Let $f,g \in \mathcal{O}(U)$. If the germs f_{x_0}, g_{x_0} are relatively prime at $x_0 \in U$ then the germs f_x, g_x are relatively prime for all

x on some neighborhood of x_0.

Proof (see [G-H].)

M is thus a sheaf space of abelian groups, a meromorphic function h is simply a section of M, M is called the sheaf of (germs of) meromorphic functions on M. Equivalently, a meromorphic function h on M is an open cover $\{U_\alpha\}$ and a collection $\{f_\alpha, g_\alpha \in \mathcal{O}(U_\alpha)\}$ of holomorphic functions with $(f_\alpha)_x, (g_\alpha)_x$ relatively prime for all $x \in U_\alpha$ and with $f_\alpha g_\beta = f_\beta g_\alpha$ on each intersection $U_\alpha \cap U_\beta$; two collections $\{U_\alpha, f_\alpha, g_\alpha\}$, $\{U'_\beta, f'_\beta, g'_\beta\}$ identified if and only if $f_\alpha g'_\beta = f'_\beta g_\alpha$ on each intersection $U_\alpha \cap U'_\beta$.

Note that setting $h = f_\alpha/g_\beta$ gives a well-defined function $h: M \setminus D \to \mathbb{C} \cup \{\infty\}$ where $D = \bigcup_\alpha \{x \in U_\alpha : f_\alpha(x) = g_\alpha(x) = 0\}$. The set D is called the indeterminacy set of h. In general $D \neq \phi$ so a meromorphic function is not a function on the whole of M. (If $\dim M = 1$, $D = \phi$ and, in this case, a meromorphic function is a function from M to $\mathbb{C} \cup \{\infty\}$.)

We can now state

(3.30) The Cousin 1-Problem. Let $U \subseteq M$ be open. Let $\{U_\alpha\}$ be an open cover of M and let $\{m_\alpha \in M(U_\alpha)\}$ be such that $m_\alpha - m_\beta \in \mathcal{O}(U_\alpha \cap U_\beta)$. Find $m \in M(U)$ such that $m - m_\alpha \in \mathcal{O}(U_\alpha)$.

To interpret this problem, say that two meromorphic functions h, k have the same principal part at x if $h_x - k_x \in \mathcal{O}_x$, i.e. their germs differ by a holomorphic function. So the $\{m_\alpha\}$ specify a "principal part" at each point $x \in M$ and the problem is to find a meromorphic function $m \in M(U)$ with the given principal part at each point $x \in M$.

To study this, we define the sheaf of principal parts to be the quotient sheaf $PP = M/\mathcal{O}$ Then by a principal part at x we mean an element of $PP_x = M_x/\mathcal{O}_x$. The principal part of a meromorphic function $m \in M(U)$ at x is then $\pi_x(m)$ where $\pi: M \to PP$ is the natural projection. A section of PP over U is a specification of principal part at each $x \in U$. We see that the $\{m_\alpha\}$ in (3.30) define a section s of PP over U and so we may reformulate (3.30) as follows:

(3.31) The Cousin 1-Problem (Reformulation). Let $U \subseteq M$ be open. Let $s \in PP(U)$. Find $m \in M(U)$ such that $\pi_U(m) = s$.

(3.32) <u>Theorem</u>: The Cousin 1-problem has a solution if $\check{H}^1(U,\mathcal{O}) = 0$.

<u>Proof</u>. We have a short exact sequence of sheaves

$$0 \to \mathcal{O} \xrightarrow{i} M \xrightarrow{\pi} PP \to 0 \quad .$$

This induces a long exact cohomology sequence

$$0 \to \mathcal{O}(U) \xrightarrow{i_u} M(U) \xrightarrow{\pi_u} PP(U) \xrightarrow{\delta^*} \check{H}^1(U,\mathcal{O}) \to \cdots$$

If $\check{H}^1(U,\mathcal{O}) = 0$, then by exactness π_u is surjective and so given $s \in PP(U)$ there exists $m \in M(U)$ such that $\pi_u(m) = s$.

(3.33) <u>Corollary</u>: The Cousin 1-problem has a solution if U is a domain of holomorphy.

<u>Proof</u>. It can be shown that $\check{H}^p(U,\mathcal{O}) = 0 \;\; \forall p \geq 1$ if and only if U a domain of holomorphy (see [G-R], [Hö]).

(3.34) <u>Remark</u>: By exactness, it is easily seen that the Cousin 1-Problem has a solution if and only if $\delta^* s = 0$. Thus, the obstruction to solving this problem lies in the sheaf cohomology group $H^1(U,\mathcal{O})$. Theorem (3.32) is a special case of this.

D. <u>Fine and Acyclic Resolutions</u>

We establish a general technique for determing cohomology groups of sheaves which will establish the important theorems of de Rham and Dolbeault.

As before, let X be a paracompact topological space.

(3.35) <u>Definition</u>: We say that a sheaf S on X is <u>fine</u> if for any locally finite open cover $\{U_\alpha\}$ of X we can find sheaf homomorphisms $\eta_\alpha : S \to S$ with the properties:

(i) $(\eta_\alpha)_x$ is the zero homomorphism for all x in some open neighbourhood of the complement of U_α,

(ii) $\sum_\alpha (\eta_\alpha)_x$ = the identity homomorphism of S_x for all x.

(3.36) Proposition: The sheaves C^∞, A^p on a paracompact C^∞ manifold M and the sheaves $A^{p,q}$ on a paracompact complex manifold M are all fine.

Proof. Given any locally finite cover $\{U_\alpha\}$ of M, there exists a partition of unity S_α subordinate to U_α (see (2.7)); set $\eta_\alpha = S_\alpha \cdot I$ where I is the identity homomorphism.

(3.37) Remark: This doesn't work for, say, $\mathcal{O}$ since $S_\alpha \cdot f$ $(f \in \mathcal{O}(U))$ is not in general holomorphic.

Now for the most important property of fine sheaves:

(3.38) Proposition: Let S be a fine sheaf on X. Then

$$\check{H}^p(X,S) = 0 \quad \forall p \geq 1 .$$

Proof. Let $\sigma \in Z^p(\mathcal{U},S)$ where $\mathcal{U} = \{U_\alpha\}_{\alpha\in I}$ is a locally finite open cover of X. Define $\tau \in C^{p-1}(\mathcal{U},S)$ by $\tau_{\alpha_0,\ldots,\alpha_{p-1}} = \sum_{\beta\in I} \eta_\beta\, \sigma_{\beta,\alpha_0,\ldots,\alpha_{p-1}}$ setting $\eta_\beta\, \sigma_{\beta,\alpha_0,\ldots,\alpha_{p-1}} = 0$ where it is undefined. Then it is easy to check that $\delta\tau = \sigma$. For example, if $p = 1$,

$$\begin{aligned}(\delta\tau)_{\alpha_0,\alpha_1} &= \tau_{\alpha_1} - \tau_{\alpha_0} \\ &= \sum \eta_\beta\, \sigma_{\beta,\alpha_1} - \sum \eta_\beta\, \sigma_{\beta,\alpha_0} \\ &= \sum \eta_\beta\, (\sigma_{\alpha_0,\alpha_1}) \text{ by the cocycle condition for } \sigma \\ &= \sigma_{\alpha_0,\alpha_1} .\end{aligned}$$

A sheaf S such that $\check{H}^p(X,S) = 0$ $\forall p \geq 1$ is called acyclic.

Fine sheaves arise naturally in C^∞ differential geometry as in (3.36). We are now going to show how such "C^∞ objects" can be used to study "topological" and "complex analytic" objects. We need:

(3.39) Definition: A fine (resp. acyclic) resolution of a sheaf S is an exact sequence $0 \to S \xrightarrow{i} S^0 \xrightarrow{d_0} S^1 \xrightarrow{d_1} S^2 \to \cdots \to S^p \to \cdots$ where each S^p is a fine (resp. acyclic) sheaf.

Fine resolutions always exist (see e.g. [G-R],[Sw]).

(3.40) Examples: (i) A fine resolution of the sheaf $\mathbb{R}$ on a C^∞ manifold is given by the Poincaré sequence

$$0 \to \mathbb{R} \xrightarrow{i} A^o \xrightarrow{d} A^1 \xrightarrow{d} \dots \to A^p \to \dots$$

where i is inclusion and d exterior differentiation (This sequence is exact by the Poincaré Lemma (2.21)).

(ii) A fine resolution of the sheaf Ω^p of holomorphic p-forms on a complex manifold is given by the Dolbeault sequence

$$0 \to \Omega^p \xrightarrow{i} A^{p,o} \xrightarrow{\bar\partial} A^{p,1} \xrightarrow{\bar\partial} \dots \xrightarrow{\bar\partial} A^{p,q} \xrightarrow{\bar\partial} \dots .$$

This sequence is exact by the $\bar\partial$-Poincaré Lemma (2.36).

Given an acyclic resolution $0 \to S \xrightarrow{i} S^o \xrightarrow{d_o} S^1 \xrightarrow{d_1} \dots \to S^p \xrightarrow{d_p} \dots$ of a sheaf S, there is induced a sequence of groups

$$0 \to S^o(X) \xrightarrow{d_o} S^1(X) \xrightarrow{d_1} \dots \to S^p(X) \xrightarrow{d_p}$$

where, for brevity, we write d_p for $(d_p)_X$ (see definition 3.10). Note that although $S(X)$ does not appear explicitly, it is contained in $S^o(X)$ as the subgroup $\ker d_o$. This sequence is clearly a complex, that is to say $d_p \circ d_{p-1} = 0$ for all p, so we may consider its cohomology groups $H^p(\{S^p,d_p\}) = \ker d_p/\mathrm{Im}\, d_{p-1}$ $(p \geq 0)$. (We set $S^{-1}(X) = 0$ $d_{-1} = 0$ for convenience). In the cases of the Poincaré (resp. Dolbeault) resolutions, these groups are the de Rham (resp. Dolbeault) cohomology groups defined in §2. The next very important result tells us how to calculate the cohomology groups of S from an acyclic resolution.

(3.41) Theorem: Let

(3.42) $$0 \to S \xrightarrow{i} S^o \xrightarrow{d_o} S^1 \xrightarrow{d_1} \dots \to S^p \xrightarrow{d_p} \dots$$

by an acyclic resolution of a sheaf S on a paracompact topological space X. Then we have a canonical isomorphism

$$\check{H}^p(X,S) \cong H^p(\{S^p,d_p\}) \;, \quad p = 0,1,\dots$$

between the Čech cohomology of S and the cohomology of the complex

$$0 \to S^0(X) \xrightarrow{d_0} \dots \to S^p(X) \xrightarrow{d_p} \dots .$$

This powerful theorem has as immediate corollaries (c.f. (2.20), (2.35)):

(3.43) <u>DeRham's Theorem</u>: Let M be a paracompact C^∞ manifold. Then

$$\check{H}^p(M,\mathbb{R}) \cong H^p_{dR}(M,\mathbb{R}) \qquad \text{for} \quad p = 0,1,\dots$$

i.e. Čech cohomology of M with coefficients in $\mathbb{R}$ and de Rham cohomology agree (Note, by (3.24), $H^p(M,\mathbb{R})$ agrees with simplicial cohomology.)

(3.44) <u>Dolbeault's Theorem</u>: Let M be a paracompact complex manifold. Then, for $p,q = 0,1,\dots$,

$$\check{H}^q(M,\Omega^p) = H^{p,q}_{\bar\partial}(M)$$

i.e. Čech cohomology of the sheaf of holomorphic forms and Dolbeault cohomology agree.

Note the special case $p = 0$: $\check{H}^q(M,\mathcal{O}) = H^{0,q}_{\bar\partial}(M)$.

<u>Proof of (3.41)</u>. From the long exact sequence (3.42) we may obtain short exact sequences of sheaves, viz:

$$0 \to S \xrightarrow{i} S^0 \xrightarrow{d_0} Z_1 \to 0$$

$$0 \to Z_{p-1} \to S^{p-1} \xrightarrow{d_{p-1}} Z_p \to 0 \qquad (p = 1,2,\dots)$$

where Z_p is the sheaf $\ker d_p$. These give rise to long exact sequences

(3.45) $0 \to S(X) \xrightarrow{i} S^0(X) \xrightarrow{d_0} \check{H}^0(X,Z_1) \to \dots \to \check{H}^q(X,S^0) \to$ and

(3.46) $\to \check{H}^q(X,S^{p-1}) \xrightarrow{d^*_{p-1}} \check{H}^q(X,Z_p) \xrightarrow{\delta^*} \check{H}^{q+1}(X,Z_{p-1}) \to \check{H}^{q+1}(X,S^p) \to \dots$

$(q \geq 0,\ p \geq 1)$

Case $p = 0$. By (3.45) and (3.23)(ii),

$$\check{H}^o(X,S) = S(X) \cong \text{Im } i = \ker d_o = H^o(\{S^p,d_p\}) \quad .$$

Case $p \geq 1$. Since $\check{H}^q(X,S^{p-1})$ and $\check{H}^{q+1}(X,S^p)$ are zero for $q \geq 1$ by (3.38), then by exactness of (3.46), $\check{H}^q(X,Z_p) \cong \check{H}^{q+1}(X,Z_{p-1})$, $(p \geq 1, q \geq 1)$, also if $q = 0$, δ^* in (3.46) is surjective, thus

$$\begin{aligned}\check{H}^p(X,S) &\cong \check{H}^{p-1}(X,Z) \cong \check{H}^{p-2}(X,Z_2) \cong \ldots \\ &\cong \check{H}^1(X,Z_{p-1}) \\ &\cong \check{H}^o(X,Z_p)/d^*_{p-1}\check{H}^o(X,S^{p-1}) && \text{by surjectivity of } \delta \\ &= \ker d_p/\text{Im } d_{p-1} && \text{by (3.23(ii))} \\ &= H^p(\{S^p,d_p\}) && \text{by definition.}\end{aligned}$$

(3.47) Remark: (i) We could have defined the cohomology group $H^p(X,S)$ of a sheaf S on a paracompact topological space by

(3.48) $\quad H^p(X,S) = H^p(\{S^p,d_p\})$

where $0 \to S \to \ldots \to S^p \xrightarrow{d_p} \ldots$ is a fine resolution. Replacing "fine" by the more general "injective" we may use (3.48) to define $H^p(X,S)$ when X is not paracompat. Techniques such as (3.41) may be used to show that there is a unique cohomology theory satisfying certain axioms (see [Sw], [Wa]). From now on we shall drop the Čech symbol ˇ and denote the cohomology group (3.48) by $H^p(X,S)$.

E. Computations

We show how the Dolbeault theorem, $\bar{\partial}$-Poincaré Lemma and the long exact sequence of cohomology can be used to calculate the cohomology groups $H^q(M,\Omega^p)$ in various important cases. We then show how this rather abstract theory leads to the solution of a very concrete problem - the Cousin II problem.

(3.49) Proposition: Let M be an m-dimensional complex manifold. Then

$$H^q(M,\mathcal{O}) = H^{o,q}_{\bar{\partial}}(M) = 0 \quad \text{for} \quad q > m.$$

Proof. There are no non-zero $(0,q)$ forms for $q > m$.

(3.50) <u>Proposition</u>: For any polydisc Δ,

$$H^q(\Delta,\Omega^p) = H^{p,q}_{\bar\partial}(\Delta) = 0 \quad \forall p \geq 0,\ q \geq 1 \quad .$$

<u>Proof</u>. By (2.36)!

In particular all the cohomology groups $H^q(\Delta,\mathcal{O})$ $q \geq 1$ vanish. Similarly $H^q(\mathbb{C}^n,\mathcal{O}) = 0$ $\forall q \geq 1$ and, more generally, $H^q(\mathbb{C}^k\times(C^*)^1,\mathcal{O}) = 0$ $\forall q \geq 1$.

These results are in accord with the statement in the proof of (3.33).

(3.51) <u>Proposition</u>: $H^q(\mathbb{C}^n,\mathcal{O}^*) = 0$ $\forall q \geq 1$.

<u>Proof</u>. From the exponential sheaf sequence (3.13) we have a long exact sequence:

$$\ldots \to H^q(\mathbb{C}^n,\mathcal{O}) \to H^q(\mathbb{C}^n,\mathcal{O}^*) \to H^{q+1}(\mathbb{C}^n,\mathbb{Z}) \to \ldots \quad .$$

Since $H^q(\mathbb{C}^n,\mathbb{Z}) = 0$ $\forall q \geq 1$ by contractibility of $\mathbb{C}^n$, and $H^q(\mathbb{C}^n,\mathcal{O}) = 0$ $\forall q \geq 1$ as above, the result follows by exactness.

(3.52) <u>Proposition</u>: $\dim H^1(\mathbb{C}^2\setminus\{0\},\mathcal{O}) = \infty$.

<u>Proof</u>. Consider the open cover of $\mathbb{C}^2\setminus\{0\}$ given by $U_1 = \{(z^1,z^2): z^1 \neq 0\}$, $U_2 = \{(z^1,z^2): z^2 \neq 0\}$, then $U_1 \cong U_2 \cong \mathbb{C}\times C^*$ and $U_1\cap U_2 = \mathbb{C}^*\times\mathbb{C}^*$, so by (3.50), $\mathcal{U} = \{U_1,U_2\}$ is an acyclic cover , and we may use Theorem (3.26) to calculate the Čech cohomology groups.

Now a 1-cochain $\in C^1(\mathcal{U},\mathcal{O})$ is simply an analytic function on $U_1\cap U_2$ and is thus given by a Laurent series:

$$f(z^1,z^2) = \sum_{m,n=-\infty}^{\infty} a_{mn}(z^1)^m(z^2)^n \quad .$$

Since $C^2(\mathcal{U},\mathcal{O})$ is trivial, all 1-cochains are cocycles.

On the other hand, a 0-cochain $\in C^0(\mathcal{U},\mathcal{O})$ consists of an analytic function on U_1, i.e. a Laurent series $g_1(z^1,z^2) = \sum_{m\geq 0} b_{mn}(z^1)^m(z^2)^n$ and an analytic function on U_2, i.e. a Laurent series $g_2(z^1,z^2) = \sum_{n\geq 0} c_{mn}(z^1)^m(z^2)^n$. The coboundary of such a 0-chain, g_2-g_1, has no terms of the form $(z^1)^m(z^2)^n$, $m,n < 0$, these functions thus give linearly independent members of $H^1(\mathcal{U},\mathcal{O}) = H^1(\mathbb{C}^2\smallsetminus\{0\},\mathcal{O})$, the proposition follows.

(3.53) <u>Proposition</u>: $H^q(\mathbb{P}^n, \Omega^p) = \begin{cases} \mathbb{C} & \text{if } p = q \leq n \\ 0 & \text{otherwise} \end{cases}$.

<u>Proof</u>. We shall prove only that (a) $H^1(\mathbb{P}^1,\mathcal{O}) = 0$, (b) $H^1(\mathbb{P}^1,\Omega^1) = \mathbb{C}$. The other cases are similar or follow from Hodge theory, see § 6.19. Consider the open cover $\mathcal{U} = \{U,V\}$ where $U = \{[z^1,z^2]: z^2 \neq 0\}$ and $V = \{[z^1,z^2]: z^1 \neq 0\}$. Since $U,V \cong \mathbb{C}$ and $U \cap V \cong \mathbb{C}^*$ the cover is acyclic so that, by (3.26), $H^1(\mathbb{P}^1,\mathcal{O}) = H^1(\mathcal{U},\mathcal{O})$ and $H^1(\mathbb{P}^1,\Omega^1) = H^1(\mathcal{U},\Omega^1)$.

(a) An element of $C^1(\mathcal{U},\mathcal{O})$ is a holomorphic function on $U \cap V$, i.e. a Laurent series $h = \sum_{n=-\infty}^{\infty} a_n u^n$ where $u = \frac{z^1}{z^2}$.

This is the coboundary of the 0-chain $(f,g) \in \mathcal{O}(U) \times \mathcal{O}(V) = C^0(\mathcal{U},\mathcal{O})$ given by $f = -\sum_{n=0}^{\infty} a_n u^n$, $g = \sum_{n=1}^{\infty} a_{-n} v^n$ where $v = \frac{1}{u}$ hence $H^1(\mathbb{P}^1,\mathcal{O}) = 0$.

(b) An element v of $C^1(\mathcal{U},\Omega^1)$ is a holomorphic (1,0)-form on $U \cap V$, so

$$v = \left(\sum_{n=-\infty}^{\infty} a_n u^n\right) du \quad .$$

Since $C^2(\mathcal{U},\Omega')$ is trivial every element of $C^1(\mathcal{U},\Omega^1)$ is a cocycle. On the other hand, an element of $C^0(\mathcal{U},\Omega^1)$ is a pair $(w,\eta) \in \Omega^1(U) \times \Omega^1(V)$, so

$$w = \left(\sum_{n=0}^{\infty} b_n u^n\right) du \quad \text{and} \quad \eta = \left(\sum_{n=0}^{\infty} a_n v^n\right) dv = -\left(\sum_{n=0}^{\infty} a_n u^{-n-2}\right) du.$$

It follows that v is the coboundary of some 0-cochain (w,η), i.e. $v = \eta - w$ on $U \cap V$, if and only if $a_{-1} = 0$. Hence $H^1(\mathcal{U},\Omega^1) \cong \mathbb{C}$ and is generated by the cohomology class of the cocycle $v = u^{-1}du$.

(3.54) <u>Corollary</u>: The Cousin 1-problem is always soluble on $\mathbb{P}^n$.

We close by considering the multiplicative version of the Cousin I problem:

(3.55) <u>The Cousin II-Problem</u>: Let $U \subseteq M$ be a domain. Let $\{U_\alpha\}$ be an open cover of U and let $\{f_\alpha \in \ (U_\alpha)\}$ be such that $f_\alpha = g_{\alpha\beta} f_\beta$ for some $\{g_{\alpha\beta} \in \mathcal{O}^*(U_\alpha \cap U_\beta)\}$. Find $f \in \mathcal{O}(U)$ such that $f = h_\alpha f_\alpha$ for some $\{h_\alpha \in \mathcal{O}^*(U_\alpha)\}$.

(3.56) Theorem: On any domain U with $H^1(U, \mathcal{O}^*) = 0$, the Cousin II problem is soluble.

Proof. Let $\{U_\alpha, f_\alpha, g_{\alpha\beta}\}$ be as above. Note that $\{g_{\alpha\beta}\}$ is a cocycle in $C^1(\mathcal{U}, \mathcal{O}^*)$. Since $H^1(U, \mathcal{O}^*) = 0$, refining $\mathcal{U}$ if necessary, this is a coboundary, i.e. there exists a 0-cochain $\{h_\alpha\} \in C^0(\mathcal{U}, \mathcal{O}^*)$ such that $h_\beta / h_\alpha = g_{\alpha\beta}$. Then $f = h_\alpha f_\alpha$ is the desired function.

(3.57) Application: The zero set $\bigcup_\alpha \{z \in U_\alpha : f_\alpha(z) = 0\}$ defined by the data $\{U_\alpha, f_\alpha, g_{\alpha\beta}\}$ of the Cousin II-problem is called an analytic hypersurface of U. The above theorem may be interpreted as saying that on any domain U with $H^1(U, \mathcal{O}^*) = 0$ (for example $U = \mathbb{C}^n$), an analytic hypersurface is the zero locus of a single holomorphic function f on U.

§4. CONNECTIONS IN VECTOR BUNDLES - KÄHLER MANIFOLDS

A. Holomorphic Vector Bundles

Recall that in (2.8), we introduced the notion of real or complex vector bundle , defined in general on a real manifold using smooth transition functions. We now consider a similar definition in the holomorphic framework.

(4.1) Definition: Let M be a complex manifold and E a complex vector bundle over M. E is called holomorphic if it has the structure of a complex manifold and if any point $z \in M$ is contained in a neighbourhood U such that there is a biholomorphic (i.e. holomorphic together with its inverse) trivialisation $\varphi_U : \pi^{-1}(U) \to U \times \mathbb{C}^k$.

Note that if φ_α and φ_β are biholomorphic trivialisations, the transition functions $g_{\alpha\beta} : U_\alpha \cap U_\beta \to GL(k;\mathbb{C})$ are holomorphic, and conversely that a bundle defined by holomorphic transition functions is holomorphic.

(4.2) Example: The holomorphic tangent bundle $T'M$ (2.25) is indeed a holomorphic bundle, since its transition functions $\left(\frac{\partial w^l}{\partial z^j}\right)$ associated to the holomorphic changes of charts $w^l(z^j)$ are matrices of holomorphic functions.

All constructions (2.9) - (2.12) made on smooth bundles carry over to holomorphic ones. In particular the $\oplus$, $\otimes$ and $\wedge$ of holomorphic bundles are holomorphic, and so is the pull-back of a holomorphic bundle by a holomorphic map.

A section or frame over U is called holomorphic if it is defined by holomorphic maps $U \to E$, and in terms of a holomorphic frame (σ_a), a section $\sigma = f^a\sigma_a$ is holomorphic iff the functions f^a are.

B. Vector Bundle Valued Forms

(4.3) Definition: Let $E \to M$ be a complex (or real) vector bundle. A p-form on M with values in E is a smooth section of the vector bundle $\Lambda^p(T^*M) \otimes E$. We denote by $A^p(E)$ the space of these sections.

In other words, an element of $A^p(E)$ associates to p vector fields a section of E in an antisymmetric way. If E is the trivial bundle $M \times \mathbb{C}$, the elements of $A^p(E)$ are "ordinary" p-forms in the sense of (2.17).

If M is complex, the p-forms can be extended to $\Lambda^p(T^{*\mathbb{C}}M) \otimes E$ by $\mathbb{C}$ linearity, then decomposed as in (2.29), so that

$$A^r(E) = \bigoplus_{p+q=r} A^{p,q}(E)$$

where $A^{p,q}(E)$ is the space of sections of $\Lambda^p T^{*\prime}M \otimes \Lambda^q T^{*\prime\prime}M \otimes E$.

On $A^*(M)$, we defined in (2.18) the exterior differential d . This definition cannot be used here, since we have not defined the action of a vector field on a section of E. This will be done in the next section, but in a non-canonical way.

However, we have the

(4.4) <u>Proposition</u>: Let E be a holomorphic vector bundle. The operator $\bar\partial: A^{p,q}(E) \to A^{p,q+1}(E)$ is well defined by

$$\bar\partial(\lambda^a \otimes e_a) = (\bar\partial\lambda^a) \otimes e_a \quad ,$$

where $(e_1,\ldots,e_k)$ is a local holomorphic frame and $\lambda^a \in A^{p,q}(U)$.

<u>Proof</u>. Let $\sigma \in A^{p,q}(E)$.
In a holomorphic frame field $(e_1,\ldots,e_k)$ over U, we have $\sigma = \lambda^a \otimes e_a$, $\lambda^a \in A^{p,q}(U)$ and we define $\bar\partial\sigma = (\bar\partial\lambda^a) \otimes e_a$.

We must check that we will get the same result using another holomorphic frame over U, say $(e_1',\ldots,e_k')$. We have $e_a = g_a^b e_b'$, where g_a^b are holomorphic functions.

Then $\sigma = \lambda^a g_a^b \otimes e_b'$, and defining $\bar\partial'$ as above in the frame (e'), we get

$$\begin{aligned} \bar\partial'\sigma &= \bar\partial(\lambda^a g_a^b) \otimes e_b' \\ &= (\bar\partial\lambda^a) g_a^b \otimes e_b' \qquad \text{since } g_a^b \text{ holomorphic} \\ &= \bar\partial\lambda^a \otimes e_a \\ &= \bar\partial\sigma \quad . \end{aligned}$$

C. Connections

Let $E \to M$, be a real or complex vector bundle, $C(E)$ its space of smooth sections (also denoted $C^\infty(E)$) and $C(M) = C(M,\mathbb{R})$ (or $C(M,\mathbb{C})$) the space of real (or complex) valued smooth functions on M.

(4.5) Definition: A connection on E is a bilinear map $\nabla: C(TM) \times C(E) \to C(E)$, written $(X,\sigma) \to \nabla_X\sigma$ and such that

$$\text{(i)}\ \nabla_{fX}\sigma = f\,\nabla_X\sigma \qquad f \in C(M,\mathbb{R})$$

$$\text{(ii)}\ \nabla_X(f\sigma) = (Xf)\sigma + f\,\nabla_X\sigma \qquad f \in C(M)$$

where $X \in C(TM)$ and $\sigma \in C(E)$.

$\nabla_X\sigma$ is called the covariant derivative of σ in the direction of X.

(4.6) Remark: If E and M are complex, we can extend ∇ to $C(T^{\mathbb{C}}M) \times C(E)$ then decompose it as $\nabla = \nabla' + \nabla''$ by restricting its action to $T'M$ and $T''M$.

(4.7) Alternatively, ∇ can be seen as a $\mathbb{C}$ - linear map (also denoted ∇)

$$\nabla: A^0(E) \to A^1(E) \quad \text{satisfying}$$

$$\nabla(f.\sigma) = df \otimes \sigma + f\cdot\nabla\sigma \qquad \forall f \in C(M).$$

This reduces to the previous definition by having T^*M (in the new definition) acting on TM (in the former one) by $\nabla\sigma\cdot X = \nabla_X\sigma$.

A connection can be described locally in the following manner. Let $e = (e_1,\ldots,e_k)$ be a frame field for E over U. Then

$$\text{(4.8)} \qquad \nabla e_a = \omega_a^b\, e_b$$

where $\omega = (\omega_a^b)$ is a matrix of complex one-forms, called the connection matrix.

The data (e) and ω determine ∇, since for $\sigma \in A^0(E)|_U$, we have

$$\sigma = \sigma^a\, e_a \quad \text{and}$$

$$\text{(4.9)} \qquad \nabla\sigma = d\sigma^a\, e_a + \sigma^a\, \nabla e_a$$

$$= (d\sigma^a + \sigma^b\, \omega_b^a)e_a \qquad \text{or}$$

$$\nabla_X\sigma = (X\,\sigma^a + \sigma^b\,\omega^a_b(X))e_a \quad .$$

If we also choose a coordinate system (x^j) on U we have $X = X^j \frac{\partial}{\partial x^j}$ and

$$\nabla_X\sigma = \left(X^j\,\frac{\partial\sigma^a}{\partial x^j} + \sigma^b\,X^j\,\omega^a_b\left(\frac{\partial}{\partial x^j}\right)\right)e_a \quad .$$

Setting

$$\nabla_{\frac{\partial}{\partial x^j}}\,e_a = \omega^b_a\left(\frac{\partial}{\partial x^j}\right)\,e_b = \Gamma^b_{ja}\,e_b \quad , \quad \text{i.e.}$$

$$\omega^b_a = \Gamma^b_{ja}\,dx^j \quad , \quad \text{we obtain}$$

(4.10) $$\nabla_X\sigma = X^j\left(\frac{\partial\sigma^a}{\partial x^j} + \sigma^b\,\Gamma^a_{jb}\right)e_a \quad .$$

Consider now a change of frame fields in $E|_U$: $e'_a = g^b_a\,e_b$.

Then $\nabla e'_a = dg^b_a\,e_b + g^c_a\,\omega^b_c\,e_b$.

Since $\nabla e'_a = \omega'^c_a\,e'_c = \omega'^c_a\,g^b_c\,e_b$, we get $\omega'^c_a = dg^b_a(g^{-1})^c_b + g^d_a\,\omega^b_d(g^{-1})^c_b$. In matrix notation, we have therefore the transformation law for ω:

(4.11) $$\omega' = dg\,g^{-1} + g\,\omega\,g^{-1} \quad .$$

Conversely, a connection can be defined by specifying in local frame fields the matrices ω, in such a way that they satisfy the transformation law (4.11).

Indeed, we can then define ∇ by (4.9) and a short calculation using (4.11) shows that ∇ is independent of the choice of frame field.

D. Exterior Differential of a Vector Valued p-Form

Using a connection ∇, we can now imitate definition (2.18).

(4.12) Definition: The exterior differential $d\colon A^p(E) \to A^{p+1}(E)$ associated to the connection ∇ is defined by

$$d\lambda(X_1,\ldots,X_{p+1}) = \sum_{j=1}^{p+1} (-1)^{j-1} \nabla_{X_j} \lambda(X_1,\ldots,\hat{X}_j,\ldots,X_{p+1})$$

$$+ \sum_{j<k} (-1)^{j+k} \lambda([X_j,X_k]X_1,\ldots,\hat{X}_j,\ldots,\hat{X}_k,\ldots,X_{p+1})$$

where $\lambda \in A^p(E)$ and $X_1,\ldots,X_{p+1} \in C(TM)$.

One can check the antiderivation rule: if in particular $\lambda \in A^p(M)$, $\sigma \in A^o(E)$, then

(4.13) $d(\lambda \wedge \sigma) = d\lambda \wedge \sigma + (-1)^p \lambda \wedge \nabla\sigma.$

This relation can in fact be used to define d, since all the operators in the right hand side are known.

For real or complex valued forms, we had $d^2 = 0$. This does not remain true here. Indeed, the calculation of d^2 involves commutators of covariant derivatives, which are not zero (as for ordinary derivatives) but are measured by the curvature of the connection.

E. Curvature

(4.14) Definition: The curvature of the connection ∇ is the map

$R: A^2(M) \times C(E) \to C(E)$ defined by

$R(X,Y)\sigma = -\nabla_X\nabla_Y\sigma + \nabla_Y\nabla_X\sigma + \nabla_{[X,Y]}\sigma$.

Remark: Our sign convention is that of [Mi]. Note that various authors (e.g. [K-N.]) use the opposite sign in the definition.

The first thing to note is that R is $C(M)$ linear in the three factors. Indeed, repeated applications of (i) and (ii) in (4.5) show that for all $f \in C(M)$:

(4.15) $R(fX,Y)\sigma = R(X,fY)\sigma = R(X,Y)f\sigma = fR(X,Y)\sigma.$

Writing vector fields and sections in terms of a basis, we deduce that the value of $R(X,Y)\sigma$ at a point x depends only of the values of X,Y and σ at that point.

The curvature is related to the exterior differential (4.12) in the following manner:

(4.16) <u>Proposition</u>: $\forall X,Y \in C(TM)$, $\sigma \in C(E)$,

$$R(X,Y)\sigma = -d^2\sigma(X,Y)$$

where $d^2 = d \circ d: A^0(E) \to A^2(E)$.

<u>Proof</u>. $d\sigma(X) = \nabla_X\sigma$

$$\begin{aligned} d(d\sigma)(X,Y) &= \nabla_X(d\sigma(Y)) - \nabla_Y(d\sigma(X)) - d\sigma[X,Y] \\ &= \nabla_X\nabla_Y\sigma - \nabla_Y\nabla_X\sigma - \nabla_{[X,Y]}\sigma \,. \end{aligned}$$

<u>Remark</u>: For $\lambda \in A^p(E)$, one can check that

$$-d^2\lambda = R \wedge \lambda \,.$$

From the fact that $(R(X,Y)\sigma)(x)$ depends only on X_x, Y_x and σ_x, we see that R is a global section of the bundle $\Lambda^2 T^*M \otimes E^* \otimes E$.

In a local frame field $e = \{e_a\}$, we can write it as $\Omega_a^b\, e^{*a} \otimes e_b$. $\Omega = \left(\Omega_a^b\right)$ is then a matrix of two-forms called the <u>curvature matrix</u> of ∇ with respect to the frame e. Then

(4.17) $$-d^2 e_a = \Omega_a^b \otimes e_b \,.$$

Using (4.15), we see that if $e_a' = g_a^b\, e_b$ is the relation between two local frames, and Ω' the curvature matrix of ∇ with respect to e' then

$$\begin{aligned} -d^2\, e_a' &= -d^2(g_a^b\, e_b) \\ &= -g_a^b\, d^2\, e_b \\ &= g_a^c\, \Omega_c^b\, e_b \,. \end{aligned}$$

Since $-d^2\, e_a' = \Omega_a'^c\, e_c' = \Omega_a'^c\, g_c^b\, e_b$, we have the transformation law:

(4.18) $$\Omega' = g\, \Omega\, g^{-1} \,.$$

In local frames, we now express the curvature in terms of the connection:

$$-d^2\, e_a = -d(\omega_a^b \otimes e_b)$$

$$= -d\omega_a^b \otimes e_b + \omega_a^c \wedge \omega_c^b \otimes e_b$$

or, in matrix notation:

(4.19) <u>The Cartan structural equation</u>

$$\Omega = -d\omega + \omega \wedge \omega .$$

Again, we can take local coordinates $\frac{\partial}{\partial x^j}$ on U, set $\Omega_a^b = R^b_{ajk} dx^j \wedge dx^k$, which means that $R\left(\frac{\partial}{\partial x^j}, \frac{\partial}{\partial x^k}\right)e_a = R^b_{ajk}\, e_b$ and get $R^b_{ajk} = \frac{\partial \Gamma^b_{ja}}{\partial x^k} - \frac{\partial \Gamma^b_{ka}}{\partial x^j} + \Gamma^c_{ja}\Gamma^b_{kc} - \Gamma^c_{ka}\Gamma^b_{jc}$.

F. <u>Tensor Product and Successive Derivations</u>

Let E and F be two vector bundles over M equipped with connections ∇_E and ∇_F. To the different bundles constructed in (2.9) - (2.10) are associated connections derived from ∇_E and ∇_F, in such a way that the expected "derivation rules" apply.

(4.20) On E^*, for $\sigma \in E$, $\tau \in E^*$ with $\langle\tau,\sigma\rangle$ the duality pairing, ∇_E^* is defined by

$$d\langle\tau,\sigma\rangle = \langle\nabla_E^*\tau,\sigma\rangle + \langle\tau,\nabla_E\sigma\rangle .$$

(4.21) On $E \oplus F$, $\nabla(\sigma \oplus \rho) = \nabla_E\sigma \oplus \nabla_F\rho$.

(4.22) On $E \otimes F$, the connection is defined as $\nabla_E \otimes 1 + 1 \otimes \nabla_F$, i.e.

$$\nabla(\sigma \otimes \rho) = \nabla_E\sigma \otimes \rho + \sigma \otimes \nabla_F\rho \quad .$$

Suppose now that $E \to M$ is a vector bundle, ∇_E a connection on E and ∇_M a connection on TM (which could be the one defined in (4.27) below). We can then define the iterated covariant differentials of a section of E as those induced by the various operations above, and we shall denote them all by ∇. We have:

$$C(E) \xrightarrow{\nabla} C(T^*M \otimes E) \xrightarrow{\nabla} C(\otimes^2 T^*M \otimes E) \xrightarrow{\nabla} \quad .$$

For instance, the second ∇ is defined by

$$\nabla(X^* \otimes \sigma) = \nabla_M^* X^* \otimes \sigma + X^* \otimes \nabla_E \sigma \ .$$

We denote by ∇^k the $k^{\underline{th}}$ iterated differential

(4.23) $\nabla^k: C(\otimes^p T^*M \otimes E) \to C(\otimes^{p+k} T^*M \otimes E)$.

G. Canonical Connections

Using partitions of unity, one can show that connections exist on all vector bundles on paracompact manifolds, but are not unique. We shall see that additional structures on E can determine a connection canonically.

We first note that the definition of a Riemannian structure on TM can be extended to any vector bundle E: the bilinear map h will now be defined on the fibres of E and be symmetric and positive definite on these.

Likewise, as on $T'M$, a Hermitian structure H on a complex vector bundle is a smooth assignment of a positive definite Hermitian product on each fibre, i.e. an $\mathbb{R}$-bilinear product with

$$H_z(a\sigma, b\rho) = a\bar{b}\, H_z(\sigma,\rho), \quad H_z(\sigma,\rho) = \overline{H_z(\rho,\sigma)}$$

and $H_z(\sigma,\sigma) > 0$ for $\sigma \neq 0$, where $\sigma,\rho \in E_z$ and $a,b \in \mathbb{C}$.

(4.24) Definition: A connection on a Riemannian bundle E, h is called a metric connection iff

$$X\, h(\sigma,\rho) = h(\nabla_X\sigma,\rho) + h(\sigma,\nabla_X\rho) \qquad \forall X \in C(TM), \ \sigma,\rho \in C(E)$$

This can be interpreted as saying that $\nabla h = 0$ or that h is covariantly constant.

We can apply the same definition to the Hermitian case:

(4.25) Definition: On a complex vector bundle E with Hermitian structure H, a connection ∇ is called a metric connection iff

$$X\, H(\sigma,\rho) = H(\nabla_X\sigma,\rho) + H(\sigma,\nabla_{\bar{X}}\rho) \qquad \forall X \in C(T^{\mathbb{C}}M), \ \sigma,\rho \in C(E) \ .$$

There may still be different metric connections associated with a Riemannian or Hermitian structure. However, we have the two following results.

(4.26) Proposition: Let E be a holomorphic vector bundle and H a Hermitian structure. There is a unique metric connection ∇ on E such that $\nabla'' = \bar{\partial}$, where $\bar{\partial}$ is the canonical operator of E (see 4.4).

Proof.

- Unicity:

Suppose that a ∇ exists. By the condition $\bar{\partial} = \nabla''$, we have in a holomorphic frame (e_a), for a section $\sigma = \sigma^a e_a$:

$$\frac{\partial \sigma^a}{\partial \bar{z}^j} e_a = \nabla_{\frac{\partial}{\partial \bar{z}^j}} (\sigma^a e_a) = \left(\frac{\partial}{\partial \bar{z}^j} \sigma^a + \sigma^b \omega^b_a\left(\frac{\partial}{\partial \bar{z}^j}\right)\right) e_a \quad .$$

Hence, $\omega^b_a\left(\frac{\partial}{\partial \bar{z}^j}\right) = 0$, which means that ω^b_a is a form of type $(1,0)$.

Therefore, the metric condition gives:

$$\begin{aligned} d\, H_{ab} &= d\, H(e_a, e_b) \\ &= \omega^c_a H_{cb} + H_{ac} \overline{\omega^c_b} \\ &= \text{type } (1,0) + \text{type } (0,1). \end{aligned}$$

Separating types, we get the two equations

$$\begin{aligned} \partial\, H_{ab} &= \omega^c_a H_{cb} \qquad \text{and} \\ \bar{\partial}\, H_{ab} &= \overline{\omega^c_b} H_{ac} \quad . \end{aligned}$$

The second equation is the complex conjugate of the first one and ω is therefore uniquely characterized by

$$\omega = \partial H \cdot H^{-1} \quad .$$

- Existence:

In a local holomorphic frame field, we set $\omega = \partial H \cdot H^{-1}$.

A short calculation shows that ω transforms as a connection form for a change of frame field, so that it defines a connection, and reading the above calculation backwards, we see that it satisfies the conditions of the proposition.

We shall not pursue the study of this connection (see [G-H], [We]) but turn to another classical situation, namely that of the tangent bundle TM.

(4.27) Definition: The torsion T of a connection ∇ on TM is defined by

$$T(X,Y) = \nabla_X Y - \nabla_Y X - [X,Y]$$

(4.28) Proposition: Let M, h be a Riemannian manifold. There is a unique torsionless $(T = 0)$ metric connection on TM. It is called the Levi-Civita connection.

Proof.

- Unicity:

Using the conditions $\nabla h = 0$ and $T = 0$, one calculates the expression $Xh(Y,Z) + Yh(Z,X) - Zh(X,Y)$ to obtain easily that

$$\begin{aligned} 2h(\nabla_X Y,Z) = {} & Xh(Y,Z) + Yh(X,Z) - Zh(X,Y) \\ & + h([X,Y],Z) + h([Z,X],Y) + h(X,[Z,Y]). \end{aligned}$$

Since this formula is valid for all vector fields X,Y,Z, it characterizes ∇.

- Existence:

Define $\nabla_X Y$ by the above formula. Short calculations show then that ∇ is a connection, that $T = 0$ and that $\nabla h = 0$.

In local coordinates (x^j), we obtain the expression for Γ by applying the above formula to the vectors $\frac{\partial}{\partial x^j}$, $\frac{\partial}{\partial x^k}$ and $\frac{\partial}{\partial x^l}$: we have

$$\Gamma^l_{jk} = \frac{1}{2} h^{lm}\left(\frac{\partial h_{mj}}{\partial x^k} + \frac{\partial h_{mk}}{\partial x^j} - \frac{\partial h_{jk}}{\partial x^m}\right)$$

where $h^{lm}h_{mj} = \delta^l_j$.

H. Kähler manifolds

We now introduce a more restricted class of Hermitian manifolds, namely the Kähler manifolds (see e.g. [Weil]). We shall give five equivalent definitions, each giving a hint of the importance of that notion in different frameworks. We shall then prove the equivalence of these definitions and at the same time give some properties.

(4.29) Definition: A Hermitian manifold M, J, h (see (2.41)) is Kähler if any of the following equivalent conditions is satisfied

(i) $\nabla J = 0$, i.e. $\nabla_X(JY) = J\nabla_X Y$, where ∇ is the Levi-Civita connection on M.

(ii) $dF = 0$, where F is the fundamental form on M (2.43).

(iii) Each point $z_o \in M$ is contained in a neighbourhood in which $F = i\partial\bar{\partial}f$ for a real function f.

(iv) The metric h osculates the Euclidean metric to order 2, i.e. for every point $z_o \in M$, there exists in a neighbourhood of z_o a system of coordinates (z^j) such that

$$ds^2 = (\delta_{jk} + O(2))dz^j \otimes d\bar{z}^k$$

where $\frac{O(2)}{|z|^2}$ is bounded as $|z| \to 0$

(v) ∇ induces on $T'M$ the canonical metric connection of Proposition (4.26) (i.e. with $\nabla'' = \bar{\partial}$).

Remarks: The condition that h be Hermitian is a pointwise condition, whereas (i) describes the variation of J from one point to another.

Condition (ii) may be quite simple to verify, and (iii) allows us to define (locally) a Kähler metric (see section I below).

We shall use condition (iv) in (6.6) to establish first order differential identities on a Kähler manifold by proving them in $\mathbb{C}^m$, then using the special charts.

Finally, (v) relates to the canonically defined operator $\bar{\partial}$ on $T'M$.

Proof of the equivalence.

(i) $\Rightarrow$ (ii):

$$\begin{aligned} dF(X,Y,Z) &= X\, F(Y,Z) + Y\, F(Z,X) + Z\, F(X,Y) \\ &\quad - F([X,Y],Z) - F([Z,X],Y) - F([Y,Z],X) && \text{by (2.18)} \\ &= X\, h(JY,Z) + \dots - h(J[X,Y],Z) - \dots && \text{by (2.43)} \end{aligned}$$

$= h(\nabla_X JY,Z) + h(JY,\nabla_X Z) + \ldots - h(J[X,Y],Z) - \ldots$ by (4.25)

$= h(\nabla_X JY - J\nabla_Y X - J[X,Y],Z)$ + circular permutations of X,Y,Z by (2.41)

$= h(\nabla_X JY - J\nabla_X Y,Z)$ + permutations since $T = 0$

$= 0$ by (i).

(ii) ⇒ (i):

Similar calculations using the same equations show that

$$2h((\nabla_X JY - J\nabla_X Y),Z) = d\,F(X,JY,JZ) - dF(X,Y,Z).$$

(ii) ⇒ (iii):

The form F is real and closed. Hence by Poincaré's lemma (2.21), there exists in a neighbourhood of z_o a real 1-form β such that $F = d\beta$.

β, being real, is equal to a (1,0)-form plus its conjugate in the decomposition dual to (2.25), i.e. $\beta = \beta_1 + \overline{\beta_1}$. Since $d\,\beta = F$ is of type (1,1), $\partial\beta_1 = \overline{\partial}\,\overline{\beta_1} = 0$ and $F = \overline{\partial}\beta_1 + \partial\overline{\beta}_1$.

Since $\overline{\partial}\,\overline{\beta}_1 = 0$, the $\overline{\partial}$-Poincaré lemma (2.36) implies that $\overline{\beta}_1 = \overline{\partial}\varphi$ in a neighbourhood, so that $\partial\overline{\varphi} = \beta_1$.

Therefore,

$$\begin{aligned} F &= \overline{\partial}\beta_1 + \partial\overline{\beta}_1 \\ &= \overline{\partial}\partial\overline{\varphi} + \partial\overline{\partial}\varphi \qquad \text{by (2.34)} \\ &= \partial\overline{\partial}(\varphi-\overline{\varphi}) \\ &= i\,\partial\overline{\partial}\,f \end{aligned}$$

where $f = -i(\varphi-\overline{\varphi})$ is a real function.

(iii) ⇒ (ii):

If $F = i\,\partial\overline{\partial}\,f$, then

$$\begin{aligned} dF &= i\,\partial\partial\overline{\partial}\,f + i\,\overline{\partial}\partial\overline{\partial}\,f \\ &= \quad 0 \quad - i\,\overline{\partial}\overline{\partial}\partial\,f \\ &= \quad 0\,. \end{aligned}$$

<u>Remark</u>: If M is a compact Kähler manifold, we cannot have $F = i\,\partial\bar\partial f$ globally.

Indeed, we would then have $F = d\beta$, with β globally defined and

$$\int_M F^m = \int_M d\beta \wedge F^{m-1} = \int_M d(\beta \wedge F^{m-1}) = 0 \quad ,$$

which is impossible since F^m is a volume form as in (2.50).

<u>(iv) ⇒ (iii)</u>:

In a neighbourhood around z_o, (iv) is satisfied and we have

$$F = i(\delta_{jk} + O(2))dz^j \wedge d\bar z^k \quad .$$

Therefore, $dF(z_o) = 0$, and this is true for all $z_o \in M$.

<u>(iii) ⇒ (iv)</u>:

We can always find coordinates such that $h_{jk}(z_o) = \delta_{jk}(z_o)$, i.e.

$$F = i(\delta_{jk} + a_{jkl} z^l + a_{jk\bar l}\bar z^l + O(2))dz^j \wedge d\bar z^k \quad .$$

Note that $h_{j\bar k} = \overline{h_{k\bar j}} \Rightarrow a_{jk\bar l} = \overline{a_{kjl}}$

$$dF = 0 \Rightarrow a_{jkl} = a_{lkj} \quad .$$

We look for a change of coordinates

$$z^l = w^l + \frac{1}{2} b_{lmn} w^m w^n \qquad (\text{with } b_{lmn} = b_{lnm})$$

such that

$$F = i(\delta_{jk} + O(2))dw^j \wedge d\bar w^k \quad .$$

Since $dz^l = dw^l + b_{lmn} w^m dw^n$, we see after a short calculation that

$$F = i(\delta_{jk} + a_{jkl} w^l + a_{jk\bar l} \bar w^l + b_{jlk} w^l + \overline{b_{jlk}} \bar w^l + O(2))dw^j \wedge dw^k$$

Setting $b_{klj} = -a_{jkl}$, we have

$$b_{klj} = -a_{jkl} = -a_{lkj} = b_{jlk} \qquad \text{and}$$

$$\overline{b_{jlk}} = -\overline{a_{kjl}} = -a_{jk\bar{l}}$$

so that F is of the right form.

(i) ⇔ (v):

We shall prove this by introducing another characterization of Kähler manifolds, in term of the Γ's.

We denote by $\frac{\partial}{\partial z^J}$, $J = 1,\dots,m,\bar{1},\dots,\bar{m}$ the quantities $\frac{\partial}{\partial z^j}$ and $\frac{\partial}{\partial \bar{z}^j}$, and set $\nabla_{\frac{\partial}{\partial z^J}} \frac{\partial}{\partial z^K} = \Gamma^L_{JK} \frac{\partial}{\partial z^L}$.

By bilinearity of ∇, $\overline{\Gamma^L_{JK}} = \Gamma^{\bar{L}}_{\bar{J}\bar{K}}$, with the convention $\bar{\bar{j}} = j$.

Since $T\left(\frac{\partial}{\partial z^J}, \frac{\partial}{\partial z^K}\right) = 0$, $\Gamma^L_{JK} = \Gamma^L_{KJ}$.

Suppose now that (i) is satisfied, so that $J\nabla_{\frac{\partial}{\partial z^j}} \frac{\partial}{\partial z^k} = \nabla_{\frac{\partial}{\partial z^j}} J\frac{\partial}{\partial z^k}$.

Since $J\nabla_{\frac{\partial}{\partial z^j}} \frac{\partial}{\partial z^k} = J\,\Gamma^l_{jk} \frac{\partial}{\partial z^l} + J\,\Gamma^{\bar{l}}_{jk}\frac{\partial}{\partial \bar{z}^l} = i\,\Gamma^l_{jk} \frac{\partial}{\partial z^l} - i\,\Gamma^{\bar{l}}_{jk}\frac{\partial}{\partial \bar{z}^l}$ and $\nabla_{\frac{\partial}{\partial z^j}} J\frac{\partial}{\partial z^k} = i\,\Gamma^l_{jk} \frac{\partial}{\partial z^l} + i\,\Gamma^{\bar{l}}_{jk}\frac{\partial}{\partial \bar{z}^l}$, this implies $\Gamma^{\bar{l}}_{jk} = 0$.

Similarly, the relation $J\nabla_{\frac{\partial}{\partial z^j}} \frac{\partial}{\partial \bar{z}^k} = \nabla_{\frac{\partial}{\partial z^j}} J\frac{\partial}{\partial \bar{z}^k}$ implies that $\Gamma^l_{j\bar{k}} = 0$.

Combining these relations, we see that if (i) is satisfied, then the only non-zero Γ's are

$$\Gamma^j_{kl} = \Gamma^j_{lk} \quad \text{and} \quad \Gamma^{\bar{j}}_{\bar{k}\bar{l}} = \Gamma^{\bar{j}}_{\bar{l}\bar{k}} = \overline{\Gamma^j_{lk}} \quad .$$

From the same calculation read backwards, we see that these conditions are in fact equivalent to (i).

Suppose now that (v) holds.

For $X,Y \in C(T'M)$, $X = X^j \frac{\partial}{\partial z^j}$, $Y = Y^k \frac{\partial}{\partial z^k}$, we have $\nabla_X Y \in C(T'M)$. Since

$$\nabla_X Y = X^j \frac{\partial Y^k}{\partial z^j} \frac{\partial}{\partial z^k} + X^j Y^k \Gamma^l_{jk} \frac{\partial}{\partial z^l} + X^j Y^k \Gamma^{\bar{l}}_{jk} \frac{\partial}{\partial \bar{z}^l} \quad ,$$

this condition is equivalent to $\Gamma^{\bar{l}}_{jk} = 0$.

For $X \in C(T''M)$, $Y \in C(T'M)$, we have $\nabla_X Y = (\bar\partial Y)X$. Since, with $X = X^{\bar j}\frac{\partial}{\partial \bar z^j}$, $\nabla_X Y = X^{\bar j}\frac{\partial Y^k}{\partial \bar z^j}\frac{\partial}{\partial z^k} + X^{\bar j}Y^k\Gamma^L_{\bar j k}\frac{\partial}{\partial z^L}$, this is equivalent to $\Gamma^L_{\bar j k} = 0$.

We see therefore that (v) is also equivalent to the condition on the Γ's expressed above.

Remark: One sees also (see e.g. [K-N]) that on a Kähler manifold, the only non-zero components of the curvature tensor (in the same basis) are given by

$$R_{i\bar j k\bar l} = -R_{i\bar j\bar l k} = -R_{\bar j i k\bar l} = R_{\bar j i\bar l k}$$
$$= \frac{1}{2}\left(\frac{\partial^2 h_{i\bar j}}{\partial z^k \partial \bar z^l} - h^{q\bar p}\frac{\partial h_{i\bar p}}{\partial z^j}\frac{\partial h_{g\bar j}}{\partial \bar z^l}\right) .$$

On a general Hermitian manifold, more components would be non-zero and all expressions would be more complicated.

I. Examples of Kähler Manifolds

(4.30) The Euclidean metric on $\mathbb{C}^m$ or $\mathbb{C}^m/\Lambda$ is Kähler, where Λ is a lattice as in (2.5).

(4.31) Any metric on a Riemann surface is Kähler. Indeed, dF is a 3-form and must be zero on a surface.

(4.32) Let M be a complex immersed submanifold of the Kähler manifold N. Then the induced metric on M is Kähler since (with $f: M \to N$),

$$\begin{aligned} dF_M &= d(f^*F_N) && (2.53)\\ &= f^*dF_N && (2.18)\\ &= 0 . \end{aligned}$$

(4.33) $\mathbb{P}^m(\mathbb{C})$, with the Fubini-Study metric is a Kähler manifold. Indeed, we have defined F in (2.56) as

$$F = -\frac{i}{2}\,\partial\bar\partial \log f_{(j)} \quad \text{on } U_j$$

so that (iii) is trivially satisfied.

(4.34) As a consequence, we get

Proposition: Any compact complex manifold that can be embedded in $\mathbb{P}^m(\mathbb{C})$ carries a Kähler metric. So, algebraic manifolds(manifolds of zeros of polynomials on $\mathbb{P}^m\mathbb{C}$) carry Kähler structures.

For a precise characterization of manifolds embeddable in $\mathbb{P}^m(\mathbb{C})$, see [We].

J. Strength of the Kähler Condition

We have seen in (2.42) that any paracompact manifold carries a Hermitian metric (and in fact many) to which a connection can be associated by (4.27). On the other hand, not all complex manifolds carry a Kähler structure. For example, the Hopf surfaces are complex manifolds homeomorphic to $S^3 \times S^1$ (see [We] for a description). Their Betti numbers are $b_o = b_1 = b_3 = b_4 = 1$ and $b_2 = 0$. Both Proposition 6.1 and Corollary 6.18 below imply that they don't carry Kähler structures.

Another phenomenon has focused attention recently, namely that of rigidity. We shall simply state without proof a recent and difficult result due to S. Mori (in a more general setting) and Y.-T. Siu and S.-T. Yau.

Definition: Let X and Y be two unit vectors in T_zM. The holomorphic bisectional curvature of X and Y is $h_z(R_z(X,JX)Y,JY)$. One observes that on $P^m(\mathbb{C})$ (with the Fubini-Study metric) all these quantities are positive. Conversely:

(4.35) Rigidity theorem [Mo][S-Y]: Let M be a compact Kähler manifold with positive holomorphic bisectional curvature. Then M is biholomorphically equivalent to $P^m(\mathbb{C})$ (i.e. there is a bijective map from M to $P^m(\mathbb{C})$, holomorphic together with its inverse.)

§5. HARMONIC THEORY ON COMPACT COMPLEX MANIFOLDS

A. Harmonic Theory on a Compact Real Manifold

Let M be a C^∞ connected compact oriented Riemannian manifold of dimension m. In §2F, we defined the de Rham cohomology group $H^r_{dR}(M) = H^r_{dR}(M,\mathbb{R})$ $(r = 0,1,2,\ldots)$ as the quotient of closed r-forms by exact r-forms; we saw in §3D that this is isomorphic to the Čech or simplicial cohomology group $H^r(M)$. Thus a cohomology class in $H^r(M)$ can be represented by a closed r-form but this closed r-form is not unique. We shall see that, given a metric, we can choose the r-form in a canonical way.

Firstly, the metric on M induces an inner product $<,>_x$ on each vector space $T^*_xM\wedge\ldots\wedge T^*_xM$ (r factors, $x\in M$), viz. if $e^1,\ldots,e^m$ is an orthonormal basis for T^*_xM, we set $<e^I,e^J>_x = \begin{cases}1 & \text{if } I = J\\ 0 & \text{if } I \neq J\end{cases}$ so that $\{e^I\colon \#I = r\}$ is an orthonormal basis for $T^*_xM\wedge\ldots\wedge T^*_xM$. (Here we use multiindex notation: if $I = (i_1,\ldots,i_r)$, $e^I = e^{i_1}\wedge\ldots\wedge e^{i_r}$.) This "pointwise" inner product induces a "global" inner product on the linear space $A^r(M)$ of r-forms by

$$<\psi,\eta> = \int_M <\psi(x),\eta(x)>_x V(x)$$

where $V(x)$ is the volume m-form on M. With this inner product, $A^r(M)$ becomes a preHilbert space. The <u>norm</u> of an r-form ψ is given by

$$\|\psi\|^2 = <\psi,\psi>.$$

Now let a de Rham cohomology class in $H^r_{DR}(M)$ be represented by a closed r-form ψ. Then the other closed r-forms in this cohomology class are the forms $\psi + d\eta$ where $\eta\in A^{r-1}(M)$ (We set $A^r(M) = \{0\}$ if $r < 0$). Thus the set of closed r-forms in a given de Rham cohomology class is an affine subspace $S = \psi + dA^{r-1}(M)$ of $A^r(M)$. A natural idea is to demand that ψ has minimum norm in S. Since $A^r(M)$ is not complete, we do not know, a priori, that such an element exists. To study this we find a criterion for ψ to have minimum norm. First note that

$$\text{(5.1)}\qquad \|\psi + d\eta\|^2 = \|\psi\|^2 + \|d\eta\|^2 + 2<\psi,d\eta>.$$

Suppose now that we can define an operator $d^*\colon A^r(M)\to A^{r-1}(M)$ which is the <u>formal adjoint</u> of $d\colon A^{r-1}(M)\to A^r(M)$, that is to say

$\langle d\psi,\eta\rangle = \langle\psi,d^*\eta\rangle$ for all $\psi \in A^{r-1}(M)$, $\eta \in A^r(M)$. Again, since $A^r(M)$ is not complete, existence of d^* is not automatic, we shall however construct d^* later. (5.1) may now be written

$$(5.2) \qquad \| \psi + d\eta \|^2 = \| \psi \|^2 + \| d\eta \|^2 + 2\langle d^*\psi,\eta\rangle .$$

We may now prove

(5.3) <u>Proposition</u>: A d-closed form ψ is of minimum norm within its de Rham cohomology class if and only if $d^*\psi = 0$.

<u>Proof</u>: <u>if</u> From (5.2), $\| \psi + d\eta \|^2 \geq \| \psi \|^2 \quad \forall\eta \in A^{r-1}(M)$.
<u>only if</u>: If ψ is of minimum norm then considering the 1-parameter variation" $t \to \psi + t\,d\eta$, from (5.2) we have

$$0 = \frac{\partial}{\partial t} \| \psi + t d\eta \|^2 = 2\langle d^*\psi,\eta\rangle \quad \forall\eta \in A^{r-1}(M) \quad .$$

Hence $d^*\psi = 0$.

(5.4) <u>Definition</u>: A d-closed form ψ such that $d^*\psi = 0$ is called <u>harmonic</u>, i.e. a <u>harmonic r-form</u> is an r-form ψ such that $d\psi = 0$, $d^*\psi = 0$.

For another formulation we make the

(5.5) <u>Definition</u>: The <u>Laplacian</u> $\Delta = \Delta_d : A^r(M) \to A^r(M)$ is the linear operator given by $\Delta\psi = (dd^* + d^*d)\psi$.

(5.6) <u>Proposition</u>: We have $\Delta\psi = 0$ if and only if $d\psi = 0$ and $d^*\psi = 0$.

<u>Proof</u>.
$$\begin{aligned}\langle\Delta\psi,\psi\rangle &= \langle(dd^*+d^*d)\psi,\psi\rangle \\ &= \langle d^*\psi,d^*\psi\rangle + \langle d\psi,d\psi\rangle \\ &= \| d\psi \|^2 + \| d^*\psi \|^2 \quad .\end{aligned}$$
So if $\Delta\psi = 0$, we must have $\| d\psi \| = \| d^*\psi \| = 0$ hence $d\psi = 0$, $d^*\psi = 0$. The converse is trivial.

Hence we have the

(5.7) <u>Equivalent Definition</u>: $\psi \in A^r(M)$ is harmonic if $\Delta\psi = 0$.

(5.8) <u>Remarks</u>: (i) A 0-form $\psi \in A^0(M)$ is just a real-valued function on M. Since $d^*(A^0(M)) = 0$, ψ is harmonic iff $d\psi$ is zero, i.e.

iff ψ is constant.

(ii) Δ is symmetric, i.e. $<\Delta\psi,\eta> = <\psi,\Delta\eta> \quad \forall\psi,\eta \in A^r(M)$.

(iii) From (5.6) it follows that a form is harmonic if and only if it gives an absolute minimum of the "Dirichlet" integral

$$D(\psi) = \frac{1}{2}\left\{\| d\psi \|^2 + \| d^*\psi \|^2\right\}$$

In fact we can say more: for a 1-parameter variation $t \to \psi+t\eta$, where $\psi,\eta \in A^r(M)$,

$$\begin{aligned}\frac{\partial}{\partial t} D(\psi+t\eta)\Big|_{t=0} &= \frac{\partial}{\partial t}\{D(\psi)+t<d\psi,d\eta>+t<d^*\psi,d^*\eta>+t^2D(\eta)\}\Big|_{t=0} \\ &= <d\psi,d\eta> + <d^*\psi,d^*\eta> \\ &= <(d^*d+dd^*)\psi,\eta> = <\Delta\psi,\eta>.\end{aligned}$$

It follows that ψ "extremises" D i.e. $\frac{\partial}{\partial t} D(\psi+t\eta)\Big|_{t=0} = 0 \quad \forall\eta \in A^r(M)$ if and only if $\Delta\psi = 0$; harmonic forms are the critical points of D.

We have seen that a d-closed form is of minimum norm within its de Rham cohomology class if and only if it is harmonic. It is clear from (5.2) that there can be at most one such form. We shall achieve our goal of finding a canonical representative if we can show that there exists such a form:

(5.9) Existence Theorem: Let $\psi \in A^r(M)$. Then there exists $\phi \in A^r(M)$ such that $\Delta\phi = \psi$ if and only if ψ is orthogonal to $\ker(\Delta)$. Further $\ker \Delta$ is finite dimensional.

Proof: only if: If $\Delta\phi = \psi$ then, for $\zeta \in \ker \Delta$,

$$<\psi,\zeta> = <\Delta\phi,\zeta> = <\phi,\Delta\zeta> = 0 \quad ,$$

if and further are part of a long story - the theory of elliptic operators, see [E], [N], [We] for expositions using different approaches.

Denote the subspace of harmonic r-forms by $H^r(M) = H^r_d(M)$. Immediate from (5.9) is the

(5.10) Theorem (Hodge Decomposition): The linear subspace $H^r(M)$ has finite dimension and there is an orthogonal direct sum decomposition

$$A^r(M) = \mathcal{H}^r(M) \oplus \mathrm{Im}(\Delta) \quad .$$

(5.11) Corollary: There is an orthogonal direct sum decomposition

(5.12) $A^r(M) = \mathcal{H}^r(M) \oplus \mathrm{Im}(d) \oplus \mathrm{Im}(d^*)$

Further

(5.13) $\ker d = \mathrm{Im}\, d \oplus \mathcal{H}^r(M)$

(5.14) $\ker d^* = \mathrm{Im}\, d^* \oplus \mathcal{H}^r(M)$

Proof. Since $\langle d\phi,\psi\rangle = \langle\phi,d^*\psi\rangle$ we see that $\mathrm{Im}\, d \perp \ker d^*$ and $\mathrm{Im}\, d^* \perp \ker d$. So $\mathrm{Im}\, d \oplus \mathrm{Im}\, d^* \subseteq \mathcal{H}^r(M)^{\perp} = \mathrm{Im}\, \Delta$. Conversely, $\mathrm{Im}\, \Delta \subseteq \mathrm{Im}\, d \oplus \mathrm{Im}\, d^*$ since $\Delta = d^*d + dd^*$. Hence $\mathrm{Im}\, \Delta = \mathrm{Im}\, d \oplus \mathrm{Im}\, d^*$. Now, since $d^2 = 0$, $\langle d\phi,d^*\psi\rangle = \langle d^2\phi,\psi\rangle = 0$, so $\mathrm{Im}\, d \perp \mathrm{Im}\, d^*$ and (5.12) is established.

Hence given $\psi \in A^r(M)$ we have an orthogonal decomposition $\psi = \psi_1 + \psi_2 + \psi_3$ where $\psi_1 \in \mathcal{H}^r(M)$, $\psi_2 \in \mathrm{Im}(d)$, $\psi_3 \in \mathrm{Im}(d^*)$. If $\psi \in \ker d$, then, since $\ker d \perp \mathrm{Im}\, d^*$, $\psi_3 = 0$, hence $\ker d \subseteq \mathrm{Im}\, d + \mathcal{H}^r(M)$. The opposite inclusion is clear since $\mathrm{Im}\, d \subseteq \ker d$ (using $d^2 = 0$) and $\mathcal{H}^r(M) \subseteq \ker d$ by definition of harmonic. (5.13) is thus proven, (5.14) follows similarly.

(5.15) Corollary: In any de Rham cohomology class, there is a unique harmonic form. In fact, we have group isomorphisms $H^r_{DR}(M) \cong \mathcal{H}^r(M)$.

We have assumed in the above the existence of an adjoint d^* to d. We now establish the existence of d^* by constructing it explicitly.

Firstly the star operator is the linear operator $*: A^r(M) \to A^{m-r}(M)$ which is defined by the requirement that $\psi(x) \wedge *\eta(x) = \langle\psi(x),\eta(x)\rangle v(x)$ $\forall x \in M$, $\psi,\eta \in A^r(M)$. So $*$ is actually an algebraic operator i.e. it can be considered as an operator $\Lambda^r T^*_x M \to \Lambda^{m-r} T^*_x M$ for each $x \in M$. Let $\phi^1,\ldots,\phi^m$ be an orthonormal oriented coframe (i.e. a local basis for the 1-forms s.t. the metric on M is $ds^2 = \phi^1 \otimes \phi^1 + \ldots + \phi^m \otimes \phi^m$ and the volume form $v = \phi^1 \wedge \ldots \wedge \phi^m$). Let

$$\eta = \sum_{\# I = r} \eta_I \phi^I \in A^r(M), \text{ where } I = (i_1,\ldots,i_r), \quad \phi^I = \phi^{i_1} \wedge \ldots \wedge \phi^{i_r}$$

Then it can be checked that

$$*\eta = \sum_{\#I=r} \varepsilon_I \eta_I \phi^{I^o}$$

where $I^o = (1,\ldots,m) \setminus I$ and $\varepsilon_I = \pm 1$ is the sign of the permutation $(1,\ldots,m) \to (i_1,\ldots,i_r,i_1^o,\ldots,i_{m-r}^o)$. Note that

(5.16) $\quad **\eta = (-1)^{r(m-r)}\eta \qquad \forall \eta \in A^r(M)$,

(5.17) $\quad *1 = v \qquad$ where 1 is considered as a 0-form.

(5.18) <u>Proposition</u>: $(-1)^{m(r+1)+1}*d*$ is the adjoint of d.

<u>Proof</u>. Let $\psi \in A^{r-1}(M)$, $\eta \in A^r(M)$. Then

$$\langle d\psi,\eta\rangle = \int_M d\psi \wedge *\eta = \int_M \{d(\psi \wedge *\eta) - (-1)^{r-1}\psi \wedge d*\eta\}$$

$$= \quad -(-1)^{(r-1)(m-r)} \int \psi\wedge*(*d*)\eta \quad \text{by (5.16) and Stokes' theorem}$$

$$= \langle \psi, -(-1)^{(r-1)(m-r)}*d*\eta\rangle \quad .$$

Hence $d^* = (-1)^{m(r+1)+1}*d* \quad (= -*d*$ if m is even, $(-1)^r*d*$ if m is odd).

B. <u>The Complex Case</u>

Let M be a compact connected Hermitian manifold of complex dimension m. We want to find canonical representatives for the Dolbeault cohomology groups $H^{p,q}_{\bar\partial}(M)$. To do this, we copy the development in §A replacing d by $\bar\partial$ and adding complex conjugate signs in various places. We shall leave proofs as exercises. See §2 especially §2J for background terminology.

Firstly the Hermitian metric on M induces a Hermitian inner product $(\ ,\ )_x$ on each space $T'_xM \wedge \ldots \wedge T'_xM \wedge T''_xM \wedge \ldots \wedge T''_xM$ and hence a "global" Hermitian inner product on $A^{p,q}(M)$, the space of forms of type (p,q), by

$$(\psi,\eta) = \int_M (\psi(x),\eta(x))_x V(x) \qquad \psi,\eta \in A^{p,q}(M)$$

making $A^{p,q}(M)$ into a complex pre-Hilbert space. The star operator $*: A^{p,q}(M) \to A^{m-p,m-q}(M)$ is defined by the requirement that $\psi(x)\wedge*\eta(x) = (\psi(x),\eta(x))_x v(x)$. If $\phi^1,\ldots,\phi^m$ is an (orthonormal) coframe for the

Hermitian metric ds^2 on M, i.e. as in §2J, $ds^2 = \sum \phi^j \otimes \overline{\phi^j}$ $(+\sum \overline{\phi^j} \otimes \phi^j)$ then if

$$\eta = \sum_{\substack{\#J=p \\ \#K=q}} \eta_{JK}\, \phi^J \wedge \overline{\phi^K} \in A^{p,q}(M)$$

where $J = (j_1,\ldots,j_p)$, $K = (k_1,\ldots,k_q)$, $\phi^J = \phi^{j_1} \wedge \ldots \wedge \phi^{j_p}$ then

$$*\eta = i^m \sum_{\substack{\#J=p \\ \#K=q}} \varepsilon_{JK}\, \overline{\eta_{JK}}\, \phi^{J^o} \wedge \overline{\phi^{K^o}}$$

where $J^o = (1,\ldots,m) \setminus J$, $K^o = (1,\ldots,m) \setminus K$ and $\varepsilon_{JK} = \pm 1$ is the sign of the permutation $(1,1',\ldots,m,m') \to (j_1,\ldots,j_p,k_1,\ldots,k_q,j_1^o,\ldots,j_{m-p}^o,k_1^o,\ldots,k_{m-q}^o)$. Then $*$ is <u>conjugate</u> linear, $*1 = v$, the volume (m,m) form and $**\eta = (-1)^{p+q}\eta$, $\forall \eta \in A^{p,q}(M)$.

(5.19) <u>Remarks</u>: The present $*$-operator can be obtained from the "real" $*$-operator $*: A^r(M) \to A^{2m-r}(M)$ of §A by extending it to complex valued r-forms by conjugate linearity, it is then clear that $*(A^{p,q}(M)) \subseteq A^{m-p,m-q}(M)$.

(5.20) <u>Proposition</u>: The formal adjoint $\bar\partial^*: A^{p,q}(M) \to A^{p,q-1}(M)$ of $\bar\partial: A^{p,q-1}(M) \to A^{p,q}(M)$ is given by $\bar\partial^* = -*\bar\partial*$ (We set $A^{p,q}(M) = \{0\}$ if p or q is negative.)

(5.21) <u>Definition</u>: The $\bar\partial$-Laplacian $\Delta_{\bar\partial}: A^{p,q}(M) \to A^{p,q}(M)$ is defined by $\Delta_{\bar\partial} = \bar\partial^*\bar\partial + \bar\partial\bar\partial^*$. A (p,q)-form ψ is said to be <u>$\bar\partial$-harmonic</u> if $\Delta_{\bar\partial}\psi = 0$.

(5.22) <u>Proposition</u>: $\psi \in A^{p,q}(M)$ is $\bar\partial$-harmonic if and only if

$$\bar\partial\psi = 0 \quad \text{and} \quad \bar\partial^*\psi = 0.$$

(5.23) <u>Proposition</u>: $\psi \in A^{p,q}(M)$ is $\bar\partial$-harmonic if and only if it minimizes $\|\psi\|$ amongst all $\bar\partial$-closed (p,q) forms in the same Dolbeault cohomology class $\psi + \bar\partial A^{p,q-1}(M)$.

(5.24) <u>Existence Theorem for $\bar\partial$-Laplacian</u>: Given $\psi \in A^{p,q}(M)$, there exists $\phi \in A^{p,q}(M)$ such that $\Delta_{\bar\partial}\phi = \psi$ if and only if ψ is orthogonal to $\ker \Delta_{\bar\partial}$. Further $\ker \Delta_{\bar\partial}$ has finite dimension.

Let $H^{p,q}_{\bar\partial}(M)$ denote the set of $\bar\partial$-harmonic (p,q) forms.

(5.25) Theorem (Hodge Decomposition): $\mathcal{H}_{\bar\partial}^{p,q}(M)$ has finite dimension and we have an orthogonal direct sum decomposition

$$A^{p,q}(M) = \mathcal{H}_{\bar\partial}^{p,q}(M) \oplus \operatorname{Im}(\Delta_{\bar\partial}) \quad .$$

(5.26) Corollary: There is an orthogonal direct sum decomposition

(5.27) $$A^{p,q}(M) = \mathcal{H}_{\bar\partial}^{p,q}(M) \oplus \operatorname{Im} \bar\partial \oplus \operatorname{Im} \bar\partial^*.$$

Further

(5.28) $\ker \bar\partial = \operatorname{Im} \bar\partial \oplus \mathcal{H}^{p,q}(M)$

(5.29) $\ker \bar\partial^* = \operatorname{Im} \bar\partial^* \oplus \mathcal{H}^{p,q}(M)$.

(5.30) Corollary: In any Dolbeault cohomology class there is a unique $\bar\partial$-harmonic (p,q)-form. In fact, we have group isomorphisms $H_{\bar\partial}^{p,q}(M) \cong \mathcal{H}_{\bar\partial}^{p,q}(M)$.

A similar development can be carried out for ∂ in place of $\bar\partial$.

C. Vector Bundle Case

(5.31) Let E be a (C^∞) Riemannian vector bundle over a compact connected C^∞ Riemannian manifold M. Suppose E is given a (C^∞) metric connection (§4G). Then as in §4D we may define the exterior derivative $d: A^r(E) \to A^{r+1}(E)$ associated to the connection. The metrics on M and E induce an inner product $\langle\ \rangle_x$ on each linear space $\Lambda^r T_x^*M \otimes E_x$ and thus a global inner product on $A^r(E)$ as in §A. We can then define the adjoint d^* of d and the Laplacian $\Delta: A^r(E) \to A^r(E)$ by $\Delta = d^*d + dd^*$. An E-valued form $\psi \in A^r(E)$ is harmonic iff $\Delta\psi = 0$ i.e. iff $d\psi = d^*\psi = 0$ and we have the existence theorem and Hodge decomposition:

$$A^r(E) = \mathcal{H}^r(E) \overset{\perp}{\oplus} \operatorname{Im}(\Delta)$$

as before. However since $d^2 \neq 0$ is general (it equals the curvature operator of E, see §4E) it is not in general true that $\operatorname{Im} d \perp \operatorname{Im} d^*$ and so (5.12) is replaced by

$$A^r(E) = \mathcal{H}^r(E) \oplus \{\operatorname{Im} d + \operatorname{Im} d^*\}$$

and (5.13) and (5.14) are, in general, false. Since $d^2 \neq 0$ we cannot define "de Rham" cohomology groups ker d/Im d.

(5.32) Application (Harmonic Maps): Let $\phi: M \to N$ be a C^∞ map between two C^∞ Riemannian manifolds. For simplicity we assume M is compact. The pull-back bundle $E = \phi^{-1}TN$ may be given a metric and a metric connection induced from the metric and Levi-Civita connection on N. The differential of ϕ, ϕ_*, is a linear map from TM to E, it is thus a 1-form with values in E. The connection on E induces an exterior derivative $d: A^0(E) \to A^1(E)$ with adjoint $d^*: A^1(E) \to A^0(E)$ called a (generalized) divergence. The map ϕ is said to be harmonic if the divergence of its differential vanishes i.e.

$$d^*(\phi_*) = 0$$

(This generalizes the familiar "div grad $\phi = 0$"). Harmonic maps include many important concepts in differential geometry e.g. geodesics and minimal surfaces; for more details see [E-L].

It is easy to see that ϕ is a harmonic map if and only if $\phi_* \in A^1(E)$ is a harmonic 1-form.

(5.33) Suppose now that E is a Hermitian holomorphic vector bundle over a Hermitian manifold. Then (§4A), $\bar\partial: A^{p,q}(E) \to A^{p,q+1}(E)$ is canonically defined by the holomorphic structure and $\bar\partial^2 = 0$. Hence we may define cohomology groups $H^{p,q}_{\bar\partial}(E) = \ker \bar\partial/\mathrm{Im}\,\bar\partial$. (If E is the trivial bundle $M \times \mathbb{C}$, these are the Dolbeault cohomology groups $H^{p,q}_{\bar\partial}(M)$.) The theory of §3 may be extended to show that $H^{p,q}_{\bar\partial}(E) = H^q(M,\Omega^p(E))$ = Čech cohomology with coefficients in the sheaf of holomorphic p-forms with values in E. The theory of §B goes through with only trivial changes, in particular $\mathcal{H}^{p,q}_{\bar\partial}(E)$, the space of harmonic (p,q)-forms with values in E is finite dimensional and we have orthogonal decompositions:

$$\begin{aligned} A^{p,q}(E) &= \mathcal{H}^{p,q}_{\bar\partial}(E) \oplus \mathrm{Im}\,\Delta_{\bar\partial} \\ &= \mathcal{H}^{p,q}_{\bar\partial}(E) \oplus \mathrm{Im}\,\bar\partial \oplus \mathrm{Im}\,\bar\partial^* \quad , \end{aligned}$$

and so each Dolbeault cohomology class in $H^{p,q}_{\bar\partial}(E)$ has a unique $\bar\partial$-harmonic representative in $\mathcal{H}^{p,q}_{\bar\partial}(E)$, hence

$$H^{p,q}_{\bar\partial}(E) \cong H^q(M,\Omega^p(E)) \cong \mathcal{H}^{p,q}_{\bar\partial}(E).$$

D. Applications

(5.34) Proposition: (i) For any compact complex manifold M,

$$\dim H^{p,q}_{\bar\partial}(M) = \dim H^q(M,\Omega^p) < \infty .$$

(ii) If E is a holomorphic vector bundle over M

$$\dim H^{p,q}_{\bar\partial}(E) = \dim H^q(M,\Omega^p(E)) < \infty.$$

Proof. (i) Choose a Hermitian metric on M. Then

$$H^q(M,\Omega^p) \cong H^{p,q}_{\bar\partial}(M) = \mathcal{H}^{p,q}_{\bar\partial}(M) .$$

Now apply (5.25). (ii) Similar.

For our next application note that $*\Delta = \Delta *$ and $*\Delta_{\bar\partial} = \Delta_{\bar\partial} *$ hence $*: A^r(M) \to A^{m-r}(M)$ (resp. $*: A^{p,q}(M) \to A^{m-p,m-q}(M)$) restricts to an isomorphism $*: \mathcal{H}^r(M) \overset{\cong}{\to} \mathcal{H}^{m-r}(M)$ (resp. $*: \mathcal{H}^{p,q}_{\bar\partial}(M) \overset{\cong}{\to} \mathcal{H}^{m-p,m-q}_{\bar\partial}(M)$) where M is any m-dimensional compact Riemannian (resp. Hermitian) manifold. Hence:

(5.35) Theorem: (i) For any m-dimensional compact connected orientable C^∞ manifold M,

$$H^r(M,\mathbb{R}) \cong H^{m-r}(M,\mathbb{R}) .$$

(ii) For any m-dimensional compact connected complex manifold M,

$$H^q(M,\Omega^p) \cong H^{p,q}_{\bar\partial}(M) \overset{\cong}{\to} H^{m-p,m-q}_{\bar\partial}(M) \cong H^{m-q}(M,\Omega^{m-p})$$

(the second isomorphism being conjugate linear).

Proof: (i) Choose a Riemannian metric and an orientation on M, then

$$H^r(M) \cong \mathcal{H}^r(M) \underset{*}{\overset{\cong}{\to}} \mathcal{H}^{m-r}(M) \cong H^{m-r}(M) .$$

(ii) similar.

Note in particular that in the real case (i)

$$H^m(M) \cong H^o(M) \cong \mathbb{R} \quad ,$$

and in the complex case (ii)

$$H^m(M,\Omega^m) \cong H^{m,m}_{\bar\partial}(M) \cong H^{o,o}_{\bar\partial}(M) \cong H^o(M,\mathcal{O}) \cong \mathbb{C} \quad ,$$

the last isomorphism holding since any holomorphic function on M is constant.

(5.36) <u>Remark</u>: The isomorphisms $*: H^m(M) \to H^o(M)$ and $*: H^{m,m}(M) \to H^{o,o}(M)$ show that (i) on a compact connected oriented m-dimensional Riemannian manifold, any harmonic m-form is a constant multiple of the volume form $v = *1$; (ii) on a compact connected m-dimensional Hermitian manifold, any harmonic (m,m)-form is a constant multiple of the volume form $v = *1$. Now define the <u>Betti numbers</u> of a paracompact manifold M by

$$b_r = \dim H^r(M) \qquad \text{(simplicial cohomology)} \quad ,$$

then we see from (5.35)(ii) that <u>if M is a compact connected orientable real manifold M of dimension m</u>,

(5.37) $b_r < \infty \qquad (r = 0,1,\ldots)$,

(5.38) $b_m = 1$,

(5.39) $b_r = b_{m-r} \qquad (r = 0,1,\ldots)$.

For a paracompact complex manifold M, we define the <u>Hodge numbers</u> by

$$h^{p,q}(M) = \dim H^{p,q}_{\bar\partial}(M) \qquad \text{(Dolbeault cohomology)}$$
$$= \dim H^q(M,\Omega^p) \qquad \text{(Čech cohomology)}$$

Then immediately from (5.35) we have

(5.40) <u>Theorem</u>: If M is a compact connected complex manifold of (complex) dimension m,

(5.41) $h^{p,q}(M) < \infty \qquad (p,q = 0,1,\ldots)$,

(5.42) $h^{m,m}(M) = 1$,

(5.43) $h^{p,q}(M) = h^{m-p,m-q}(M)$.

(5.44) Remarks: We have developed the real and complex theories separately. On a Hermitian manifold M we have the two notions: "d-harmonic" and "$\bar{\partial}$-harmonic". These are, in general, unrelated. However, we shall see in the next section that many special relations exist in the Kähler case.

§6. COHOMOLOGY OF KÄHLER MANIFOLDS

We shall establish some special relations on the Betti and Hodge numbers of compact Kähler manifolds.

A. Betti numbers

(6.1) Proposition: The even Betti numbers $b_{2p} = \dim H^{2p}(M,\mathbb{R})$ $(0 \le p \le$ $\le m = \dim M)$ of a compact Kähler manifold are positive.

Proof. We show that $F^p = F \wedge \ldots \wedge F$ represents a non-zero element of $H^{2p}(M,\mathbb{R})$, i.e. that it is closed and not exact.

F^p is closed since F is, and if $F^p = d\psi$, then

$$\int F^m = \int d\psi \wedge F^{m-p} = \int d(\psi \wedge F^{m-p}) = 0 \quad ,$$

which is impossible since F^m is a volume form (2.50).

B. The Hodge identities

On a compact Hermitian manifold, we have defined a number of operators on the space A^*M, such as $d, \partial, \bar\partial$, their adjoints d^*, ∂^* and $\bar\partial^*$ and the associated Laplacians $\Delta_d = dd^* + d^*d$, Δ_∂ and $\Delta_{\bar\partial}$.

We define three more operators

(6.2) $$d^c = \frac{i}{4\pi}(\bar\partial - \partial)$$

(6.3) $L: A^{p,q}(M) \to A^{p+1,q+1}(M)$

$\eta \to \eta \wedge F$, where F is the fundamental form.

(6.4) $\Lambda = L^*: A^{p,q}(M) \to A^{p-1,q-1}(M)$ the adjoint of L.

Note that d^c (like d) is a real operator and that

(6.5) $$\begin{aligned} dd^c &= \frac{i}{4\pi}(\partial+\bar\partial)(\bar\partial-\partial) \\ &= \frac{i}{2\pi}\partial\bar\partial \qquad \text{by (2.34)} \\ &= -\frac{i}{2\pi}\bar\partial\partial \\ &= -d^c d. \end{aligned}$$

On a general Hermitian manifold, there are no simple relations between these operators. In the Kähler case, however, we shall now establish some of the Hodge identities relating them.

(6.6) Lemma: i) $[\Lambda,d] = \Lambda \circ d - d \circ \Lambda = -4\pi\, d^{c*}$

ii) $[L,d^*] = 4\pi\, d^c$

iii) $[\Lambda,\bar\partial] = -i\,\partial^*$

iv) $[\Lambda,\partial] = i\,\bar\partial^*$.

Note first that ii) is the dual of i), and that iii) and iv) come from the decomposition of i) in complex types. Since Λ, d and d^c are real, we have iii) $\Leftrightarrow$ iv) and therefore the four equations are equivalent. We shall prove iv).

The idea is to use the osculation property of the Kähler metric h (def. (4.29) iv) to reduce the calculation at one point to a calculation in $\mathbb{C}^m$.

To proceed with the latter, we calculate "in coordinates" by introducing operators associated to the variables $z^1 \ldots z^m$. We shall not go into full details (see e.g. [G-H.]) but rather try to show how to handle the adjoint operators.

Let $e^k \colon A_c^{p,q}(\mathbb{C}^m) \to A_c^{p+1,q}(\mathbb{C}^m)$ ((p+1,q)-forms with compact support) be defined by

$$e^k(\varphi) = dz^k \wedge \varphi,$$

and $\bar e^k$ by

$$\bar e^k(\varphi) = d\bar z^k \wedge \varphi$$

and call i_k and $\bar i_k$ their adjoints.

Let also ∂_k and $\bar\partial_k$ be defined on $A_c^{p,q}(\mathbb{C}^m)$ by

$$\partial_k(\varphi_{JK}\, dz^J \wedge d\bar z^K) = \frac{\partial \varphi_{JK}}{\partial z^k}\, dz^J \wedge d\bar z^K$$

(with the notations of (2.31)) and

$$\bar\partial_k(\varphi_{JK}\, dz^J \wedge d\bar z^K) = \frac{\partial \varphi_{JK}}{\partial \bar z^k}\, dz^J \wedge d\bar z^K \quad .$$

We first note that ∂_k and $\bar{\partial}_l$ commute with each other and with e^j and $\bar{e}^j$. From the description of i_j and $\bar{i}_j$ which we shall outline below, it will appear that ∂_k and $\bar{\partial}_l$ also commute with these.

The operators under consideration can then be written as

$$\partial = \partial_k\, e^k = e^k\, \partial_k$$

$$\bar{\partial} = \bar{\partial}_k\, \bar{e}^k = \bar{e}^k\, \bar{\partial}_k$$

$$L = i\, e^k\, \bar{e}^k$$

$$\Lambda = -i\, \bar{i}_k\, i_k \quad .$$

We now calculate the adjoint of ∂_k and relate e^k and i_k, in order to calculate $[\Lambda,\partial]$.

We first note that the adjoint of ∂_k is $-\bar{\partial}_k$. Indeed, for $\varphi = \varphi_{JK}\, dz^J \wedge d\bar{z}^K$ and ψ a C^∞ function:

$$(-\bar{\partial}_k\varphi, \psi\, dz^M \wedge d\bar{z}^N) = \left(-\frac{\partial\varphi_{JK}}{\partial\bar{z}^k}\, dz^J \wedge d\bar{z}^K,\ \psi\, dz^M \wedge d\bar{z}^N\right)$$

$$= 2^{\#M+\#N} \int_{\mathbb{C}^m} -\frac{\partial\varphi_{MN}}{\partial\bar{z}^k}\,\bar{\psi}\,.\,V(z) \quad (\text{recall } |dz^k|^2 = 2)$$

$$= 2^{\#M+\#N} \int_{\mathbb{C}^m} \varphi_{MN}\,\frac{\partial\bar{\psi}}{\partial\bar{z}^k}\,.\,V(z)$$

$$= 2^{\#M+\#N} \int_{\mathbb{C}^m} \varphi_{MN}\,\overline{\frac{\partial\psi}{\partial z^k}}\,.\,V(z)$$

$$= (\varphi_{JK}\, dz^J \wedge d\bar{z}^K, \partial_k(\psi\, dz^M \wedge d\bar{z}^N))$$

$$= (\varphi, \partial_k(\psi\, dz^M \wedge d\bar{z}^N))$$

One proves also that

$$i_k e^k + e^k i_k = 2$$

$$i_k e^l + e^l i_k = 0 \qquad \text{for } k \neq l$$

This is done by evaluating i_k on every (p,q) form of the basis $dz^J \wedge d\bar{z}^K$ and checking the relation on these forms.

For instance, we show that $i_k \ (dz^k \wedge dz^M \wedge d\bar{z}^N) = 2\, dz^M \wedge d\bar{z}^N$. Indeed, $\forall (J,K)$:

$$\begin{aligned}(i_k(dz^k \wedge dz^M \wedge d\bar{z}^N), dz^J \wedge d\bar{z}^K) \\ = (dz^k \wedge dz^M \wedge d\bar{z}^N, dz^k \wedge dz^J \wedge d\bar{z}^K) \\ = 2(dz^M \wedge d\bar{z}^N, dz^J \wedge d\bar{z}^K).\end{aligned}$$

Using all these relations, one gets

$$-i\, \overline{i_k}\, i_k\, \partial_l\, e^l = -i\, \partial_l\, e^l\, \overline{i}_k\, i_k + i\, \partial_k\, \overline{i_k}$$

that is $\Lambda\partial = \partial\Lambda - i\bar{\partial}^*$, which establishes lemma (6.6) in $\mathbb{C}^m$.

For any point z_o of a Kähler manifold M, we can by (4.29)(iv) find in a neighbourhood a local coframe $\varphi^1,\dots,\varphi^m$ for the metric such that $d\varphi^j(z_o) = 0$. The expression for Λ holds with dz^J replaced by φ^J, and the calculation of $[\Lambda,\bar{\partial}]$ involving only first derivatives yields the same answer as in $\mathbb{C}^m$, except for terms involving $\bar{\partial}\varphi^J$, which cancel at z_o. Likewise, $\partial^* = -{*}\partial{*}$ and only terms in $\bar{\partial}\varphi^j$ will be introduced on M.

The identity is therefore satisfied at each point.

We draw some important consequences from lemma (6.6).

(6.7) <u>Proposition</u>: On a compact Kähler manifold,

$[L,\Delta_d] = 0$ or equivalently

$[\Lambda,\Delta_d] = 0$.

<u>Proof</u>. Since F is closed,

$$d(F \wedge \eta) = F \wedge d\eta \quad , \quad \text{or} \quad [L,d] = 0 \ \text{ and } \ [\Lambda,d^*] = 0.$$

Then:

$$\begin{aligned}\Lambda(dd^*+d^*d) &= (d\,\Lambda\, d^* - 4\pi\, d^{c*}\, d^*) + d^*\Lambda\, d && (6.6)\text{i} \\ &= d\Lambda\, d^* + (4\pi\, d^* d^{c*} + d^*\Lambda\, d) && (6.5) \\ &= (dd^* + d^*d)\Lambda && (6.6)\text{i}\end{aligned}$$

(6.8) <u>Theorem</u>: On a compact Kähler manifold: $\Delta_d = 2\,\Delta_\partial = 2\,\Delta_{\bar{\partial}}$.

Proof. By (6.6)iv, we have:

$$\begin{aligned} i(\partial\bar\partial^* + \bar\partial^*\partial) &= \partial(\Lambda\partial - \partial\Lambda) + (\Lambda\partial - \partial\Lambda)\partial \\ &= \partial\Lambda\partial \qquad - \qquad \partial\Lambda\partial \\ &= 0. \end{aligned}$$

Then:

$$\begin{aligned} \Delta_d &= (\partial+\bar\partial)(\partial^*+\bar\partial^*) + (\partial^*+\bar\partial^*)(\partial+\bar\partial) \\ &= (\partial\partial^*+\partial^*\partial) + (\bar\partial\bar\partial^*+\bar\partial^*\bar\partial) + (\partial\bar\partial^*+\bar\partial\partial^*+\partial^*\bar\partial+\bar\partial^*\partial) \\ &= \Delta_\partial + \Delta_{\bar\partial} \quad . \end{aligned}$$

Finally, we show that $\Delta_\partial = \Delta_{\bar\partial}$:

$$\begin{aligned} -i\ \Delta_\partial &= \partial(\Lambda\bar\partial-\bar\partial\Lambda) + (\Lambda\bar\partial - \bar\partial\Lambda)\partial && \text{by (6.6)iii} \\ &= \partial\Lambda\bar\partial - \partial\bar\partial\Lambda + \Lambda\bar\partial\partial - \bar\partial\Lambda\partial \end{aligned}$$

and

$$\begin{aligned} i\ \Delta_{\bar\partial} &= (\bar\partial(\Lambda\partial-\partial\Lambda) + (\Lambda\partial-\partial\Lambda)\bar\partial) && \text{by (6.6)iv} \\ &= \bar\partial\Lambda\partial - \bar\partial\partial\Lambda + \Lambda\partial\bar\partial - \partial\Lambda\bar\partial \\ &= -\partial\Lambda\bar\partial + \partial\bar\partial\Lambda - \Lambda\bar\partial\partial + \bar\partial\Lambda\partial && \text{since } \partial\bar\partial = -\bar\partial\partial \\ &= i\ \Delta_\partial . \end{aligned}$$

From this, we obtain immediately:

(6.9) Corollary: On a compact Kähler manifold, Δ_d preserves the bidegree, or, $[\Delta_d, \pi^{p,q}] = 0$.

Indeed, $\Delta_{\bar\partial}$ does.

Let us now see what this implies on cohomology. To avoid confusion, set

$$H_d^{p,q}(M) = Z_d^{p,q}(M)/B_d^{p,q}(M)$$

$$\mathcal{H}_d^{p,q}(M) = \{\eta \in A^{p,q}(M) : \Delta_d\eta = 0\}$$

$$\mathcal{H}_d^{r}(M) = \{\eta \in A^{r}(M) : \Delta_d\eta = 0\}$$

and similarly for ∂ and $\bar\partial$.

Since $\Delta_d = 2\Delta_{\bar\partial}$, we see immediately that

(6.10) $\mathcal{H}_d^{p,q}(M) = \mathcal{H}_{\bar\partial}^{p,q}(M)$.

We also have

(6.11) $\mathcal{H}_d^r(M) = \bigoplus_{p+q=r} \mathcal{H}_d^{p,q}(M)$.

Indeed, all (p,q) components of a harmonic form are harmonic since Δ_d preserves bidegrees.

(6.12) $\mathcal{H}_d^{p,q}(M) = \overline{\mathcal{H}_d^{q,p}(M)}$.

Indeed, Δ_d is real

(6.13) $H_d^{p,q} \cong \mathcal{H}_d^{p,q}$.

Indeed, the projection of a (p,q)-form on its harmonic component is of type (p,q) since it is the same for Δ_d or $\Delta_{\bar\partial}$.

(6.14) $H_{dR}^r(M) \cong \mathcal{H}_d^r(M)$.

Combining these identities and isomorphisms, we get the

(6.15) Hodge decomposition: For a compact Kähler manifold:

$$\begin{aligned} H^r(M,\mathbb{C}) &\cong \bigoplus_{p+q=r} H_d^{p,q}(M) \\ &\cong \bigoplus_{p+q=r} H_{\bar\partial}^{p,q}(M) \\ &\cong \bigoplus_{p+q=r} H^q(M,\Omega^p) \\ H_d^{p,q}(M) &= \overline{H_d^{q,p}(M)} \end{aligned}$$

Note that (6.10)-(6.12) are equalities whereas other relations (e.g. the first one in (6.15)) are isomorphisms depending on the Kähler metric.

As a special case of (6.15), we see that

$$H^{p,o}(M) \cong H^o(M,\Omega^p) \quad ,$$

the space of holomorphic p-forms. In fact:

(6.16) Proposition: Given any Kähler metric on a compact complex manifold (if it exists), the harmonic (p,0)-forms are precisely the holomorphic p-forms.

Remark: This illustrates the relationship between a complex notion (holomorphic forms) and a Riemannian one (harmonic forms) appearing when the structures are related by the Kähler condition.

Proof. We have to prove equality between spaces - not only isomorphism.

Since $\Delta_d = 2\Delta_{\bar\partial}$, we have $H^{p,o}_d = H^{p,o}_{\bar\partial}$. In the Hodge decomposition we have in general

$$Z^{p,q}_{\bar\partial}(M) = H^{p,q}_{\bar\partial} + B^{p,q}_{\bar\partial}(M) \quad ,$$

with $B^{p,q}_{\bar\partial}(M) = \bar\partial\, A^{p,q-1}(M)$.

For $q = 0$, we have therefore $Z^{p,o}_{\bar\partial}(M) = H^{p,o}_{\bar\partial}$. And $Z^{p,o}_{\bar\partial}(M)$ is precisely the space of holomorphic p-forms.

Recall from §5 that the Hodge numbers are defined by $h^{p,q}(M) = \dim H^{p,q}_{\bar\partial}(M)$ and that they satisfy various relations on a Hermitian manifold. On a Kähler manifold, we can say more, using (6.15) and the fact that in (6.1) w is a (1,1)-form:

(6.17) Proposition: On a compact Kähler manifold,

$$b_r(M) = \sum_{p+q=r} h^{p,q}(M)$$

$$h^{p,q}(M) = h^{q,p}(M)$$

$$h^{p,p}(M) \geq 1 \qquad (p \leq m) \quad .$$

(6.18) Corollary: The odd Betti numbers of a compact Kähler manifold are even.

Proof. $b_{2s+1}(M) = \sum_{p=0}^{2s+1} h^{p,2s+1-p}(M) = 2 \sum_{p=0}^{s} h^{p,2s+1-p}(M)$.

As mentioned in (4.J), this shows in particular that some complex manifolds do not carry any Kähler metric.

(6.19) <u>Cohomology of the complex projective space</u>. Using the Hodge identities in the case of $P^m(\mathbb{C})$, we can of course deduce b_r from the knowledge of $h^{p,q}$, and get $b_{2k+1}(P^m(\mathbb{C})) = 0$

$$b_{2k}(P^m(\mathbb{C})) = 1 \qquad (k \leq m)$$

(see 3.53).

Conversely, if we know the Betti numbers, we can in this case deduce what the Hodge numbers must be.

Indeed, since $b_{2k+1} = 0$, $h^{p,q} = 0$ for $p+q$ odd, and since $b_{2k} = 1$, for $p \neq k$ we have

$$\begin{aligned} 1 &\geq h^{p,2k-p} + h^{2k-p,p} \\ &= 2h^{p,2k-p} \end{aligned}$$

so that $h^{p,2k-p} = 0$.

Hence, $h^{p,p} = b_{2p} = 1$.

References

[dR] G. de Rham: Variétés différentiables. Hermann (1955).

[D] J. Dieudonné: Eléments d'analyse. Gauthier Villars (1969-) Treatise on analysis, Academic Press (1969-).

[E] J. Eells: Elliptic operators on manifolds. Complex analysis and its applications, Trieste, 1975, IAEA, Vienna (1976), vol. I pp. 95-152.

[E-L] J.Eells and L. Lemaire: A report on harmonic maps. Bull. London Math. Soc. 10 (1978) 1-68.

[F] M.J. Field: Several complex variables. Complex analysis and its applications, Trieste, 1975, IAEA, Vienna (1976), vol. I, pp. 153-234.

[Fu] B.A. Fuks: Introduction to the theory of analytic functions of several complex variables. Translations of mathematical monographs no. 8, American Mathematical Society (1963).

[G] C. Godbillon: Géometrie différentielle et mécanique analytique. Hermann (1969).

[Go] R. Godement: Topologie algébrique et théorie des faisceaux. Hermann (1958).

[G-H] P. Griffiths and J. Harris: Principles of algebraic geometry. Wiley-Interscience (1978).

[G-R] R. Gunning and H. Rossi: Analytic functions of several complex variables. Prentice Hall (1965).

[Gr-R] H. Grauert and R. Remmert: The theory of Stein spaces. Grundlagen der Mathematischen Wissenschaft 236, Springer-Verlag (1979).

[Ha] F.R. Harvey: Integral formulae connected with Dolbeault's isomorphism. Rice Studies (1970), pp. 77-97.

[He] S. Helgason: Differential geometry and symmetric spaces. Academic Press (1962); Differential geometry, Lie groups and symmetric spaces, Academic Press (1978).

[Hö] L. Hörmander: An introduction to complex analysis in several variables. Van Nostrand Reinhold (1966).

[K-M] K. Kodaira and J. Morrow: Complex manifolds. Holt, Rinehart and Winston (1971).

[K-N] S. Kobayashi, K. Nomizu: Foundations of differential geometry. Wiley-Interscience (vol. 1, 1963, vol. 2, 1969).

[Ma] B. Malgrange: Sur l'intégrabilité des structures presque complexes. Symposia Math., Istituto di Alta Matematica, Academic Press (1969), pp. 289-296.

[Mi] J. Milnor: Morse theory. Annals of Mathematical Studies, 51, Princeton University Press (1963).

[Mo] S. Mori: Projective manifolds with ample tangent bundles. Ann. Math. 110 (1979) 593-606.

[N] R. Narasimhan: Analysis on real and complex manifolds. Masson & Cie/North-Holland (1973).

[Na] L. Nachbin: Holomorphic functions, domains of holomorphy and local properties. North Holland (1970).

[N-N] A. Newlander and L. Nirenberg: Complex analytic coordinates in almost complex manifolds. Ann. Math. 65 (1957)391-404.

[Sp] M. Spivak: A comprehensive introduction to differential geometry (5 volumes), Publish or Perish (1975).

[Sw] R.G. Swan: The theory of sheaves. Univ. of Chicago Press (1964).

[S-Y] Y.T. Siu and S.-T. Yau: Compact Kähler manifolds of positive bisectional curvature. Invent. Math. 59 (1980) 189-204.

[T] B.R. Tennison: Sheaf theory. London Math. Soc. Lecture notes no. 20, Cambridge University Press (1975).

[Wa] F. Warner: Introduction to manifolds. Scott-Foresman (1971).

[Weil] A. Weil: Introduction à l'étude des variétés Kählériennes. Hermann (1958).

[We] R.O. Wells: Differential analysis on complex manifolds. Prentice Hall (1973) and Springer-Verlag (1979).

COMPLEX ANALYSIS AND COMPLEXES OF DIFFERENTIAL OPERATORS

Mauro Nacinovich
Istituto Matematico " L. Tonelli"
Via Buonarroti,2
Università di Pisa (Italy)

Introduction. Complex analysis and the theory of complexes of differential operators are two closely related subjects. First we notice that the study of the Dolbeault complex is essential to understand holomorphic functions of several complex variables: this complex is a particular example of a complex of differential operators and thus many results in complex analysis could be considered as particular instances of the more general theory of differential complexes. Our knowledge of the Dolbeault complex is a powerful source of intuition to forecast the behaviour of general differential complexes. On the other hand, complex analysis and the Dolbeault complex play a very peculiar rôle because they are also an essential tool for the study of differential operators. In my lectures, I will try to give an idea of the close relationship of the two fields. The arguments that I will discuss will be the following:

1) the theory of Ehrenpreis-Malgrange for differential operators with constant coefficients.
2) Differential equations with constant coefficients in the class of real--analytic functions (analytic convexity).
3) The theory of convexity and the theorem of Cartan Thullen for general operators.
4) Boundary complexes and boundary values of pluriharmonic functions.
5) The Lemma of Poincaré for complexes of differential operators with smooth (non constant) coefficients.

LECTURE 1. DIFFERENTIAL OPERATORS WITH CONSTANT COEFFICIENTS.

1. We denote by $\mathcal{E}(\Omega)$ the space of complex valued, C^{∞} functions defined on an open set Ω of $\mathbb{R}^n$. To a matrix $A(\xi) = \left(a_{ij}(\xi)\right)_{1\leq i\leq p, 1\leq j\leq q}$ of polynomials (with complex coefficients) in the indeterminates $\xi_1, \ldots, \xi_n$ we associate the differential operator

$$A(D) : \mathcal{E}(\Omega)^q \longrightarrow \mathcal{E}(\Omega)^p$$

obtained by substituting $D_j = \dfrac{\partial}{\partial x_j}$ to ξ_j $(j = 1, \ldots, n)$.

We consider, for $f \in \mathcal{E}(\Omega)^p$, the differential equation:

$$(*) \qquad \begin{cases} u \in \mathcal{E}(\Omega)^q \\ A(D)u = f \quad \text{on} \quad \Omega \end{cases} .$$

If this equation is solvable and $Q(\xi) = (Q_1(\xi), \ldots, Q_p(\xi))$ is a vector with polynomial components such that

$$Q(\xi)\ A(\xi) = 0 ,$$

then we must have $Q(D)f = 0$ on Ω (integrability condition).

On the other hand, if $S(\xi) = \begin{pmatrix} S_1(\xi) \\ \vdots \\ S_q(\xi) \end{pmatrix}$ is a polynomial vector such that

$$A(\xi)\ S(\xi) = 0$$

and $u \in \mathcal{E}(\Omega)^q$ solves $(*)$, then for every $v \in \mathcal{E}(\Omega)$ also $u+S(D)v$ is a solution of $(*)$ (S is a cointegrability condition).

These facts can be taken into account by inserting $A(D)$ into a complex :

$$\mathcal{E}^s(\Omega) \xrightarrow{C(D)} \mathcal{E}^q(\Omega) \xrightarrow{A(D)} \mathcal{E}^p(\Omega) \xrightarrow{B(D)} \mathcal{E}^r(\Omega)$$

where the lines of the matrix B form a basis (over $\mathcal{P} = \mathbb{C}[\xi_1, \ldots, \xi_n]$) of the integrability conditions and the columns of C form a basis for the cointegrability conditions (it is obvious that the integrability and cointegrability conditions form modules over $\mathcal{P}$).

Thus the study of an over determined system led us naturally to consider complexes of differential operators.

2. General Results

Notations. $\mathcal{E}(\Omega) = C^\infty$ functions on Ω

$\mathcal{E}'(\Omega)$ = distributions with compact support in Ω

$\mathcal{D}(\Omega) = C^\infty$ functions with compact support in Ω

$\mathcal{D}'(\Omega)$ = distributions on Ω

$\mathcal{D}'_F(\Omega)$ = distributions of finite type on Ω

$\mathcal{E}_{x_0}$ = space of germs of C^∞ functions at $x_0 \in \mathbb{R}^N$

$\mathcal{A}_{x_0}$ = space of germs of real-analytic functions at $x_0 \in \mathbb{R}^n$

$\mathcal{S}$ = Schwartz space of rapidly decreasing functions on $\mathbb{R}^n$

$\mathcal{S}'$ = slowly increasing generalized functions on $\mathbb{R}^n$

$\mathbb{F}$ = space of (finite) linear combinations of exponential-polynomials

$\mathcal{P} = \mathbb{C}[\xi_1, \ldots, \xi_n]$

$\Phi_{x_0} = \mathbb{C}\{\{x_1 - x_1^\circ, \ldots, x_n - x_n^\circ\}\}$ formal power series centered at $x_0 \in \mathbb{R}^n$.

Theorem a.

Let $A(\xi)$, $B(\xi)$ be respectively $q \times p$ and $r \times q$ matrices of polynomials. Then the following statements 1,2,3,5,6,7,8,9,10,11 are equivalent and imply 4) and 12) .

1) The sequence $\mathcal{P}^p \xleftarrow{{}^tA(\xi)} \mathcal{P}^q \xleftarrow{{}^tB(\xi)} \mathcal{P}^r$ is exact.

2) The sequence $\mathcal{E}^p(\Omega) \xrightarrow{A(D)} \mathcal{E}^q(\Omega) \xrightarrow{B(D)} \mathcal{E}^r(\Omega)$ is exact for Ω convex and non empty.

3) The sequence $\mathbb{F}^p \xrightarrow{A(D)} \mathbb{F}^q \xrightarrow{B(D)} \mathbb{F}^r$ is exact.

4) The sequence $\mathcal{S}'^p \xrightarrow{A(D)} \mathcal{S}'^q \xrightarrow{B(D)} \mathcal{S}'^r$ is exact.

5) The sequence $\mathcal{D}'^p_F(\Omega) \xrightarrow{A(D)} \mathcal{D}'^q_F(\Omega) \xrightarrow{B(D)} \mathcal{D}'^r_q(\Omega)$ is exact, for Ω

convex $\neq \emptyset$.

6) The sequence $\mathcal{D}'^p(\Omega) \xrightarrow{A(D)} \mathcal{D}'^q(\Omega) \xrightarrow{B(D)} \mathcal{D}'^r(\Omega)$ is exact, for Ω convex $\neq \emptyset$.

7) The sequence $\mathcal{E}^p_{x_0} \xrightarrow{A(D)} \mathcal{E}^q_{x_0} \xrightarrow{B(D)} \mathcal{E}^r_{x_0}$ is exact for $x_0 \in \mathbb{R}^n$.

8) The sequence $\mathcal{A}^p_{x_0} \xrightarrow{A(D)} \mathcal{A}^q_{x_0} \xrightarrow{B(D)} \mathcal{A}^r_{x_0}$ is exact for $x_0 \in \mathbb{R}^n$.

9) The sequence $\Phi^p_{x_0} \xrightarrow{A(D)} \Phi^q_{x_0} \xrightarrow{B(D)} \Phi^r_{x_0}$ is exact for $x_0 \in \mathbb{R}^n$.

10) The sequence $\mathcal{E}'^p(\Omega) \xleftarrow{{}^tA(D)} \mathcal{E}'^q(\Omega) \xleftarrow{{}^tB(D)} \mathcal{E}'^r(\Omega)$ is exact for Ω open, convex $\neq \emptyset$.

11) The sequence $\mathcal{D}^p(\Omega) \xleftarrow{{}^tA(D)} \mathcal{D}^q(\Omega) \xleftarrow{{}^tB(D)} \mathcal{D}^r(\Omega)$ is exact for Ω open, convex, $\neq \emptyset$.

12) The sequence $\mathcal{S}^p \xleftarrow{{}^tA(D)} \mathcal{S}^q \xleftarrow{{}^tB(D)} \mathcal{S}^r$ is exact.

Remark. On all the spaces listed at the beginning, we can let $\mathcal{P}$ operate (to the left) by the rule $p(\xi)\cdot f = p(D)f \quad \forall p \in \mathcal{P}$. This gives to all those spaces a structure of $\mathcal{P}$- modules. The implication 1) $\Longrightarrow$ 2),3),...,12) can be expressed in a more algebraic way by saying that:
for Ω open and convex :

$\mathcal{E}(\Omega), \mathcal{D}'(\Omega), \mathcal{D}'_F(\Omega), \mathcal{E}_{x_0}, \mathcal{A}_{x_0}, \Phi_{x_0}, \mathcal{S}', \mathbb{F}$ are injective $\mathcal{P}$-modules

and $\mathcal{E}'(\Omega), \mathcal{D}(\Omega), \mathcal{S}$ are flat $\mathcal{P}$- modules.

Indeed the complexes in 2),...,12) are obtained from the complex in 1) by application of the functions $\mathrm{Hom}_{\mathcal{P}}$ (from 2 to 9) and $\otimes_{\mathcal{P}}$ (from 10 to 12).

Theorem b (Approximation theorem)

The restriction to Ω (open and convex) of solutions $u \in \mathbb{F}^p$ of $A(D)u = 0$ are dense in the space of solutions $u \in \mathcal{E}^p(\Omega)$ (risp. $u \in \mathcal{D}'_F(\Omega)^p$, $u \in \mathcal{D}'(\Omega)^p$) of $A(D)u = 0$ on Ω .

The proof of these theorems is done by means of Fourier analysis, reducing to the cases of the spaces $\mathcal{S}$, $\mathcal{D}(\Omega)$, $\mathcal{E}'(\Omega)$ (that, when Ω is convex, are characterized by the theorem of Paley - Wiener) and duality. We will obtain most of these results as consequences of the "Fundamental Principle" of Ehrenpreis.

3. Characteristic Variety

In the statement of Theorem a, we see that the study of the differential operator $A(D) : \mathcal{E}^p(\Omega) \longrightarrow \mathcal{E}^q(\Omega)$ is reduced to the study of the homomorphism of polynomials: ${}^tA(\xi): \mathcal{P}^q \longrightarrow \mathcal{P}^p$. Let us set I = Image $({}^tA(\xi): \mathcal{P}^q \longrightarrow \mathcal{P}^p)$ and $M = \mathcal{P}^p/_I$ = cokernel$({}^tA(\xi): \mathcal{P}^q \longrightarrow \mathcal{P}^p)$. The first goal in our study will be to obtain a suitable characterization of the module I.

To this aim we begin by recalling some well known facts from algebra:

a) Reduced primary decomposition.

An ideal $\wp \subset \mathcal{P}$ is prime if $p_1, p_2 \in \mathcal{P}$ and $p_1 p_2 \in \wp$ implies that either p_1 or p_2 belongs to $\wp$.

An ideal $\mathfrak{b}$ is primary if $p_1, p_2 \in \mathcal{P}$ and $p_1 p_2 \in \mathfrak{b}$ implies that either p_1 or p_2 belongs to $\sqrt{\mathfrak{b}} = \{p \in \mathcal{P} \mid p^k \in \mathfrak{b}$ for some integer $k \geq 1\}$ = the radical of $\mathfrak{b}$. When $\mathfrak{b}$ is primary $\sqrt{\mathfrak{b}}$ is prime and is called the associated prime to $\mathfrak{b}$.

Given a $\mathcal{P}$ - module M, for $m \in M$ we define $\mathrm{Ann}(m) = \{p \in \mathcal{P} \mid p\, m = 0\}$. This is an ideal of $\mathcal{P}$. The set of ideals $\mathrm{Ann}(m)$, for $m \in M$, that are prime is the set of associated ideals to M and is denoted by $\mathrm{Ass}(M)$. If M is of finite type (has a finite number of generators), then $\mathrm{Ass}(M)$ is a finite set of prime ideals (is $\{0\}$ if M is torsion-free, i.e. if M is a submodule of a free module).

We also define: $\mathrm{Ann}(M) = \bigcap_{m \in M} \mathrm{Ann}(m)$ and supp $M = \{\wp$ prime ideal of $\mathcal{P} \mid \wp \supset \mathrm{Ann}(M)\}$.

Then $\mathrm{Ass}(M)$ and $\mathrm{supp}(M)$ have the same minimal elements and thus

$$\sqrt{\mathrm{Ann}(M)} = \bigcap \{\wp \mid \wp \in \mathrm{Ass}(M)\}.$$

The module M is co-primary if $M \neq 0$ and, $\forall m \in M - \{0\}$, $\mathrm{Ann}(m) \subset \sqrt{\mathrm{Ann}\, M}$.

In this case $\wp = \sqrt{\mathrm{Ann}\, M}$ is a prime ideal, $\mathrm{Ass}(M) = \{\wp\}$ contains a single element, and we say that M is $\wp$ - coprimary. A submodule N of M is $\wp$ - primary in M if $M/_N$ is $\wp$ -coprimary.

We have the following (theorem of Lasker-Noether):

If M is a $\mathcal{P}$ - module of finite type, N a submodule of M, for every prime ideal $\mathfrak{p} \in \mathrm{Ass}(M/N)$, we can find a $\mathfrak{p}$ - primary submodule $Q(\mathfrak{p})$ of M, in such a way that

i) $N = \bigcap_{\mathfrak{p} \in \mathrm{Ass}(M/N)} Q(\mathfrak{p})$

ii) $\forall \mathfrak{p} \in \mathrm{Ass}(M/N), \quad \bigcap_{\substack{\mathfrak{q} \in \mathrm{Ass}(M/N) \\ \mathfrak{q} \neq \mathfrak{p}}} Q(\mathfrak{q}) \not\subset Q(\mathfrak{p})$

The datum of the $\mathcal{P}$-modules $Q(\mathfrak{p})$ with the properties listed above is called a reduced primary decomposition of N in M.

Remark. The definitions and statements above still hold if we substitute to the ring of polynomials any noetherian integral domain with a unit and we use the word "module" meaning "unitary module".

b. Given the operator $A(D): \mathcal{E}^p(\Omega) \to \mathcal{E}^q(\Omega)$, with I and M defined as at the beginning, we consider a reduced primary decomposition $I = I_1 \cap \ldots \cap I_k$ of I in $\mathcal{P}^p$. Let $\mathfrak{p}_1, \ldots, \mathfrak{p}_k$ be the associated prime ideals and $V_1, \ldots, V_k$ the associated irreducible algebraic varieties: $V_j = \{\xi \in \mathbb{C}^n | p(\xi) = 0 \ \forall p \in \mathfrak{p}_j\}$. Then the set $\{V_1, V_2, \ldots, V_k\}$ is called the characteristic variety associated to $A(D)$.

Note that $V_1 \cup \ldots \cup V_k$ is the set of points $\xi \in \mathbb{C}^n$ such that the matrix $A(\xi)$ has rank (over $\mathbb{C}$) less than p.

4. Characterization of the Module I.

a. The case in which I is a primary submodule of $\mathcal{P}^p$.

Let in this case V denote the characteristic variety of $A(D)$. By a linear change of coordinates in $\mathbb{R}^n$ we can assume that the projection in the first d coordinates : $\pi: V \to \mathbb{C}^d$, d being the complex dimension of V, is generally biholomorphic and onto. Then the module $M = \mathcal{P}^p/I$, as a $\mathcal{P}_d = \mathbb{C}[\xi_1, \ldots, \xi_d]$-module, is of finite type and torsion free. With $\mathbb{C}(\xi_1, \ldots, \xi_d)$ = the field of rational functions of $\xi_1, \ldots, \xi_d$, the space $M \otimes_{\mathcal{P}_d} \mathbb{C}(\xi_1, \ldots, \xi_d)$ is a finite dimensional $C(\xi_1, \ldots, \xi_d)$ - vector space.

We choose for it a basis $\mu_1, \ldots, \mu_r$ of elements of M and we denote by N

the $\mathcal{P}_d$ - module generated by $\mu_1, \ldots, \mu_r$. Then N is free $\cong \mathcal{P}_d^r$. Moreover, for some $\psi \in \mathcal{P}_d - \{0\}$, we have $\psi M \subset N$, while obviously $N \subset M$.

If we consider $\mathcal{P}$ as a $\mathcal{P}_d$ - module, we realize that the inclusions $N \hookrightarrow M$ and $\psi : M \hookrightarrow N$ factor through $\mathcal{P}_d$ - homomorphism:

$$\rho : \mathcal{P}_d^r \longrightarrow \mathcal{P}^p$$

$$\tau : \mathcal{P}^p \longrightarrow \mathcal{P}_d^r$$

in such a way that we obtain the commutative diagrams with exact rows:

$$\begin{array}{ccccccc}
\mathcal{P}^q & \xrightarrow{\ {}^tA(\xi)\ } & \mathcal{P}^p & \longrightarrow & M & \longrightarrow & 0 \\
 & & \rho\uparrow & & \uparrow & & \\
0 & \longrightarrow & \mathcal{P}_d^r & \longrightarrow & N & \longrightarrow & 0 \\
 & & & & \uparrow & & \\
 & & & & 0 & &
\end{array}$$

$$\begin{array}{ccccccc}
 & & & & 0 & & \\
 & & & & \downarrow & & \\
\mathcal{P}^q & \xrightarrow{\ {}^tA(\xi)\ } & \mathcal{P}^p & \longrightarrow & M & \longrightarrow & 0 \\
 & & \downarrow\tau & & \downarrow\psi & & \\
0 & \longrightarrow & \mathcal{P}_d^r & \longrightarrow & N & \longrightarrow & 0
\end{array} \quad .$$

<u>Remark</u>. From the diagrams above one obtains at once:

i) $X \in I$ if and only if $\tau(X) = 0$

ii) $\forall X \in \mathcal{P}^p$, $\psi X - \rho \circ \tau X \in I$.

For $\eta^\circ \in \mathbb{C}^d$, we denote by $\mathcal{O}_{\eta^\circ, d}$ the ring of germs of holomorphic functions at η° in $\mathbb{C}^d$. This is flat over $\mathcal{P}_d$: this means that the tensor product by $\mathcal{O}_{\eta^\circ, d}$ of an exact sequence of $\mathcal{P}_d$ - modules and $\mathcal{P}_d$ - homomorphisms yields an exact sequence. In particular, for $X \in \mathcal{O}_{\eta^\circ, d}[\xi_{d+1}, \ldots, \xi_n]$ one obtains :

$$X \in I \otimes_{\mathcal{P}_d} \mathcal{O}_{\eta^\circ, d} \iff \tau(X) = 0$$

$$\psi X - \rho \circ \tau X \in I \otimes_{\mathcal{P}_d} \mathcal{O}_{\eta^\circ, d} .$$

By a classical preparation theorem after a real linear change of coordinates in $\mathbb{C}^n$ we can assume that the image of ξ_{d+1} in $\mathcal{P}/\wp$ generates the quotient field of $\mathcal{P}/\wp$ (that is an integral domain because $\wp$ is prime) over the field $C(\xi_1, \ldots, \xi_d)$ of rational functions.

Let P be the monic generator of the prime ideal $\mathbb{C}[\xi_1, \ldots, \xi_{d+1}] \cap \wp \subset \mathbb{C}[\xi_1, \ldots, \xi_{d+1}]$ and $\Delta \in \mathcal{P}_d$ its discriminant with respect to ξ_{d+1}. Then for $i = 2, \ldots, n-d$, $\wp$ contains a polynomial of the form

$$\Delta \xi_{d+i} - \gamma_i(\xi_1, \ldots, \xi_{d+1}) .$$

Let $s_\circ$ be the smallest integer such that $\wp^{s_\circ+1} M = 0$.

Then $\forall \alpha = (\alpha_1, \ldots, \alpha_{n-d}) \in \mathbb{N}^{n-d}$ with $\alpha_1 + \ldots + \alpha_{n-d} \geq s_\circ + 1$ we have a $\mathcal{P}$-homomorphism $T_\alpha : \mathcal{P}^p \longrightarrow \mathcal{P}^q$ such that

$$\circledast \quad P^{\alpha_1}(\Delta \xi_{d+2} - \gamma_2(\xi_1, \ldots, \xi_{d+1}))^{\alpha_2} \ldots (\Delta \xi_n - \gamma_{n-d}(\xi_1, \ldots, \xi_{d+1}))^{n-d} \, \mathrm{id}_{\mathcal{P}^p} = {}^t A(\xi) T_\alpha(\xi).$$

Let us fix now a point $\theta \in V$ where $\Delta(\theta) = \Delta(\theta_1, \ldots, \theta_d) \neq 0$ (V is certainly smooth at θ and π is biholomorphic when restricted to a neighborhood of θ in V).

We choose local coordinates at θ by :

$$\begin{aligned} z_i &= \xi_i - \theta_i \quad \text{for } 1 \leq i \leq d \\ z_{d+1} &= P(\xi_1, \ldots, \xi_{d+1}) \\ z_{d+2} &= \Delta \xi_{d+2} - \gamma_2 \\ &\cdots\cdots\cdots\cdots \\ z_n &= \Delta \xi_n - \gamma_{n-d} \end{aligned}$$

so that locally V has the equation $z_{d+1} = \ldots = z_n = 0$.

We set $\mathfrak{z}' = (\mathfrak{z}_1, \ldots, \mathfrak{z}_d)$, $\mathfrak{z}'' = (\mathfrak{z}_{d+1}, \ldots, \mathfrak{z}_n)$.

Note that $\mathfrak{z}''^{\alpha} X = {}^tA\, T_\alpha X \quad \forall X \in \mathcal{O}^p_\theta$ by ⊛ if $|\alpha| \geq s_\circ + 1$.

Given $X \in \mathcal{O}^p_\theta$, we can write X as a convergent power series:

$$X = \sum_\alpha \frac{1}{\alpha!} z''^{\alpha} \frac{\partial^{|\alpha|} X}{\partial z''^{\alpha}} (z',0) .$$

We set

$$\omega_\theta(X) = \sum_{|\alpha| \leq s_\circ} \frac{1}{\alpha!} z''^{\alpha} \frac{\partial^{|\alpha|} X}{\partial z''^{\alpha}} (z',0)$$

and we have :

$$X \in I \otimes_{\mathcal{P}} \mathcal{O}_\theta \iff \omega_\theta(X) \in I \otimes_{\mathcal{P}_d} \mathcal{O}_{\theta',d} \iff \tau \circ \omega_\theta(X) = 0 .$$

The last formula gives the condition as a differential equation for X on the germ of V at θ. Going back to the coordinates $\xi_1, \ldots, \xi_n$, and taking into account that the determinant of the jacobian matrix $\partial \mathfrak{z}/\partial \xi$ is $\delta(\xi) = \frac{\partial P_1}{\partial \xi_{d+1}}(\xi) \cdot \Delta^{n-d-1}$, since $\Delta(\xi) = q_1 P + q_2 \frac{\partial P}{\partial \xi_1}$ for suitable $q_1, q_2 \in \mathcal{P}$, we obtain

$$\tau \circ \omega_\theta(\Delta^{s_\circ(n-d)} X) = \sum_{|\alpha| \leq s_\circ} \lambda_\alpha(\xi_1, \ldots, \xi_d, \varphi_{d+1}, \ldots, \varphi_n) \frac{\partial^{|\alpha|} X}{\partial \xi^\alpha} (\xi_1, \ldots, \xi_d, \varphi_{d+1}, \ldots, \varphi_n)$$

where $\xi_j = \varphi_j(\xi_1, \ldots, \xi_d)$ $j = d+1, \ldots, n$ for $\xi \in V$ in a neighborhood of θ. ($\xi_{d+1}, \ldots, \xi_n$ have a unique branch as algebraic functions of $\xi_1, \ldots, \xi_d$ on a neighborhood of θ).

Then we define the differential operator with polynomial coefficients:

$$\mathcal{L}(\xi, \frac{\partial}{\partial \xi}) = \sum_{|\alpha| \leq s_\circ} \lambda_\alpha(\xi) \frac{\partial^{|\alpha|}}{\partial \xi^\alpha} .$$

Then the equation $\tau \circ \omega_\theta(X) = 0$ can be written as:

$$\mathcal{L}(\xi, \frac{\partial}{\partial \xi}) X = 0 \text{ on the germ of } V \text{ at } \theta .$$

Now we make the following remarks: ($\mathcal{H}$ denoting the ring of entire functions on $\mathbb{C}^n$) :

1) A submodule I of $\mathcal{P}^p$ is primary if and only if $I \otimes \mathcal{H}$ is a primary submodule of $\mathcal{H}^p$.

2) $X \in \mathcal{H}^p$ (resp. $X \in \mathcal{P}^p$) belongs to I (primary) if and only if for a $\theta \in V$ the holomorphic germ defined by X at θ, say X_θ, belongs to $I \otimes \mathcal{O}_\theta$.

(In particular, if V is irreducible, $\theta \in V$ and a holomorphic function f on $\mathbb{C}^n$ vanishes at the germ (V,θ), then f vanishes on V).

Therefore we have obtained the following statement:

$X \in \mathcal{H}^p$ (resp. $X \in \mathcal{P}^p$) belongs to $I \otimes \mathcal{H}$ (resp. I) if and only if

$$\mathcal{L}(\xi, \frac{\partial}{\partial \xi})X = 0 \quad \text{on } V.$$

b. The general case.

If we drop the assumption that I is primary, we will obtain for each associated characteristic variety $V_j \in \{V_1,\dots,V_k\}$ a differential operator $\mathcal{L}_i(\xi, D_\xi)$ with polynomial coefficients such that:

$X \in \mathcal{H}^p$ (resp. $X \in \mathcal{P}^p$) belongs to $I \otimes \mathcal{H}$ (resp. I) if and only if

$$\mathcal{L}_j(\xi, \frac{\partial}{\partial \xi})X = 0 \quad \text{on } V_j \quad \text{for } j = 1,\dots,k.$$

These operators $\mathcal{L}_j(\xi, \frac{\partial}{\partial \xi})$ where first defined by V. Palamodov that called them "Noetherian operators".

5. The extension Theorem.

The fundamental result is the following:

Theorem I

There are constants $C,K > 0$ and an integer $N > 0$ such that: if φ is a plurisubharmonic function in $\mathbb{C}^n$, Lipschitz continuous with Lipschitz constant L: $|\varphi(\xi') - \varphi(\xi'')| \le L|\xi' - \xi''|$, then for every $X \in \mathcal{H}^p$ with

$$|\mathcal{L}_j(\xi, \frac{\partial}{\partial \xi})X| \le \text{const}(X)\, e^{\varphi(\xi)} \quad \text{on } V_j \quad \text{for } j = 1,\dots,k$$

there exists $Y \in \mathcal{H}^p$ with

$$X - Y = {}^tA(\xi)Z \quad \text{for some } Z \in \mathcal{H}^q$$

<u>and</u>

$$|Y(\xi)| \leq C\cdot \text{const}(X)e^{KL}(1+|\xi|)^N e^{\varphi(\xi)} \qquad \forall \xi\in\mathbb{C}^n .$$

We have also:

<u>Theorem</u> II

<u>Same assumptions on</u> φ . <u>If</u> $X \in \mathcal{H}^p$ <u>and</u> $|X(\xi)| \leq c(1+|\xi|)^m e^{\varphi(\xi)}$ <u>on</u> $\mathbb{C}^n$ <u>and</u> $X = {}^tA(\xi)Y$ <u>for some</u> $Y \in \mathcal{H}^q$, <u>then there is</u> $Z \in \mathcal{H}^q$ <u>such that</u>

$${}^tA(\xi)Z = X \quad \text{and} \quad |Z(\xi)| \leq C\cdot c\cdot(1+|\xi|)^{m+N} e^{KL} e^{\varphi(\xi)} \quad \text{on} \quad \mathbb{C}^n .$$

This statement is a consequence of the previous one, taking into account that, if $\mathcal{P}^p \xleftarrow{{}^tA(\xi)} \mathcal{P}^q \xleftarrow{{}^tB(\xi)} \mathcal{P}^r$ is exact, then for $J = \text{Ker}\ {}^tA(\xi) = \text{Im}\ {}^tB(\xi)$, J is a primary submodule of $\mathcal{P}^q$ and ${}^tA(\xi)$ is the associated noetherian operator to J on $V = \mathbb{C}^n$. Actually we use theorem 2 to prove theorem 1 .

The proof of this theorem uses "cohomology with bounds" , that is discussed by Hörmander and some extensions of his results due to Petersen.

I will not go into details: first one makes local extensions by using the maps $\rho\circ\tau\circ\omega_\theta$ at points θ where $\Delta(\theta) \neq 0$. Then by Riemann extension theorem and Hermite's interpolation formula one obtains an extension to a neighborhood of V (first the case of a primary I is considered) that satisfies the estimate required for Y on an open neighborhood of V in $\mathbb{C}^n$. The global Y on $\mathbb{C}^n$ is obtained using Hörmander's results on cohomology with bounds.

6. <u>The Retard Formula</u>

The importance of the characterization of modules given above is shown by the "retard formula": we have the following identity:

if P and Q are polynomials in n variables, ξ and z are variables in $\mathbb{C}^n$, we have:

$$P(\tfrac{\partial}{\partial z})\ (Q(z)\ \exp\langle z,\xi\rangle) = Q(\tfrac{\partial}{\partial\xi})(P(\xi)\ \exp\langle z,\xi\rangle),$$

where $\langle z,\xi\rangle = z_1\xi_1 + \ldots + z_n\xi_n$. (Indeed both sides equal $\left(\sum_\alpha 1/_{\alpha!}\ Q^{(\alpha)}(z)P^{(\alpha)}(\xi)\right)\exp\langle z,\xi\rangle$, with $Q^{(\alpha)} = \dfrac{\partial^{|\alpha|}Q}{\partial z^\alpha}$, $P^{(\alpha)} = \dfrac{\partial^{|\alpha|}P}{\partial\xi^\alpha}$).

When P is a $p \times q$ matrix and Q a $q \times r$ matrix of polynomials the identity becomes:

$$ {}^t\{P(\tfrac{\partial}{\partial z})(Q(z)\exp\langle z,\xi\rangle)\} = {}^tQ(\tfrac{\partial}{\partial \xi})\{{}^tP(\xi)\exp\langle z,\xi\rangle\} \qquad \forall z,\xi \in \mathbb{C}^n . $$

If $\mathcal{L}_j(\xi,D_\xi)$ $(j = 1,\dots,k)$ are the matrices of differential operators with polynomial coefficients considered in the previous sections, then

$$ \mathcal{L}_j(\xi,D_\xi)\{{}^tA(\xi)X(\xi)\} = 0 \qquad \forall \xi \in V_j, \forall\ X \in \mathcal{H}^q . $$

In particular, one obtains

$$ A(\tfrac{\partial}{\partial z})\{{}^t\mathcal{L}_j(\xi,z)\exp\langle \xi,z\rangle\} = 0 \quad \forall z \in \mathbb{C}^n,\ \forall \xi \in V_j : $$

this proves that the columns of $(\exp\langle \xi,z\rangle)\,{}^t\mathcal{L}_j(\xi,z)$ are, for any fixed $\xi \in V_j$, exponential polynomial solutions of $A(D)u = 0$.

7. <u>Functionals</u>.

We consider the space $\mathcal{H}_A = \{u \in \mathcal{H}^p \mid A(\frac{\partial}{\partial z})u = 0\}$ (this is a Fréchet space being a closed subspace of $\mathcal{H}^p$ with the customary Fréchet topology of uniform convergence on compact sects) and for Ω open in $\mathbb{R}^n$ we consider the space $\mathcal{E}_A(\Omega) = \{u \in \mathcal{E}^p(\Omega) \mid A(D)u = 0\}$ (this is also Fréchet, being a closed subspace of $\mathcal{E}^p(\Omega)$ for the Schwartz topology of uniform convergence of the function and all partial derivatives on compact subsets of Ω).

Let

$$ \mathcal{H}'_A = \text{Homcont}\ (\mathcal{H}_A,\mathbb{C}) $$

$$ \mathcal{E}'_A(\Omega) = \text{Homcont}\ (\mathcal{E}_A(\Omega),\mathbb{C}) $$

be the dual spaces.

By Hahn-Banach theorem any functional $\mu \in \mathcal{H}'_A$ has a representative $\eta \in \mathcal{E}'^p(\mathbb{C}^n)$ and thus we can consider the Fourier-Laplace transform of η :

$$ \tilde{\eta}(\xi) = \eta(\exp\langle z,\xi\rangle) \in \mathcal{H}^p $$

$(\tilde{\eta}(\xi) = (\tilde{\eta}_1(\xi),\dots,\tilde{\eta}_p(\xi))$ with $\tilde{\eta}_j(\xi) = \eta_j(\exp\langle z,\xi\rangle)$, being $\eta = (\eta_1,\dots,\eta_p)$, $\eta_j \in \mathcal{E}'(\mathbb{C}^n))$.

We note that for $j = 1,\dots,k$, the restriction to V_j of $\mathcal{L}_j(\xi, \frac{\partial}{\partial \xi})\tilde{\eta}(\xi)$ depends only on μ and not on the extension we have chosen.

For K compact $\subset \mathbb{C}^n$ we set

$$\|u\|_{K,m} = \sup_{z \in K} \sum_{|\alpha| \leq m} \left| \frac{\partial^{|\alpha|}}{\partial x^\alpha} u(z) \right| \qquad (z = x+iy,\ x,y \in \mathbb{R}^n)$$

$$H_K(\xi) = \sup_{z \in K} \operatorname{Re} \langle z, \xi \rangle .$$

We have the following :

Theorem

If $\mu \in \mathcal{H}'_A$ and $\eta \in \mathcal{E}'^p(\mathbb{C}^n)$ is an extension of μ to $\mathcal{E}^p(\mathbb{C}^n)$, if K is a compact set in $\mathbb{C}^n$ then we have:

1) If $|\mathcal{L}_j(\xi, \partial/\partial\xi)\tilde{\eta}(\xi)| \leq C(1+|\xi|)^m \exp H_K(\xi) \quad \forall \xi \in V_j \quad \forall j = 1,\dots,p$, then

$$|\mu(u)| \leq c \cdot c_0 \|u\|_{K,m+N} \quad \forall u \in \mathcal{H}_A ,$$

where $c_0 = c_0(m,K,A)$, and $N = N(A)$.

2) If $|\mu(u)| \leq c \|u\|_{K,m} \quad \forall u \in \mathcal{H}_A$, then

$$|\mathcal{L}_j(\xi, \partial/\partial\xi)\tilde{\eta}(\xi)| \leq c \cdot c_1 \cdot (1+|\xi|)^{m+N_1} \exp H_K(\xi) \quad \forall \xi \in V_j, \ \forall j = 1,\dots,k$$

with $c_1 = c_1(m,K,A)$ and $N_1 = N_1(A)$.

8. The Approximation theorem

We prove the statement made at the beginning that the restrictions of elements of $\mathbb{F}_A$ are dense in $\mathcal{E}_A(\Omega)$ when Ω is open and convex.

By the theorem of Hahn-Banach this is equivalent to prove that if $\eta \in \mathcal{E}'^p(\Omega)$ and $\eta(u) = 0 \quad \forall u \in \mathbb{F}_A$, then $\eta(u) = 0 \quad \forall u \in \mathcal{E}_A(\Omega)$.

But $\eta(u) = 0 \quad \forall u \in \mathbb{F}_A$ implies that $\mathcal{L}_j(\xi, \partial/\partial\xi)\tilde{\eta}(\xi) = 0$ on V_j for $j = 1,\dots,k$ by the retard formula. Then $\tilde{\eta}(\xi) = {}^tA(\xi)G(\xi)$ for some $G \in \mathcal{H}^q$. But $\tilde{\eta}(\xi)$ satisfies some estimate of the form

$$|\tilde{\eta}(\xi)| \leq c(1+|\xi|)^m \exp H_k(\xi)$$

with K compact convex $\subset \Omega$. By Oka's theorem with bounds G H^q can be chosen to satisfy the estimate

$$|G(\xi)| \leq c'(1+|\xi|)^s \exp H_K(\xi)$$

and thus, by Paley-Wiener, to be the Fourier-Laplace transform of a distribution χ with support in Ω : $\eta = {}^tA(D)\chi$, supp $\chi \subset \Omega$

and thus : $\eta(u) = \chi(A(D)u) = 0 \quad \forall u \in \mathcal{E}_A(\Omega)$.

Analogously one proves that the restrictions to Ω of elements of $\mathbb{F}_A$ are dense in $\mathcal{D}'_A(\Omega)$ and $\mathcal{D}'_{F_A}(\Omega)$.

Remark. In proving the approximation theorem one uses only the "easy part" of the Ehrenpreis-Palamodov theory outlined in the previous sections, namely the characterization of the modules, and not the extension theorem.

Remark. We have proved above the following.

Theorem

If Ω is open and convex and $A(D)$ is a differential operator with constant coefficients, then ${}^tA(\ )\ \mathcal{E}'^q(\Omega)$ is closed in $\mathcal{E}'^p(\Omega)$.

Indeed:

$${}^tA(D)\ \mathcal{E}'^q(\Omega) = \{\eta \in \mathcal{E}'^p(\Omega) \mid \eta(u) = 0 \quad \forall\, u \in \mathbb{F}_A\} .$$

From the extension theorem we obtain at once that:

i) ${}^tA(D)\ \mathcal{E}'^q(\Omega)$ is a closed subspace of $\mathcal{E}'^p(\Omega)$ for Ω open and convex

ii) If $\mathcal{P}^p \xleftarrow{{}^tA(\xi)} \mathcal{P}^q \xleftarrow{{}^tB(\xi)} \mathcal{P}^r$ is exact, then

$$\mathcal{E}'(\Omega)^r \xrightarrow{B(D)} \mathcal{E}'(\Omega)^q \xrightarrow{{}^tA(D)} \mathcal{E}'(\Omega)^p$$

is exact for Ω open and convex.

Then by duality we obtain:

i') $A(D)\,\mathcal{E}^p(\Omega)$ is a closed subspace of $\mathcal{E}^q(\Omega)$ for Ω open and convex.

ii') If $\mathcal{P}^p \xleftarrow{{}^tA(\xi)} \mathcal{P}^q \xleftarrow{{}^tB(\xi)} \mathcal{P}^r$ is exact, then

$\mathcal{E}^p(\Omega) \xrightarrow{A(D)} \mathcal{E}^q(\Omega) \xrightarrow{B(D)} \mathcal{E}^r(\Omega)$ is exact for Ω open and convex.

We note also that the extension theorem yields the following statements: there is N integer ≥ 0 such that :

a) $\forall\, m \geq N$, if $u \in C_0^m(\Omega)^p$ and $\int f\, u\, dx = 0 \;\; \forall f \in \mathbb{F}_A$, then $u = {}^tA(D)v$ for some $v \in C_0^{m-N}(\Omega)^q$.

b) If $\mathcal{P}^p \xleftarrow{{}^tA(\xi)} \mathcal{P}^q \xleftarrow{{}^tB(\xi)} \mathcal{P}^r$ is exact, then for $m \geq N : \forall u \in C_0^m(\Omega)^q$ satisfying

${}^tA(D)u = 0$ there is $v \in C_0^{m-N}(\Omega)^r$ such that ${}^tB(D)v = u$ (Ω is always assumed to be open and convex).

Denoting by $\mathcal{D}'_m(\Omega)$ the space of distributions of order m in Ω , i.e. distributions that extend continuously to $C_0^m(\Omega)$, we notice that $\mathcal{D}'_m(\Omega)$ is in a natural way a Fréchet space and thus by standard duality arguments one obtains: under the assumptions in b) above:

$$\forall f \in \mathcal{D}'_m(\Omega)^q \text{ with } B(D)f = 0 \text{ on } \Omega$$

$$\exists\, u \in \mathcal{D}'_{m+N}(\Omega)^p \text{ such that } A(D)u = f .$$

This proves that we have an exact sequence

$$\mathcal{D}'_F(\Omega)^p \xrightarrow{A(D)} \mathcal{D}'_F(\Omega)^q \xrightarrow{B(D)} \mathcal{D}'_F(\Omega)^r ,$$

where moreover all maps have a closed image.

Let now $u \in \mathcal{D}^p(\Omega)$ be such that $\int u\, f\, dx = 0 \;\; \forall f \in \mathbb{F}_A$. Then we define a linear functional on the space $A(D)\,\mathcal{D}'_F(\Omega)^p \subset \mathcal{D}'_F(\Omega)^q$ by

$$A(D)f \;\longrightarrow\; f(u)$$

This is well defined and is continuous for the topology of $\mathcal{D}'_F(\Omega)^q$.

By Hahn-Banach, because $\mathcal{D}(\Omega)$ is the dual space of $\mathcal{D}'_F(\Omega)$, there is $v \in \mathcal{D}^q(\Omega)$ such that

$$f(u) = (A(D)f)(v) \quad \forall f \in \mathcal{D}'_F(\Omega)^p .$$

But this gives $u = {}^tA(D)v$ as we wanted.

An analogous argument proves the result for $\mathcal{D}'(\Omega)$ starting from the result for $\mathcal{D}(\Omega)$.

We end up by noticing that the results on $\mathcal{S}$ and $\mathcal{S}'$ need some additional considerations besides the arguments of complex analysis used in the other points.

For the case of $\mathcal{E}_{x_0}$, $\mathcal{A}_{x_0}$ we reduce to the case of $\mathcal{E}(\Omega)$, Ω open and convex, while for Φ_{x_0} the relevant consideration is that, considering on formal power series the natural Fréchet topology, differential operators have always closed range on spaces of formal power series. The case of $\mathbb{F}$ reduces to linear algebra and the necessity of the algebraic condition can be reduced to the following observation: if $\mathcal{P}^p \xleftarrow{{}^tA(\xi)} \mathcal{P}^q \xleftarrow{{}^tB(\xi)} \mathcal{P}^r$ is not exact, one can find $f \in \mathbb{F}^q$ and ${}^tQ(\xi) = {}^t(Q_1(\xi),\dots,Q_q(\xi)) \in \mathcal{P}^q$ such that

$$B(D)f = 0 ,$$

$${}^tA(\xi)\,{}^tQ(\xi) = 0 \implies Q(D)\,A(D) = 0$$

but $Q(D)f \neq 0$.

9. Resolutions and Hilbert Complexes.

We have the following theorem, due to Hilbert:

Theorem.

Given any $\mathcal{P}$-homomorphism $S(\xi): \mathcal{P}^{p_1} \to \mathcal{P}^{p_0}$ we can construct a finite exact sequence

$$(*) \qquad 0 \to \mathcal{P}^{p_d} \xrightarrow{S_{d-1}(\xi)} \mathcal{P}^{p_{d-1}} \xrightarrow{S_{d-2}(\xi)} \cdots \to \mathcal{P}^{p_2} \xrightarrow{S_1(\xi)} \mathcal{P}^{p_1} \xrightarrow{S(\xi)} \mathcal{P}^{p_0}$$

of lenght $d \leq n+1$.

Remark.

Let $M = \operatorname{coker}\left(S(\xi): \mathcal{P}^{p_1} \to \mathcal{P}^{p_0}\right)$. Choosing a different presentation

$M = \mathrm{coker}\left(R(\xi): \mathcal{P}^{q_1} \longrightarrow \mathcal{P}^{q_o} \right)$ we can obtain for the homomorphism $R(\xi)$ a resolution of lenght $\leq n$. Recall that, in the study of the solvability of systems of equations associated to the operators in (*) , only the module M matters ; indeed the complexes of differential operators are obtained by using the functors Hom and $\otimes$ and then the obstruction to solvability is expressed by the groups Ext and Tor, which only depend on the module M .

Definition.

A complex of differential operators with constant coefficients of the form

$$\mathcal{E}^{p_o}(\Omega) \xrightarrow{{}^tS(D)} \mathcal{E}^{p_1}(\Omega) \xrightarrow{{}^tS_1(D)} \mathcal{E}^{p_2}(\Omega) \longrightarrow \ldots$$

(**)

$$\ldots \longrightarrow \mathcal{P}^{p_{d-1}}(\Omega) \xrightarrow{{}^tS_{d-1}(D)} \mathcal{P}^{p_d}(\Omega) \longrightarrow 0$$

obtained from an exact sequence of $\mathcal{P}$-modules and $\mathcal{P}$-homomorphisms (*) is called a Hilbert complex.

Hilbert complexes are characterized by the condition that (**) is exact for Ω open and convex.

LECTURE 2.

ANALYTIC CONVEXITY FOR DIFFERENTIAL OPERATORS WITH CONSTANT COEFFICIENTS .

The study of analytic convexity originated from the following remarks (Cattabriga, De Giorgi, Piccinini):

1) If $P(D)$ is a differential operator with constant coefficients in $\mathbb{R}^2$, then the equation $P(D)u = f \in \mathcal{A}(\mathbb{R}^2)$ (real-analytic functions of two real variables) has always a global solution $u \in \mathcal{A}(\mathbb{R}^2)$.

2) The equation $(\partial/_{\partial x_1} + i\, \partial/_{\partial x_2})u = f \in \mathcal{A}(\mathbb{R}^3)$ has no global real analytic solution u on $\mathbb{R}^3$ for many $f \in \mathcal{A}(\mathbb{R}^3)$.

More in general, we consider a Hilbert complex of differential operators with constant coefficients

$$\mathcal{E}(\Omega)^{p_0} \xrightarrow{A_0(D)} \mathcal{E}^{p_1}(\Omega) \xrightarrow{A_1(D)} \mathcal{E}^{p_2}(\Omega) \rightarrow \ldots$$

defined on all open subsets Ω of $\mathbb{R}^n$. This defines Hilbert complexes of sheaves:

$$0 \rightarrow \mathcal{E}_{A_0} \rightarrow \mathcal{E}^{p_0} \xrightarrow{A_0(D)} \mathcal{E}^{p_1} \xrightarrow{A_1(D)} \mathcal{E}^{p_2} \longrightarrow \ldots\ldots$$

$$0 \rightarrow \mathcal{A}_{A_0} \rightarrow \mathcal{A}^{p_0} \xrightarrow{A_0(D)} \mathcal{A}^{p_1} \xrightarrow{A_1(D)} \mathcal{A}^{p_2} \longrightarrow \ldots\ldots$$

where

$\mathcal{E}$ = sheaf of germs of C^∞ functions on $\mathbb{R}^n$

$\mathcal{A}$ = sheaf of germs of real analytic functions on $\mathbb{R}^n$.

By the results given in the previous lecture, these complexes of sheaves are acyclic, and then we have

$$H^j(\Omega, \mathcal{E}_{A_\circ}) = \mathrm{Ext}^j(M, \mathcal{E}(\Omega)) = \frac{\mathrm{Ker}\left(A_j(D): \mathcal{E}^{p_j}(\Omega) \to \mathcal{E}^{p_{j+1}}(\Omega)\right)}{\mathrm{Im}\left(A_{j-1}(D): \mathcal{E}^{p_{j-1}}(\Omega) \to \mathcal{E}^{p_j}(\Omega)\right)}$$

$$H^j(\Omega, \mathcal{A}_{A_\circ}) = \mathrm{Ext}^j(M, \mathcal{A}(\Omega)) = \frac{\mathrm{Ker}\left(A_j(D): \mathcal{A}^{p_j}(\Omega) \to \mathcal{A}^{p_{j+1}}(\Omega)\right)}{\mathrm{Im}\left(A_{j-1}(D): \mathcal{A}^{p_{j-1}}(\Omega) \to \mathcal{A}^{p_j}(\Omega)\right)},$$

where $M = \mathrm{coker}\ {}^tA_\circ(\xi): \mathcal{P}^{p_1} \to \mathcal{P}^{p_\circ}$, and where $H^j(\Omega, \mathcal{S})$ denote the j-th cohomology group with coefficients in the sheaf $\mathcal{S}$.

We say that an open set $\Omega \subset \mathbb{R}^n$ is C^∞ (resp. Analytically) convex if $H^j(\Omega, \mathcal{E}_{A_\circ}) = 0$ (resp. $H^j(\Omega, \mathcal{A}_{A_\circ}) = 0$) $\forall_j \geq 1$.

By the results given in the first lecture, $H^j(\Omega, \mathcal{E}_{A_\circ}) = 0\ \forall_j \geq 1$ if Ω is open and convex.

The example of De Giorgi and Piccinini (2) shows that $H^1(\mathbb{R}^3, \mathcal{A}_{\frac{\partial}{\partial x_1} + i\frac{\partial}{\partial x_2}}) \neq 0$, and hence C^∞ and analytic convexity are different.

Elliptic Operators.

Let $S(D): \mathcal{E}^{s_\circ}(\mathbb{R}^N) \to \mathcal{E}^{s_1}(\mathbb{R}^N)$ be a differential operator with constant coefficients in $\mathbb{R}^N$. We say that $S(D)$ is elliptic if every solution $u \in \mathcal{E}^{s_\circ}(\mathbb{R}^N)$ of $S(D)u = 0$ is real-analytic on $\mathbb{R}^N$. This is equivalent to say that for every open set $\Omega \subset \mathbb{R}^N$ and every $u \in \mathcal{E}^{s_\circ}(\Omega)$ that solves $S_\circ(D)u = 0$ on Ω, $u \in \mathcal{A}^{s_\circ}(\Omega)$. The condition of being elliptic can be characterized algebraically, by saying that, if $\{V_1, \dots, V_k\}$ is the characteristic variety of $S(D)$, then

$$|\mathrm{Im}\,\xi| \leq c(1 + |\mathrm{Re}\,\xi|)\quad \forall \xi \in V_1 \cup \dots \cup V_k$$

for some constant $c > 0$.

Remark. Let (Ω open in $\mathbb{R}^N$)

$$(*)\qquad \mathcal{E}^{s_0}(\Omega) \xrightarrow{S_0(D)} \mathcal{E}^{s_1}(\Omega) \xrightarrow{S_1(D)} \mathcal{E}^{s_2}(\Omega) \longrightarrow \ldots$$

be a Hilbert complex where $S_0(D)$ is an elliptic operator. Then C^∞ and analytic convexity are the same: indeed $\mathcal{E}_{S_0} = \mathcal{A}_{S_0}$ and hence

$$H^j(\Omega, \mathcal{E}_{S_0}) = H^j(\Omega, \mathcal{A}_{S_0}) \ \forall j .$$

The Cauchy Problem.

We consider a Hilbert complex $(*)$ and let $\mathfrak{M} = \operatorname{coker}({}^tS_0(\xi,\eta) : \mathcal{P}_N^{s_1} \longrightarrow \mathcal{P}_N^{s_0})$, so that

$$(**)\qquad 0 \leftarrow \mathfrak{M} \leftarrow \mathcal{P}_N^{s_0} \xleftarrow{{}^tS_0(\xi,\eta)} \mathcal{P}_N^{s_1} \xleftarrow{{}^tS_1(\xi,\eta)} \mathcal{P}_N^{s_2} \longleftarrow \ldots$$

is a Hilbert resolution of the $\mathcal{P}_N$-module $\mathfrak{M}(\xi_1,\ldots,\xi_n,\eta_1,\ldots,\eta_m$, $m+n = N$ are indeterminates).

We consider the embedding $\mathbb{R}^n \longrightarrow \mathbb{R}^N$ where $N = n+m$, $x_1 \ldots x_n$, $y_1 \ldots y_m$ are cartesian coordinates on $\mathbb{R}^N$ and $\mathbb{R}^n = \{(x,y) \in \mathbb{R}^N \mid y = 0\}$.

Algebraically, the Cauchy problem with initial data on $\mathbb{R}^n$ consists essentially in determining a solution of $S_0(D)u = 0$ on a neighborhood of $\mathbb{R}^n$ in $\mathbb{R}^N$ by the restrictions of a finite number of transversal derivatives of u on $\mathbb{R}^n$.

The necessary and sufficient condition for this to happen is that $\mathfrak{M}$, as a $\mathcal{P}_n = \mathbb{C}[\xi_1,\ldots,\xi_n]$ - module, is of finite type. Let us denote by $(\mathfrak{M})_n$ the module $\mathfrak{M}$ thought as a $\mathcal{P}_n$-module.

If we take a Hilbert resolution:

$$0 \leftarrow (\mathfrak{M})_n \leftarrow \mathcal{P}_n^{r_0} \xleftarrow{{}^tR_0(\xi)} \mathcal{P}_n^{r_1} \xleftarrow{{}^tR_1(\xi)} \mathcal{P}_m^{r_2} \longleftarrow \ldots .$$

we notice that also $(**)$ is a free $\mathcal{P}_n$ - resolution (by infinite $\mathcal{P}_n$-modules) of $(\mathfrak{M})_n$ and hence there are maps

$$ {}^t\tau_j(\xi,\eta) : \mathcal{P}_n^{r_j} \longrightarrow \mathcal{P}_N^{s_j} $$

$$ {}^t\rho_j \quad : \mathcal{P}_N^{s_j} \longrightarrow \mathcal{P}_n^{r_j} $$

that extend the identity map $(\mathcal{N})_n \cong (\mathcal{N})_n$, making the following diagrams commute:

$$
\begin{array}{ccccccccc}
0 \leftarrow & \mathcal{N} & \leftarrow & \mathcal{P}_N^{s_o} & \xleftarrow{{}^tS_o} & \mathcal{P}_N^{s_1} & \xleftarrow{{}^tS_1} & \mathcal{P}_N^{s_2} & \leftarrow \dots \\
 & \| & & \downarrow {}^t\rho_o & & \downarrow {}^t\rho_1 & & \downarrow {}^t\rho_2 & \\
0 \leftarrow & \mathcal{N} & \leftarrow & \mathcal{P}_n^{r_o} & \xleftarrow{{}^tR_o} & \mathcal{P}_n^{r_1} & \xleftarrow{{}^tR_1} & \mathcal{P}_n^{s_2} & \leftarrow \dots
\end{array}
$$

$$
\begin{array}{ccccccccc}
0 \leftarrow & \mathcal{N} & \leftarrow & \mathcal{P}_N^{s_o} & \xleftarrow{{}^tS_o} & \mathcal{P}_N^{s_1} & \xleftarrow{{}^tS_1} & \mathcal{P}_N^{s_2} & \leftarrow \dots \\
 & \| & & \uparrow {}^t\tau_o & & \uparrow {}^t\tau_1 & & \uparrow {}^t\tau_2 & \\
0 \leftarrow & \mathcal{N} & \leftarrow & \mathcal{P}_n^{r_o} & \xleftarrow{{}^tR_o} & \mathcal{P}_n^{r_1} & \xleftarrow{{}^tR_1} & \mathcal{P}_n^{r_2} & \leftarrow \dots
\end{array}
$$

The Hilbert complex:

$$ \mathcal{E}^{r_o}(\omega) \xrightarrow{R_o(D)} \mathcal{E}^{r_1}(\omega) \xrightarrow{R_1(D)} \mathcal{E}^{r_2}(\omega) \rightarrow \dots $$

is called « a complex of Cauchy data for the complex (*) on $\mathbb{R}^n$ ».

The reason for that is the following:

let $\mathcal{F}$ denote any of the differential $\mathcal{P}_n$-modules: $\mathcal{E}(\omega)$, $\mathcal{A}(\omega)$, $\mathcal{E}_{x^\circ}$, $\mathcal{A}_{x^\circ}$, $\mathbb{C}_{x^\circ}\{\{x - x^\circ\}\} = \Phi_{x^\circ,n}$.

(ω open in $\mathbb{R}^n$, $x_\circ \in \mathbb{R}^n$, $\Phi_{x^\circ,n}$ = formal power series of n variables centered at x°).

If $H^j(\mathcal{F}^*, R_*)$ denotes the j-th cohomology group of the complex

$$(\mathcal{F}^*, R_*) = \{ \mathcal{F}^{r_\circ} \xrightarrow{R_\circ(D)} \mathcal{F}^{r_1} \xrightarrow{R_1(D)} \mathcal{F}^{r_2} \longrightarrow \dots \}$$

and $H^j(\mathcal{F}\{\{y\}\}^*, S_*)$ the j-th cohomology group of the complex:

$$(\mathcal{F}\{\{y\}\}^*, S_*) = \{ \mathcal{F}\{\{y\}\}^{s_\circ} \xrightarrow{S_\circ(D)} \mathcal{F}\{\{y\}\}^{s_1} \xrightarrow{S_1(D)} \mathcal{F}\{\{y\}\}^{s_2} \longrightarrow \dots \}$$

then under the assumptions above:

$$H^j(\mathcal{F}^*, R_*) \cong H^j(\mathcal{F}\{\{y\}\}^*, S_*) \quad \forall j$$

(the isomorphism is given explicitly by the maps: $\tau_j : \mathcal{F}\{\{y\}\}^{s_j} \longrightarrow \mathcal{F}^{r_j}$, $\rho_j : \mathcal{F}^{r_j} \longrightarrow \mathcal{F}\{\{y\}\}^{s_j}$).

($\mathcal{F}\{\{y\}\}$ denotes the space of formal power series of $y = (y_1, \dots, y_m)$ with coefficients in $\mathcal{F}$).

One obtains the following generalization of the theorem of Cauchy-Kowalewska: The necessary and sufficient condition to have an isomorphism

$$\tau_\circ : \mathcal{A}_{S_\circ} \xrightarrow{\sim} \mathcal{A}_{R_\circ}$$

(that implies the isomorphism $H^j(\omega, \mathcal{A}_{S_\circ}) \cong H^j(\omega, \mathcal{A}_{R_\circ})$ $\forall j$, $\forall \omega$ open $\subset \mathbb{R}^n$) is that, $\{V_1, \dots, V_k\}$ denoting the characteristic variety of $S_\circ(D)$, one has:

$$|\eta| \leq c(1 + |\xi|) \quad \forall (\xi, \eta) \in V_1 \cup \dots \cup V_k$$

Suspensions

An elliptic complex (i.e. with $S_\circ(D)$ elliptic)

$$(*) \qquad \mathcal{E}^{s_\circ}(\Omega) \xrightarrow{S_\circ(D)} \mathcal{E}^{s_1}(\Omega) \xrightarrow{S_1(D)} \mathcal{E}^{s_2}(\Omega) \longrightarrow \dots \quad \text{on } \mathbb{R}^N$$

is a suspension to $\mathbb{R}^N$ of the complex

$$(**) \qquad \mathcal{E}^{p_0}(\omega) \xrightarrow{A_0(D)} \mathcal{E}^{p_1}(\omega) \xrightarrow{A_1(D)} \mathcal{E}^{p_2}(\omega) \xrightarrow{A_2(D)} \dots$$ <u>on $\mathbb{R}^n$ if this is a complex of Cauchy data for</u> $(*)$.

<u>Remark</u>. For elliptic complexes the Cauchy-Kowalewska condition is automatical. Thus, if $(*)$ is an elliptic suspension of $(**)$ we have (reduction of analytic to C^∞ convexity):

$$H^j(\omega, \mathcal{A}_{A_0}) = H^j(\omega, \mathcal{E}_{S_0}) \quad \forall j \ .$$

<u>Examples</u> 1. Let $\Delta = -\sum_1^n \frac{\partial^2}{\partial x_i^2} - \frac{\partial^2}{\partial y^2}$ be the Laplace operator in $n+1$ variables.

We consider the complex: $\mathcal{E}(\Omega) \xrightarrow{\Delta} \mathcal{E}(\Omega) \longrightarrow 0$.
By tensor product with the complex (**) we obtain a complex:

$$\begin{array}{ccccccc} & & \mathcal{E}^{p_0}(\Omega) & \xrightarrow{A_0} & \mathcal{E}^{p_1}(\Omega) & & \\ & \nearrow & & & & & \\ \mathcal{E}^{p_0}(\Omega) & & \oplus & -\Delta \nearrow & \oplus & \longrightarrow & \dots \\ & \underset{A_0}{\searrow} & & & & & \\ & & \mathcal{E}^{p_1}(\Omega) & \xrightarrow[A_1]{} & \mathcal{E}^{p_2}(\Omega) & & \end{array}$$

which is a suspension of the complex obtained taking two copies of the complex $(**)$: if $B_0(D)\colon \mathcal{E}^{p_0}(\Omega) \xrightarrow{\Delta \oplus A_0} \mathcal{E}^{p_0+p_1}(\Omega)$, we have:

$$H^j(\omega, \mathcal{E}_{B_0}) = \bigoplus_1^2 H^j(\omega, \mathcal{A}_{A_0}) \quad \forall j \ .$$

<u>Remark</u>. The complex $\mathcal{E}(\Omega) \xrightarrow{\Delta} \mathcal{E}(\Omega) \to 0$ is acyclic on all open sets Ω in $\mathbb{R}^N = \mathbb{R}^{n+1}$. It follows that, $\mathcal{H}(\Omega)$ denoting the space of harmonic functions on Ω , the cohomology groups $H^j(\Omega, \mathcal{E}_{B_0})$ are isomorphic to the cohomology groups of the complex:

$$\mathcal{H}^{p_0}(\Omega) \xrightarrow{A_0(D)} \mathcal{H}^{p_1}(\Omega) \xrightarrow{A_1(D)} \mathcal{H}^{p_2}(\Omega) \longrightarrow \dots$$

2. Let $\mathcal{E}^{(0,0)}(\Omega) \xrightarrow{\bar\partial} \mathcal{E}^{(0,1)}(\Omega) \xrightarrow{\bar\partial} \dots \rightarrow \mathcal{E}^{(0,n)}(\Omega) \rightarrow 0$ be the Dolbeault complex on Ω open $\subset \mathbb{C}^n$ ($\mathcal{E}^{(r,s)}(\Omega)$ are the alternated forms of type r,s with smooth coefficients on Ω). By tensor product with the complex (✱✱) ($\mathbb{R}^n \subset \mathbb{C}^n$ as the subspace $\{ \operatorname{Im} z = 0 \}$) we obtain an elliptic complex:

$$
(\ddagger) \qquad (\mathcal{E}^{(0,0)}(\Omega))^{p_0}
\begin{array}{c} \overset{\bar\partial}{\nearrow} \\ \underset{A_0}{\searrow} \end{array}
\begin{array}{c} (\mathcal{E}^{(0,1)}(\Omega))^{p_0} \\ \oplus \\ (\mathcal{E}^{(0,0)}(\Omega))^{p_1} \end{array}
\begin{array}{c} \xrightarrow{\bar\partial} \\ \underset{A_0}{\searrow} \; \overset{\bar\partial}{\nearrow} \\ \underset{A_1}{\searrow} \end{array}
\begin{array}{c} (\mathcal{E}^{(0,2)}(\Omega))^{p_0} \\ \oplus \\ (\mathcal{E}^{(0,1)}(\Omega))^{p_1} \\ \oplus \\ (\mathcal{E}^{(0,0)}(\Omega))^{p_2} \end{array}
\longrightarrow \dots
$$

for which the given complex (✱✱) is a complex of Cauchy data.

If Ω is an open set of holomorphy, the cohomology of the complex (‡) is isomorphic to the cohomology of the complex:

$$
\Gamma(\Omega,\mathcal{O})^{p_0} \xrightarrow{A_0(D)} \Gamma(\Omega,\mathcal{O})^{p_1} \xrightarrow{A_1(D)} \Gamma(\Omega,\mathcal{O})^{p_2} \rightarrow \dots
$$

($\mathcal{O}$ = sheaf of holomorphic functions in $\mathbb{C}^n$, $\Gamma(\Omega,\mathcal{O})$ = sections of $\mathcal{O}$ over Ω) .

3. We consider $\mathbb{R}^3$ as the subspace $\{(z_1,z_2) \in \mathbb{C}^2 \mid \operatorname{Im} z_2 = 0\}$ of $\mathbb{C}^2$. Then the Dolbeault complex:

$$
\mathcal{E}^{(0,0)}(\Omega) \xrightarrow{\bar\partial} \mathcal{E}^{(0,1)}(\Omega) \xrightarrow{\bar\partial} \mathcal{E}^{(0,2)}(\Omega) \rightarrow 0
$$

is a suspension of the complex

$$
\mathcal{E}(\omega) \xrightarrow{\partial/\partial x_1 + i\,\partial/\partial x_2} \mathcal{E}(\omega) \longrightarrow 0 \qquad (\,\omega \text{ open in } \mathbb{R}^3\,).
$$

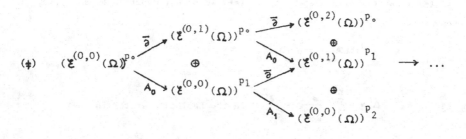

Sufficient Conditions for Analytic Convexity. ((*) is an elliptic suspension of (**))

Let $j \geq 1$. The following are sufficient conditions for $H^j(\omega, \mathcal{A}_{A_\circ}) = 0$:

i) There is a fundamental system of open neighborhoods Ω of ω in $\mathbb{R}^N$ such that

$$H^j(\Omega, \mathcal{E}_{S_\circ}) = 0 .$$

ii) $\forall$ open neighborhood Ω of ω in $\mathbb{R}^N$ there is a neighborhood Ω' of ω in Ω such that the natural map induced by the restriction: $H^j(\Omega, \mathcal{E}_{S_\circ}) \longrightarrow H^j(\Omega', \mathcal{E}_{S_\circ})$ is the zero map.

One proves that condition (ii) is independent of the suspension considered (this remark will be useful in the sequel). Probably the same is not true of condition (i).

Two open sets $\Omega' \subset \Omega$ such that $H^j(\Omega, \mathcal{E}_{S_\circ}) \to H^j(\Omega', \mathcal{E}_{S_\circ})$ is the zero map are said to be j-compatible.

First results on open convex sets.

Staircases: $\Omega \subset \mathbb{R}^N$ is a staircase if it is a reunion of countably many convex open sets $U_1, U_2, \ldots$ such that $U_i \cap U_j \subset U_h \cap U_k$ if $i \leq h \leq k \leq j$.

Example. Let ω be open in $\mathbb{R}^n$ and convex. Let $\omega_1 \subset\subset \omega_2 \subset\subset \omega_3 \subset\subset \ldots$ be an increasing sequence of convex open sets with $\cup \omega_j = \omega$. Let $\{T_j\}$ be a decreasing sequence of strictly positive numbers. Then, with

$$U_j = C(\omega_j, T_j) = \{(x,y) \in \mathbb{R}^N \mid x \in \omega_j,\ |y| < T_j\} ,$$

$$\Omega = \cup U_j \quad \text{is a staircase.}$$

When $\{T_j\}$ varies over all decreasing sequence of positive numbers, Ω describes a fundamental system of open neighborhoods of ω in $\mathbb{R}^N$.

We have the following

Theorem

If Ω is a staircase, then $H^j(\Omega, \mathcal{E}_{S_\circ}) = 0 \quad \forall\, j \geq 2$.

<u>Proof</u>. To compute $H^j(\Omega, \mathcal{E}_{S_\circ})$ we use a flabby resolution:

$$0 \longrightarrow \mathcal{E}_{S_\circ} \longrightarrow \mathcal{C}^{(\circ)} \xrightarrow{\delta_\circ} \mathcal{C}^{(1)} \xrightarrow{\delta_1} \mathcal{C}^{(2)} \longrightarrow \cdots$$

of the sheaf $\mathcal{E}_{S_\circ}$: we have

$$H^j(\Omega, \mathcal{E}_{S_\circ}) = \frac{\operatorname{Ker}\{\delta_j : \mathcal{C}^{(j)}(\Omega) \to \mathcal{C}^{(j+1)}(\Omega)\}}{\operatorname{Im}\{\delta_{j-1} : \mathcal{C}^{(j-1)}(\Omega) \to \mathcal{C}^{(j)}(\Omega)\}},$$

where $\mathcal{C}^{(r)}(\Omega) = \Gamma(\Omega, \mathcal{C}^{(r)})$.

Let $f \in \mathcal{C}^{(j)}(\Omega)$ $(j \geq 2)$ satisfy $\delta_j f = 0$. For each U_i of a representation of Ω as a staircase, because $H^j(U_i, \mathcal{E}_{S_\circ}) = 0$, there is $u_i \in \mathcal{C}^{(j-1)}(U_i)$ such that $\delta_{j-1} u_i = f$ on U_i . On $U_1 \cap U_2$ we have $\delta_{j-1}(u_1 - u_2) = 0$ and hence, being $j - 1 \geq 1$, there is $\nu_2 \in \mathcal{C}^{(j-2)}(U_1 \cap U_2)$ such that $u_1 - u_2 = \delta_{j-2} \nu_2$. Because $\mathcal{C}^{(j-2)}$ is flabby, ν_2 is the restriction to $U_1 \cap U_2$ of a section that we still denote by ν_2 on $\mathbb{R}^N$. Hence $v_2 = u_2 + \delta_{j-2} \nu_2$ still satisfies $\delta_{j-1} v_2 = f$ on U_2 and moreover $v_2 = u_1$ on $U_1 \cap U_2$.

Thus we obtain a section $g_2 = \begin{cases} u_1 & \text{on } U_1 \\ u_2 + \delta_{j-2} \nu_2 & \text{on } U_2 \end{cases} \in \mathcal{C}^{(j-2)}(U_1 \cup U_2)$ such that $\delta_{j-1} g_2 = f$ on $U_1 \cup U_2$.

On $(U_1 \cup U_2) \cap U_3 = U_2 \cap U_3$ we have $\delta_{j-1}(u_3 - g_2) = 0$. Thus $g_2 - u_3 = \delta_{j-2} \nu_3$ on $U_2 \cap U_3$ for some $\nu_3 \in C^{(j-2)}(\mathbb{R}^N)$ and

$$g_3 = \begin{cases} g_2 & \text{on } U_1 \cup U_2 \\ u_3 + \delta_{j-2} \nu_3 & \text{on } U_3 \end{cases}$$ satisfies $g_3 \in \mathcal{C}^{(j-1)}(U_1 \cup U_2 \cup U_3)$, solves

$\delta_{j-1} g_3 = f$ on $U_1 \cup U_2 \cup U_3$ and coincides with g_2 on $U_1 \cup U_2$. So we obtain by iteration $g_2, g_3, \ldots$ such that $g_s \in \mathcal{C}^{(j-1)}(U_1 \cup \ldots \cup U_s)$,
$\delta_{j-1} g_s = f$ on $U_1 \cup \ldots \cup U_s$, $g_s = g_{s+1}$ on $U_1 \cup \ldots \cup U_s$.
Then the sequence g_s defines $g \in \mathcal{C}^{(j-1)}(\Omega)$ with $\delta_{j-1} g = f$ on Ω , q.e.d.

Corollary.

If ω is an open convex set in $\mathbb{R}^n$, then $H^j(\omega, \mathcal{A}_{A_\circ}) = 0 \ \forall\ j \geq 2$.

Moreover one easily proves the following:

Proposition.

Let Ω be a staircase in $\mathbb{R}^N$, $Z(\Omega) = \{ f \in \mathcal{E}^{s_1}(\Omega) \mid S_1(D)f = 0 \}$; then $B(\Omega) = \{ S_\circ(D)u \mid u \in \mathcal{E}^{s_\circ}(\Omega) \}$ is a dense subspace of $Z(\Omega)$ (topology of $\mathcal{E}^{s_1}(\Omega)$).

Corollary.

Let $\Omega' \subset \Omega$ be open sets in $\mathbb{R}^N$, Ω a staircase. Then either image $(H^1(\Omega, \mathcal{E}_{S_\circ}) \to H^1(\Omega', \mathcal{E}_{S_\circ}))$ is infinite dimensional or is zero.

A general result (functional analysis).

Let ω be an open set in $\mathbb{R}^n$ (not necessarily convex). Let $\omega_1 \subset\subset \omega_2 \subset\subset \omega_3 \subset\subset \dots$ be an increasing sequence of relatively compact open substs such that $\omega = \bigcup \omega_j$. An application of Baire's cathegory argument yields:

Proposition.

Let Ω be a neighborhood of ω in $\mathbb{R}^N$. If $\dim_{\mathbb{C}} H^q(\omega, \mathcal{E}_{S_\circ}) = d < +\infty$, then we can find a sequence $\{T_j\}$ of positive numbers such that, setting

$B_k = \bigcup_{j=1}^{k} \{(x,y) \mid x \in \omega_j,\ |y| < T_j\}$, one has

$$\dim_{\mathbb{C}} \text{Image}(H^q(\Omega, \mathcal{E}_{S_\circ}) \to H^q(B_k, \mathcal{E}_{S_\circ})) \leq d \quad \forall\ k .$$

We want to prove the following statement:

"A necessary and sufficient condition in order that $H^q(\omega, \mathcal{A}_{A_\circ}) = 0$ is that $\forall$ open neighborhood Ω of ω in $\mathbb{R}^N$ there is a neighborhood B of ω in Ω such that

$$\text{Image}(H^q(\Omega, \mathcal{E}_{S_\circ}) \to H^q(B, \mathcal{E}_{S_\circ})) = 0 \text{ ."}$$

We will prove this in the cases: a) ω is convex (only the case $q = 1$ is of interest then) b) ω is star-shaped and $A_\circ$ is homogeneous.

Formally, to obtain the statement above from the previous proposition one only has to pass to the limit for $k \to \infty$: $B = \bigcup B_k = \lim_{k\to\infty} B_k$. This requires a "Runge type" approximation theorem:

Let Ω be open in $\mathbb{R}^N$, $B_1 \subset B_2 \subset \dots$ be an increasing sequence of open subsets of Ω such that for some integer $q \geq 1$ we have

$$\text{Image}\left(H^q(\Omega, \mathcal{E}_{S_\circ}) \to H^q(B_h, \mathcal{E}_{S_\circ})\right) = 0 \ (\forall h) .$$

Let $B = \bigcup B_k$ and assume that for every K compact $\subset B$ there is an index $h(K) \geq 1$ such that $K \subset B_{h(K)}$ and the restriction map

$$Z^{q-1}(B, \mathcal{E}_{S_\circ}) \to Z^{q-1}(B_{h(K)}, \mathcal{E}_{S_\circ})_K$$

has a dense image. Then

$$\text{Image}\left(H^q(\Omega, \mathcal{E}_{S_\circ}) \to H^q(B, \mathcal{E}_{S_\circ})\right) = 0$$

We are able to prove Runge's type theorems that are sufficient to discuss the two cases mentioned above. We have:

Proposition α .

If $S_\circ(D)$ is an elliptic operator in $\mathbb{R}^{n+1}$ and Ω is an open set of the form

$$\Omega = \{(x,y) \in \mathbb{R}^{n+1} \mid x \in \omega \ , |y| < \rho(x)\}$$

for ω open in $\mathbb{R}^n$ and ρ upper semicontinuous and strictly positive on ω , then the restrictions to Ω of functions in $\mathcal{E}_{S_\circ}(\mathbb{R}^{n+1})$ are dense in $\mathcal{E}_{S_\circ}(\Omega)$.

In the proof of this proposition one uses essentially the fact that we are dealing with neighborhoods of open sets ω in $\mathbb{R}^n$ in a euclidean space of dimension greater by one unit.

A sketch of the proof: set $\Omega^+ = \{(x,y) \in \mathbb{R}^{n+1} \mid x \in \omega, \ y > -\rho(x)\}$, $\Omega^- = \{(x,y) \mid x \in \omega \ , \ y < \rho(x)\}$.

Then $\Omega^+ \cup \Omega^- = \omega \times \mathbb{R}$ is convex and $\Omega^+ \cap \Omega^- = \Omega$.

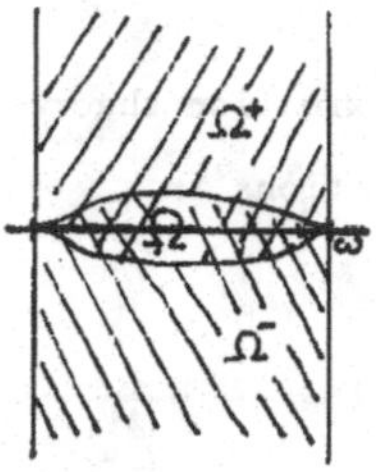

We have the Mayer-Victoris exact sequence:

$$0 \to \mathcal{E}_{S_\circ}(\omega\times\mathbb{R}) \to \mathcal{E}_{S_\circ}(\Omega^+) \oplus \mathcal{E}_{S_\circ}(\Omega^-) \to \mathcal{E}_{S_\circ}(\Omega) \to 0$$

where the last zero is $H^1(\omega\times\mathbb{R}, \mathcal{E}_{S_\circ}) = 0$ ($\omega\times\mathbb{R}$ is convex)

Because functions on $\mathcal{E}_{S_\circ}(\Omega)$ are sums of functions in $\mathcal{E}_{S_\circ}(\Omega+)$ and $\mathcal{E}_{S_\circ}(\Omega^-)$, it is sufficient to approximate such functions by functions in $\mathcal{E}_{S_\circ}(\mathbb{R}^{n+1})$.

Assume $f \in \mathcal{E}_{S_\circ}(\Omega^-)$ and let K be a compact subset of Ω^-. Let $e = (0,\dots,0,1)$.

For some $\lambda_\circ > 0$ we have $K - \lambda_\circ e = \{(x,y-\lambda_\circ) \mid (x,y) \in K\} \subset \{x \in \omega\ ,\ y < 0\}$.

Then one can prove the following:

$$\text{let } F = \bigcup_{0 \leq \lambda \leq \lambda_\circ} K - \lambda e .$$

Then F is a compact subset of Ω^-. The function f is real analytic and thus extends holomorphically to an open neighborhood of F in $\mathbb{C}^{n+1}$, that we can assume to contain the balls $\overline{B(x,y;4\varepsilon)}$ for some $\varepsilon > 0$ and $\forall (x,y) \in F$.

<u>If</u> k <u>is an integer sufficiently large, in such a way that</u> $\sigma = \frac{\lambda_\circ}{k} \leq \varepsilon$, <u>then for any</u> $\delta > 0$ <u>one can find integers</u> $m_1(\delta),\dots,m_k(\delta)$ <u>such that, with</u>

$$f_\delta(z,w) = \sum_{\substack{0 \leq s_1 \leq m_1(\delta) \\ \cdots\cdots\cdots \\ 0 \leq s_k \leq m_k(\delta)}} \frac{\sigma^{s_1 \cdots s_k}}{s_1! \dots s_k!} D_w^{s_1+\dots+s_k} f(z,w-\lambda_\circ)$$

((z,w) <u>complex variables extending</u> (x,y) <u>to</u> $\mathbb{C}^{n+1}$; $f(z,w)$ <u>extension of</u> f <u>as a holomorphic function</u>) <u>one has</u>

$$\sup_{\text{dist}(z,w);K) \leq \varepsilon} |f_\delta(z,w) - f(z,w)| < \delta .$$

This proves that, on K, f can be approximated by functions in $\mathcal{E}_{S_\circ}(\{(x,y) \mid x \in \omega,\ y < \lambda_\circ\})$.

Because the set $x \in \omega$, $y < \lambda_\circ$ is convex, by the approximation theorem given in the first lecture f can be approximated on K by functions in $\mathcal{E}_{S_\circ}(\mathbb{R}^{n+1})$.

Consequence.

If ω is an open convex set in $\mathbb{R}^n$, in order that $H^1(\omega, \mathcal{A}_{A_\circ}) = 0$ it is necessary and sufficient that for every open neighborhood Ω of ω in $\mathbb{R}^N$ there is an open neighborhood Ω' of ω in Ω such that

$$\operatorname{Image}\left(H^1(\Omega, \mathcal{E}_{S_\circ}) \to H^1(\Omega', \mathcal{E}_{S_\circ})\right) = 0 .$$

Homogeneous Differential Operators.

Let $S_\circ(D) = (S^\circ_{ij}(D))_{\substack{i=1,\dots,s_1\\ j=1,\dots,s_\circ}}$ be a $s_1 \times s_\circ$ matrix of differential operators with constant coefficients. We say that $S_\circ$ is homogeneous if there are integers $a_1 \dots a_{s_\circ}$ $b_1,\dots,b_{s_1}$ such that $\forall\, i,j$ either $S^\circ_{ij}(D) = 0$ or $S^\circ_{ij}(D)$ is a homogeneous differential operator of order $a_j - b_i$.

Proposition β.

If $S_\circ(D)$ is homogeneous, then any real analytic solution f of $S_\circ(D)f = 0$ on a star shaped open domain $\Omega \subset \mathbb{R}^N$ is limit in $\mathcal{E}^{s_\circ}(\Omega)$ of restrictions to Ω of elements of $\mathcal{E}_{S_\circ}(\mathbb{R}^N)$.

Sketch of the proof.

One can assume Ω starshaped with respect to 0. Because f is real analytic, it extends holomorphically to a star-shaped (with respect to 0) open neighborhood $\tilde{\Omega}$ of Ω in $\mathbb{C}^N$. Let us write

$$f = \sum_{0}^{\infty} f_h$$

in a neighborhood of 0, where $f_h = (f_h^1, \dots, f_h^{s_\circ})$ with f_h^j a homogeneous polynomial of degree $h+a_j$ (we assume as we can that $a_j \le 0\ \forall j$).

Then one easily verifies that

$$f_\varepsilon = \sum_{0}^{\infty} \frac{f_h}{\Gamma(1+\varepsilon h)}$$

are entire functions that solve $S_\circ(D)f_\varepsilon = 0$ and that $f_\varepsilon \to f$ for $\varepsilon \to 0^+$ in $\mathcal{E}^{S_\circ}(\Omega)$.

Consequence.

If ω is star-shaped in $\mathbb{R}^n$, if $A_\circ(D)$ is homogeneous, then $H^q(\omega, \mathcal{A}_{A_\circ}) = 0$ if and only if $\forall\,\Omega$ open neighborhood of ω in $\mathbb{R}^N$ there is a neighborhood Ω' of ω in Ω such that

Image $(H^q(\Omega, \mathcal{E}_{S_\circ}) \to H^q(\Omega', \mathcal{E}_{S_\circ})) = 0$.

Indeed one uses the facts: for a homogeneous operator there are homogeneous Hilbert resolutions and homogeneous suspensions.

Remark.

By the same approximation theorem given above (props. β) one also obtains that "boundaries are dense in the cycles" also in the case of homogeneous operators on star-shaped domains. Thus we have the alternative:

$$\text{either } H^q(\omega, \mathcal{A}_{A_\circ}) = 0 \quad \text{or} \quad \dim_{\mathbb{C}} H^q(\omega, \mathcal{A}_{A_\circ}) = +\infty$$

(ω star shaped open set in $\mathbb{R}^n$).

Examples.

1. We consider the example of De Giorgi and Piccinini, i.e. the complex :

$$(*) \qquad \mathcal{A}(\omega) \xrightarrow{\partial/\partial x_1 + i\, \partial/\partial x_2} \mathcal{A}(\omega) \longrightarrow 0 \ ,$$

ω star-shaped open domain of $\mathbb{R}^3$.

We noticed before that the Dolbeault complex in $\mathbb{C}^2$ is a suspension of $(*)$.

For the Dolbeault complex in $\mathbb{C}^2$ we have the following:

Proposition.

Let $\Omega' \subset \Omega$ be open sets in $\mathbb{C}^2$, and let $(\tilde{\Omega}', \pi)$ be the envelope of holomorphy of Ω'. Then a necessary and sufficient condition in order that $\text{Image}(H^1(\Omega,\mathcal{O}) \to H^1(\Omega',\mathcal{O})) = 0$ is that $\pi(\tilde{\Omega}')\subset\Omega$.

Proof. If $0\notin\Omega$, then $U_j = \{(z_1,z_2)\in\Omega \mid z_j \neq 0\}$ $j = 1,2$ is an open covering of Ω and then the datum of the holomorphic function $\frac{1}{z_1 z_2}$ on on $U_1\cap U_2$ defines a cohomology class in $H^1(\Omega,\mathcal{O})$. Because in dimension 1 the natural map from the Čech cohomology relative to every open covering into the first cohomology group with values in the sheaf is injective, if $\text{Image}(H^1(\Omega,\mathcal{O}) \to H^1(\Omega',\mathcal{O})) = 0$ there are $u_j\in\mathcal{O}(V_j)$ where $V_j = \{(z_1,z_2)\in\Omega' \mid z_j \neq 0\}$, such that $\frac{1}{z_1 z_2} = u_1 - u_2$ on $V_1\cap V_2$. Then we obtain two holomorphic functions g_1, g_2 on Ω' by setting:

$$g_1 = \begin{cases} z_1 u_1 & \text{on } V_1 \\ \frac{1}{z_2} + z_1 u_2 & \text{on } V_2 \end{cases} \qquad g_2 = \begin{cases} z_2 u_2 & \text{on } V_2 \\ z_2 u_1 - \frac{1}{z_1} & \text{on } V_1 \end{cases}$$ that satisfy

$$z_1 g_2 - z_2 g_1 = 1 \quad \text{on } \Omega' .$$

Ig $\tilde{g}_1, \tilde{g}_2$ are the extensions of g_1, g_2 to $\tilde{\Omega}'$, we must have by analytic continuation $\pi_1(p)\tilde{g}_2(p) - \pi_2(p)\tilde{g}_1(p) = 1$ on $\tilde{\Omega}'$, where $(\pi_1(p),\pi_2(p)) = \pi(p)$ is the projection $\pi:\tilde{\Omega}'\to\mathbb{C}^2$ and hence $\pi(p)\neq 0\ \forall p\in\tilde{\Omega}'$, i.e. $\pi(\tilde{\Omega}')\not\ni 0$. Repeating this argument substituting for 0 any point of $\mathbb{C}^2 - \Omega$, we obtain the statement of the proposition (the sufficiency part is trivial because $H^1(\tilde{\Omega}',\mathcal{O}) = 0$).

Let ω be a star-shaped open domain in $\mathbb{R}^3$. Let Ω be an open neighborhood of ω in $\mathbb{C}^2$. We have seen before that, if $H^1(\omega,\mathcal{A}_{\bar\partial'}) = 0$ ($\bar\partial' = \frac{\partial}{\partial x_1} + i\frac{\partial}{\partial x_2}$), then we can find an increasing sequence $\{B_h\}$ of star-shaped open sets of $\mathbb{C}^2$ such that $\omega\subset\bigcup B_h\subset\Omega$ and $\text{Image}(H^1(\Omega,\mathcal{O}) \to H^1(B_h,\mathcal{O})) = 0\ \forall h$. Because B_h is star - shaped, $\tilde{B}_h$ is an open subset of $\mathbb{C}^2$ and thus we should have by the previous proposition $\tilde{B}_h\subset\tilde{B}_{h+1}\subset\Omega\ \forall h$.

Hence $\bigcup_h \tilde{B}_h = \tilde{B}$ is a domain of holomorphy (by the theorem of Behnke-Tullen) and $\omega\subset\tilde{B}\subset\Omega$.

Thus, if $H^1(\omega,\mathcal{A}_{\bar\partial'}) = 0$, then ω has in $\mathbb{C}^2$ a fundamental system of neighborhoods that are open sets of holomorphy. This is not true (as can be proved by the logarithmic convexity of the domains of convergence of holomorphic power series) and hence this proves that $\dim_{\mathbb{C}} H^1(\omega,\mathcal{A}_{\bar\partial'}) = +\infty$ $\forall$ star-shaped open domain $\omega \subset \mathbb{R}^3$.

2. This example can be generalized to the following situation: let $\mathbb{R}^{n+k} = \mathbb{C}^k \times \mathbb{R}^{n-k} \subset \mathbb{C}^n$ $(n \geq 2,\ k < n,\ k \geq 1)$, $\mathbb{C}^n$ being the complex span of $\mathbb{R}^{n+k}$. We consider the Dolbeault complex along the fibers $\mathbb{C}^k$ of $\mathbb{R}^{n+k}$:

$$\mathcal{E}^{(0,0)}(\omega) \xrightarrow{\bar\partial'} \mathcal{E}^{(0,1)}(\omega) \xrightarrow{\bar\partial'} \mathcal{E}^{(0,2)}(\omega) \longrightarrow \dots \xrightarrow{\bar\partial'} \mathcal{E}^{(0,k)}(\omega) \to 0 .$$

Again we have $H^j(\omega,\mathcal{A}_{\bar\partial'}) \cong H^j(\omega,\mathcal{O})$ $\forall j$.

<u>Proposition.</u>

<u>If ω is a convex non empty open set in $\mathbb{R}^{n+k}$, then $H^1(\omega,\mathcal{A}_{\bar\partial'})$ is infinite dimensional.</u>

<u>Proof.</u> Assume that $0 \in \omega$. The proof is done by induction on n, as the statement has been proved true in the case $n = 2$. Assume that the statement is true for $n-1$ $(n \geq 3)$ and let us prove that is true for n. Let $\mathbb{R}^{n+k} = \{\operatorname{Im} z_{k+1} = \dots = \operatorname{Im} z_n = 0\}$.

If $k = n-1$ we set $\mathbb{C}^{n-1} = \{z_1 = 0\} \subset \mathbb{C}^n$, if $1 \leq k < n-1$ we set $\mathbb{C}^{n-1} = \{z_n = 0\} \subset \mathbb{C}^n$.

We set $\zeta = z_1$ in the first case, $\zeta = z_n$ in the second. We have the exact sequence of sheaves: $0 \to \mathcal{O}_{\mathbb{C}^n} \xrightarrow{\zeta} \mathcal{O}_{\mathbb{C}^n} \to \mathcal{O}_{\mathbb{C}^{n-1}} \to 0$ and thus we obtain the exact cohomology sequence:

$$H^1(\omega,\mathcal{O}_{\mathbb{C}^n}) \longrightarrow H^1(\omega',\mathcal{O}_{\mathbb{C}^{n-1}}) \to H^2(\omega,\mathcal{O}_{\mathbb{C}^n}).$$

The last group is zero, as was proven before and thus $H^1(\omega,\mathcal{O}_{\mathbb{C}^n})$ is infinite dimensional because $H^1(\omega',\mathcal{O}_{\mathbb{C}^{n-1}})$ is infinite dimensional.

3. Let $\omega = (\mathbb{C}^2 - \{0\}) \times \mathbb{C} \times \mathbb{R} \subset \mathbb{C}^4$, $\bar\partial'$ the exterior antiholomorphic differential on the fibers $\mathbb{C}^3$ of $\mathbb{R}^7$. Then one can prove that for $j \geq 2$ one has $H^j(\omega,\mathcal{E}_{\bar\partial'}) = 0$, while $H^2(\omega,\mathcal{A}_{\bar\partial'}) \cong H^2(\omega,\mathcal{O})$ is infinite dimensional.

LECTURE 3.

ANALYTIC CONVEXITY FOR CONVEX OPEN SETS.

(a) The Splitting condition.

In the following we will consider a fixed Hilbert complex on $\mathbb{R}^n$

$$\mathcal{E}^{p_0}(\omega) \xrightarrow{A_0(D)} \mathcal{E}^{p_1}(\omega) \xrightarrow{A_1(D)} \mathcal{E}^{p_2}(\omega) \longrightarrow \ldots$$

and an elliptic suspension of it in $\mathbb{R}^N$:

$$\mathcal{E}^{s_0}(\Omega) \xrightarrow{S_0(D)} \mathcal{E}^{s_1}(\Omega) \xrightarrow{S_1(D)} \mathcal{E}^{s_2}(\Omega) \longrightarrow \ldots$$

We denote by $x = (x_1,\ldots,x_n)$ cartesian coordinates in $\mathbb{R}^n$, by $(x,y) = (x_1,\ldots,x_n,y_1,\ldots,y_m)$ cartesian coordinates in $\mathbb{R}^N$ and set

$C(\omega,T) = \{(x,y) \in \mathbb{R}^N \mid x \in \omega \ , \ |y| < T\}$.

Let ω be a nonempty open convex set in $\mathbb{R}^n$.

We say that ω satisfies the splitting condition (S) if :

> $\forall \ \emptyset \neq \omega_1 \subset\subset \omega$ we can find $\delta > 0$ and an open convex set ω_2 with $\omega_1 \subset \omega_2 \subset\subset \omega$ such that:
>
> $$\forall \ 0 < T < \delta$$
>
> we can find an open neighborhood $\Omega(T) \subset C(\omega,T)$ of ω in $\mathbb{R}^N$ such that $\forall \ u \in \Gamma(C(\omega_2,T), \mathcal{E}_{s_0})$
>
> $$\exists \ u_1 \in \Gamma(C(\omega_1,\delta), \mathcal{E}_{s_0}) \text{ and } u_2 \in \Gamma(\Omega(T), \mathcal{E}_{s_0})$$
>
> such that $u_1 - u_2 = u$ on $\Omega(T) \cap C(\omega_1,T)$.

The last three lines can be substituted by the condition:

$$\text{Image } (H^1(C(\omega,T) \cup C(\omega_2,\delta), \mathcal{E}_{S_\circ}) \to H^1(\Omega(T) \cup C(\omega_1,\delta), \mathcal{E}_{S_\circ}) = 0$$

Remark. This δ can be substituted by any larger real number.

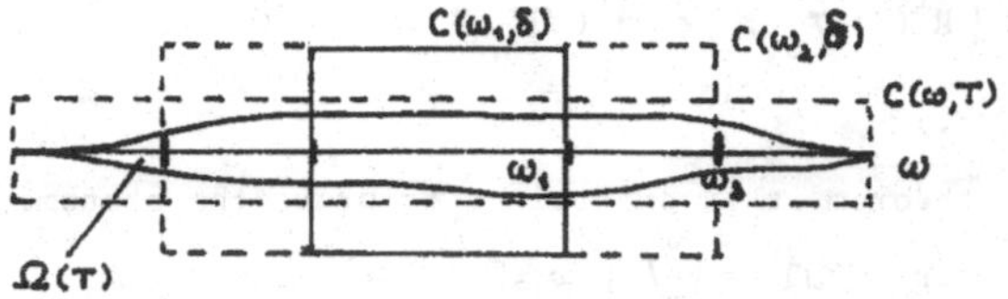

Remarks. 1) Condition (S) is indipendent of the suspension.

2) If $N = n+1$, then one can choose $\Omega(T) = C(\omega,T)$ in the condition above:

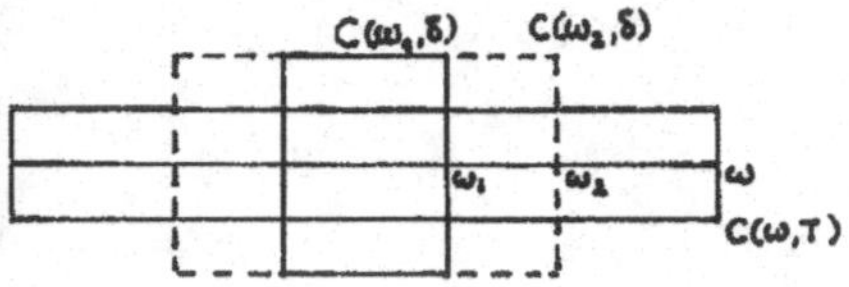

$\forall$ convex $\omega_1 \subset\subset \omega$
$\exists \omega_2$ convex $\omega_1 \subset \omega_2 \subset\subset \omega$, $\delta > 0$
such that $\forall\ 0 < T < \delta$

$$\text{Image}\{H^1(C(\omega,T) \cup C(\omega_2,\delta), \mathcal{E}_{S_\circ}) \to H^1(C(\omega,T) \cup ((\omega_1,\delta), \mathcal{E}_{S_\circ})\} = 0.$$

We have the following:

Theorem.

Let ω be an open convex set $(\neq \emptyset)$ in $\mathbb{R}^n$. Then $H^1(\omega, \mathcal{A}_{A_\circ}) = 0$ if and only if ω satisfies the splitting condition (S) .

(By remark 2.2. we work with $N = n+1$)

Proof. (sufficiency). Let $\omega_1 \subset\subset \omega_2 \subset\subset \ldots$ be a sequence of convex open sets such that $\omega = \cup \omega_j$ and such that there exists a sequence $\delta_1 > \delta_2 > \delta_3 > \ldots > 0$ of positive numbers such that

$$\forall\ 0 < T < \delta_i \ , \ \text{Image}\{H^1(C(\omega,T) \cup C(\omega_{i+1},\delta_i), \mathcal{E}_{S_\circ}) \to H^1(C(\omega,T) \cup C(\omega_i,\delta_i), \mathcal{E}_{S_\circ})\} = 0$$

With any sequence $T_1 \geq T_2 \geq T_3 \geq \ldots > 0$ with $0 < T_j < \delta_j$

$$\Omega = \bigcup C(\omega_i, T_i)$$

$$\Omega' = \bigcup C(\omega_i, T_{i+1}) .$$

We claim that

$$\text{Image}\,\{ H^1(\Omega, \mathcal{E}_{S_0}) \to H^1(\Omega', \mathcal{E}_{S_0}) \} = 0$$

(and this yields sufficiency).

By the approximation theorem it is sufficient to prove the statement with Ω'_ν instead of Ω', where $\Omega'_\nu = \bigcup_1^\nu C(\omega_i, T_{i+1})$.

Every cohomology class in $H^1(\Omega'_\nu, \mathcal{E}_{S_0})$ can be represented by a 1-cycle in $Z^1(\{C(\omega_i, T_{i+1})\}_{1 \le i \le \nu}, \mathcal{E}_{S_0})$, say $\{f_{ij}\}$ with $f_{ij} \in \Gamma(C(\omega_i, T_{i+1}) \cap C(\omega_j, T_{j+1}), \mathcal{E}_{S_0})$

By the cycle condition:

$$f_{ij} = \sum_{h=i}^{j-1} f_{h,h+1} \quad \text{on } C(\omega_i, T_j) \text{ for } i < j ,$$

and it follows that this is a coboundary if and only if

$$f_{h,h+1} = u_h - u_{h+1} \quad (1 \le h \le \nu-1) \text{ with } u_h \in \mathcal{E}_{S_0}(C(\omega_h, T_{h+1})) .$$

Being $f_{h,h+1} \in \mathcal{E}_{S_0}(C(\omega_h, T_{h+1}))$, by the splitting condition one can find

$$u_{h-1}^{(h)} \in \Gamma(C(\omega_{h-1}, \delta_{h-1}), \mathcal{E}_{S_0}) \text{ and } v_{h-1}^{(h)} \in \Gamma(C(\omega, T_{h+1}), \mathcal{E}_{S_0})$$

such that

$$f_{h,h+1} = u_{h-1}^{(h)} - v_{h-1}^{(h)} \quad \text{on} \quad C(\omega_{h-1}, T_{h+1}) .$$

By the equality $u_{h-1}^{(h)} = f_{h,h+1} + v_{h-1}^{(h)}$, one sees that

$u_{h-1}^{(h)}$ extends to an element of $\Gamma(C(\omega_{h-1}, \delta_{h-1}) \cup C(\omega_h, T_{h+1}), \mathcal{E}_{S_0})$.

If $h-1 > 1$, $u_{h-1}^{(h)} \in \Gamma(C(\omega_{h-1}, \delta_{h-1}), \mathcal{E}_{S_0})$ and by condition (S) there are

$$u_{h-2}^{(h)} \in \Gamma(C(\omega_{h-2}, \delta_{h-2}), \mathcal{E}_{S_0}) \text{ and } v_{h-2}^{(h)} \in \Gamma(C(\omega, \delta_{h-1}), \mathcal{E}_{S_0})$$

such that

$$u_{h-1}^{(h)} = u_{h-2}^{(h)} - v_{h-2}^{(h)} \quad \text{on} \quad C(\omega_{h-2},\delta_{h-1}) .$$

By the equality

$$u_{h-2}^{(h)} =: u_{h-1}^{(h)} + v_{h-2}^{(h)}$$

We see that $u_{h-2}^{(h)}$ extends to an element of $\Gamma(C(\omega_{h-2},\delta_{h-2}) \cup C(\omega_{h-1},\delta_{h-1}) \cup C(\omega_h,T_{h+1}), \mathcal{E}_{S_o})$.

Iterating this procedure, we obtain

$$u_{h-2}^{(h)}, u_{h-3}^{(h)},\dots,u_1^{(h)}$$

$$v_{h-2}^{(h)}, v_{h-3}^{(h)},\dots,v_1^{(h)}$$

with

$$u_j^{(h)} \in \Gamma(C(\omega_j,\delta_j) \cup C(\omega_{j+1},\delta_{j+1}) \cup \dots \cup C(\omega_{h-1},\delta_{h-1}) \cup C(\omega_h,T_{h+1}),\mathcal{E}_{S_o})$$

$$v_j^{(h)} \in \Gamma(C(\omega_j,\delta_{j+1}),\mathcal{E}_{S_o})$$

such that

$$u_{j+1}^{(h)} = u_j^{(h)} - v_j^{(h)} \quad \text{on} \quad C(\omega_j,\delta_{j+1}) \ (1 \leq j \leq h-2).$$

Hence for $h \geq 2$ one obtains:

$$f_{h,h+1} = u_{h-1}^{(h)} - v_{h-1}^{(h)} = u_{h-2}^{(h)} - v_{h-2}^{(h)} - v_{h-1}^{(h)} =$$

$$= \dots = u_1^{(h)} - (v_{h-1}^{(h)} + v_{h-2}^{(h)} + \dots + v_1^{(h)})$$

Set $u^{(h)} = u_1^{(h)}$, $v^{(h)} = v_1^{(h)} + \dots + v_{h-1}^{(h)}$, so that

$$u^{(h)} \in \Gamma(C(\omega_1,\delta_1) \cup \dots \cup C(\omega_{h-1},\delta_{n-1}) \cup C(\omega_h,T_{h+1}),\mathcal{E}_{S_o})$$

$$v^{(h)} \in \Gamma(C(\omega,T_{h+1}), \mathcal{E}_{S_o})$$

$$f_{h,h+1} = u^{(h)} - v^{(h)} \quad \text{on} \quad C(\omega_h, T_{h+1}) .$$

With $u^{(1)} = f_{12}$, $v^{(1)} = 0$, we set

$$u_1 = u^{(1)} + u^{(2)} + \ldots + u^{(\nu-1)}$$

$$u_2 = u^{(2)} + u^{(3)} + \ldots + u^{(\nu-1)} + v^{(1)}$$

....

$$u = v^{(1)} + \ldots + v^{(\nu-1)} .$$

Then $u_i \in \Gamma(C(\omega_i, T_{i+1}), \mathcal{E}_{S_\circ})$ and $f_{h,h+1} = u_h - u_{h+1}$ on $C(\omega_h, T_{h+1})$.

Necessity.

Let $\omega_1 \subset\subset \omega_2 \subset\subset \omega_3 \subset\subset \ldots$ be as at the beginning.

Let $T_1 > T_2 > \ldots > 0$ with $T_j < 1/_j \; \forall_j$, and let U_j denote any open neighborhood of ω in $\mathbb{R}^N$ containing $C(\omega_1, 1/_j)$.

We can assume that the splitting condition does not hold with ω_1 and then we can choose T_j with $0 < T_j < \frac{1}{j}$, T_{j-1} in such a way that for every choice of $U_j \subset C(\omega, T_j) \cup C(\omega_j, 1)$

$$H^1(C(\omega, T_j) \cup C(\omega_j, 1), \mathcal{E}_{S_\circ}) \to H^1(U_j, \mathcal{E}_{S_\circ})$$

has an image $\neq 0$.

If $H^1(\omega, \mathcal{A}_{A_\circ}) = 0$, we have proved that there exists for every open neighborhood Ω of ω in $\mathbb{R}^N$ a neighborhood B of ω in Ω such that

$$H^1(\Omega, \mathcal{E}_{S_\circ}) \to H^1(B, \mathcal{E}_{S_\circ}) \quad \text{has zero image.}$$

We can assume that, for every j , $\Omega \subset C(\omega, T_j) \cup C(\omega_j, 1)$. But for large j we have $B \supset C(\omega_1, 1/_j)$ and this gives a contradiction.

b. $S_\circ$ - analytic functionals and the co-splitting or carrier condition.

We denote by $\mathcal{E}_{S_\circ}(\Omega)$ the space of solutions $u \in \mathcal{E}^{S_\circ}(\Omega)$ of $S_\circ(D)u = 0$, when $S_\circ(D)$ is an elliptic differential operator. Let $\mathcal{E}'_{S_\circ}(\Omega)$ denote the strong dual of $\mathcal{E}_{S_\circ}(\Omega)$ and

let $\mathcal{E}'_{S_\circ}(K) = \underset{\Omega \supset K}{\underrightarrow{\lim}} \mathcal{E}'_{S_\circ}(\Omega)$. By the approximation theorem given in the first lecture, for K compact convex and Ω convex open there are natural inclusions (dual of the restriction maps):

$$\lambda_K : \mathcal{E}'_{S_\circ}(K) \to \mathcal{E}'_{S_\circ}(\mathbb{R}^N), \quad \lambda_\Omega : \mathcal{E}'_{S_\circ}(\Omega) \to \mathcal{E}'_{S_\circ}(\mathbb{R}^N) .$$

Let $\mu \in \mathcal{E}'_{S_\circ}(\mathbb{R}^n)$. We say that the convex compact set K is a carrier for μ if $\mu \in \lambda_K(\mathcal{E}'_{S_\circ}(K))$. This fact is equivalent to the following $(K_\varepsilon = \{r \in \mathbb{R}^N \mid \mathrm{dist}(r,K) < \varepsilon\})$: $\forall \varepsilon > 0$ $\exists$ a constant $C_\varepsilon > 0$ such that $|\mu(f)| \leq C_\varepsilon \|f\|_{K_\varepsilon} = C_\varepsilon \sup_{K_\varepsilon} |f|$ $\forall f \in \mathcal{E}_{S_\circ}(\mathbb{R}^N)$.

Let $A_\circ$, $S_\circ$ be as in the preceding sections. We say that an open convex set $\omega \subset \mathbb{R}^n$ satisfies the "carrier condition" if :

(C) $\forall$ K convex compact $\subset \omega$, $\exists$ $\delta > 0$ and K' convex compact with $K \subset K' \subset \omega$ such that every $\mu \in \mathcal{E}'_{S_\circ}(\mathbb{R}^N)$ that is carried by K_δ and by a compact $K'' \subset \omega$ is carried by K' .

Remark.

The "carrier condition" is essentially obtained from the "splitting condition" by duality. One proves that this condition is independent of the particular suspension $S_\circ(D)$.

Reducing to the splitting condition one proves:

Theorem.

If ω is an open convex set in $\mathbb{R}^n$, then $H^1(\omega, \mathcal{A}_{A_\circ}) = 0$ if and only if the carrier condition (C) holds for ω .

Remark.

It follows from the discussion above that informations about the characterization of carries of $S_\circ$ - analytic functionals will translate by means of the carrier condition into conditions for the analytic convexity of convex sets in $\mathbb{R}^n$.

In the following, we will do it for the case of the $\bar{\partial}$ - suspension, using the Ehrenpreis fundamental principle and a generalization of the gauge function of Martineau.

c. Let V be an algebraic variety in $\mathbb{C}^n$. Let $\mathfrak{b}$ denote the ideal of polynomials vanishing on V . We denote by $\mathfrak{N}$ the homogeneous ideal in the graded ring of homogenous polynomials generated by the principal parts of polynomials in $\mathfrak{b}$. The homogeneous cone W of common zeros of polynomials in $\mathfrak{N}$ is called the asymptotic variety, or the asymptotic cone of V . It defines an algebraic variety W° in $\mathbb{P}_{n-1}(\mathbb{C})$ (n-1 dimensional complex projective space).

Given a differential operator $A_\circ(D)$, let $\{V_1,\dots,V_k\}$ denote its characteristic variety. We call the <u>asymptotic variety</u> of $A_\circ(D)$ the set of homogeneous algebraic cones $\{W_1,\dots,W_s\}$ that are irreducible components of the asymptotic cone of V_j for some $1 \le j \le k$.

In the following we will show how to obtain the following characterization of analytic convexity for convex open sets due to Hörmander: (Phragmèn-Lindelöff principle : result of Hörmander for a single scalar operator).

Theorem.

<u>A necessary and sufficient condition for an open convex set</u> $\omega \subset \mathbb{R}^n$ <u>to have</u> $H^1(\omega,\mathcal{A}_{A_\circ}) = 0$ <u>is that:</u>

$\forall K$ <u>convex compact</u> $\subset \omega \ \exists \delta > 0$ <u>and</u> K' <u>convex compact</u> $\subset \omega$ <u>such that</u> $\forall \varphi$ <u>plurisubharmonic on</u> $\mathbb{C}^n$ <u>that satisfies:</u>

$$\left.\begin{array}{ll} \varphi(\xi) \le H_K(\xi) + \delta|\xi| & \forall \xi \in \mathbb{C}^n \\ \varphi(\xi) \le 0 & \forall \xi \in i\,\mathbb{R}^n \cap W_j \end{array}\right\} \ \underline{\text{one has also}} \quad \varphi(\xi) \le H_{K'}(\xi) \quad \forall \xi \in W_j \ \ \underline{\text{for}}\ j = 1,\dots,s$$

Notation:

$$H_K(\xi) = \sup_{x \in K} \operatorname{Re} \langle \xi, x \rangle \quad .$$

d. Gauge Function for an Entire Function of Exponential type on an Algebraic Variety.

We recall that an entire function F on $\mathbb{C}^n$ is of exponential type if $|F(z)| \leq c_1 e^{c|z|}$ $\forall z \in \mathbb{C}^n$ for some constants c_1, $c > 0$. Entire functions of exponential type are Fourier-Laplace transforms of analytic functionals in $\mathbb{C}^n$.

α. The classical theory of Martineau.

The gauge function is defined by:

$$g(z) = \limsup_{\substack{\mathbb{R} \ni \lambda \to +\infty \\ w \to z}} \frac{\ln |F(\lambda w)|}{\lambda} \quad \forall z \in \mathbb{C}^n \ (\ln 0 = -\infty) \ .$$

This is a function $g: \mathbb{C}^n \longrightarrow \mathbb{R} \cup \{-\infty\}$. The following holds:

1) g is a plurisubharmonic function on $\mathbb{C}^n$ and $g(z) \leq c|z|$, $\forall z \in \mathbb{C}^n$.

2) F is the Fourier Laplace transform of an analytic functional in $\mathbb{C}^n$ carried by the compact convex set $K \subset \mathbb{C}^n$ if and only if

$$g(\theta) \leq H_K(\theta) = \sup_{z \in K} \operatorname{Re}\langle \theta, z \rangle \ \forall \theta \in C^n \ .$$

β. Let V be an irreducible algebraic variety in $\mathbb{C}^n$ and let W denote its asymptotic cone. The gauge function of F on V is a function $g: W \longrightarrow [-\infty, \infty)$ defined in the following way :

for $\theta \in W$, $|\theta| = 1$, we set:

$$g(\theta) = \limsup_{\substack{z \in V \\ z \to \infty \\ z/|z| \to \theta}} \frac{\ln |F(z)|}{|z|}$$

and we extend this definition by setting:

$$g(\theta) = |\theta|\, g(\frac{\theta}{|\theta|}) \quad \text{if} \quad \theta \in W - \{0\}$$

$$g(0) = \limsup_{\theta \to 0} g(\theta) \quad .$$

The statement 2) in α is then generalized by the following

Proposition. Let F, g be as above and let K be a compact convex set in $\mathbb{C}^n$. Then the following statements are equivalent:

(i) $g(\theta) \leq H_K(\theta) \quad \forall \theta \in W$

(ii) $|F(z)| \leq c_\varepsilon \exp(H_K(z) + \varepsilon|z|) \quad \forall \varepsilon > 0$

Statement 1 extends as follows:

Proposition. The function $g: W \to \mathbb{R} \cup \{-\infty\}$ defined above has the properties:

1) $g(\theta) = \limsup_{\substack{\eta \in W \\ \eta \to \theta}} g(\eta) \quad \forall \theta \in W$. (Thus is upper semicontinuous)

2) Let $S(W)$ denote the set of singular points of W. Then g is plurisubharmonic on $W - S(W)$ (and locally bounded on W)(such a function will be called « weakly plurisubharmonic on W »).

The proof of this proposition requires the preparation theorem for irreducible algebraic varieties adn the Riemann extension theorem for plurisubharmonic functions. Technically it is a little complicated and hence we refer the reader to a forthcoming paper by Andreotti and Nacinovich on the Quaderni della Scuola Normale Superiore. We prefer to give some more comments on the geometrical interpretation of the gauge function.

Let $W = W_1 \cup \ldots \cup W_k$ be the decomposition of W into irreducible components(*).

(*) We recall a few facts about normalizations and normal spaces: Let X be a (reduced) complex space, $S(X)$ the set of singular points of X. A weakly holomorphic function on an open subset U of X is a holomorphic function f on $U - S(X)$ that is locally bounded, i.e. such that any point $p \in U - S(X)$ has a neighborhood V in U such that f is bounded on $V - S(X)$.

./.

We consider the restriction of g to a component W_j. Then by Riemann extension theorem, g extends to a plurisubharmonic function on $W_j - S(W_j)$ ($S(W_j)$ = singular points of W_j), because $(\bigcup_{i \neq j} W_i) \cap W_j$ is a proper analytic subvariety of W_j.

One can ask wheter g is plurisubharmonic on the whole of W_j, in the sense that for any holomorphic map $\sigma : D = \{ t \in \mathbb{C} ; |t| < 1 \} \rightarrow W_j$ the function $\sigma^* g = g \circ \sigma$ is subharmonic on D. The answer is negative if W_j is not normal.

Let us consider the normalization $\tilde{W}_j \xrightarrow{\nu_j} W_j$ of W_j. (The disjoint union of $\tilde{W}_1, \ldots, \tilde{W}_k$ is then the normalization of W).

The function $g^{(j)} = \nu_j^* g$ on $\tilde{W}_j - \nu_j^{-1}(S(W_j))$ is plurisubharmonic and then by a theorem of Grauert and Remmert extends to a plurisubharmonic function on $\tilde{W}_j$. The collection of $g^{(1)} \ldots g^{(k)}$ defines a plurisubharmonic function on $\tilde{W} = \tilde{W}_1 \dot{\cup} \tilde{W}_2 \dot{\cup} \ldots \dot{\cup} \tilde{W}_k$: $\tilde{g} : \tilde{W} \rightarrow \mathbb{R} \cup \{-\infty\}$ and we have :

$$g(z) = \sup \{ \tilde{g}(p) \mid p \in \tilde{W} , \nu(p) = z \}$$

($\nu : \tilde{W} \rightarrow W$ is the normalization of W).

Thus, properly speaking, we should say that <u>the gauge function of an entire function of exponential growth over an irreducible algebraic variety V is a plurisubharmonic function on the normalization $\tilde{W}$ of the asymptotic cone of V</u>.

e. <u>The Principle of Phragmèn-Lindeloff as a sufficient condition.</u>

Let $S_o(D) = A_o(D) \oplus \tilde{\mathcal{B}}$ be the $\tilde{\mathcal{B}}$ suspension of $A_o(D)$.

(*) Weakly holomorphic functions form a sheaf $\tilde{\mathcal{O}}$. <u>A complex space</u> X <u>is normal if</u> $\forall\, p \in X$ <u>one has</u> $\mathcal{O}_p = \tilde{\mathcal{O}}_p$.

<u>The normalization of a (reduced) complex space</u> X <u>is the datum of a normal space</u> $\tilde{X}$ <u>and a holomorphic map</u> $\nu : \tilde{X} \rightarrow X$ <u>such that</u>

a) ν <u>is finite and surjective</u>

b) <u>If</u> $N(X)$ <u>denote the set of points</u> $p \in X$ <u>where</u> $\mathcal{O}_p \neq \tilde{\mathcal{O}}_p$, <u>then</u> $\nu^{-1}(N(X))$ <u>is analytically rare and the restriction</u> $\nu' : X - \nu^{-1}(N(X)) \rightarrow X - N(X)$ <u>is biholomorphic</u>.

(The normalization is defined up to isomorphism).

By the fundamental principle explained in the first lecture, we have the following : let $\{V_1,\dots,V_k\}$ be the characteristic variety of $A_\circ(D)$, let $\mathcal{L}_1(\xi,\partial/\partial\xi),\dots,\mathcal{L}_k(\xi,\partial/\partial\xi)$ be the differential operators with polynomial coefficients introduced in lecture 1 .

Let μ be an $S_\circ$ - analytic functional. Let $\eta\in\mathcal{E}'^{P_\circ}(\mathbb{C}^n)$ be an extension of μ and let $\tilde{\eta}(\xi) = \eta(\exp\langle z,\xi\rangle)$ denote its Fourier-Laplace transform.

Let K be a compact convex subset of $\mathbb{C}^n$. Then a necessary and sufficient condition for μ to be carried by K is that: $\forall\varepsilon>0$ there is $C_\varepsilon>0$ such that

$$|\mathcal{L}_j(\xi,\partial/\partial\xi)\tilde{\eta}(\xi)| \leq C_\varepsilon \exp(H_{K_\varepsilon}(\xi)) \quad \forall\xi\in V_j\ ,\quad j=1,\dots,k\ .$$

We note that for every j $\mathcal{L}_j(\xi,\partial/\partial\xi)\tilde{\eta}(\xi)$ is a vector whose components are entire functions of exponential growth. If g_j , defined on the asymptotic cone $W_{(j)}$ of V_j is the gauge function of $\mathcal{L}_j(\xi,\partial/\partial\xi)\tilde{\eta}(\xi)$ on $W_{(j)}$, we obtain

$$g_j(\theta) \leq H_K(\theta) \quad \forall\,\theta \quad W_{(j)},\quad j=1,\dots,k$$

as a necessary and sufficient condition for μ to be carried by K .

We obtain the sufficiency of the Phragmèn-Lindeloff principle in the (apparently) much stronger form:

(Ph. L.) $\forall W_j$ <u>in the asymptotic variety of</u> $A_\circ(D)$

$\forall K$ <u>convex compact</u> $\subset\omega$ $\exists K'$ <u>compact</u> $\subset\omega$ <u>and</u> $\delta>0$ <u>such that</u>

$\forall\varphi$ <u>weakly plurisubharmonic on</u> W_j <u>that satisfies:</u>

$$\varphi(\xi) \leq H_K(\xi) + \delta|\xi| \quad \underline{\text{on}} \quad W_j$$

$$\varphi(\xi) \leq 0 \quad \underline{\text{on}} \quad i\,\mathbb{R}^n \cap W_j$$

<u>one has also</u>

$$\varphi(\xi) \leq H_{K'}(\xi) \quad \underline{\text{on}} \quad W_j\ .$$

Then the proof of the theorem stated in section c) (sufficiency part) follows by showing that this condition is equivalent (with different δ's) to the condition involving plurisubharmonic functions on $\mathbb{C}^n$ of the statement of the theorem.

f. The difficulty in proving that the condition is also necessary comes out from the fact that only the particular weakly plurisubharmonic functions that are gauge functions of functions of the form $\mathcal{L}_j(\xi,\partial/_{\partial\xi})\tilde{\eta}(\xi)$ are used above. The proof then is done by first reducing to inequalities of Phragmèn-Lindelöff for a particular class of entire functions and then constructing from these particular $S_\circ$ - analytic functionals to which we apply the carrier condition. I will not discuss this in details, as complete proofs will be found in the above mentioned paper.

g. Application.

The criterion for analytic convexity given by the Phragmèn-Lindeloff principle is very implicit and it is therefore of the greatest interest to show how the principle can be applied to get more explicit answers.

Some easy consequences of the Ph. L. principle are the following:

Proposition A.

If $\{\omega_m\}_{m\in N}$ is an increasing sequence of open convex sets and $H^1(\omega_m,\mathcal{A}_{A_\circ}) = 0 \ \forall m$, then $H^1(\bigcup\omega_m,\mathcal{A}_{A_\circ}) = 0$.

As a consequence we obtain: if for ω open in $\mathbb{R}^n$, $x_\circ\in\mathbb{R}^n$ we set $\Gamma(x_\circ,\omega) = x_\circ + \bigcup_{\lambda>0}\lambda(\omega - x_\circ)$, then :

Proposition B.

If ω is an open convex set with $H^1(\omega,\mathcal{A}_{A_\circ}) = 0$, then

$$H^1(\Gamma(x_\circ,\omega),\mathcal{A}_{A_\circ}) = 0 \quad \forall x_\circ\in\mathbb{R}^n .$$

In particular:

Corollary.

If $H^1(\omega,\mathcal{A}_{A_\circ}) = 0$ for some open convex set $\omega\neq\emptyset$, then $H^1(\mathbb{R}^n,\mathcal{A}_{A_\circ}) = 0$.

Given an open half space Σ in $\mathbb{R}^n$, we say that it is a tangent subspace to an open convex set ω in $\mathbb{R}^n$ if $\Sigma = \Gamma(x_\circ,\omega)$ for some $x_\circ\in\partial\Omega$.

Every open convex set is the interior of the intersection of its tangent half spaces.

If $A_\circ(D)$ is a "terminal operator", i.e. if we have a Hilbert complex of the form $\mathcal{E}^p(\omega) \xrightarrow{A_\circ(D)} \mathcal{E}^q(\omega) \longrightarrow 0$, then we obtain the following criterion: $H^1(\omega, \mathcal{A}_{A_\circ}) = 0$ if and only if $H^1(\Sigma, \mathcal{A}_{A_\circ}) = 0$ for every tangent open half space to ω. (Here ω is assumed to be open and convex).

From the corollary above it follows that if any non void convex open set is analytically convex for $A_\circ(D)$, then $\mathbb{R}^n$ is analytically convex. Then we start by studying analytic convexity for $\mathbb{R}^n$. Because the Ph. L. condition is about the asymptotic variety $\{W_1,\ldots,W_s\}$ of $A_\circ(D)$, it will be sufficient to study conditions under which the Ph. L. principle holds on an irreducible algebraic cone W in $\mathbb{C}^n$.

We first obtain the following:

Theorem.

Let W be an irreducible algebraic cone in $\mathbb{C}^n$ on which the Ph. L. principle (for $\mathbb{R}^n$) holds. Then $\forall\, a \in W \cap \mathbb{R}^n$, $a \neq 0$, $\dim_{\mathbb{C}}(W,a) = \dim_{\mathbb{R}}(W \cap \mathbb{R}^n, a)$.

This theorem gives an explanation for the fact that $H^1(\mathbb{C}^m \times \mathbb{R}^k, \mathcal{A}_{\bar\partial'})$ is infinite dimensional in the case $m,k > 0$ and $\bar\partial'$ is the Cauchy Riemann operator along the fibers $\mathbb{C}^m$. Indeed in this case the asymptotic variety has dimension $k+m$ while the real dimension of the real part is $k < k+m$.

Using as plurisubharmonic test functions the real parts of coordinates of a suitable embedding in some $\mathbb{C}^N$ of the normalization $\tilde W$ of W one also obtains the following:

let $a \in W \cap \mathbb{R}^n$, $a \neq 0$ and let (W_α, a) be an irreducible germ of W at a and let $(\tilde W_\alpha, b) \longrightarrow (W_\alpha, a)$ be the normalization of it.

Assume that the cone W has the property of Ph. L. then:

if the germ $(\tilde W_\alpha, b)$ is non-singular at b, then the germ (W_α, a) is non-singular at a.

Corollary.

Let $a \in W_{\mathbb{R}} = W \cap \mathbb{R}^n$, $a \neq 0$, let (W_α, a) be any irreducible germ of W at a. Let $S(W_\alpha)_{\mathbb{R}}$ denote the set of points of $W_{\alpha,\mathbb{R}}$ which are singular points of

W_α . Then the real codimension of $(S(W_\alpha)_{\mathbb{R}}, a)$ in $(W_{\alpha\,\mathbb{R}}, a)$ is ≥ 2 .

Corollary.

If $\dim_{\mathbb{C}} W = 2$, then W has the Phragmèn Lindelöff property if and only if every irreducible germ (W_α, a) of W at $a \in W_{\mathbb{R}} - \{0\}$ is non-singular.

Definition.

W is locally hyperbolic at $a \in W_{\mathbb{R}} \setminus \{0\}$ if and only if , d denoting the complex dimension of W , we can find a real linear projection $\pi_a : \mathbb{C}^n \to \mathbb{C}^d$ inducing a proper map of a neighborhood $U(a)$ of a in W onto an open neighborhood Ω of $\pi(\Omega)$ in $\mathbb{C}^d$ such that

$$W \cap \pi_a^{-1}(\Omega \cap \mathbb{R}^d) = U(a) \cap W_{\mathbb{R}}$$

Then one proves that an irreducible algebraic cone W that is locally hyperbolic admits the Phragmèn-Lindelöff principle.

The case of a differential operator in $\mathbb{R}^2$, originally treated by De Giorgi and Cattabriga, extends by the following statement:

The Ph. L. principle holds on all complex lines through the origin.

Indeed it is trivial for complex lines that have no real point $\neq 0$.

When the complex line is the complexification of a real line, the statement reduces to the classical Ph. L. principle for functions of one complex variable:

if φ is subharmonic on $\mathbb{C}$ and

$$\varphi(z) \leq A\,|z| \quad \forall z \in \mathbb{C}$$

$$\varphi(z) \leq 0 \quad \forall z \in \mathbb{R}$$

Then $\varphi(z) \leq A\,|\mathrm{Im}\, z| \;\; \forall z \in \mathbb{C}$.

LECTURE 4.

ENVELOPE OF REGULARITY FOR HOMOGENEOUS SYSTEMS OF P.D.E.

1. Convexity Theory.

Let X be a topological space with a countable topology. Denoting by $\mathcal{C}(X)$ the space of real valued continuous functions on X, compact subsets C of X are characterized as those closed sets $C \subset X$ such that

$\sup_C |f| \quad +\infty \quad \forall f \in \mathcal{C}(X)$. (Weierstrass theorem).

In general, given a subset $S \subset \mathcal{C}(X)$, one can consider the set

$$C(X,S) = \{ C \subset X \text{ , closed} \mid \forall f \in S \quad \sup_C |f| \; +\infty \} .$$

This set contains all compact subsets of X and , in a certain sense, convexity theory consider the problem of deciding when $C(X,S)$ = compact subsets of X .

A very important instance of this question is the case in which $X = \Omega$ is an open set in $\mathbb{C}^n$ and $S = \mathcal{O}(\Omega)$ is the space of holomorphic functions on Ω .

We have that $C(\Omega, \mathcal{O}(\Omega)) = \{$ compact subsets of $\Omega \}$ if and only if is an open set of holomorphy: this is the contents of the theorem of Cartan Thullen that in the following we will generalize to the case in which $\mathcal{O}(\Omega)$ is substituted by the space of solutions $u \in \mathcal{E}^p(\Omega)$ of a homogeneous system of differential equations $A(x,D)u = 0$ on Ω ; where $A(x,D): \mathcal{E}^p(\Omega) \to \mathcal{E}^q(\Omega)$ is a suitable differential operator with smooth coefficients on Ω , open in $\mathbb{R}^n$

On $A(x,D): \mathcal{E}^p(\Omega) \to \mathcal{E}^q(\Omega)$ we make the crucial assumption : $A(x,D)$ <u>is elliptic, in the sense that all solutions</u> $u \in \mathcal{D}'(\Omega)$ <u>of</u> $A(x,D)u = 0$ <u>are real analytic on</u> Ω (sometimes in the letterature such operators are called analytic-hypoelliptic).

2. Envelopes of Regularity.

<u>Let</u> $A(x,D)$ <u>be defined on</u> $\mathbb{R}^n$ <u>and elliptic</u>. We denote by $\mathcal{E}_A$ the sheaf

of germs of solutions of $A(x,D)u = 0$ in $\mathbb{R}^n$. By assumption the map

$$\mathcal{T}_{x_0} : \mathcal{E}_{A,x_0} \longrightarrow \Phi^p_{x_0} \quad (\Phi_{x_0} = \text{formal power series centered at } 0)$$

described by the Taylor series at x_0 is injective and also the restriction map $\Gamma(\Omega,\mathcal{E}_A) \longrightarrow \mathcal{E}_{A,x_0}$ is injective if Ω is connected and $x_0 \in \Omega$ (unique continuation).

The topology of the sheaf $\mathcal{E}_A$ is defined in the usual way : a basis for the open sets of $\mathcal{E}_A$ is given by the sets of the form:

$$V(\Omega,u) = \{ u_x \mid u \in \mathcal{E}_A(\Omega), x \in \Omega \}$$

for Ω open in $\mathbb{R}^n$. The unique continuation property tells that the topology of $\mathcal{E}_A$ is of Hausdorff.

α . <u>Partial Completion.</u>

Let Ω be a connected open set in $\mathbb{R}^n$.

Given $u \in \mathcal{E}_A(\Omega) = \Gamma(\Omega, \mathcal{E}_A)$, a function $F_u : \Omega \longrightarrow \mathcal{E}_A$ is defined, associating to $x \in \Omega$ the germ u_x of u at x. We denote by $\tilde{\Omega}_u$ <u>the connected component</u> of $F_u(\Omega)$ in $\mathcal{E}_A$, and let

$\omega : \tilde{\Omega}_u \longrightarrow \mathbb{R}^n$ be the natural projection induced by the projection $\pi : \mathcal{E}_A \longrightarrow \mathbb{R}^n$.

<u>Proposition.</u>

$\tilde{\Omega}_u$, <u>with the topology induced by</u> $\mathcal{E}_A$, <u>is a connected Hausdorff topological space and</u> ω <u>is a local homeomorphism.</u>

Then, via ω, $\tilde{\Omega}_u$ inherits a differentiable (real analytic) structure (which makes of ω a local diffeomorphism). For $\alpha \in \tilde{\Omega}_u$, we set $U(\alpha) = \alpha(\omega(\alpha)) \in \mathbb{C}^p$, obtaining a smooth function on $\tilde{\Omega}_u$ that extends the function $\omega^* u$ defined on $F_u(\Omega)$ and that satisfies the differential equation $(\omega^* A)U = 0$ on $\tilde{\Omega}_u$.

The space $\omega : \tilde{\Omega}_u \longrightarrow \mathbb{R}^n$ is a Riemann domain and it represents a maximal completion of u in the following sense:

Proposition.

Let $\hat{\Omega} \xrightarrow{\hat{\omega}} \mathbb{R}^n$ be a connected Riemann domain and assume that there is a section $\hat{F}_u : \Omega \to \hat{\Omega}$ and a function $\hat{U} : \hat{\Omega} \to \mathbb{C}^p$ such that

$$(\hat{\omega}^* A)\hat{U} = 0 \quad \text{on} \quad \hat{\Omega}$$

$$\hat{U}\Big|_{\hat{F}_u(\Omega)} = \hat{\omega}^* u$$

(we say that $(\hat{\Omega}, \hat{\omega}, \hat{F}_u)$ is a completion of u). Then there is a unique natural map $\lambda : \hat{\Omega} \to \tilde{\Omega}_u$ that makes the following diagram commute:

$$\begin{array}{ccc} \hat{\Omega} & \xrightarrow{\lambda} & \tilde{\Omega}_u \\ \hat{\omega} & & \omega \\ & \mathbb{R}^n & \end{array}$$

The map λ induces an isomorphism $\hat{F}_u(\Omega) \cong F_u(\Omega)$ for which $\hat{U} = \lambda^* U$.

β. Envelope.

If $\mathcal{S}$ is a subset of $\Gamma(\Omega, \mathcal{E}_A)$, we replace in the previous considerations the sheaf $\mathcal{E}_A$ by the sheaf $\mathcal{E}_A^{\mathcal{S}} \subset \prod_{\sigma \in \mathcal{S}} (\mathcal{E}_A)_\sigma$.

Note that an element $\varphi \in \mathcal{E}_A^{\mathcal{S}}$ is now a collection $\varphi = (f_{\sigma, x_0})_{\sigma \in \mathcal{S}}$ of germs $f_{\sigma, x_0} \in \mathcal{E}_{A, x_0}$ all defined in a sufficiently small but common neighborhood of x_0 in $\mathbb{R}^n$. Because the inclusion $\mathcal{E}_A^{\mathcal{S}} \hookrightarrow \prod_{\sigma \in \mathcal{S}} (\mathcal{E}_A)_\sigma$ is continuous, $\mathcal{E}_A^{\mathcal{S}}$ has a Hausdorff topology.

We consider the map $F_{\mathcal{S}} : \Omega \to \mathcal{E}_A^{\mathcal{S}}$ given by $F_{\mathcal{S}}(x) = \{\sigma_x \mid \sigma \in \mathcal{S}\}$.

Let $\tilde{\Omega}_{\mathcal{S}}$ be the connected component of $F_{\mathcal{S}}(\Omega)$ in $\mathcal{E}_A^{\mathcal{S}}$. Denoting by ω the restriction to $\tilde{\Omega}_{\mathcal{S}}$ of the natural projection $\pi : \mathcal{E}_A^{\mathcal{S}} \to \mathbb{R}^n$ we obtain a Riemann domain $\omega : \tilde{\Omega}_{\mathcal{S}} \to \mathbb{R}^n$ with the property:

there exists a section $F_{\mathcal{S}} : \Omega \to \tilde{\Omega}_S$ such that

$\forall g \in \mathcal{S}$ there exists a real analytic function $G = G_g$ on $\tilde{\Omega}_{\mathcal{S}}$ such that:

(i) $(\omega^* A)G = 0$ on $\tilde{\Omega}_{\mathcal{S}}$

(ii) $G\big|_{F_{\mathcal{S}}(\Omega)} = \omega^* g$

A Riemann domain $\hat{\Omega}_{\mathcal{S}} \xrightarrow{\hat{\omega}} \mathbb{R}^n$ endowed with a section $\hat{F}_{\mathcal{S}} : \Omega \to \hat{\Omega}_{\mathcal{S}}$ such that $\forall \; g \in \mathcal{S}$ there is a function $\hat{G}_g : \hat{\Omega}_{\mathcal{S}} \to \mathbb{C}^p$ such that (i) and (ii) are satisfied is called an $\mathcal{S}$ - completion of Ω.

<u>Proposition.</u>

<u>For every</u> $\mathcal{S}$ - <u>completion</u> $(\hat{\Omega}_{\mathcal{S}} \to \mathbb{R}^n, \hat{F}_u, \{\hat{G}_g\}_{g \in \mathcal{S}})$ <u>of</u> Ω <u>there exists a uniquely defined natural map</u> $\lambda : \hat{\Omega}_{\mathcal{S}} \to \tilde{\Omega}_{\mathcal{S}}$ <u>such that the diagram</u>

$$\begin{array}{ccc} \hat{\Omega}_{\mathcal{S}} & \longrightarrow & \tilde{\Omega}_{\mathcal{S}} \\ & {\scriptstyle \hat{\omega}} \searrow \quad \swarrow {\scriptstyle \omega} & \\ & \mathbb{R}^n & \end{array}$$

<u>commutes and moreover</u> $\lambda^* G_g = \hat{G}_g \;\; \forall \, g \in \mathcal{S}$.

In other words: $(\tilde{\Omega}_{\mathcal{S}} \xrightarrow{\omega} \mathbb{R}^n, F_{\mathcal{S}}, \{G_g\}_{g \in \mathcal{S}})$ <u>is the maximal</u> $\mathcal{S}$ -<u>completion of</u> Ω.

When $\mathcal{S} = \Gamma(\Omega, \mathcal{E}_A)$ we say that $\tilde{\Omega} = \tilde{\Omega}_{\Gamma(\Omega, \mathcal{E}_A)} \xrightarrow{\omega} \mathbb{R}^n$ is the "<u>envelope of regularity</u>" of Ω.

If $\mathcal{S} \subset \mathcal{S}' \subset \Gamma(\Omega, \mathcal{E}_A)$ we have a natural map $\mathcal{E}_A^{\mathcal{S}'} \to \mathcal{E}_A^{\mathcal{S}}$ that induces a morphism of domination between the $\mathcal{S}$ and $\mathcal{S}'$ envelopes of Ω:

$$\begin{array}{ccc} \tilde{\Omega}_{\mathcal{S}'} & \longrightarrow & \tilde{\Omega}_{\mathcal{S}} \\ & {\scriptstyle \omega'} \searrow \quad \swarrow {\scriptstyle \omega} & \\ & \mathbb{R}^n & \end{array}$$

Hence the envelope of regularity dominates every $\mathcal{S}$ - envelope of Ω, and every $\Gamma(\Omega, \mathcal{E}_A)$ - completion of Ω factors through it.

<u>Definition.</u>

$\mathbb{R}^n$ <u>is a domain of regularity</u> (<u>for</u> $A(x,D)$) <u>if the map</u> $F_{\Gamma(\Omega, \mathcal{E}_A)} : \Omega \to \tilde{\Omega}$ <u>is an isomorphism.</u>

Let Ω be an open set in $\mathbb{R}^n$ and let D be a connected open set $\subset \Omega$. We say that a domain $D' \subset R^n$ is a $\mathcal{E}_A(\Omega)|D$ - completion of D if

$$\text{Image} \; \{ r_D^{D'} : \mathcal{E}_A(D') \to \mathcal{E}_A(D) \} \supset \mathcal{E}_A(\Omega)|D \, .$$

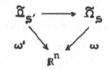

Proposition.

$\Omega \subset \mathbb{R}^n$ is a domain of regularity if and only if $\forall$ domain $D \subset \Omega$ and every $\mathcal{E}_A(\Omega)|D$ completion D' of D, we have $D' \subset \Omega$.

Remark.

Completions for differential operators other than the Cauchy Riemann operators have not been studied very much till now. It seems however that this theory presents very interesting problems. The case of harmonic functions should be of relevant phisical significance (electromagnetic stationary fields, stationary currents, periods, etc.)

Remark.

In the following we will assume that the coefficients of the operators considered are real analytic and therefore it will be possible, by adding to the operator $A(x,D)$ the Cauchy Riemann equations, to reduce to the situation in which $\mathbb{R}^n$ is substituted by $\mathbb{C}^n$ and we are dealing with spaces of holomorphic functions satisfying additional partial differential equations (with holomorphic coefficients).

3. Convexity Theory for operators with Constant Coefficients.

We treat first the case of constant coeffcients, in which our results are better formulated.

Let $A(D)$ be an elliptic differential operator $A(D): \mathcal{E}^p(\Omega) \to \mathcal{E}^q(\Omega)$ (Ω open $\subset \mathbb{R}^n$) with constant coefficients. We set for K compact in Ω and $u \in \mathcal{E}^s(\Omega)$

$$\|u\|_K = \sup_K |u| \quad \text{where } |u| \text{ denotes any fixed norm in } \mathbb{C}^s .$$

Given K compact $\subset \Omega$, $c \geq s$, we set

$$\hat{K}_\Omega(c) = \{x \in \Omega \mid |u(x)| \leq c\|u\|_K \quad \forall u \in \mathcal{E}_A(\Omega)\} .$$

Remark.

If $\mathcal{E}_A(\Omega)$ is an algebra ($p = 1$) then $\hat{K}_\Omega(c) = \hat{K}_\Omega(1)$ $\forall c \geq 1$.
Indeed, if $x \in \hat{K}_\Omega(c)$, then $\forall u \in \mathcal{E}_A(\Omega)$ we have:

$$|u^\ell(x)| \leq c\|u^\ell\|_K \iff |u(x)| \leq c^{1/\ell}\|u\|_K \quad \forall u \in \mathcal{E}_A(\Omega)$$

letting $\ell \to +\infty$ we obtain $|u(x)| \leq \|u\|_K \; \forall u \in \mathcal{E}_A(\Omega)$, i.e. $x \in \hat{K}_\Omega(1)$. Hence $\hat{K}_\Omega(1) \subset \hat{K}_\Omega(c)$ and because the opposite inclusion is trivial we have equality.

Let us consider the following conditions:

$(K)_\Omega$: $\forall$ K compact $\subset \Omega$, $\forall\, c \geq 1$, $\hat{K}_\Omega(c)$ is compact

$(D)_\Omega$: $\forall$ divergent sequence $\{x_n\} \subset \Omega$ there is $u \in \mathcal{E}_A(\Omega)$ such that

$$\sup_n |u(x_n)| = +\infty .$$

Theorem.

If A(D) is elliptic, $(K)_\Omega$ and $(D)_\Omega$ are equivalent.

Proof.

It is obvious that $(D)_\Omega \Rightarrow (K)_\Omega$. We prove the opposite implication by contradiction, assuming that for a divergent sequence $\{x_n\} \subset \Omega$ we have $\sup_n |u(x_n)| < +\infty \;\; \forall u \in \mathcal{E}_A(\Omega)$. Then we set $\mathbb{A} = \{u \in \mathcal{E}_A(\Omega) \mid |u(x_n)| \leq 1 \;\forall n\}$

This is a closed convex circled subset of $\mathcal{E}_A(\Omega)$ and under the assumption of the theorem $\mathcal{E}_A(\Omega) = \bigcup_1^\infty m\mathbb{A}$. By Baire's cathegory argument it follows that $\mathbb{A}$ is a neighborhood of 0 in $\mathcal{E}_A(\Omega)$ and hence contains a set of the form $\{u \in \mathcal{E}_A(\Omega) \mid \|u\|_K \leq \varepsilon\}$ for some $\varepsilon > 0$, K compact $\subset \Omega$. Then one readly verifies that $\hat{K}_\Omega(\frac{1}{\varepsilon}) \supset \{x_n\}$, contradicting $(K)_\Omega$.

Calling "A-convex" open sets Ω for which $(K)_\Omega$ holds, the theorem above says that "A-convex" open sets are regularity domains.

The converse is not true: for $A = \begin{pmatrix} \partial/\partial x_1 \\ \vdots \\ \partial/\partial x_n \end{pmatrix}$ we have that $\mathbb{R}^n$, which is obviously a regularity domain, is not A- convex.

We have:

Theorem.

If $\mathbb{R}^n$ is A-convex, then for Ω to be a domain of regularity it is necessary and sufficient that Ω is A - convex.

Remark.

Let $\{V_1,\dots,V_k\}$ be the characteristic varieties of A, $V = V_1 \cup \dots \cup V_k$. Then $\mathbb{R}^n$ is A-convex if the convex cone generated by $\{\mathrm{Re}\, a | a \in V\}$ is $\mathbb{R}^n$. Viceversa, for $\mathbb{R}^n$ to be A - convex, it is necessary that $\{\mathrm{Re}\, a \mid a \in V\}$ contains a basis of $\mathbb{R}^n$ as a vector space when we assume that $p = 1$ and the ideal generated by by the symbols of the entires of A coincides with its radical.

(Easy examples show that no better conditions involving only V can be given).

The proof of the theorem above is close to the proof of the analogous theorem in complex analysis: first of all we notice that for an elliptic operator $A(x,D)$ (even with variable coefficients) we can find a function

$\rho_\Omega : \Omega \to \mathbb{R}^+ = \{t > 0\}$, lower semicontinuous and with the properties:

1) $\rho_\Omega(x) \le \mathrm{dist}(x,\partial\Omega) \quad \forall x \in \Omega$

2) $\forall x_0 \in \Omega$ and $\forall u \in \mathcal{E}_A(\Omega)$ the Taylor series of u centered at x_0 converges in the polycilinder centered at x_0 of radius $\rho_\Omega(x_0)$.

Such a function will be called a "pseudodistance from the boundary".

Then one proves (same proof of the analogous statement in complex analysis):

Theorem.

If Ω is a regularity domain, then $\rho_\Omega(K) = \rho_\Omega(\hat{K}_\Omega(c))$ $\forall K$ compact $\forall c > 1$. $\left(\rho_\Omega(C) = \inf_C \rho_\Omega(x)\right)$.

We can use pseudo-distances to obtain a characterization of domains of regularity without the restriction of $\mathbb{R}^n$ being A - convex: first we notice that the conditions:

(i) $\forall K$ compact $\subset \Omega$, $\forall c \ge 1$, $\rho_\Omega(\hat{K}_\Omega(c)) > 0$.

(ii) $\forall \{x_n\} \subset \Omega$ such that $\lim_n \rho_\Omega(x_n) = 0$ there is $u \in \mathcal{E}_A(\Omega)$ with $\sup_n |u(x_n)| = \infty$ are equivalent.

Then one obtains:

Theorem.

Ω is a domain of regularity iff the equivalent conditions (i) and (ii) hold.

4. Convexity Theory for Operators with Variable Coefficients.

In the case of variable coefficients we need to introduce: for $C \geq 1$, $L \geq 1$

$$\hat{K}_\infty(L,c) = \{x \in \Omega \mid |u(x)|_m \leq L\, c^m \|u\|_{K,m}\ \forall u \in \mathcal{E}_A(\Omega), \forall m \in \mathbb{N}\}$$

where $|u(x)|_m = \sum_{|\alpha| \leq m} |D^\alpha u(x)|$

$$\|u\|_{K,m} = \sup_K |u(x)|_m .$$

Remark.

If $A(x,D)$ has constant coefficients, then $\hat{K}_\infty(L,c) = \hat{K}_\infty(L,1) = \hat{K}_\Omega(L)\ \forall L,c \geq 1$

With a pseudodistance ρ_Ω we set $\delta_\Omega(x) = \inf\{1, \rho_\Omega(x)\}$. Then we obtain:

Proposition.

If Ω is a domain of regularity for $A(x,D)$, then:

$\forall K$ compact $\subset \Omega$, $\forall L,\ c \geq 1$ we have:

$$\delta_\Omega(\hat{K}_\infty(L,c)) \geq \frac{1}{c}\, \delta_\Omega(K) > 0 .$$

and hence:

Theorem.

If $\mathbb{R}^n$ is $A(x,D)$ - convex (i.e. $\hat{K}_\infty(L,c) \subset\subset \mathbb{R}^n\ \forall K \subset\subset \mathbb{R}^n$)

then $\Omega \subset \mathbb{R}^n$ is a domain of regularity if and only if is $A(x,D)$-convex.

Set $\tau(\varepsilon,x,u) = \sup_\alpha \varepsilon^{|\alpha|} \dfrac{|D^\alpha u(x)|}{\alpha!}$, $u \in \mathcal{E}_A(\Omega)$, $\varepsilon > 0$

Then the following conditions are equivalent:

1) Ω is a domain of regularity

2) $\forall K \subset\subset \Omega$, $\forall L,\ c \geq 1$, $\rho_\Omega(\hat{K}_\infty(L,c)) > 0$.

3) $\forall$ sequence $\{x_n\} \subset \Omega$ such that $\rho_\Omega(x_n) \to 0$, $\forall \varepsilon > 0$ there is $u \in \mathcal{E}_A(\Omega)$ such that

$$\sup_n \tau(\varepsilon, x_n, u) = +\infty .$$

(Generalized Cartan-Thullen theorem).

Remark.

It is an open question whether these statements can be improved to obtain statements closer to those obtained for operators with constant coefficients.

5. Geometrical Properties of Regularity Domains.

α. Let $A(D)$ be a differential operator with constant coefficients in $\mathbb{R}^n$; we let $A(D)$ operate on holomorphic (vector valued) functions on $\mathbb{C}^n$ and we denote by $\mathcal{O}_A$ the sheaf of germs of holomorphic solutions $u \in \mathcal{O}^p$ of $A(D)u = 0$.

Let $W = W_1 \cup \ldots \cup W_k$ the asymptotic variety of $A(D)$ $(W \subset \mathbb{C}^n)$, and denote by $\mathcal{W}$ the associated algebraic variety in $\mathbb{P}_{n-1}(\mathbb{C})$.

Let Ω be an open set in $\mathbb{C}^n$. Let $z_\circ \in \partial\Omega$. If the boundary is of class C^1 at $z_\circ$ and φ is a differentiable function at $z_\circ$ such that locally Ω has the equation $\varphi < 0$, we notice that the vector:

$$\frac{\partial\varphi}{\partial z}(z_\circ) = \left(\frac{\partial\varphi}{\partial z_1}(z_\circ), \ldots, \frac{\partial\varphi}{\partial z_n}(z_\circ)\right) \neq 0$$

is uniquely determined by Ω and $z_\circ$ up to multiplication by a positive constant and thus defines a unique element $\theta_{z_\circ} \in \mathbb{P}_{n-1}(\mathbb{C})$.

We say that $\partial\Omega$ is noncharacteristic for A at $x_\circ$ if $\theta_{x_\circ} \notin \mathcal{W}$, and is characteristic if $\theta_{z_\circ} \in \mathcal{W}$.

Extension lemma.

Assume $\partial\Omega$ is C^2 at $z_\circ$ and that $\theta_{z_\circ} \notin \mathcal{W}$, i.e. that $\partial\Omega$ is noncharacteristic for A at $z_\circ$. Then there is a neighborhood $V(z_\circ)$ of $z_\circ$ in $\mathbb{C}^n$ such that every $u \in \Gamma(\Omega, \mathcal{O}_A)$ extends uniquely to an element $\tilde{u} \in \Gamma(\Omega \cup V(z_\circ), \mathcal{O}_A)$.

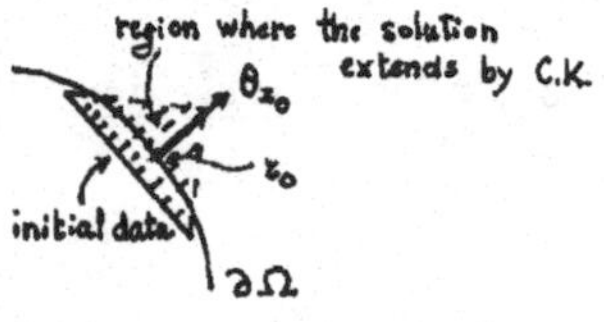

The proof uses essentially the theorem of Cauchy Kowalewska (see preceding lectures) using the fact that the domains to which the solution extends have the form shown in the figure.

Corollary.

Let Ω be a domain of regularity for $A(D) \oplus \bar\partial$. Then at every point $z_\circ \in \partial\Omega$ where $\partial\Omega \in C^2$, $\theta_{z_\circ} \in W$.

Corollary.

$\Omega \subset \mathbb{C}^n$, bounded regularity domain, cannot have a boundary every-where of class C^2 unless $A(D) = 0$ and the system reduces to $\bar\partial$.

β. Envelopes of Regularity of convex sets.

For $a \in \mathbb{C}^n$, we set $\pi_a : \mathbb{C}^n \to \mathbb{C}$ by $\pi_a(z) = \sum_1^n a_j z_j$.

We have the following:

Theorem.

Let Ω be a convex open set $\subset \mathbb{C}^n$. Then $\tilde{\Omega} = \overbrace{\bigcap_{a \in W-\{0\}} \pi_a^{-1}(\pi_a(\Omega))}^{\circ}$ (*).

Proof.

First of all we notice that $\tilde{\Omega} \subset \mathbb{C}^n$ by the density theorem for convex sets given in the first lecture.

Then we notice that the right hand side of (*) is a regularity domain containing Ω and hence is larger than $\tilde{\Omega}$. This fact follows because a half space $\Sigma = \{\mathrm{Re}\langle z,a\rangle < c\}$ is either of regularity or its envelope $\tilde{\Sigma}$ is $\mathbb{C}^n$.

Then we notice that the set of points $z \in \partial\tilde{\Omega}$ such that $\tilde{\Omega}$ contains a small ball B with $z \in \partial B$ is dense in $\partial\tilde{\Omega}$ and that by the extension lemma the holomorphic components of the exterior normal to B at z must belong to $W - \{0\}$. From this it follows that $\tilde{\Omega}$ is intersection of half spaces of the form $\mathrm{Re}\langle a,z\rangle < \mathrm{const}$ with $a \in W - \{0\}$ and hence must contain the set in the right hand side of (*), from which one obtains equality.

γ. Properties of the distance from the boundary.

Remark 1.

For $B = \{|z| < 1\}$, we have $\tilde{B} = \overbrace{\{|\sum a_i z_i| < |a| \ \forall\ a \in W - \{0\}\}}^{\circ}$.

Remark 2.

Let $f: \Omega \to \mathbb{R}$, Ω open $\subset \mathbb{R}^N$. Let $x_\circ \in \Omega$ and let f be differentiable at $x_\circ$. If $f(x) - f(x_\circ) \leq \langle \xi, x - x_\circ\rangle + \sigma(|x-x_\circ|)$ in a nbd of $x_\circ$ in

Ω, then $\xi = \text{grad}_x f(x_0)$.

Remark 3.

For Ω open in $\mathbb{C}^n = \mathbb{R}^N$, the function $\text{dist}(z, C\Omega)$ is Lipschitz continuous on Ω .

Remark 4. (Rademaker's theorem)

A Lipschitz continuous function is differentiable almost everywhere.

From these remarks we obtain the following:

Corollary.

Let Ω be a domain of regularity; then

$$\text{grad}_z(\text{dist}(z, C\Omega)) \in W \quad \text{almost everywhere.}$$

Indeed: let $z_0 \in \Omega$, $\delta = \text{dist}(z_0, C\Omega)$ and let $w \in \overline{B(z_0,\delta)} \cap \partial\Omega$. For $z \in B(z_0,\delta)$ we have :

$$d(z,\partial\Omega)^2 \leq d(z,w)^2 = |z-z_0|^2 + |w-z_0|^2 + 2\,\text{Re}\langle z-z_0, w-z_0\rangle .$$

Thus

$$d(z,\partial\Omega)^2 - d(z_0,\partial\Omega)^2 \leq 2\,\text{Re}\langle w-z_0, z-z_0\rangle + |z-z_0|^2$$

and hence from remark 2 we obtain that $\overline{w-z_0} = \text{grad}_z\, d(z,\partial\Omega)\big|_{z=z_0}$ if $d(z,\partial\Omega)$ is differentiable at z_0 .

Corollary.

A regularity domain Ω in $\mathbb{C}^n$ is a domain of holomorphy on which we have moreover

$$\text{grad}_z(\text{dist}(z, C\Omega) \in W \quad \text{a.e. on } \Omega .$$

Remark.

On easily obtains, with $\mathfrak{A}$ = ideal of polynomials vanishing on W, $\forall g \in \mathfrak{A}$, $\forall \mu_1,\dots,\mu_n \in \mathbb{C}$, setting

$$L = \sum_{\alpha=1}^{n} \frac{\partial g}{\partial \xi_\alpha} (\text{grad}_z\, d(z_0, C\Omega)) \frac{\partial}{\partial z_\alpha} + \sum_{\beta=1}^{n} \mu_\beta \frac{\partial}{\partial \bar{z}_\beta}$$

one has:

$$L\bar{L}\, \text{dist}(z, C\Omega)\big|_{z=z_0} \geq 0$$

(where $\text{dist}(z, C\Omega) \in C^2$) that generalizes the Levi condition for $\bar{\partial}$.

Levi problem for system of operators with constant coefficients.

Are the necessary conditions of the corollary above also sufficient for to be a domain of regularity?

This is an open (and apparently difficult) question.

* * *

LECTURE 5.

BOUNDARY COMPLEXES.

1. The Local Situation.

Let $A(x,D): \mathcal{E}^p(\Omega) \to \mathcal{E}^q(\Omega)$ be a differential operator with C^∞ coefficients on Ω open $\subset \mathbb{R}^n$.

Assume that we have chosen integers $a_1,\dots,a_p$ for $\mathcal{E}^p(\Omega)$, $b_1,\dots,b_q$ for $\mathcal{E}^q(\Omega)$ such that $A(x,D) = (a_{ij}(x,D))_{1\le i\le q, 1\le j\le p}$ with $a_{ij}(x,D) = \sum_{|\alpha|\le a_j-b_i} a_{ij\alpha}(x)D^\alpha$ an operator of order $\le a_j - b_i$ (note that the integers a_j,b_i are defined modulo the addition of an integer k).

Then we say that (a_j,b_i) is a multigrading for $A(x,D)$ and define the "symbol of $A(x,D)$ for the multigrading (a_j,b_i)" by

$$\hat{A}(x,\xi) = (\hat{a}_{ij}(x,\xi)) \quad \text{where} \quad \hat{a}_{ij}(x,\xi) = \sum_{|\alpha|=a_j-b_i} a_{ij\alpha}(x)\,\xi^\alpha$$

(matrix of polynomials in the variables $\xi = (\xi_1,\dots,\xi_n) \in \mathbb{C}^n$ and coefficients in $\mathcal{E}(\Omega)$).

Given a second differential operator $B(x,D): \mathcal{E}^q(\Omega) \to \mathcal{E}^r(\Omega)$, we fix a third sequence of integers $c_1,\dots,c_r$ for $\mathcal{E}^r(\Omega)$ in such a way that $B(x,D)$ has multigrading (b_i,c_h). Then $B(x,D)\circ A(x,D): \mathcal{E}^p(\Omega) \to \mathcal{E}^r(\Omega)$ has multigrading (a_j,c_h) and we have (multiplicative property of symbols):

$$\widehat{B\circ A}(x,\xi) = \hat{B}(x,\xi)\hat{A}(x,\xi) .$$

2. Differential Operators between vector bundles.

Let X be a differentiable manifold of pure dimension n and let $E \xrightarrow{\pi} X$, $F \xrightarrow{\mu} X$ be vector bundles with fibres modeled respectively on $\mathbb{C}^p$, $\mathbb{C}^q$. We say that E (F) is a vector bundle on X of rank p (q) .

Let $\mathfrak{U} = \{U_i\}$ be an atlas on X such that $E|_{U_i}$, $F|_{U_i}$ are trivial $\forall$ i. We fix trivializations: $E|_{U_i} = U_i \times \mathbb{C}^p$, $F|_{U_i} = U_i \times \mathbb{C}^q$ and consequently transition functions

$$e_{ij} : U_i \cap U_j \to GL(p,\mathbb{C}), \quad f_{ij} : U_i \cap U_j \to GL(q,\mathbb{C})$$

for the bundles E and F respectively:

$$e_{ij}\, e_{jk} = e_{ik} \quad \text{and} \quad f_{ij}\, f_{jk} = f_{ik} \quad \text{on} \quad U_i \cap U_j \cap U_k .$$

A section $s: X \to E$ is represented in the local trivializations by $(x, s_i(x))$, $x \in U_i$, $s_i(x) \in \mathbb{C}^p$, so that $s_i \in \mathcal{E}^p(U_i)$ and

$$s_i = e_{ij}\, s_j \quad \text{on} \quad U_i \cap U_j .$$

A differential operator between the bundles E and F is a linear map

$$A(x,D) : \Gamma(X,E) \to \Gamma(X,F)$$

(where $\Gamma(X,E)$, $\Gamma(X,F)$ represent C^∞ sections of E,F respectively) such that

1) $A(x,D) : \Gamma(X,E) \to \Gamma(X,F)$ is continuous for the Schwartz topologies.

2) $A(x,D)$ is local, i.e. $\forall\, s \in \Gamma(X,E)$, $\operatorname{supp} A(x,D)s \subset \operatorname{supp} s$.

By a theorem of Peetre this is equivalent to give, for every i , a differential operator

$$A^{(i)}(x,D) : \mathcal{E}^p(U_i) \to \mathcal{E}^q(U_i)$$

such that the diagrams

$$\begin{array}{ccc}
\mathcal{E}^p(U_i) & \xrightarrow{A^{(i)}(x,D)} & \mathcal{E}^q(U_i) \\
\downarrow e_{ji} & & \downarrow f_{ji} \\
\mathcal{E}^p(U_j) & \xrightarrow{A^{(j)}(x,D)} & \mathcal{E}^q(U_j)
\end{array}$$

commute where they are defined. If $A^{(i)} = \sum a_\alpha^{(i)}(x)\, D^\alpha$, from the identity $A^{(i)} e_{ij}\, s_j = f_{ij}\, A^{(j)} s_j$ on $U_i \cap U_j$ $\forall\, s_j \in \mathcal{E}^p(U_j)$ one obtains:

$$\sum a^{(i)}_{\alpha}(x) \binom{\alpha}{\beta} D^{\alpha-\beta} e_{ij}(x) = f_{ij}\, a^{(j)}_{\beta}(x).$$

Gradings. A grading on the bundle E is by definition the assignement, for a given covering $\{U_i\}=\mathcal{U}$ and trivializations $E|U_i \cong U_i \times \mathbb{C}^p$, of a grading $a_1^{(i)},\dots,a_p^{(i)}$ for $\mathcal{E}^p(U_i) \cong \Gamma(U_i,E)$ such that the transition functions $e_{ij}=(e_{ij,rs}(x))_{r,s=1,\dots,p}$ satisfy $e_{ij,rs}=0$ whenever $a_s^{(j)} - a_r^{(i)} \neq 0$.

Example: when $a_j^{(i)} = a$ is the same for $j=1,\dots,p$, we say that the grading is classical: in this case the condition on the transition functions is empty.

One proves the following statement about the structure of graded bundles:

Proposition. Let X be a connected manifold and let E be a graded vector bundle on X. Then E splits into a direct sum $E = E_1 \oplus E_2 \oplus \dots \oplus E_h$ of vector bundles on each of which a classical grading is given.

Let $A(x,D): \Gamma(X,E) \longrightarrow \Gamma(X,F)$ be a differential operator between vector bundles E and F of rank p and q respectively and let $a_1^{(i)},\dots,a_p^{(i)}$ be a grading on E and let $b_1^{(i)},\dots,b_q^{(i)}$ be a grading on F, compatible with the operator A(x,D). Then for each index i we can consider the symbol $\hat{A}^{(i)}(x,\xi)$. We have the formula:

$$(*) \qquad \hat{A}^{(i)}(x,\xi)\, e_{ij}(x) = f_{ij}\, \hat{A}^{(j)}(x,\xi)\ ,$$

while a change of the x coordinates affects $\xi=(\xi_1,\dots,\xi_n)$ as the components of a covariant vector. Thus (x,ξ) can be tought of as a point in the cotangent bundle $T^*(X)$. We also consider the bundle $\mathrm{Hom}_X(E,F)$: a section $\sigma \in \mathrm{Hom}_X(E,F)$ defines on each U_i a matrix $M_i(x)$, of type $q \times p$, with C^∞ entries, such that

$$f_{ij}(x)\, M_j(x) = M_i(x)\, e_{ij}(x) \quad \text{on} \quad U_i \cap U_j .$$

Hence by (*) we can define the symbol of the differential operator A(x,D) for the given grading as a map $\hat{A}(x,\xi): T^*(X) \longrightarrow \mathrm{Hom}_X(E,F)$ making the following diagram commute:

$$\begin{array}{ccc} T^*(X) & \xrightarrow{\ \hat{A}(x,\xi)\ } & \mathrm{Hom}_X(E,F) \\ & \searrow \quad \swarrow & \\ & X & \end{array}$$

Let us denote by $\mathcal{P}(X)$ the ring of C^∞ functions on $T^*(X)$ that are polynomials along the fibres: if $T^*(X)|U_i \cong U_i \times \mathbb{R}^n$ is a trivialization of $T^*(X)$, denoting by $x^{(i)}=(x_1^{(i)},\dots,x_n^{(i)})$ coordinates on U_i and $\xi^{(i)} = (\xi_1^{(i)},\dots,\xi_n^{(i)})$ coordinates along the fibres, an element $p(x,\xi) \in \mathcal{P}(X)$ is a collection of polynomials $p_i(x^{(i)},\xi^{(i)})$ with smooth coefficients on U_i and such that on $U_i \cap U_j$ one has:

$$p_j(x^{(j)},\xi^{(j)}) = p_i(x^{(i)}(x^{(j)}),\ (\partial x^{(j)} / \partial x^{(i)})\, \xi^{(j)}).$$

The space $\mathcal{P}(X)$ is "the ring of codifferential symmetric forms". When X is parallelizable, $T^*(X) \cong X \times \mathbb{R}^n$ and $\mathcal{P}(X) \cong \mathcal{E}(X)[\xi_1,\ldots,\xi_n]$ (polynomials with coefficients in $\mathcal{E}(X)$).

Given a vector bundle E on X, the tensor product $\mathcal{P}(X) \otimes_{\mathcal{E}(X)} \Gamma(X,E)$ defines a $\mathcal{P}(X)$-module that we denote by $\mathcal{P}(X,E)$. An element $\varphi \in \mathcal{P}(X,E)$ is defined in a trivialization of E by a collection $\varphi_i(x,\xi) = {}^t(\varphi_i^1(x,\xi),\ldots,\varphi_i^p(x,\xi))$ where the $\varphi_i^j(x,\xi)$ are codifferential symmetric forms on U_i, and $e_{ij}\varphi_j = \varphi_i$ on $U_i \cap U_j$.

Given gradings on two bundles E and F over X and a differential operator compatible with the gradings: $A(x,D)\colon \Gamma(X,E) \longrightarrow \Gamma(X,F)$, we can consider its symbol as a $\mathcal{P}(X)$-linear map $\quad \hat{A}(x,\xi)\colon \mathcal{P}(X,E) \longrightarrow \mathcal{P}(X,F)$.

Let G be a third vector bundle of rank r over X, with grading $c_1^{(i)},\ldots,c_r^{(i)}$ and let $B(x,D)\colon \Gamma(X,F) \longrightarrow \Gamma(X,G)$ be a differential operator compatible with the gradings of F and G. Then the symbol $\hat{B}(x,\xi)\colon \mathcal{P}(X,F) \longrightarrow \mathcal{P}(X,G)$ and that of the composed map $B(x,D)\circ A(x,D)\colon \Gamma(X,E) \longrightarrow \Gamma(X,G)$ satisfy the multiplicative property that the following diagram commutes:

$$\begin{array}{ccc} \mathcal{P}(X,E) & \xrightarrow{\hat{A}(x,\xi)} & \mathcal{P}(X,F) \\ & \searrow^{\widehat{B\circ A}(x,\xi)} \quad \swarrow_{\hat{B}(x,\xi)} & \\ & \mathcal{P}(X,G) & \end{array}$$

<u>Complexes.</u> On X we give a (finite) sequence of vector bundles $E^0, E^1, E^2, \ldots$ of rank $p_0, p_1, p_2, \ldots$ respectively and a sequence of differential operators

$$A^j(x,D)\colon \Gamma(X,E^j) \longrightarrow \Gamma(X,E^{j+1}) \quad \text{for } j=0,1,2,\ldots$$

We say that the sequence (here $\mathcal{E}^{(j)}(X) = \Gamma(X,E^j)$)

$$\mathcal{E}^{(0)}(X) \xrightarrow{A^0(x,D)} \mathcal{E}^{(1)}(X) \xrightarrow{A^1(x,D)} \mathcal{E}^{(2)}(X) \longrightarrow \ldots$$

forms a complex if $A^{j+1}\circ A^j = 0$ for $j=0,1,2,\ldots$

Suppose we have given gradings $a_1^{(i)},\ldots,a_{p_0}^{(i)}$ on E^0, $b_1^{(i)},\ldots,b_{p_1}^{(i)}$ on E^1, $c_1^{(i)},\ldots$ $\ldots,c_{p_2}^{(i)}$ on E^2 for a covering $\mathcal{U} = \{U_i\}$ such that for every i and j $E^j|U_i$ is trivial.

Assuming that all gradings are compatible with the differential operators, taking symbols we obtain a complex of $\mathcal{P}(X)$-homomorphisms:

$$\mathcal{P}(X,E^0) \xrightarrow{\hat{A}^0(x,\xi)} \mathcal{P}(X,E^1) \xrightarrow{\hat{A}^1(x,\xi)} \mathcal{P}(X,E^2) \longrightarrow \ldots \qquad \text{(\underline{symbolic complex})}.$$

Given $x_0 \in X$, let $\mathfrak{m}_{x_0}$ be the ideal of $\mathcal{E}(X)$ of functions $f \in \mathcal{E}(X)$ vanishing at x_0; tensoring the sequence above by $\mathcal{E}(X)/\mathfrak{m}_{x_0}$ over the ring $\mathcal{E}(X)$, we obtain a complex of $\mathcal{P} = \mathbb{C}[\xi_1,\ldots,\xi_n]$- homomorphisms:

$$\mathcal{P}^{p_0} \xrightarrow{\hat{A}^0(x_0,\xi)} \mathcal{P}^{p_1} \xrightarrow{\hat{A}^1(x_0,\xi)} \mathcal{P}^{p_2} \longrightarrow \ldots \qquad \text{(\underline{symbolic complex at } x_0)}$$

and finally, for fixed $x_0 \in X$ and $\xi^0 \in \mathbb{R}^n-\{0\}$ on the fiber of $T^*(X)$ over x_0 we obtain another complex

$$0 \longrightarrow \mathbb{C}^{p_0} \xrightarrow{\hat{A}^0(x_0,\xi^0)} \mathbb{C}^{p_1} \xrightarrow{\hat{A}^1(x_0,\xi^0)} \mathbb{C}^{p_2} \longrightarrow \dots$$

We say that the given complex is <u>elliptic</u> at x_0 if this sequence is exact for any choice of ξ^0 in $\mathbb{R}^n-\{0\}$. We say that the given complex is <u>correct</u> at the point x_0 if the complex

$$\mathcal{E}^{p_0}(\mathbb{R}^n) \xrightarrow{\hat{A}^0(x_0,D)} \mathcal{E}^{p_1}(\mathbb{R}^n) \xrightarrow{\hat{A}^1(x_0,D)} \mathcal{E}^{p_2}(\mathbb{R}^n) \longrightarrow \dots$$

or equivalently if the complex

$$\mathcal{P}^{p_0} \xleftarrow{{}^t\hat{A}^0(x_0,\xi)} \mathcal{P}^{p_1} \xleftarrow{{}^t\hat{A}^1(x_0,\xi)} \mathcal{P}^{p_2} \longleftarrow \dots$$

is acyclic. Correctedness is a very important notion in connection with the lemma of Poincaré (local solvability), as we will show in the next lecture.

3. <u>Fiber Transformations and Change of Gradings.</u>

Let M be a differential operator acting from a vector bundle E into itself. Let $a_1^{(i)},\dots,a_p^{(i)}, \alpha_1^{(i)},\dots,\alpha_p^{(i)}$ be gradings on E, thought as source and target space respectively and assume that M is compatible with these gradings. We have the following:

<u>Proposition</u>. <u>If</u> M(x,D) <u>has total degree</u> 0 (i.e. $a_1^{(i)}+\dots+a_p^{(i)}=\alpha_1^{(i)}+\dots+\alpha_p^{(i)}$ $\forall$ i) <u>and on each</u> U_i $\det \hat{M}^{(i)}(x,\xi) = \det \hat{M}^{(i)}(x,0) \neq 0$, <u>then there exists a unique differential operator</u> N(x,D) <u>from</u> E <u>to</u> E, <u>compatible with the gradings</u> $\alpha_1^{(i)},\dots,\alpha_p^{(i)}$, <u>on</u> E <u>as source and</u> $a_1^{(i)},\dots,a_p^{(i)}$ <u>on</u> E <u>as target space, such that</u>

$$N \circ M = M \circ N = \text{identity on } \Gamma(X,E).$$

We call such an M a "fiber transform" of type $(a_j^{(i)}, \alpha_h^{(i)})$ on E. Note that when the gradings on E are classical, then M is simply a matrix of functions in each coordinate patch.

Let us suppose now that we have a complex of differential operators:

$$\mathcal{E}^{(0)}(X) \xrightarrow{A^0(x,D)} \mathcal{E}^{(1)}(X) \xrightarrow{A^1(x,D)} \mathcal{E}^{(2)}(X) \longrightarrow \dots = (\mathcal{E}^*(X), A^*)$$

with $\mathcal{E}^{(j)}(X) = \Gamma(X,E^j)$ for some bundle E^j of rank p_j over X, and that we have gradings $a_1^{(i)},\dots,a_{p_0}^{(i)}$ on E^0, $b_1^{(i)},\dots,b_{p_1}^{(i)}$ on $E^1,\dots$ compatible with the differential operators of the complex.

Given other gradings $\alpha_1^{(i)},\dots,\alpha_{p_0}^{(i)}$ on E^0 with $a_1^{(i)}+\dots+a_{p_0}^{(i)} = \alpha_1^{(i)}+\dots+\alpha_{p_0}^{(i)}$ $\forall$ i

$\beta_1^{(i)},\dots,\beta_{p_1}^{(i)}$ on E^1 with $b_1^{(i)}+\dots+b_{p_1}^{(i)} = \beta_1^{(i)}+\dots+\beta_{p_1}^{(i)}$ $\forall$ i,...

and fiber transforms $M_i:\Gamma(X,E^j) \longrightarrow \Gamma(X,E^j)$ compatible with the gradings introduced above, we set $B^j(x,D) = M_{j+1}(x,D)\circ A^j(x,D)\circ M_j^{-1}(x,D)$.

Then $B^j(x,D)\colon \mathcal{E}^{(j)}(X) \longrightarrow \mathcal{E}^{(j+1)}(X)$ is a differential operator and we have a commutative diagram:

$$\begin{array}{ccccccc}
\mathcal{E}^{(0)}(X) & \xrightarrow{A^0(x,D)} & \mathcal{E}^{(1)}(X) & \xrightarrow{A^1(x,D)} & \mathcal{E}^{(2)}(X) & \longrightarrow & \dots \\
M_0(x,D)\downarrow & & M_1(x,D)\downarrow & & M_2(x,D)\downarrow & & \\
\mathcal{E}^{(0)}(X) & \xrightarrow{B^0(x,D)} & \mathcal{E}^{(1)}(X) & \xrightarrow{B^1(x,D)} & \mathcal{E}^{(2)}(X) & \longrightarrow & \dots
\end{array}$$

in which the vertical arrows represent isomorphisms. The complex in the bottom line, whose operators are compatible with the gradings $\alpha_1^{(i)},\dots,\alpha_{p_0}^{(i)}$ on E^0, $\beta_1^{(i)},\dots,\beta_{p_1}^{(i)}$ on E^1, ..., is called the transformed of the complex in the upper line by the fiber transformations $M_0, M_1, M_2, \dots$

Let us set: $$H^j(X;\mathcal{E}^*(X),A^*) = \frac{\operatorname{Ker}(A^j(x,D)\colon \mathcal{E}^{(j)}(X) \longrightarrow \mathcal{E}^{(j+1)}(X))}{\operatorname{Image}(A^{j-1}(x,D)\colon \mathcal{E}^{(j-1)}(X) \longrightarrow \mathcal{E}^{(j)}(X))}$$

for j=0,1,2,...(the space at the quotient should be interpreted as 0 for j=0). Defining analogous groups $H^j(X;\mathcal{E}^*(X),B^*)$ we obtain:

<u>Proposition.</u> <u>Under the above assumptions</u> $H^j(X;\mathcal{E}^*(X),A^*) \cong H^j(X;\mathcal{E}^*(X),B^*)\ \forall\, j \geq 0$.

Let S be a closed subset of X. We denote by $\mathcal{F}_S^{(j)}(X)$ the space of sections in $\Gamma(X,E^j)$ that are flat on S. We obtain a complex

$$\mathcal{F}_S^{(0)}(X) \xrightarrow{A^0(x,D)} \mathcal{F}_S^{(1)}(X) \xrightarrow{A^1(x,D)} \mathcal{F}_S^{(2)} \longrightarrow \dots$$

whose cohomology groups will be denoted by $H^j(X;\mathcal{F}_S^*,A^*)$ ($j\geq 0$). Taking the quotient we obtain a new complex:

$$\frac{\mathcal{E}^{(0)}(X)}{\mathcal{F}_S^{(0)}(X)} \xrightarrow{A^0(x,D)} \frac{\mathcal{E}^{(1)}(X)}{\mathcal{F}_S^{(1)}(X)} \xrightarrow{A^1(x,D)} \frac{\mathcal{E}^{(2)}(X)}{\mathcal{F}_S^{(2)}(X)} \longrightarrow \dots$$

whose cohomology groups are denoted by $H^j(X;\mathcal{E}^*/\mathcal{F}_S^*, A^*)$.

With obvious notations, we obtain the isomorphisms:

$$H^j(X;\mathcal{F}_S^*,A^*) = H^j(X;\mathcal{F}_S^*, B^*) \qquad (j\geq 0)$$

$$H^j(X;\mathcal{E}^*/\mathcal{F}_S^*,A^*) = H^j(X;\mathcal{E}^*/\mathcal{F}_S^*,B^*) \qquad (j\geq 0).$$

4. <u>Noncharacteristic Hypersurfaces</u>.

Let $\rho\colon X \longrightarrow \mathbb{R}$ be a C^∞ function on X, and let $S = \{x\in X \mid \rho(x)=0\}$. We say that S is a smooth orientable hypersurface if ρ can be chosen in such a way that $d\rho(x) \neq 0$ on S. Then X is divided into two closed distinct regions:

$X^+ = \{x \in X \mid \rho(x) \geq 0\}$ and $X^- = \{x \in X \mid \rho(x) \leq 0\}$.

We say that S is noncharacteristic at $x_0 \in S$ for the complex $(\mathcal{E}^*(X), A^*)$ if the sequence

$$0 \longrightarrow \mathbb{C}^{p_0} \xrightarrow{\hat{A}^0(x_0, \mathrm{grad}\,\rho(x_0))} \mathbb{C}^{p_1} \xrightarrow{\hat{A}^1(x_0, \mathrm{grad}\,\rho(x_0))} \mathbb{C}^{p_2} \longrightarrow \cdots$$

is exact and we say that S is noncharacteristic if it is noncharacteristic at every point. We make the following remarks:

1) The notion of "noncharacteristic" does not depend on the choice of the defining function ρ for S.

2) If the complex $(\mathcal{E}^*(X), B^*)$ is obtained from $(\mathcal{E}^*(X), A^*)$ by fiber transformations, then S is non-characteristic at x_0 for the first complex if and only if S is non-characteristic at x_0 for the second one: "the notion of being non-characteristic is invariant by fiber transformations".

3) If the given complex is elliptic on X, then any given smooth oriented hypersurface is non-characteristic.

5. Formally Non-characteristic Hypersurfaces.

a. Natural Boundary conditions (local situation: Ω open in $\mathbb{R}^n = X$)

One can prove that, by substituting to Ω a smaller open neighborhood of the smooth oriented hypersurface S in Ω one can assume that the defining function ρ satisfies also the condition $|\mathrm{grad}\,\rho(x)| = 1$ on Ω.

(Example: $S = \{\Sigma x_i^2 - 1 = 0\}$ in $\mathbb{R}^n$ is described in $\mathbb{R}^n - 0$ by $\rho(x) = (\Sigma x_i^2)^{1/2} - 1 = 0$).

Then we introduce the differential operators:

$$\partial/\partial\rho = \Sigma\, \partial\rho(x)/\partial x_i \cdot \partial/\partial x_i \qquad \text{(normal derivative)}$$

$$D_{t_j} = \partial/\partial x_j - \partial\rho(x)/\partial x_j \cdot \partial/\partial\rho \ \text{(tangential derivative)} \quad \text{for } j=1,\ldots,n,$$

for which the following relations hold:

$$D_{t_j}(\rho h) = \rho D_{t_j} h \qquad \forall\, h \in \mathcal{E}(\Omega),$$

$$[D_{t_j}, \partial/\partial\rho] = D_{t_j}\, \partial/\partial\rho - \partial/\partial\rho\, D_{t_j} = \sum_i \partial^2\rho/\partial x_i \partial x_j \cdot D_{t_i}$$

$$\Sigma\, \partial\rho/\partial x_j \cdot D_{t_j} = 0$$

and for the (oriented) Euclidean element of hypersurface area dS we have

$$\partial\rho/\partial x_j\, dS = (-1)^{j-1} dx_1 \wedge dx_2 \wedge \ldots \wedge \widehat{dx_j} \wedge \ldots \wedge dx_n \mid S$$

$$dS = \Sigma\, (-1)^{j-1} \partial\rho/\partial x_j \; dx_1 \wedge dx_2 \wedge \ldots \wedge \widehat{dx_j} \wedge \ldots \wedge dx_n \mid S.$$

Given a differential operator $A(x,D) = \sum a_\alpha D^\alpha : \mathcal{E}^p(\Omega) \longrightarrow \mathcal{E}^q(\Omega)$ we define the (formal) adioint operator $A^*(x,D): \mathcal{E}^q(\Omega) \longrightarrow \mathcal{E}^p(\Omega)$ by

$$A^*(x,D) = \sum (-1)^{|\alpha|} D^\alpha({}^t a_\alpha(x)\, D^\alpha \cdot\,) .$$

One has

$$\int_\Omega {}^t v \cdot A(x,D)u \, dx = \int_\Omega {}^t(A^*(x,D)\, v)\cdot u \, dx \qquad \forall\, u\in\mathcal{D}^p(\Omega),\ \forall\, v\in\mathcal{D}^q(\Omega).$$

For the adjoint of $\partial/\partial\rho$ we have:

$$\nabla_\rho = -(\partial/\partial\rho)^* = \partial/\partial\rho + \Delta\rho \quad \text{with } \Delta\rho = \sum \partial^2\rho/\partial x_i^2 .$$

We have also

$${}^t(\partial v/\partial\rho)\cdot u = \sum \partial/\partial x_i (\partial\rho/\partial x_i \; {}^t v\cdot u) - {}^t v\cdot\nabla_\rho u \quad \text{and}$$

$$(D_{t_j})^* = - D_{t_j} .$$

Any differential operator $A(x,D)$ can be written in the form:

$$\begin{aligned} A(x,D) &= A_0(x,D_t) + A_1(x,D_t)\,\partial/\partial\rho + \dots + A_k(x,D_t)\,\partial^k/\partial\rho^k \\ &= A_0'(x,D_t) + \nabla_\rho A_1'(x,D_t) + \dots + \nabla_\rho^k A_k'(x,D_t) \end{aligned}$$

where by $P(x,D_t)$ we denote an operator containing only the tangential derivatives D_t.

With $C_j(x,D)$ defined by

$$A(x,D) = A_0'(x,D_t) + \dots + \nabla_\rho^{j-1} A_{j-1}'(x,D_t) + \nabla_\rho^j C_j(x,D),$$

i.e. $\quad C_j(x,D) = A_j'(x,D_t) + \dots + \nabla_\rho^{k-j} A_k'(x,D_t)$

and $\Omega^- = \{x\in\Omega \mid \rho(x) \leq 0\}$, we obtain (Green's formula):

$$\int_{\Omega^-} {}^t v\cdot A(x,D)u\,dx = \int_{\Omega^-} {}^t(A^*(x,D)\,v)\cdot u\,dx + \sum_{i=0}^{k-1} (-1)^i \int_S {}^t((\partial/\partial\rho)^i v)\; C_{i+1} u\, dS.$$

6. The Sheaf $\mathcal{S}_A(S)$.

Let E,F be vector bundles on X (of rank p and q respectively) and let $A(x,D)$ be a differential operator between the bundles E and F. We set $X = \{x\in X \mid \rho(x)\leq 0\}$ and $X^+ = \{x\in X \mid \rho(x) \geq 0\}$ for a smooth real valued function ρ on X defining a smooth oriented closed hypersurface in X. Let E^* and F^* be the dual bundles of E and F respectively (they are the bundles defined by the transition functions ${}^t e_{ij}^{-1}$ and ${}^t f_{ij}^{-1}$ resp.)

With $n=\dim_{\mathbb{R}} X$, we denote by Ω^n the bundle of differential n-forms on X: we have $\Omega^n = \Lambda^n T^*X$ where T^*X is the cotangent bundle of X. Then the (formal) adjoint of $A(x,D)$ is a differential operator $A^*(x,D) : \Gamma(X,F^*\otimes\Omega^n) \longrightarrow \Gamma(X,E^*\otimes\Omega^n)$ such that for any $v\,dx \in\Gamma(X,F^*\otimes\Omega^n)$ and any $u\in\Gamma(X,E)$, both with compact support, we have:

$$\int_X \langle v, A(x,D)u\rangle\,dx = \int_X \langle A^*(x,D)v, u\rangle\,dx$$

where $\quad \langle\cdot\, ,\, \cdot\rangle\,dx : \Gamma(X,E)\times\Gamma(X,E^*\otimes\Omega^n) \longrightarrow \Gamma(X,\Omega^n)$

and $\quad \langle\cdot\, ,\, \cdot\rangle\,dx : \Gamma(X,F)\times\Gamma(X,F^*\otimes\Omega^n) \longrightarrow \Gamma(X,\Omega^n)$

are the natural pairings.

Let U be an open set in X and let $u \in \Gamma(U,E)$. We say that u is in the domain $\mathcal{S}_A(S,U)$ of A(x,D) along S (or that u has zero Cauchy data for A(x,D) on S) if for every v dx in $\Gamma(X,F^* \otimes \Omega^n)$ with compact support contained in U we have

$$\int_{X^-} \langle v, A(x,D)u \rangle \, dx = \int_{X^-} \langle A^*(x,D)v, u \rangle \, dx.$$

One verifies that $U \longrightarrow \mathcal{S}_A(S,U)$ is a sheaf, that we denote by $\mathcal{S}_A(S)$.

Remark. In the local situation considered in the previous section, we have that $u \in \mathcal{S}_A(S,\Omega)$ if and only if $C_i(x,D)u \,|\, S = 0$ for $i=1,\dots,k$.

Let $\mathcal{F}_S(U)$ denote the space of sections of E on U that are flat on S. Then $U \longrightarrow \mathcal{F}_S(U)$ is a sub-sheaf of the sheaf $\mathcal{S}_A(S)$, that we will denote by $\mathcal{F}_S$. The quotient sheaf $\mathcal{S}_A(S)/\mathcal{F}_S$ is defined by the exact sequence:

$$0 \longrightarrow \mathcal{F}_S \longrightarrow \mathcal{S}_A(S) \longrightarrow \mathcal{S}_A(S)/\mathcal{F}_S \longrightarrow 0 .$$

7. Formally Noncharacterisitic Hypersurfaces.

Given a complex $(\mathcal{E}^*(X), A^*)$ of differential operators on X, we have $A^j(x,D)\,\mathcal{S}_{A^j}(S,X) \subset \mathcal{S}_{A^{j+1}}(S,X)$ and also, denoting by $\mathcal{F}_S^j$ the sub-sheaf of sections of $\mathcal{E}^{(j)}$ that are flat on S, we have $A^j(x,D)\,\mathcal{F}_S^j(X) \subset \mathcal{F}_S^{j+1}(X)$. Therefore we obtain subcomplexes:

$$\left\{ \mathcal{S}_{A^0}(S,X) \xrightarrow{A^0(x,D)} \mathcal{S}_{A^1}(S,X) \xrightarrow{A^1(x,D)} \mathcal{S}_{A^2}(S,X) \longrightarrow \dots \right\} = (\mathcal{S}^*(S,X), A^*)$$

and

$$\left\{ \mathcal{F}_S^0(X) \xrightarrow{A^0(x,D)} \mathcal{F}_S^1(X) \xrightarrow{A^1(x,D)} \mathcal{F}_S^2(X) \longrightarrow \dots \right\} = (\mathcal{F}_S^*(X), A^*).$$

We say that S is formally noncharacteristic (for the complex $(\mathcal{E}^*(X), A^*)$ if the quotient complex, augmented by zero:

$$(*) \left\{ 0 \longrightarrow \frac{\mathcal{S}_{A^0}(S,X)}{\mathcal{F}_S^0(X)} \xrightarrow{A^0(x,D)} \frac{\mathcal{S}_{A^1}(S,X)}{\mathcal{F}_S^1(X)} \xrightarrow{A^1(x,D)} \frac{\mathcal{S}_{A^2}(S,X)}{\mathcal{F}_S^2(X)} \longrightarrow \dots \right\} = \left(\frac{\mathcal{S}^*}{\mathcal{F}^*}(S,X), A^* \right)$$

is acyclic. We have the following:

Proposition. Let us assume that classical gradings are defined on the bundles E^j, compatible with the operators of $(\mathcal{E}^*(X), A^*)$. If the complex $(\mathcal{E}^*(X), B^*)$ is obtained from the first one by classical fiber transformations, then S is formally noncharacteristic for the first complex if and only if is formally noncharacterisic for the second one.

When the gradings are nonclassical, the statement of the proposition may be false. Take for instance $X = \mathbb{R}$, $S = \{t=0\}$ (t coordinate in $\mathbb{R}$). We have a commutative diagram of differential operators:

$$\begin{array}{ccccccc}
\mathcal{E}(\mathbb{R}) & \xrightarrow{d/dt} & \mathcal{E}(\mathbb{R}) & \xrightarrow{0} & \mathcal{E}^2(\mathbb{R}) & \xrightarrow{\begin{pmatrix}1 & d/dt\\ 0 & t\,d/dt\end{pmatrix}} & \mathcal{E}^2(\mathbb{R}) \\
\uparrow 1 & & \uparrow 1 & & \uparrow \begin{pmatrix}1 & -d/dt\\ 0 & 1\end{pmatrix} & & \uparrow \begin{pmatrix}1 & 0\\ 0 & 1\end{pmatrix} \\
\mathcal{E}(\mathbb{R}) & \xrightarrow[d/dt]{} & \mathcal{E}(\mathbb{R}) & \xrightarrow[0]{} & \mathcal{E}^2(\mathbb{R}) & \xrightarrow[\begin{pmatrix}1 & 0\\ 0 & t\,d/dt\end{pmatrix}]{} & \mathcal{E}^2(\mathbb{R})
\end{array}$$

The complexes in the two lines can be thought as obtained one from the other by fiber transformations corresponding to suitable nonclassical gradings. One has:

$$\mathcal{J}_{d/dt}(0,\mathbb{R}) = \{u \mid u(0) = 0\} = t\,\mathcal{E}(\mathbb{R}) \qquad \mathcal{J}_0(0,\mathbb{R}) = \mathcal{E}(\mathbb{R})$$

$$\mathcal{J}_{\begin{pmatrix}1 & d/dt\\ 0 & t\,d/dt\end{pmatrix}}(0,\mathbb{R}) = \{(u,v) \mid v(0)=0\} = \mathcal{E}(\mathbb{R}) \oplus t\,\mathcal{E}(\mathbb{R}) \qquad \mathcal{J}_{\begin{pmatrix}1 & 0\\ 0 & t\,d/dt\end{pmatrix}}(0,\mathbb{R}) = \mathcal{E}^2(\mathbb{R}).$$

Denoting by Φ_0 the space of formal power series of t, the sequence

$$0 \longrightarrow t\,\Phi_0 \xrightarrow{d/dt} \Phi_0 \xrightarrow{0} \Phi_0^2 \xrightarrow{\begin{pmatrix}1 & d/dt\\ 0 & t\,d/dt\end{pmatrix}} \Phi_0^2$$

is exact and thus $\{0\}$ is formally noncharacteristic for the complex in the upper line, while for the complex in the bottom line we obtain the non exact sequence:

$$0 \longrightarrow t\,\Phi_0 \xrightarrow{d/dt} \Phi_0 \xrightarrow{0} \Phi_0^2 \xrightarrow{\begin{pmatrix}1 & 0\\ 0 & t\,d/dt\end{pmatrix}} \Phi_0^2$$

and hence $\{0\}$ is formally characteristic for that complex.

We can conclude that the notion of "formally noncharacteristic" is not invariant under fiber transformations of general type.

8. The Mayer Vietoris Sequence.

The importance of the notion of formally noncharacteristic that has been introduced in the previous section depends on the following considerations.

Let us set $Q^{(j)}(S) = \mathcal{E}^{(j)}(X)/\mathcal{J}_{A^j}(S,X)$. We derive a quotient complex:

$$\{Q^{(0)}(S) \xrightarrow{A^0_S} Q^{(1)}(S) \xrightarrow{A^1_S} Q^{(2)}(S) \longrightarrow \dots\} = (Q^*(S), A^*_S)$$

where the maps A^j_S are induced by the differential operators $A^j(x,D)$ but are not necessarily differential operators on S. We denote by $H^j(S;Q^*(S),A^*_S)$ the cohomology groups of this complex.

<u>Proposition.</u> <u>If</u> S <u>is formally noncharacteristic for the complex</u> $(\mathcal{E}^*(X),A^*)$, <u>then we have a long exact sequence (Mayer Vietoris sequence):</u>

$$0 \longrightarrow H^0(X;\mathcal{E}^*(X),A^*) \longrightarrow H^0(X^+;\mathcal{E}^*(X^+),A^*) \oplus H^0(X^-;\mathcal{E}^*(X^-),A^*) \longrightarrow H^0(S;Q^*(S),A^*_S)$$

$$\longrightarrow H^1(X;\mathcal{E}^*(X),A^*) \longrightarrow H^1(X^+;\mathcal{E}^*(X^+),A^*) \oplus H^1(X^-;\mathcal{E}^*(X^-),A^*) \longrightarrow H^1(S;Q^*(S),A^*_S)$$

$$\longrightarrow \dots$$

$$\dots \longrightarrow H^j(X;\mathcal{E}^*(X),A^*) \longrightarrow H^j(X^+;\mathcal{E}^*(X^+),A^*) \oplus H^j(X^-;\mathcal{E}^*(X^-),A^*) \longrightarrow H^j(S;Q^*(S),A^*_S)$$
$$\longrightarrow H^{j+1}(X;\mathcal{E}^*(X),A^*) \longrightarrow \dots$$

(The exactness of this sequence follows from the isomorphism

$$H^j(S;\mathcal{E}^*(X)/\mathcal{J}^*_S(X),A^*) \cong H^j(S;Q^*(S),A^*_S) \quad).$$

The importance of the exact sequence established in the proposition above obviously depends on the interpretation of the cohomology groups $H^j(S;Q^*(S),A^*_S)$.

9. A Preparation Lemma for Differential Operators.

Theorem. Let Ω be an open set in $\mathbb{R}^n$ and let

$$\mathcal{E}^{p_0}(\Omega) \xrightarrow{A^0(x,D)} \mathcal{E}^{p_1}(\Omega) \xrightarrow{A^1(x,D)} \mathcal{E}^{p_2}(\Omega) \longrightarrow \dots$$

be a graded finite complex of differential operators. Let S be an oriented smooth hypersurface in Ω and let us assume that S is noncharacteristic at x_0 ($\in S$) for the given complex. Then we can find an open neighborhood ω of x_0 in Ω and for each j a graded fiber transformation $M_j(x,D)$ on $\mathcal{E}^{p_j}(\omega)$ in such a way that the complex obtained by these fiber transformations:

$$(*) \qquad \mathcal{E}^{p_0}(\omega) \xrightarrow{B^0(x,D)} \mathcal{E}^{p_1}(\omega) \xrightarrow{B^1(x,D)} \mathcal{E}^{p_2}(\omega) \longrightarrow \dots$$

has the properties:

i) $S_\omega = S \cap \omega$ is formally noncharacterisitic for (*)

ii) The boundary complex $(\mathcal{E}^*(\omega)/\mathcal{J}_{B^*}(S_\omega,\omega),B^*_S)$ is such that
$$C^j(S_\omega) = \mathcal{E}^{p_j}(\omega)/\mathcal{J}_{B^j}(S_\omega,\omega) \cong \mathcal{E}^{q_j}(S) \text{ for some } q_j \text{ for } j=0,1,2,\dots$$
and $B^j_S:\ \mathcal{E}^{q_j}(S_\omega) \longrightarrow \mathcal{E}^{q_{j+1}}(S_\omega)$ is a differential operator on S_ω $\forall j$.

iii) The sheaf on ω : $U \longrightarrow \mathcal{J}_{B^j}(S,U)$, and thus also the sheaf $U \longrightarrow \mathcal{J}_{B^j}(S,U)/\mathcal{J}^j_S(U)$, are soft.

Remark. When the gradings considered are all classical, properties i),ii)and iii) hold also for the given complex.

In the proof of the theorem above, we can assume that $(y,t)=(y_1,\dots,y_{n-1},t)$ are cartesian coordinates in $\mathbb{R}^n$ and that $S=\{t=0\}$. The proof follows then from a lemma that states that, under the assumptions of the theorem, the given complex can be reduced, by means of graded fiber transformations, to a canonical form that we describe now: we have:

$$0 \longrightarrow \mathcal{E}^{r_0}(\omega) \xrightarrow{A^0} \mathcal{E}^{r_1}(\omega) \oplus \mathcal{E}^{r_0}(\omega) \xrightarrow{A^1} \mathcal{E}^{r_2}(\omega) \oplus \mathcal{E}^{r_1}(\omega) \longrightarrow \dots$$

where $A^j = A^j(y,t;\partial/\partial y,\partial/\partial t): \mathcal{E}^{r_j}(\omega) \oplus \mathcal{E}^{r_{j-1}}(\omega) \longrightarrow \mathcal{E}^{r_{j+1}}(\omega) \oplus \mathcal{E}^{r_j}(\omega)$ is written in block notations as

$$A^j = \begin{pmatrix} A^j_0 & A^j_1 \\ A^j_2 & A^j_3 \end{pmatrix}$$

with $A_2^{(j)}$ of type $r_j \times r_j$ of the form

$$A_2^{(j)}(y,t;\partial/\partial y,\partial/\partial t) = \mathrm{diag}(\partial^{k_1^{(j)}}/\partial t^{k_1^{(j)}},\ldots,\partial^{k_{r_j}^{(j)}}/\partial t^{k_{r_j}^{(j)}}) + R^{(j)}$$

where the entry $R_{hs}^{(j)}$ of $R^{(j)}$ has order $< k_s^{(j)}$ with respect to $\partial/\partial t$ and the entries a_{hs} of $A_0^{(j)}$ have order $< k_s^{(j)}$ with respect to $\partial/\partial t$. With $q_j = k_1^{(j)}+\ldots+k_{r_j}^{(j)}$ it is easy to establish the isomorphism $\mathcal{E}^{r_j}(\omega)/\mathcal{G}_{A_2^{(j)}}(S_\omega,\omega) \cong \mathcal{E}^{q_j}(S_\omega)$

and then the statement follows from the isomorphism $\dfrac{\mathcal{E}^{r_j}(\omega)}{\mathcal{G}_{A_2^{(j)}}(S_\omega,\omega)} \cong \dfrac{\mathcal{E}^{r_j}(\omega)\oplus\mathcal{E}^{r_{j+1}}(\omega)}{\mathcal{G}_{A^{(j)}}(S_\omega,\omega)}$

that is induced by the map $w \xrightarrow{\sigma} w \oplus 0$.

The fact that the sheaf $U \longrightarrow \mathcal{G}_{A^j}(S,U)$ is soft (when the complex is in canonical form) follows from the identity:

$$\mathcal{G}_{A^{(j)}}(S,U) = A^{(j-1)}\sigma\mathcal{G}_{A_2^{(j-1)}}(S,U) \oplus \sigma\mathcal{G}_{A_2^{(j)}}(S,U) + \mathcal{F}_S^j(U)$$

and the fact that the sheaves $U \longrightarrow \mathcal{G}_{A_2^{(r)}}(S,U)$ are soft and $\mathcal{F}_S^j$ is soft.

Consequence. a.) For the complex $(\mathcal{E}^*(\omega),B^*)$, we have the Mayer Vietoris sequence for every paracompactifying family of supports.

b.) When we are in the general situation, but the grading is classical, then the statements of the theorem are true globally: for every j the space $C^{(j)}(S)$ is the space of sections of a vector bundle over S and the Mayer Vietoris sequence is exact for every paracompactifying family of supports in X.

LECTURE 6. BOUNDARY VALUES OF PLURIHARMONIC FUNCTIONS.

1. Let X be a complex manifold of pure dimension n. For Ω open $\subset$ X we denote by $A^{(r,s)}(\Omega)$ the space of C^∞ forms of type r,s in Ω, and we set $A^{(j)}(\Omega) = \bigoplus_{r+s=j} A^{r,s}(\Omega)$ (differential forms of total degree j). Let d denote the exterior differential: $d = \partial + \bar\partial$ where ∂ (resp. $\bar\partial$) is the differential with respect to holomorphic (resp. antiholomorphic) coordinates. We consider the complex of differential operators:

$$(*) \quad A^{0,0}(\Omega) \xrightarrow{\partial\bar\partial} A^{1,1}(\Omega) \xrightarrow{d} A^{1,2}(\Omega) \oplus A^{2,1}(\Omega) \longrightarrow \dots$$
$$\dots \xrightarrow{d} \bigoplus_{j=1}^{n-1} A^{j,n-j}(\Omega) \xrightarrow{d} A^{(n+1)}(\Omega) \longrightarrow \dots \xrightarrow{d} A^{(2n)}(\Omega) \longrightarrow 0.$$

Denoting by $\mathcal{H}(\Omega)$ the space of (complex valued) pluriharmonic functions on the open set Ω, we obtain an augmentation of (*) by

$$0 \longrightarrow \mathcal{H}(\Omega) \longrightarrow A^{0,0}(\Omega) \xrightarrow{\partial\bar\partial} A^{1,1}(\Omega).$$

We denote by $\mathcal{H}$ the shaf of germs of pluriharmonic function on X and by $\mathcal{O}$ the sheaf of germs of holomorphic function on X. With $\underset{\sim}{\mathbb{C}}$ denoting the constant sheaf, we have the exact sequence of sheaves:

$$0 \longrightarrow \underset{\sim}{\mathbb{C}} \longrightarrow \mathcal{O} \oplus \bar{\mathcal{O}} \longrightarrow \mathcal{H} \longrightarrow 0.$$

We also note that the complex (*) is a complex of differential operators with constant coefficients in any holomorphic coordinate patch.

2. Let us look at the complex (*) in the way indicated in the previous lecture: the bundle E^0 is the trivial bundle, E^1 is the bundle $\mathcal{J}^*(X) \oplus \overline{\mathcal{J}^*(X)}$ (where $\mathcal{J}^*(X)$ is the holomorphic cotangent bundle), E^2 is the bundle $\mathcal{J}^*(X) \otimes \overset{2}{\Lambda}\mathcal{J}^*(X) \oplus \overset{2}{\Lambda}\mathcal{J}^*(X) \otimes \overline{\mathcal{J}^*(X)}$ etc. Gradings are classical, so that there is a jump of two units from E^0 to E^1 and of one unit from E^j to E^{j+1} for j=1,2,...

To write the symbolic complex we introduce the notation $\mathbb{P} = \mathbb{C}[\xi_1,\dots,\xi_n,\bar\xi_1,\dots,\bar\xi_n]$ (the ring of polynomials with complex coefficients in the indeterminates $\xi, \bar\xi$),

$\mathbb{P}^{r,s}$ = space of exterior forms of type r in $d\xi_1,\dots,d\xi_n$ and of type s in $d\bar\xi_1,\dots,d\bar\xi_n$

$\mathbb{P}^{(j)} = \bigoplus_{h=0}^{j} \mathbb{P}^{h,j-h}$

$\alpha = \sum \xi_j d\xi_j$, $\bar\alpha = \sum \bar\xi_j d\bar\xi_j$.

Then the symbolic complex of (*) at any point x_0 of X is the complex:

$$\mathbb{P}^{0,0} \xrightarrow{\alpha\wedge\bar\alpha} \mathbb{P}^{1,1} \xrightarrow{\wedge(\alpha+\bar\alpha)} \mathbb{P}^{1,2} \oplus \mathbb{P}^{2,1} \xrightarrow{\wedge(\alpha+\bar\alpha)} \dots$$
$$\dots \longrightarrow \bigoplus_{j=1}^{n-1} \mathbb{P}^{j,n-j} \xrightarrow{\wedge(\alpha+\bar\alpha)} \mathbb{P}^{(n+1)} \xrightarrow{\wedge(\alpha+\bar\alpha)} \dots \longrightarrow \mathbb{P}^{(2n)} \longrightarrow 0.$$

We know that the sequence (*) is exact on any open set of holomorphy and then the sequence obtained from the sequence above by applying the functor $\mathrm{Hom}_{\mathbb{P}}(\cdot,\mathbb{P})$ and the isomorphism $\mathrm{Hom}_{\mathbb{P}}(\mathbb{P}^{r,s},\mathbb{P}) = \mathbb{P}^{n-r,n-s}$ is also exact. In particular we note that (*) is a Hilbert complex.

3. Given an oriented smooth hypersurface $S = \{\rho = 0\}$ in X, having fixed holomorphic coordinates $z_j = x_j + i\, x_{n+j}$ we set

$$\operatorname{grad}\rho(x) = (\nabla\rho, \bar\nabla\rho) = (\partial\rho/\partial z_1,\dots,\partial\rho/\partial z_n, \partial\rho/\partial\bar z_1,\dots,\partial\rho/\partial\bar z_n).$$

Note that $\alpha(\operatorname{grad}\rho) = \partial\rho = \sum \partial\rho/\partial z_j\, d\xi_j$ and $\bar\alpha(\operatorname{grad}\rho) = \bar\partial\rho = \sum \partial\rho/\partial\bar z_j\, d\bar\xi_j$.

Denote by $\mathbb{C}^{r,s}$ the space of exterior forms of degree r in $d\xi_1,\dots,d\xi_n$ and of degree s in $d\bar\xi_1,\dots,d\bar\xi_n$ with coefficients in $\mathbb{C}$ and set $\mathbb{C}^{(j)} = \bigoplus \mathbb{C}^{h,j-h}$. Then the sequence

$$0 \longrightarrow \mathbb{C}^{0,0} \xrightarrow{\partial\rho\wedge\bar\partial\rho} \mathbb{C}^{1,1} \xrightarrow{\wedge(\partial\rho+\bar\partial\rho)} \mathbb{C}^{1,2} \oplus \mathbb{C}^{2,1} \longrightarrow \dots$$

is always exact because the complex (*) is elliptic. Thus all hypersurfaces S are noncharacteristic and formally noncharacteristic for the complex (*).

4. To construct the boundary complex, we compute the domains of the operators of the complex (*):

$$\mathfrak{J}_{\partial\bar\partial}(S,\Omega) = \rho^2 A^{0,0}(\Omega)$$
$$\mathfrak{J}_d(S,\Omega) = \rho A^{1,1}(\Omega) + \partial\rho\wedge\bar\partial\rho\, A^{0,0}(\Omega)$$
$$\mathfrak{J}_d(S,\Omega) = \rho(A^{1,2}(\Omega) + A^{2,1}(\Omega)) + d\rho\wedge A^{1,1}(\Omega)$$
$$\dots\dots\dots\dots\dots\dots$$
$$\mathfrak{J}_d(S,\Omega) = \sum_{j=1}^{n-1} \rho\, A^{j,n-j}(\Omega) + d\rho\wedge \sum_{j=1}^{n-2} A^{j,n-j-1}(\Omega)$$
$$\dots\dots\dots\dots\dots\dots$$
$$\mathfrak{J}_d(S,\Omega) = \rho\, A^{(2n)}(\Omega) + d\rho\wedge A^{(2n-1)}(\Omega).$$

The computation is straightforward: as an example we compute $\mathfrak{J}_{\partial\bar\partial}(S,\Omega)$. A function u in $A^{0,0}(\Omega)$ belongs to $\mathfrak{J}_{\partial\bar\partial}(S,\Omega)$ iff, for every $v \in \mathcal{D}^{n-1,n-1}(\Omega)$ we have:

$$\int_{\Omega^-} \partial\bar\partial u\wedge v = \int_{\Omega^-} u\,\partial\bar\partial\, v.$$

We have: $\displaystyle\int_{\Omega^-} \partial\bar\partial u\wedge v = \int_{\Omega^-} d(\bar\partial u\wedge v) + \int_{\Omega^-} \bar\partial u\wedge\partial v = \int_S \bar\partial u\wedge v + \int_{\Omega^-} d(u\,\partial v) - \int_{\Omega^-} u\,\partial\bar\partial v$

$$= \int_S \bar\partial u\wedge v + \int_S u\,\partial v + \int_{\Omega^-} u\,\partial\bar\partial v\,.$$

This means that $u \in \mathfrak{J}_{\partial\bar\partial}(S,\Omega)$ iff

$$\int_S \bar\partial u\wedge v + \int_S u\,\partial v = 0 \qquad \text{for every } v\in\mathcal{D}^{n-1,n-1}(\Omega).$$

Taking $v = \rho\, w$ for some $w \in \mathcal{D}^{n-1,n-1}(\Omega)$ we deduce that $u = \rho h$ for some $h \in A^{0,0}$. Substituting ρh for u in the above equation, we deduce that $h = \rho k$ for some k in $A^{0,0}(\Omega)$, i.e. $u = \rho^2 k \in \rho^2 A^{0,0}(\Omega)$.

Then we define:

$$Q^{(0)}(S_\Omega) = A^{0,0}(\Omega)/\rho^2 A^{0,0}(\Omega) = A^{(0)}(S_\Omega) + \rho A^{(0)}(S_\Omega) \cong A^{(0)}(S_\Omega)^2,$$

$$Q^{(1)}(S_\Omega) = A^{1,1}(\Omega)/(\rho A^{1,1}(\Omega) + \partial\rho\wedge\bar\partial\rho A^{0,0}(\Omega))$$

. . . .

$$Q^{(k)}(S_\Omega) = \sum_{\substack{r+s=k+1\\ r,s\geq 1}} A^{r,s}(\Omega) \Big/ \rho \sum_{\substack{r+s=k+1\\ r,s\geq 1}} A^{r,s}(\Omega) + d\rho \sum_{\substack{r+s=k\\ r,s\geq 1}} A^{r,s}(\Omega) \qquad 2\leq k\leq n-1$$

. . . .

$$Q^{(k)}(S_\Omega) = A^{(k+1)}(S_\Omega) \qquad \text{for } k=n,\dots,2n-2.$$

The boundary complex takes the form

$$Q^{(0)}(S) \xrightarrow{(\partial\bar\partial)_S} Q^{(1)}(S) \xrightarrow{d_S} Q^{(2)}(S) \longrightarrow \dots \xrightarrow{d_S} Q^{(n)}(S) \xrightarrow{d} Q^{(n+1)}(S) \longrightarrow \dots$$
$$\dots \xrightarrow{d} Q^{(2n-2)}(S) \longrightarrow 0$$

where, because the gradings of the complex (*) are classical, by the results of the previous lecture we have $Q^{(j)}(S) \cong \mathcal{E}^{q_j}(S)$ for some integer q_j for all $j=0,\dots,2n-2$ and all operators of the complex are differential operators.

5. <u>Explicit Expression of the Operator</u> $(\partial\bar\partial)_S$.

We take a coordinate patch Ω for a point $z_0 \in S$. We can assume that on Ω the function ρ defining S takes the form $\rho = y_n - \sigma(z_1,\dots,z_{n-1},x_n)$ where $z_j = x_j + iy_j$ are holomorphic coordinates on Ω and σ vanishes at $z_0=0$ with all first order derivatives.

Then in a small neighborhood of 0 we have

$$dz_n = a\,\partial\rho + \sum_1^{n-1} \alpha_j dz_j\ , \quad d\bar z_n = \bar a\,\bar\partial\rho + \sum_1^{n-1} \bar\alpha_j d\bar z_j \quad \text{with} \quad a = -2(i+\partial\sigma/\partial x_n)^{-1},\ \alpha_j = a\,\partial\sigma/\partial z_j$$

and
$$\partial\bar\partial\rho = \sum_1^{n-1} \ell_{ij}\, dz_i d\bar z_j + \partial\rho \sum_1^{n-1} \bar\beta_i d\bar z_i - \bar\partial\rho \sum_1^{n-1} \beta_i dz_i + \gamma\,\partial\rho\,\bar\partial\rho$$

where the first summand, computed in 0, is the Levi form at 0 restricted to the tangent analytic space to S. Taking $z_1,\dots,z_{n-1},x_n$ as local coordinates near z_0, we write

$$u = u_0 + \rho u_1 \in A^{(0)}(S) + \rho A^{(0)}(S) \text{ with } u_h = u_h(z_1,\dots,z_{n-1},x_n),\ h=0,1$$

Then the equation $(\partial\bar\partial)_S u=0$ can be written as

$$\partial\bar\partial(u_0+\rho u_1)\wedge\partial\rho|S = 0, \quad \partial\bar\partial(u_0+\rho u_1)\wedge\bar\partial\rho|S = 0.$$

Setting $\mathcal{L}_{ij} = \partial^2/\partial z_i\partial\bar z_j + (1/2)\,\bar\alpha_j\,\partial^2/\partial z_i\partial x_n + (1/2)\,\alpha_i\,\partial^2/\partial\bar z_j\partial x_n + (1/4)\alpha_i\bar\alpha_j\partial^2/\partial x_n^2$

$S_i = a\,(\partial^2/\partial\bar z_i\partial x_n + (1/4)\,\bar\alpha_i\,\partial^2/\partial x_n^2)$ and $T_i = \partial/\partial\bar z_i + (1/2)\,\bar\alpha_i\,\partial/\partial x_n$, this system takes the explicit form:

$$\mathcal{L}_{ij}u_0 + \ell_{ij}u_1 = 0 \quad 1\leq i,j\leq n-1\ ; \quad S_iu_0 + T_iu_1 = 0 \text{ and } \bar S_iu_0 + \bar T_iu_1 = 0 \text{ for } 1\leq i\leq n-1.$$

<u>Remark</u>. If the Levi form of ρ along the analytic tangent space to S is $\neq 0$, then the first of the above equations uniquely determines u_1 in function of u_0. Computing u_1 from u_0 and substituting in the last two equations, one obtains a system of third order differential equations for u_0.

6. Hartogs Type Theorems. Notations: $X^- = \{x \in X \mid \rho(x) \leq 0\}$,

$H^0(X^-, \mathcal{H})$ = C^∞ functions on X^- that are pluriharmonic on $\overset{\circ}{X}{}^-$.

$H^0(S, \mathcal{H}_S^{(1)})$ = space of pairs of functions $u_0+\rho u_1 \in A^{(0)}(S)+\rho A^{(0)}(S)$ satisfying $(\partial\bar\partial)_S(u_0+\rho u_1) = 0$.

We have a natural map $r: H^0(X^-, \mathcal{H}) \longrightarrow H^0(S, \mathcal{H}_S^{(1)})$ ($r(u) = (u|S, \partial u/\partial\rho|S)$.

Theorem. Assume that X^- is compact and all components of X^+ are not compact. If X is n-2 - complete and $H^2(X,\mathbb{C})=0$, then the natural map r is an isomorphism.

Indeed: from the Mayer Vietoris sequence with compact supports we obtain the exact sequence: $0 \longrightarrow H^0(X^-, \mathcal{H}) \xrightarrow{r} H^0(S, \mathcal{H}_S^{(1)}) \longrightarrow H^1_k(X, \mathcal{H})$.

On the other hand, from the exact sequence $0 \longrightarrow \underset{\sim}{\mathbb{C}} \longrightarrow \mathcal{O}\oplus\bar{\mathcal{O}} \longrightarrow \mathcal{H} \longrightarrow 0$ we obtain the exact sequence $H^1_k(X, \mathcal{O}) \oplus H^1_k(X, \bar{\mathcal{O}}) \longrightarrow H^1_k(X, \mathcal{H}) \longrightarrow H^2_k(X, \mathbb{C})$. Under the assumptions of the theorem $H^1_k(X, \mathcal{O})=0$, $H^1_k(X, \bar{\mathcal{O}})=0$, $H^2_k(X, \underset{\sim}{\mathbb{C}})=0$ and thus $H^1_k(X, \mathcal{H})=0$, q.e.d.

Denote by $\mathcal{H}_S$ the sheaf of germs of C^∞ function u on S for which we can find a germ of C^∞ function v on S such that $(\partial\bar\partial)_S(u+\rho v)=0$. We have a natural surjective map $\mathcal{H}_S^{(1)} \longrightarrow \mathcal{H}_S \longrightarrow 0$. If the Levi form on the analytic tangent space to S is always different from 0, this map is an isomorphism. Therefore, under the same assumptions of the previous theorem we find that also $H^0(X^-, \mathcal{H}) = H^0(S, \mathcal{H}_S)$.

7. For Ω open in X, $S_\Omega = S \cap \Omega$, we set $A^{0,0}(S_\Omega) = A^{0,0}(\Omega)/\rho A^{0,0}(\Omega) = A^{(0)}(S_\Omega)$,

$A^{0,1)}(S_\Omega) = A^{0,1}(\Omega)/\rho A^{0,1}(\Omega) + \bar\partial\rho\, A^{0,0}(\Omega)$, ...,

$A^{0,j}(S_\Omega) = A^{0,j}(\Omega)/\rho\, A^{0,j}(\Omega) + \bar\partial\rho\, A^{0,j-1}(\Omega)$, ...

The Dolbeault complex $A^{0,0}(\Omega) \xrightarrow{\bar\partial} A^{0,1}(\Omega) \xrightarrow{\bar\partial} \dots \longrightarrow A^{0,n}(\Omega) \longrightarrow 0$ induces a boundary complex $A^{0,0}(S_\Omega) \xrightarrow{\bar\partial_S} A^{0,1}(S_\Omega) \xrightarrow{\bar\partial_S} \dots \longrightarrow A^{0,n-1}(S_\Omega) \longrightarrow 0$.

We denote by $\mathcal{O}_S$ the sheaf of germs of smooth function on S satisfying $\bar\partial_S u=0$. Passing to the conjugates, with obvious notations we obtain a complex

$$0 \longrightarrow \bar{\mathcal{O}}_S \longrightarrow A_S^{0,0} \xrightarrow{\partial_S} A_S^{1,0} \xrightarrow{\partial_S} \dots \xrightarrow{\partial_S} A_S^{n-1,0} \longrightarrow 0.$$

By a version in formal power series of the theorem of Cauchy-Kowalewska and an extension theorem of Whitney one obtains:

Lemma. a) $\forall\, u \in \Gamma(S_\Omega, \mathcal{O}_S)$ there is $\tilde{u} \in A^{0,0}(\Omega)$ with $\tilde{u}|S_\Omega = u$ and $\bar\partial\tilde{u} \in \mathcal{F}_S^{0,1}(\Omega)$

b) $\forall\, u_0+\rho u_1 \in \Gamma(S_\Omega, \mathcal{H}_S^{(1)})$ there is $\tilde{u} \in A^{0,0}(\Omega)$ with $\tilde{u}|S_\Omega=u_0$ $\partial\tilde{u}/\partial\rho|S_\Omega = u_1$ and such that $\partial\bar\partial\tilde{u} \in \mathcal{F}_S^{1,1}(\Omega)$.

Notation: $\mathcal{F}_S^{r,s}(\Omega)$=forms of type r,s on Ω with smooth coefficients, that are flat on S_Ω. We also denote by $\mathcal{F}_S^{(r)}$ exterior differential forms of total degree r that are flat on S. We set $W_S^{(r)} = A^{(r)}/\mathcal{F}_S^{(r)}$. The exterior differential induces a map $d: W_S^{(r)} \longrightarrow W_S^{(r+1)}$ and we have (formal Cauchy Problem):

Lemma. The following is an exact sequence of sheaves:

$$0 \longrightarrow \underset{\sim}{C} \longrightarrow W_S^{(0)} \xrightarrow{d} W_S^{(1)} \xrightarrow{d} W_S^{(2)} \xrightarrow{d} \cdots \longrightarrow W_S^{(n)} \longrightarrow 0$$

We also have the following:

Lemma. With $\mathcal{L} = \mathcal{O}_S \cap \overline{\mathcal{O}}_S$, we have the exact sequence of sheaves:

$$0 \longrightarrow \mathcal{L} \longrightarrow \mathcal{O}_S \oplus \overline{\mathcal{O}}_S \longrightarrow \mathcal{H}_S \longrightarrow 0 .$$

Remark.

On open sets S_Ω where the Levi form restricted to the analytic tangent space is $\neq 0$ at every point, $\mathcal{L} \cong \underset{\sim}{C}$.

Corollary. Assume that $H^1(S,\underset{\sim}{C}) = 0$ and that the Levi form restricted to the analytic tangent space to S is always $\neq 0$ on S. Then we have an exact sequence:

$$0 \longrightarrow \Gamma(S,\underset{\sim}{C}) \longrightarrow \Gamma(S,\mathcal{O}_S) \oplus \Gamma(S,\overline{\mathcal{O}}_S) \longrightarrow \Gamma(S,\mathcal{H}_S) \longrightarrow 0 .$$

8. The case of a Levi form of rank ≥ 2.

In this subsection we will prove that in this case the trace of a pluriharmonic function is characterized by a set of second order differential equations and that the cohomology at the boundary can be computed by means of a somehow simpler complex. (The generic situation for $n > 3$ is simpler than that for $n = 2$).

We set:

$$\mathcal{J}^0(S,\Omega) = \rho A^{0,0}(\Omega)$$
$$\mathcal{J}^1(S,\Omega) = \rho A^{1,1}(\Omega) + \partial\rho\wedge A^{0,1}(\Omega) + \bar\partial\rho\wedge A^{1,0}(\Omega) + \partial\bar\partial\rho\, A^{0,0}(\Omega)$$
$$\mathcal{J}^\mu(S,\Omega) = \mathcal{J}_d(S,\Omega) \quad \text{for} \quad \mu \geq 2 .$$

Then one realizes that

$$(\underset{*}{*}) \qquad \mathcal{J}^0(S,\Omega) \xrightarrow{\partial\bar\partial} \mathcal{J}^1(S,\Omega) \xrightarrow{d} \mathcal{J}^2(S,\Omega) \xrightarrow{d} \cdots$$

is a subcomplex of the complex $(\underset{*}{*})$ of $\partial\bar\partial$.

Moreover the sheaves $\Omega \longrightarrow \mathcal{J}^j(S,\Omega)$ are fine (and thus soft) sheaves.

The subcomplex of $(\underset{*}{*})$

$$\mathcal{J}_{\partial\bar\partial}(S,\Omega) \longrightarrow \mathcal{J}_d(S,\Omega) \xrightarrow{d} \cdots$$

is a subcomplex of (*) . We set

$$C^{(0)}(S_\Omega) = A^{00}(\Omega)/_{\mathcal{J}^0(S,\Omega)} = A^{00}(S_\Omega)$$

$$C^{(1)}(S_\Omega) = A^{11}(\Omega)/_{\mathcal{J}^1(S,\Omega)}$$

$$C^{(\mu)}(S_\Omega) = Q^{(\mu)}(S_\Omega) \quad \mu \geq 2 .$$

At the sheaf level, we get a commutative diagram of sheaves and linear maps:

$$\begin{array}{ccccccccc} 0 \longrightarrow & \mathcal{H}_S^{(1)} & \longrightarrow & Q^{(0)} & \xrightarrow{(\partial\bar\partial)_S} & Q^{(1)} & \xrightarrow{d_S} & Q^{(2)} & \xrightarrow{d_S} \cdots \\ & \downarrow \lambda & & \downarrow & & \downarrow & & \downarrow & \\ 0 \longrightarrow & \mathcal{T}_S & \longrightarrow & C^{(0)} & \xrightarrow{(\partial\bar\partial)^R} & C^{(1)} & \xrightarrow{d^R} & Q^{(2)} & \xrightarrow{d_S} \cdots \end{array}$$

where by definition $\mathcal{T}_S = \mathrm{Ker}(\partial\bar\partial)^R$, $(\partial\bar\partial)^R$ and d^R being the operators obtained from $\partial\bar\partial$ and d passing to the quotient.

We note that $C^{(0)}$, $C^{(1)}$ are sheaves of $\mathcal{E}_S$ - modules.

Proposition.

Let S_Ω be the open subset of S where the Levi form restricted to the analytic tangent space is $\neq 0$. On S_Ω , $C^{(1)}$ is a locally free sheaf of $\mathcal{E}_S$ - modules of rank $(n-1)^2 - 1$.

Remark.

From proposition above and Peetrés theorem, the operators $(\partial\bar\partial)^R$ and d^R are differential operators on S_Ω . We note that $(\partial\bar\partial)^R$ is a differential operator of second order, that is 0 for $n = 2$.

Note that we have $\lambda(\mathcal{H}_S^{(1)}) = \mathcal{H}_S \subset \mathcal{T}_S$.

Let us consider now the complex of (soft) sheaves:

$$(**) \qquad 0 \rightarrow \frac{\mathcal{J}^0}{\mathcal{F}_S^0} \longrightarrow \frac{\mathcal{J}^1}{\mathcal{F}_S^1} \xrightarrow{d} \frac{\mathcal{J}^2}{\mathcal{F}_S^2} \xrightarrow{d} \cdots$$

Proposition.

a) If the Levi form restricted to the tangent analytic hyperplane is $\neq 0$, then (**) is exact at $\mathcal{J}^0 / \mathcal{J}_S^0$.

b) If the Levi form restricted to the tangent analytic hyperplane has rank ≥ 2, then (**) is exact at $\mathcal{J}^1 / \mathcal{J}_S^1$ and hence everywhere.

Proof of a) Let $w = \rho u \in \mathcal{J}^0$ and assume that $\partial\bar\partial(\rho u) \in \mathcal{J}_S^1$. From

$$\partial\bar\partial\rho \cdot u + \partial\rho \wedge \bar\partial u - \bar\partial\rho \wedge \partial u + \rho\partial\bar\partial u \in \mathcal{J}_S^1$$

it follows that $\partial\rho \wedge \bar\partial\rho \wedge \partial\bar\partial\rho \cdot u = 0$ on S and then by the assumption that $\partial\rho \wedge \bar\partial\rho \wedge \partial\bar\partial\rho \neq 0$ it follows that $u = 0$ on S. Thus $w = \rho^2 v \in \mathcal{J}_{\partial\bar\partial}$ and hence $\partial\bar\partial w \in \mathcal{J}_S^1$ implies that $w \in \mathcal{J}_S^0$ because S is formally noncharacteristic for the complex (*).

Proof of b) Let $g^{11} = \rho\alpha^{11} + \partial\rho\wedge\beta^{01} + \bar\partial\rho\wedge\gamma^{10} + \partial\bar\partial\rho\cdot\sigma^{00} \in \mathcal{J}^1$.
Then $\rho\sigma^{00} \in \mathcal{J}^0$ and

$$g^{11} - \partial\bar\partial(\rho\sigma^{00}) = \rho\,\theta^{11} + \partial\rho\wedge\theta^{01} + \bar\partial\rho\wedge\theta^{10} \in \mathcal{J}^1 .$$

Assume that $dg^{11} \in \mathcal{J}_S^2$: then $d(g^{11} - \partial\bar\partial(\rho\sigma^{00})) \in \mathcal{J}_S^2$ and this gives:

$$\rho\,\partial\theta^{11} + \partial\rho\wedge(\theta^{11} - \partial\theta^{01}) - \bar\partial\rho\,\partial\theta^{10} + \partial\bar\partial\rho\wedge\theta^{10} \in \mathcal{J}_S^2$$

$$\rho\,\bar\partial\theta^{11} + \bar\partial\rho\wedge(\theta^{11} - \bar\partial\theta^{10}) - \partial\rho\wedge\bar\partial\theta^{01} + \partial\bar\partial\rho\wedge\theta^{01} \in \mathcal{J}_S^2$$

We have

$$\partial\rho\wedge\bar\partial\rho\wedge\partial\bar\partial\rho\wedge\theta^{10}\big|_S = 0$$

$$\partial\rho\wedge\bar\partial\rho\wedge\partial\bar\partial\rho\wedge\theta^{01}\,{}_S = 0 .$$

By the assumption that $\partial\bar\partial\rho\wedge\partial\rho\wedge\bar\partial\rho$ has rank 2 it follows that:

$$\theta^{10} = \rho\,\lambda^{10} + \partial\rho\,\mu^{00}$$

$$\theta^{01} = \rho\,\lambda^{01} + \bar\partial\rho\,\nu^{00}$$

Thus $g^{11} = \partial\bar\partial(\rho\sigma^{00}) + \rho(\theta^{11} + \partial\rho\,\lambda^{01} + \bar\partial\rho\,\lambda^{10}) + \partial\rho\wedge\bar\partial\rho\,(\nu^{00} - \mu^{00})$.

Therefore $g^{11} - \partial\bar\partial(\rho\sigma^{00}) \in \mathcal{J}_d$ and thus because S is formally noncharacteristic one deduces that $g^{11} - \partial\bar\partial(\rho\sigma^{00}) \in \mathcal{J}_S^1$.

Let Σ open $\subset$ S and $\emptyset$ a paracompactifying family of supports on Σ.
We consider:

$$1)\quad \Gamma_\emptyset(\Sigma, A^{00}/\mathcal{F}_S^{00}) \xrightarrow{\partial\bar\partial} \Gamma_\emptyset(\Sigma, A^{11}/\mathcal{F}_S^{11}) \xrightarrow{d} \Gamma_\emptyset(\Sigma,(A^{12}\oplus A^{21})/(\mathcal{F}_S^{12}\oplus\mathcal{F}_S^{21})) \xrightarrow{d} \cdots$$

$$2)\quad \Gamma_\emptyset(\Sigma, Q^{(0)}) \xrightarrow{(\partial\bar\partial)_S} \Gamma_\emptyset(\Sigma, Q^{(1)}) \xrightarrow{d_S} \Gamma_\emptyset(\Sigma, Q^{(2)}) \xrightarrow{d_S} \cdots$$

$$3)\quad \Gamma_\emptyset(\Sigma, C^{(0)}) \xrightarrow{(\partial\bar\partial)^R} \Gamma_\emptyset(\Sigma, C^{(1)}) \xrightarrow{d^R} \Gamma_\emptyset(\Sigma, Q^{(2)}) \xrightarrow{d_S} \cdots$$

We set
$$\hat{\mathcal{H}} = \operatorname{Ker} \partial\bar\partial : A^{00}/\mathcal{F}_S^0 \to A^{11}/\mathcal{F}_S^{11},$$
$$\mathcal{H}_S^{(1)} = \operatorname{Ker}\{Q^{(0)} \xrightarrow{(\partial\bar\partial)_S} Q^{(1)}\}$$
$$\mathcal{T}_S = \operatorname{Ker}\{C^{(0)} \xrightarrow{(\partial\bar\partial)^R} C^{(1)}\}$$

and we denote the cohomology groups of these complexes by:

$$H^j_\emptyset(\Sigma,[\hat{\mathcal{H}}]),\quad H^j_\emptyset(\Sigma,[\mathcal{H}_S^{(1)}]),\quad H^j_\emptyset(\Sigma,[\mathcal{T}_S]).$$

The brackets are used to recall that these are not cohomology groups with values in a sheaf, because the Poincaré lemma may fail to hold for the complexes 1), 2), 3).

Set $S^{(2)}$ for the set of points of S where the Levi form restricted to the tangent analytic space has rank $\geqslant 2$.

<u>Proposition. On $S^{(2)}$ we have</u>

(i) $\mathcal{H}_S^{(1)} \cong \mathcal{H}_S \cong \mathcal{T}_S$.

(ii) $\forall \Sigma$ open $\subset S^{(2)}$:

$$H^j_\emptyset(\Sigma,[\hat{\mathcal{H}}]) \cong H^j_\emptyset(\Sigma,[\mathcal{H}_S^{(1)}]) \cong H^j_\emptyset(\Sigma,[\mathcal{T}_S]).$$

<u>Corollary. If $S^{(2)} = S$ we have the Mayer-Vietoris exact sequence:</u>

$$0 \to H^0_\emptyset(X,\mathcal{H}) \to H^0_\emptyset(X^+,\mathcal{H}) \oplus H^0_\emptyset(X^-,\mathcal{H}) \to H^0_\emptyset(S,[\mathcal{T}_S])$$
$$\to H^1_\emptyset(X,\mathcal{H}) \to H^1_\emptyset(X^+,\mathcal{H}) \oplus H^1_\emptyset(X^-,\mathcal{H}) \to H^1_\emptyset(S,[\mathcal{T}_S]) \to \cdots$$

Remarks on Poincaré Lemma for the complex of $(\partial\bar\partial)_S$.

We make the following observation:

Proposition. $H^j(X^-,\mathbb{C})$ is the j-th cohomology group of the complex:

$$A^0(X^-) \xrightarrow{d} A^1(X^-) \xrightarrow{d} A^2(X^-) \longrightarrow \ldots$$

where $A^j(X^-)$ is the space of C^∞ forms of degree j defined on X^- up to the boundary (but not beyond it) and d is exterior differentiation.

From this it follows:

Proposition. If $H^j(X^-,\mathbb{C}) = 0 = H^{j+1}(X,\mathbb{C})$ $(j \geq 1)$ we have an isomorphism

$$H^j(X^-,\mathcal{O}) \oplus H^j(X^-,\bar{\mathcal{O}}) \cong H^j(X^-,\mathcal{H}) .$$

Let $z_0 \in S$ and let Ω be a coordinate patch at z_0 in X . Let $z_1 \ldots z_n$ be coordinates on Ω and assume $z_j = 0$ at z_0 . Denote by B_n the ball of radius $\frac{1}{n}$ about z_0 in this system of coordinates.

If the Levi form at z_0 , restricted to the analytic tangent space has p positive and q negative eigenvalues, with $p+q = n-1$, then Andreotti and Hill proved that

$$\lim_n H^p(B_n^+,\mathcal{O}) \text{ is infinite dimensional}$$

$$\lim_n H^q(B_n^-,\mathcal{O}) \text{ is infinite dimensional}$$

$$\lim_n H^r(B_n^+,\mathcal{O}) = 0 \text{ if } r \neq p,0$$

$$\lim_n H^s(B_n^-,\mathcal{O}) = 0 \text{ if } s \neq q,0 .$$

Then it follows:

Proposition. $\lim_n H^j(S_{B_n},[\mathcal{H}^{(1)}]) = 0$ if $j \neq p,q,0$ and is infinite dimensional for $j = 0,p,q$.

LECTURE 7. THE POINCARE' LEMMA.

In the first lecture we have seen that the Poincaré lemma always holds for complexes of differential operators with constant coefficients and at the end of last lecture we showed that the (C^∞) Poincaré lemma does not hold for the boundary complex of $\partial\bar\partial$ (and for the boundary Cauchy Riemann complex) in dimensions p,q, if the Levi form restricted to the analytic tangent plane has signature (p,q) $(p+q$ = complex dimension of the analytic tangent plane).

The problem to establish the validity of the Lemma of Poincaré is a very central one in the theory of complexes of differential operators with variable coefficients.

Remark. Because the validity of the Poincaré lemma is a local question, we will only consider operators defined on open subsets Ω of a euclidean space $\mathbb{R}^n$ (in the case of differential operators between vector bundles we reduce to a fixed local trivialization).

a) Let $$\mathcal{E}^{p_0}(\Omega) \xrightarrow{A^0(x,D)} \mathcal{E}^{p_1}(\Omega) \xrightarrow{A^1(x,D)} \mathcal{E}^{p_2}(\Omega) \longrightarrow \dots \qquad (*)$$

be a complex of differential operators with smooth coefficients in an open set $\Omega \subset \mathbb{R}^n$. Let $x_0 \in \Omega$: we denote :

$\mathcal{E}_{x_0}$ = germ of C^∞ functions at x_0

Φ_{x_0} = formal power series in $x - x_0$

$\mathcal{A}_{x_0}$ = germs of (real) analytic, complex valued functions at x_0

We say that the complex (*) admits the Poincaré lemma at x_0 if the sequence:

$$\mathcal{E}^{p_0}_{x_0} \xrightarrow{A^0(x,D)} \mathcal{E}^{p_1}_{x_0} \xrightarrow{A^1(x,D)} \mathcal{E}^{p_2}_{x_0} \longrightarrow \dots \text{ is exact.}$$

(This means: the integrability conditions $A^{j+1}(x,D)f = 0$ are sufficient for the local solvability at x_0 of the equation $A^j(x,D)u = f$).

Substituting to the coefficients of $A^j(x,D)$ the corresponding Taylor series at x_0 one consider the complex

$$\Phi_{x_0}^{p_0} \xrightarrow{A^0(x,D)} \Phi_{x_0}^{p_1} \xrightarrow{A^1(x,D)} \Phi_{x_0}^{p_2} \longrightarrow \;\cdots\cdots$$

If this sequence is exact, we say that the given complex admits the formal Poincaré lemma.

<u>Remark</u>. The Poincaré lemma for the complex (*) can be false for somehow trivial reasons, i.e. in the case, for instance, that $A^1(x,D)$ does not contain sufficiently many integrability conditions for $A^0(x,D)$. In this sense, the validity of the formal Poincaré lemma is a good algebraic condition that says that all integrability conditions at x_0 have been taken into account.

<u>Remark</u>. (C^∞ vs. formal Poincaré lemma).

A differential operator $A(x,D)\colon \mathcal{E}^p(\Omega) \longrightarrow \mathcal{E}^q(\Omega)$ is honest at x_0 if the following is true:

$\forall$ distribution T with support reduced to the point x_0, $T \;\;'^{p}_{x_0}$, such that neighborhood B of x_0 in there is a distribution $S \;\;'^{q}(B)$ such that ${}^tA(x,D)U = T$.

We have the following

<u>Theorem</u>. <u>If the operators of the complex</u> (*) <u>are honest at</u> x_0 <u>and if the Poincaré lemma for the complex</u> (*) <u>holds at</u> x_0, <u>then also the formal Poincaré lemma holds at</u> x_0 .

Let us assume that all operators of the complex have real analytic coefficients in Ω . We say that the complex (*) admits the analytic Poincaré lemma at x_0 if the sequence:

$$\mathcal{A}_{x_0}^{p_0} \xrightarrow{A^0(x,D)} \mathcal{A}_{x_0}^{p_1} \xrightarrow{A^1(x,D)} \mathcal{A}_{x_0}^{p_2} \longrightarrow \;\ldots \;\text{ is exact.}$$

<u>Examples</u>.

1) Let $A^{(j)}(\Omega)$ be the space of exterior differential forms of degree j with C^∞ coefficients in Ω . We consider the complex (d = exterior differentiation):

$$A^{(0)}(\Omega) \xrightarrow{d} A^{(1)}(\Omega) \xrightarrow{d} A^{(2)}(\Omega) \rightarrow \;\ldots$$

This is a complex of differential operators with constant coefficients and then admits the C^∞, analytic and formal Poincaré lemma (the proof of local solvability was first given for this complex by Vito Volterra and then by Poincaré, from which the lemma is named).

2) (Non-honest differential operators).

With Ω open in $\mathbb{R}^2$, where x,y are cartesian coordinates, we consider the complex:

$$A^{(0)}(\Omega) \xrightarrow{d} A^{(1)}(\Omega) \xrightarrow{\exp(-1/(x^2+y^2))\ d} A^{(2)}(\Omega) .$$

This complex admits the C^∞ but not the formal Poincaré lemma at 0. (From the remarks above it follows that the last operator is not "honest").

3) Consider the complex:

$$\mathcal{E}(\Omega) \xrightarrow{x^2(d/dx)+1} \mathcal{E}(\Omega) \to 0 , \quad \Omega \text{ open } \subset \mathbb{R}^1, \text{ x cartesian coordinate on } \mathbb{R}.$$

This complex admits the formal but not the analytic Poincaré lemma at 0 : the equation $x^2 u' + u = x^2$ admits the unique formal series solution

$$u = \sum_1^\infty (-1)^{h+1}\ h!\ x^{h+1}$$

that is not convergent.

4) Let us consider the complex:

$$\mathcal{E}(\mathbb{R}^3) \xrightarrow{\partial/\partial x_1 + i\ \partial/\partial x_2 - 2i(x_1+ix_2)\partial/\partial x_3} \mathcal{E}(\mathbb{R}^3) \to 0$$

(Hans Lewy's example). This has analytic coefficients and admits at every point the formal and analytic Poincaré lemma, but not the Poincaré lemma. This example has been the starting point of many investigations on local solvability.

b. Formal and Analytic Poincaré lemma.

Assume that multigrading have been introduced as in Lecture 4, so that (principal) symbols are well defined. We can also reduce to the case of short complexes:

(**) $$\mathcal{E}^p(\Omega) \xrightarrow{A(x,D)} \mathcal{E}^q(\Omega) \xrightarrow{B(x,D)} \mathcal{E}^r(\Omega) .$$

We consider the complex

(*) $$\mathcal{P}^p \xleftarrow{{}^t\hat{A}(x_0,\xi)} \mathcal{P}^q \xleftarrow{{}^t\hat{B}(x_0,\xi)} \mathcal{P}^r .$$

We have the following:

Theorem. If the complex (*) is exact, then the complex (**) admits the formal Poincaré lemma at x_0.

Essentially by means of the preparation lemma for differential operators of lecture (4) and a variant of Cauchy-Kowalewska theorem, we obtain

Theorem. If $A(x,D)$ and $B(x,D)$ have real analytic coefficients and the sequence (*) is exact, then the complex (**) admits the analytic Poincaré lemma at x_0.

c. C^∞ - Poincaré lemma for Elliptic Complexes with analytic coefficients.

Theorem. Assume:

i) The sequence (*) is exact.

ii) $A(x,D)$ and $B(x,D)$ have real-analytic coefficents.

iii) The complex (**) is elliptic at x_0 : i.e. $\forall\ \xi^0 \in \mathbb{R}^n \sim \{0\}$ the sequence:

$$\mathbb{C}^p \xrightarrow{\hat{A}(x_0,\,\xi^0)} \mathbb{C}^q \xrightarrow{\hat{B}(x_0,\,\xi^0)} \mathbb{C}^p \quad \text{is exact.}$$

Then the complex (**) admits the Poincaré lemma at x_0.

Proof. Let $a_1,\dots,a_p$; $b_1,\dots,b_q$; $c_1,\dots,c_p$ all $\geqslant 0$ be the multigradings on $\mathcal{E}^p(\Omega)$, $\mathcal{E}^q(\Omega)$, $\mathcal{E}^r(\Omega)$ respectively, and let $m \in \mathbb{Z}$ be an upper bound for them. Set $\Delta = -\sum_1^n \partial^2/\partial x_i^2$ for the Laplace operator and define:

$$\Delta^{2m-a} : \mathcal{E}^p(\Omega) \to \mathcal{E}^p(\Omega) \quad \text{by} \quad \Delta^{2m-a} = \operatorname{diag} < \Delta^{2m-a_1},\dots, \Delta^{2m-a_p} >$$

$$\Delta^{m-b} : \mathcal{E}^q(\Omega) \to \mathcal{E}^q(\Omega) \quad \text{by} \quad \Delta^{m-b} = \operatorname{diag} < \Delta^{m-b_1},\dots, \Delta^{m-b_q} >$$

$$\Delta^{c} : \mathcal{E}^p(\Omega) \to \mathcal{E}^r(\Omega) \quad \text{by} \quad \Delta^{c} = \operatorname{diag} < \Delta^{c_1},\dots, \Delta^{c_r} > .$$

Let $\square(x,D) : \mathcal{E}^q(\Omega) \to \mathcal{E}^q(\Omega)$ be defined by :

$$\square(x,D) = A(x,D)\,\Delta^{2m-a}\,A^*(x,D) + \Delta^{m-b}\,B^*(x,D)\Delta^c\,B(x,D)\Delta^{m-b}.$$

This differential operator corresponds to multigradings $(2m-b_j, b_i-2m)$ and is elliptic (in the sense of Douglis and Nirenberg), with real analytic coefficients

($A^*(x,D)$ and $B^*(x,D)$ are the formal adjoints conjugated).

Let $f \in \mathcal{E}^q_{x_0}$ be such that $B(x,D)f = 0$. Because $\Box$ is elliptic, we can find $w \in \mathcal{E}^q_{x_0}$ such that $\Box w = f$. Replacing u by $u - \Delta^{2m-a} A^*(x,D)w = v$ the equation $A(x,D)u = f$ becomes $A(x,D)v = \Delta^{m-b}\mu$ with

$$\mu = B^*(x,D)\Delta^c B(x,D)\Delta^{m-b} w .$$

But we have $\Box\mu = 0$. Hence $\mu \in \mathcal{A}^q_{x_0}$ and thus we are reduced to the analytic Poincaré lemma.

d. Operators with Analytic Coefficients.

The problem of constructing "resolutions" of a given differential operator is solved by the syzygies theorem of Hilbert (as we have seen in the first lecture). We have seen that complexes of differential operators with variable coefficients arise as boundary complexes of complexes of operators with constant coefficients in the last lecture. However, the problem that faces us is the existence of sufficiently many complexes of differential operators with variable coefficients. In this section we will show that resolutions exist in the case of real analytic coefficients and that the condition (exactness of (*)) that was the essential assumption to obtain formal and analytic Poincaré lemma is a very natural one.

1. Let $\mathcal{F}(\Omega)$ denote any subring of the ring $\mathcal{E}(\Omega)$ of C^∞ functions on Ω, stable by differentiation.

Examples: $\mathcal{F}(\Omega) = \mathcal{E}(\Omega)$, $\mathcal{F}(\Omega) = \mathcal{A}(\Omega)$, $\mathbb{R}^n = \mathbb{C}^m$ and $\mathcal{F}(\Omega) = \mathcal{O}(\Omega) = \Gamma(\Omega,\mathcal{O})$, holomorphic functions on Ω. We consider a complex of differential operators

$$(\#) \qquad \mathcal{F}^{p_0}(\Omega) \xrightarrow{A^0(x,D)} \mathcal{F}^{p_1}(\Omega) \xrightarrow{A^1(x,D)} \mathcal{F}^{p_2}(\Omega) \longrightarrow \dots$$

corresponding to multigradings $a_1,\dots,a_{p_0}$ on $\mathcal{F}^{p_0}(\Omega)$, $b_1,\dots,b_2$, on $\mathcal{F}^{p_1}(\Omega)$, $c_1,\dots,c_{p_2}$ on $\mathcal{F}^{p_2}$ ....

Let $\mathcal{H}(\Omega)$ denote the graded ring of homogeneous polynomials in $\xi_1,\dots,\xi_n$ with coefficients in $\mathcal{F}(\Omega)$: $\mathcal{H}(\Omega) = \mathcal{F}(\Omega)_0[\xi_1,\dots,\xi_n]$ and we consider the complex of multigraded $\mathcal{H}(\Omega)$-homomorphisms:

$$(\#) \qquad \mathcal{H}^{p_0}(\Omega) \xrightarrow{{}^t\hat{A}^0(x,\xi)} \mathcal{H}^{p_1}(\Omega) \xrightarrow{{}^t\hat{A}^1(x,\xi)} \mathcal{H}^{p_2}(\Omega) \longrightarrow \dots$$

Let $N_\Omega = \operatorname{coker} {}^t\hat{A}^0(x,\xi) : \mathcal{H}^{p_1}(\Omega) \to \mathcal{H}^{p_0}$.

When $(\#)$ is an exact sequence, we say that the complex $(\#)$ with the given multigrading is a <u>correct complex</u>.

Let $\mathfrak{m}_{x_0}(\Omega)$ denote the ideal of $\mathcal{F}(\Omega)$ of functions vanishing at x_0. We assume that the linear functional $\delta_{x_0} : \mathcal{F}(\Omega) \longrightarrow \mathbb{C}$ (Dirac's measure) is onto, so that $\mathbb{C} = \mathcal{F}(\Omega)/\mathfrak{m}_{x_0}(\Omega)$ can be considered as an $\mathcal{F}(\Omega)$ - module.

Tensoring the sequence $(\#)$ by $\mathbb{C}$ we then obtain a complex :

$$\mathcal{H}^{p_0} \xleftarrow{{}^t\hat{A}^0(x_0,\xi)} \mathcal{H}^{p_1} \xleftarrow{{}^t\hat{A}^1(x_0,\xi)} \mathcal{H}^{p_2} \longleftarrow \ldots.$$

where $\mathcal{H} = \mathbb{C}_0[\xi_1,\ldots,\xi_n]$ is the ring of homogeneous polynomials in n variables $\xi_1,\ldots,\xi_n$.

If the complex $(\#)$ is correct, the homology of $(\#)$ is given by the $\mathcal{F}(\Omega)$- - modules:

$$\mathrm{Tor}^j_{\mathcal{F}(\Omega)}(N_\Omega, \mathcal{F}(\Omega)/\mathfrak{m}_{x_0}(\Omega)) \qquad (j \geqslant 1) :$$

i.e. given a correct complex of differential operators with coefficients in $\mathcal{F}(\Omega)$ the condition $0 = \mathrm{Tor}^j_{\mathcal{F}(\Omega)}(N_\Omega, \mathcal{F}(\Omega)/\mathfrak{m}_{x_0}(\Omega))$ $\forall j \geqslant 1$ is necessary and sufficient for the symbolic complex to be exact at x_0 .

We have also: <u>If $(\#)$ is a correct complex of differential operators</u>, $\forall\, x_0 \in \Omega$:

i) $\mathrm{Tor}^j_{\mathcal{F}(\Omega)}(N_\Omega, \mathcal{F}(\Omega)/\mathfrak{m}_{x_0}(\Omega)) = 0$ <u>for</u> $j > n$ <u>if</u> $\mathcal{F}(\Omega) = \mathcal{E}(\Omega)$ or $\mathcal{F}(\Omega) = \mathcal{A}(\Omega)$

ii) $\mathrm{Tor}^j_{\mathcal{F}(\Omega)}(N_\Omega, \mathcal{F}(\Omega)/\mathfrak{m}_{x_0}(\Omega)) = 0$ <u>for</u> $j > m = \frac{n}{2}$ if $\mathbb{R}^n = \mathbb{C}^m$ and $\mathcal{F}(\Omega) = \mathcal{O}(\Omega)$.

2. The following is a very important result:

<u>Proposition</u>. <u>Let Ω be an open connected set, and let $\mathcal{F}(\Omega) = \mathcal{A}(\Omega)$ or $\mathcal{F}(\Omega) = \mathcal{O}(\Omega)$ (in the last case $\mathbb{R}^n = \mathbb{C}^m$). If the complex $(\#)$ is correct, then</u> $\forall\, j \geqslant 1$

$$A_j = \{ x \in \Omega \mid \mathrm{Tor}^j_{\mathcal{F}(\Omega)}(N_\Omega, \mathcal{F}(\Omega)/\mathfrak{m}_x(\Omega)) \neq 0 \}$$

<u>is a proper analytic subset of Ω</u> .

This statement is a consequence of a lemma on homogeneous Hilbert resolutions for matrices of polynomials with complex coefficients in n indeterminates:

Lemma. Given n,p,q and integers $a_1,\dots,a_p, b_1,\dots,b_q$ we can find an integer $\sigma \geq 0$ with the property:

$\forall$ homogeneous $q \times p$ matrix $M(\xi)$ of gradings (a_j,b_i), $\operatorname{Ker} M(\xi)$ has a set of generators that are polynomials with homogeneous components of degree $\leq \sigma$.

From this lemma it follows that, if we have a complex

$$(\#) \qquad \mathcal{H}^p \xrightarrow{M(\xi)} \mathcal{H}^q \xrightarrow{N(\xi)} \mathcal{H}^r$$

with M homog. of deg. (a_j,b_i) and N homog. of deg. (b_j,c_i) and we set

$$(\mathcal{H}^p)_h = \{v = \begin{pmatrix} v_1 \\ \vdots \\ v_p \end{pmatrix} \in \mathcal{H}^p \mid v_j \text{ is homogeneous of deg. } -a_j+h \quad j = 1\dots p\}$$

$$(\mathcal{H}^q)_h = \{w = \begin{pmatrix} w_1 \\ \vdots \\ w_q \end{pmatrix} \in \mathcal{H}^q \mid w_j \text{ is homogeneous of deg. } h-b_j \quad j = 1\dots q\}$$

$$(\mathcal{H}^r)_h = \{z = \begin{pmatrix} z_1 \\ \vdots \\ z_r \end{pmatrix} \in \mathcal{H}^r \mid z_j \text{ is homogeneous of deg. } h-c_j \quad j = 1\dots r\}$$

Then we can find an integer h_0 such that the complex $(\#)$ is acyclic if and only if

$$(\#)_h \qquad (\mathcal{H}^p)_h \xrightarrow{M(\xi)} (\mathcal{H}^q)_h \xrightarrow{N(\xi)} (\mathcal{H}^r)_h$$

is exact $\forall\, h \leq h_0$.

These statements reduce the proof of the proposition above to linear algebra (matrices operating on finite dimensional vector spaces over $\mathbb{C}$ with coefficients in $\mathcal{F}(\Omega)$).

Remark. When $\mathcal{F}(\Omega) = \mathcal{E}(\Omega)$, the same argument shows that the sets A_j are closed (intersection of the set of zeros of finitely many functions in $\mathcal{E}(\Omega)$).

3. Let K be a compact set and let $\mathcal{F}(K) = \varinjlim_{\omega \text{ open} \supset K} \mathcal{F}(\omega)$. We have also in this case analogous statements to those given in the previous subsection. The advantages of taking a compact K are given by the following statements, that are consequences of a theorem of Frish:

α) <u>If</u> K <u>is a compact semianalytic subset of</u> $\mathbb{R}^n$, <u>then</u> $\mathcal{A}(K)$ <u>is noetherian.</u>

β) <u>If</u> $K \subset \mathbb{C}^m$ <u>is a compact semianalytic set of holomorphy , then</u> $\mathcal{O}(K)$ <u>is noetherian.</u>

4. With $\mathcal{F}$ equals to $\mathcal{A}$ or $\mathcal{O}$, and K satisfying the corresponding assumptions α) or β) above, we denote by $\mathcal{D}(K)$ the ring of differential operators with coefficients in $\mathcal{F}(K)$.

Setting $\mathcal{D}_h(K) = \{$diff. operators in $\mathcal{D}(K)$ of order $\leqslant h\}$ we obtain a filtration of $\mathcal{D}(K) : \mathcal{D}_0(K) \subset \mathcal{D}_1(K) \subset \mathcal{D}_2(K) \subset \dots$. The associated graded ring is isomorphic to the graded ring $\mathcal{H}(K)$ of homogeneous polynomials with coefficients in $\mathcal{F}(K)$:

$$\mathcal{H}(K) = \mathcal{F}(K)_0 [\xi_1, \dots, \xi_n] .$$

(The isomorphism is obtained by associating to the element
$a(x,D) = \sum_{|\alpha| \leqslant h} a_\alpha(x) D^\alpha \in \mathcal{D}_h(K)$ the symbol $\hat{a}(x,\xi) = \sum_{|\alpha|=h} a_\alpha(x) \xi^\alpha$).

Given a $p_1 \times p_0$ matrix $A^0(x,D)$ with entries in $\mathcal{D}(K)$, we associate to it the $\mathcal{D}(K)$ - homomorphism:

$$\mathcal{D}(K)^{p_1} \xrightarrow{A^0(x,D)} \mathcal{D}(K)^{p_0}$$

that sends the row vectors $v = (v_1(x,D), \dots, v_q(x,D)) \in \mathcal{D}(K)^{p_1}$ into the row vector $v(x,D)\, A^0(x,D)$.

Given a multigrading (a_j, b_j) for $A^0(x,D)$, we set $\forall\, h \in \mathbb{Z}$:

$$(\mathcal{D}(K)^{p_0})_h = \{(r_1, \dots, r_{p_0}) \in \mathcal{D}(K)^{p_0} \mid \text{order of } r_j \leqslant a_j + h \quad j=1\dots p_0\}$$

$$(\mathcal{D}(K)^{p_1})_h = \{(s_1, \dots, s_{p_1}) \in \mathcal{D}(K)^{p_1} \mid \text{order of } s_j \leqslant b_j + h \quad j=1,\dots p_1\} .$$

Then $A^0(x,D) : (\mathcal{D}(K)^{p_1})_h \to (\mathcal{D}(K)^{p_0})_h \quad \forall\, h \in \mathbb{Z}$.

Taking the associated graded rings we obtain the maps:

$$\mathcal{H}(K)^{p_1} \xrightarrow{{}^t\hat{A}^0(x,\xi)} \mathcal{H}(K)^{p_0} \quad \text{and} \quad (\mathcal{H}(K)^{p_1})_h \xrightarrow{{}^t\hat{A}^0(x,\xi)} (\mathcal{H}(K)^{p_0})_h \quad .$$

We always have: $\operatorname{Im} {}^t\hat{A}^0(x,\xi) \subset {}^t\,\widehat{\operatorname{Im} A^0(x,D)}$.

<u>Definition</u>. <u>We say that the operator</u> $A^0(x,D)$, <u>with multigradings</u> (a_j, b_i) , <u>is involutive if</u>

$$\operatorname{Im} {}^t\hat{A}^0(x,\xi) = {}^t\widehat{\operatorname{Im} A^0(x,D)} .$$

Proposition. Under the above specified assumption on K and $\mathcal{F}$:

1) Given any differential operator $A^0(x,D): \mathcal{D}^{p_1}(K) \to \mathcal{D}^{p_0}(K)$ one can always find a matrix $M(x,D): \mathcal{D}^{p_1}(K) \to \mathcal{D}^{q_1}(K)$ such that

   i) for a suitable grading $\beta_1,\dots,\beta_{q_1}$ of $\mathcal{F}^{q_1}(K)$ the operator $M(x,D)\,A^0(x,D)$ is of type (a_j,β_i) and is involutive on K.

   ii) There exists a matrix $N(x,D) : \mathcal{D}^{q_1}(K) \to \mathcal{D}^{p_1}(K)$ such that

$$N(x,D)\,M(x,D)\,A^0(x,D) = A^0(x,D) = A^0(x,D) \text{ on } K .$$

2) If $A^0(x,D)$ is involutive on K, then:

$$\widehat{{}^t\operatorname{Ker} A^0(x,D)} = \operatorname{Ker} {}^t\hat{A}^0(x,\xi) .$$

This proposition reduces the proof of the existence of "good resolutions" for differential operators to resolutions of homogeneous operators on graded rings.

The following is the final result:

Theorem. Let K be a connected semianalytic compact set, which when $\mathbb{R}^n = \mathbb{C}^m$ and $\mathcal{F}(\Omega) = \mathcal{O}(\Omega)$ is assumed to be of holomorphy, having a fundamental sequence of topologically contractible open neighborhoods.

Let $A^0(x,D): \mathcal{D}(K)^{p_1} \longrightarrow \mathcal{D}(K)^{p_0}$ be a multigraded involutive operator with coefficients in $\mathcal{F}(K)$.

Then $A^0(x,D)$ admits a correct involutive resolution:

$$\mathcal{F}^{p_0}(K) \xrightarrow{A^0(x,D)} \mathcal{F}^{p_1}(K) \xrightarrow{A^1(x,D)} \mathcal{F}^{p_2}(K) \to \dots \to \mathcal{F}^{p_d}(K) \to 0$$

of lenght $d \leqslant 2n+1$.

* * *

References

L. HÖRMANDER, "*An introduction to complex analysis in several variables*" . North Holland Math. Lib. 7 (North-Holland, Amsterdam, 1973)

J. BJÖRK, "*Rings of Differential Operators*" , North Holland Math. Lib 21 (North Holland, Amsterdam, 1979)

L. EHRENPREIS, "*Fourier Analysis in Several variables*" , Pure and Appl. Math. (Wiley-Intersci., New York) 17, 1970

V. PALAMODOV, "*Linear differential operators with constant coefficients*", Grundlehren 168 (Springer-Verlag, Berlin) 1969.

A. ANDREOTTI - M. NACINOVICH, "*Complexes of Partial Differential Operators*" , Ann. Scuola Norm. Sup. Pisa Cl.Sci.(4) 3 (1976) n. 4 pp.553-621.

- "*Analytic convexity: some comments on an example of De Giorgi and Piccinini*" Complex Analysis and its applications, v. III, pp. 25-37 Vienna, 1976.
- "*Some remarks on formal Poincaré lemma*", Complex Analysis and algebraic geometry pp. 245-305. Iwanami Shoten, Tokyo, 1977.
- "*Théorie élementaire de la convexité* " IRMA, Strasbourg RCP 25, v. 24 (1977) pp. 1-12.
- "*Domains de Regularité pour les operateurs elliptiques à coefficients constant*" IRMA, Strasbourg, RCP 25, v. 26 (1978) pp.1-13 .
- "*Convexité analitique*" Springer Lecture notes 670, pp.354-364 (1979).
- "*Convexité analitique II*" Springer Lecture notes 670, pp. 410-415 (1979).
- "*On the envelope of regularity for solutions of homogeneous systems of Linear Partial Differential Operators*" Ann. Scuola Norm. Sup. Pisa cl. Sci. (4) 6 (1979) pp. 69-141.
- "*Analytic Convexity*" to app. Ann. Sc. Norm. Sup. Pisa .
- "*Analytic Convexity and the Phragmén-Lindeloff Principle*" to app. Quaderni della Scuola Normale.
- "*On the theory of analytic convexity*" to app. Atti Accad. Naz. Lincei.
- "*On analytic and* C^{∞} *Poincaré lemma*" to app. in Advances in Math.
- "*Sopra il lemma di Poincaré su complessi di operatori differenziali*" to app. Rendiconti Torino.
- "*Noncharacteristic hypersurfaces for complexes of differential operators*" to app. in Annali di Matematica.

- "*Some remarks on complexes of differential operators with constant coefficients*" preprint Istituto di Matematica L. Tonelli - Pisa.

A. ANDREOTTI- D. HILL, "*E.E. Levi convexity and the Hans Lewy problem, I and II*" Ann. Scuola Norm. Sup. Pisa (3) 26 (1972) pp. 325-363, pp. 747-806.

A. ANDREOTTI, "*Complexes of Partial Differential Operators*" , Yale Univ. Press. 1975 , pp. 49 .

A. ANDREOTTI-D. HILL-S. ŁOJASIEWICZ-B. MACKICKAN, "*Complexes of differential operators. The Mayer-Vietoris sequence*" , Invent. Math. 35 (1976) pp. 43-86.

L. HÖRMANDER, "*On the existence of analytic solutions of partial differential equations with constant coefficients*" Inv. Math. vol. 21 (1973)pp. 151-182.

A. ANDREOTTI, "*Complessi di Operatori Differenziali*", Boll. Un. Mat. Ital. A (5) 13 (1976), n. 2, pp. 273-281 .

* * *

DEFORMATIONS OF COMPLEX STRUCTURES AND HOLOMORPHIC VECTOR BUNDLES

M.S. Narasimhan
Tata Institute of Fundamental Research

Homi Bhabha Road, Colaba, Bombay 400005, India

The aim of these lectures is to give an introduction to the theory of deformations of complex structures as developed by Kodaira and Spencer. The theory studies complex structures which are near a given complex structure on a compact differentiable manifold. One also has an analogous theory of deformations of holomorphic vector bundles on a compact complex manifold.

It is well known that the real two dimensional torus carries a one parameter family of distinct complex structures. The complex structures on a compact connected orientable (real) 2-manifold of genus $g \geq 2$ depend on (3g-3) complex parameters, so called moduli. In this theory, the number of moduli, 3g-3, appears as the dimension of the first cohomology group of a compact Riemann surface with coefficients in its tangent sheaf.

1. Families of complex structures

Definition 1.1. Let T be a connected complex manifold. A holomorphic family of compact complex manifolds parametrised by T is a (paracompact) complex manifold $\mathcal{V}$ together with a proper surjective holomorphic map $p: \mathcal{V} \to T$ which is of maximal rank.

Remark 1.1. 'Maximal rank' means that for every $x \in \mathcal{V}$ the differential of p is surjective from the tangent space to $\mathcal{V}$ at x to that of T at $p(x)$. 'Proper' means that the inverse image by p of every compact set of T is compact in $\mathcal{V}$.

Remark 1.2. By the implicit function theorem, for $t \in T$, $p^{-1}(t)$ has a natural structure of compact complex manifold, which we shall denote by V_t. If $t_o \in T, V_t$ is said to be a deformation of V_{t_o}. If t is near t_o we may say that the complex structure V_t is near V_{t_o}. We shall assume that each V_t is connected.

Remark 1.3. We shall see in the next proposition that for $t_1, t_2 \in T, V_{t_1}$ and V_{t_2} are diffeomorphic. But they need not be complex analytically isomorphic. The ideas used in the proof of this proposition would also be useful later.

Proposition 1.1. Let $p: \mathcal{V} \to T$ be a proper, surjective smooth (= C^∞) map of maximal rank between two smooth manifolds. Then $\mathcal{V}$ is a locally trivial fibre space. That is, every point t in T has a neighbourhood N such that there exists a diffeomorphism $\Phi : p^{-1}(N) \to N \times V_t$ with commutativity in the diagram

$$\begin{array}{ccc} p^{-1}(N) & \xrightarrow{\ \Phi\ } & N \times V_t \\ \downarrow p & & \downarrow p_N \\ N & \xrightarrow{\text{Identity}} & N \end{array}$$

where p_N denotes the projection from $N \times V_t$ onto the factor N.

Lemma 1.1. Let X be a smooth vector field on T. Then there exists a (smooth) vector field $\tilde{X}$ on V such that for any $x \in V$ we have $dp(\tilde{X}(x)) = X(p(x))$, where dp denotes the differential map of p.

Remark 1.4. We shall say that $\tilde{X}$ is a projectable vector field on V and that X is its projection on T.

Proof of Lemma 1.1. By the implicit function theorem we may assume that there is an open covering $\{U_\alpha\}$ of V such that there is a commutative diagram

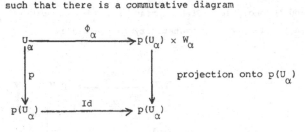

$$\begin{array}{ccc} U_\alpha & \xrightarrow{\ \Phi_\alpha\ } & p(U_\alpha) \times W_\alpha \\ \downarrow p & & \downarrow \text{projection onto } p(U_\alpha) \\ p(U_\alpha) & \xrightarrow{\ \text{Id}\ } & p(U_\alpha) \end{array}$$

where Φ_α is a diffeomorphism and W_α are open in $\mathbb{R}^n$. ($p(U_\alpha)$ is open in T). Let $X_\alpha = X|p(U_\alpha)$. Using the product structure on U_α we can find a projectable vector field $\tilde{X}_\alpha$ on U_α projecting into X (For example we may take the horizontal lift with respect to the product decomposition: given by $(X,0)$). Consider in $U_\alpha \cap U_\beta$ the vector field $\Theta_{\alpha\beta} = \tilde{X}_\beta - \tilde{X}_\alpha$, which is tangential to the fibres. By using a partition of unity we can find a vector field Y_α in U_α, tangential to the fibres, such that $Y_\beta - Y_\alpha = \Theta_{\alpha\beta}$. (A tangent vector in V is said to be tangential to the fibres if its image by the differential map is zero. Such vectors build a vector subbundle of the tangent bundle of V). Then $\tilde{X}_\alpha - Y_\alpha = \tilde{X}_\beta - Y_\beta$ in $U_\alpha \cap U_\beta$ and hence define a globally defined vector field $\tilde{X}$ on V projecting into X.

Proof of Proposition 1.1. We shall give the proof in the case $\dim T = 1$. The general case can be proved by an easy induction using Lemma 1.1. We may assume that T is an interval in $\mathbb{R}$ containing 0. Take $X = \frac{d}{dt}$ in Lemma 1.1. Let N be a neighbourhood 0 with $\bar{N}$ compact and $\bar{N} \subset T$. Then $p^{-1}(\bar{N})$ is compact. The local one parameter group of transformations $\exp(t\tilde{X})$ generated by the vector field $\tilde{X}$ is defined on the compact set $p^{-1}(\bar{N})$ for $|t|<\varepsilon$ for some $\varepsilon > 0$. It is immediate that $\exp(-t\tilde{X})$ maps V_t diffeomorphically onto V_o for small t and that the map

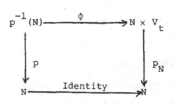

$x \to (p(x) = t, \exp(-t\tilde{X})x)$ gives a diffeomorphism of $p^{-1}(N')$ onto $N' \times V_o$ for some neighbourhood N' of 0.

2. The infinitesimal deformation map

Let $p : \mathcal{V} \to T$ be a holomorphic family of complex structures. Let $t \in T$ and T_t denote the holomorphic tangent space to T at t. We shall define a linear map

$$\rho_t : T_t \to H^1(V_t, \Theta)$$

where Θ denotes the sheaf (of germs) of holomorphic vector fields on V_t and $H^1(V_t, \Theta)$ denotes the first cohomology group of V_t with coefficients in the sheaf Θ. The map ρ_t will be called the infinitesimal deformation or the Kodaira-Spencer map.

By the implicit function theorem we can assume that there is an open coverg $\{U_\alpha\}$ of $\mathcal{V}$ and holomorphic diffeomorphism $\Phi_\alpha : U_\alpha \to p(U_\alpha) \times W_\alpha$ for each α, where W_α is open in $\mathbb{C}^n$, with commutativity in the diagram

$$\begin{array}{ccc} U_\alpha & \xrightarrow{\Phi_\alpha} & p(U_\alpha) \times W_\alpha \\ {\scriptstyle p}\downarrow & & \downarrow{\scriptstyle \text{projection}} \\ p(U_\alpha) & \longrightarrow & p(U_\alpha) \end{array}$$

Let v be a tangent vector at t. We can lift v into a holomorphic section of $\mathcal{T}|U_\alpha \times V_t$ projecting into v, where $\mathcal{T}$ is the holomorphic tangent bundle of $\mathcal{V}$. Let us call this section $\tilde{v}_\alpha$. In $U_\alpha \cap U_\beta \cap V_t$, $\tilde{v}_\beta - \tilde{v}_\alpha$ is a vector field tangent to V_t, as $\tilde{v}_\beta$ and $\tilde{v}_\alpha$ have the same projection v. Then elements $\theta_{\alpha\beta} = \tilde{v}_\beta - \tilde{v}_\alpha$ clearly satisfy the cocycle condition and hence define an element of $H^1(V_t, \Theta)$. One can check that this element in $H^1(V_t, \Theta)$ depends only on v and not on the other choices. We define this element to be $\rho_t(v)$.

It is clear that if $\mathcal{V} = V \times T$, then the map ρ is zero.

3. Deformations of holomorphic Vector bundles

Let V be a compact complex manifold and T a complex manifold, both connected.

Definition 3.1. A (holomorphic) family of holomorphic vector bundles on V parametrised by T is a holomorphic vector bundle $\mathcal{W}$ on $T \times V$.

Remark 3.2. If $t \in T$, the restriction of $\mathcal{W}$ to $t \times V$ can be considered as a holomorphic vector bundle on V, say W_t. We thus have a family $\{W_t\}$ of holomorphic bundles on V parametrised by T.

If $t_o \in T$, we have in this case, an infinitesmial deformation map

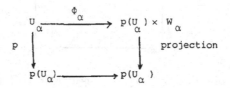

$$\eta_{t_o} : T_{t_o} \longrightarrow H^1(V, \underline{\text{End}}\, W_{t_o})$$

where $\underline{\text{End}}\, W_{t_o}$ denote the sheaf of germs of endomorphisms of the vector bundle W_{t_o}. To define this map, note that we can find an open cover $\{U_\alpha\}$ of V such that for t in a neighbourhood U of t_o the transition functions of W_t are given by $g_{\alpha\beta}(x,t)$, where $g_{\alpha\beta}$ are holomorphic in $(U_\alpha \cap U_\beta) \times U$. $[x \in U_\alpha \cap U_\beta]$. The matrix valued functions

$$v(g_{\alpha\beta}(x,t))\, g^{-1}_{\alpha\beta}(x,t_o),$$

considered as an endomorphism of $W_{t_o} | U_\alpha \cap U_\beta$ by using the trivialisation with respect to U_α, define a 1-cocycle of the sheaf $\underline{\text{End}}\, W_{t_o}$ and hence an element of $H^1(V_{t_o}, \underline{\text{End}}\, W_{t_o})$. This is the element $\eta_{t_o}(v)$.

4. Basic Theorems

Theorem 4.1 (Rigidity Theorem).

If $H^1(V_{t_o}, \Theta) = 0$ there exists a neighbourhood U of t_o such that $p^{-1}(U)$ is (holomorphically) a product i.e, we have a holomorphic isomorphism $\Phi : p^{-1}(U) \to U \times V_{t_o}$ with the commutative diagram

$$\begin{array}{ccc} p^{-1}(U) & \xrightarrow{\ \Phi\ } & U \times V_{t_o} \\ \downarrow{\scriptstyle p} & & \downarrow{\scriptstyle p_U} \\ U & \xrightarrow{\ Id\ } & U \end{array}$$

Theorem 4.2. (Existence of deformations of complex structures). Let V be a compact complex manifold with $H^2(V, \Theta) = 0$. Then there exists a family $\mathcal{V}$ of complex manifolds parametrised by a complex manifold T such that

i) there exists $t_o \in T$ such that V_{t_o} is isomorphic to V.

ii) the infinitesimal deformation map $\rho_{t_o} : T_{t_o} \longrightarrow H^1(V_{t_o}, \Theta)$ is an isomorphism.

Before we state the next theorem we make

Definition 4.1. Let $\mathcal{V} \to T$ be a family of (compact) complex manifolds. Let $t_o \in T$. We say that the family $\mathcal{V}$ is complete at t_o if the following condition is satisfied. Let $\mathcal{M} \xrightarrow{q} S$ be a family of complex manifolds with M_{s_o} isomorphic holomorphically to V_{t_o} for some $s_o \in S$. Then there exists a neighbourhood U of s_o in S and a holomorphic map $\varphi : U \to T$ with $\varphi(s_o) = t_o$ such that the pull-back of the family $\mathcal{V} \to T$ by φ is isomorphic to the restriction, $\mathcal{M}|U$, of $\mathcal{M}$ to U. (This means that there exists a holomorphic map $\tilde{\varphi} : q^{-1}(U) \to \mathcal{V}$ with a commutative diagram

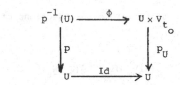

$$\begin{array}{ccc} q^{-1}(U) & \xrightarrow{\widetilde{\varphi}} & V \\ \downarrow q & & \downarrow p \\ U & \xrightarrow{\varphi} & T \end{array}$$

and such that $\widetilde{\varphi}\,|M_s : M_s \to V_{f(s)}$ is an isomorphism for $s \in U$).

The definition means that in some sense all small deformations of V_{t_o} are found in the family $\{V_t\}$.

<u>Theorem 4.3</u>. (Theorem of Completeness)

Let $V \to T$ be a family of complex structures such that $\rho_{t_o} : T_{t_o} \to H^1(V_{t_o}, \Theta)$ is surjective for $t_o \in T$. Then the family V is complete at t_o.

<u>Theorem 4.4</u>. (Semi-Continuity Theorem. Invariance of the Euler characteristic).

Let W be a holomorphic vector bundle on V and let W_t denote the holomorphic bundle on V_t obtained by restricting W to V_t. Then

i) For any $t_o \in T$ there exists a neighbourhood U of t_o such that

$$\dim_{\mathbb{C}} H^q(V_t, \underline{W}_t) \leq \dim_{\mathbb{C}} H^q(V_{t_o}, \underline{W}_{t_o})$$

for any q, where $\underline{W}_t$ denotes the sheaf of holomorphic sections of W_t; i.e., the set $S_k = \{t \in T \mid \dim_{\mathbb{C}} H^q(V_t, W_t) \geq k\}$ is closed.

ii) The Euler characteristic

$$\chi(V_t, \underline{W}_t) = \sum_{i=0} (-1)^i \dim_{\mathbb{C}} H^i(V_t, \underline{W}_t)$$

is independent of t.

<u>Remark, 4.1</u>. The set S_k is in fact an analytic set [7].

Theorem 4.1 is proved in [1], Th.4.2 in [3], Th.4.3 in [2] and Th.4.4 in [4].

By Theorem 4.3, the family constructed in Theorem 4.2 is complete. In 4.2, the question arises as to when, given $t_1, t_2 \in T$, the manifolds V_{t_1} and V_{t_2} are isomorphic. In some special cases one has a satisfactory answer (see [6]). For instance if V is a compact Riemann surface of genus $g \geq 2$ which has no non-trivial holomorphic automorphism, one can choose a neighbourhood U of t_o such that V_{t_1} and V_{t_2} are isomorphic if and only if $t_1 = t_2$ for $t_1, t_2 \in U$.

Note that the analogues of Theorems 4.1 ~ 4.3 have obvious formulations in the case of deformations of holomorphic vector bundles on a compact complex manifold, with $H^q(V, \underline{\text{End}}\ W)$ playing the role of $H^q(V, \Theta)$. These analogous theorems are also true.

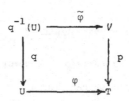

In the case of bundles there is a fairly general theorem answering the question corresponding to the above question [5, § 2].

Definition 4.2. Let W be a holomorphic vector bundle on a compact (connected) complex manifold V. We say that W is a simple bundle if the only global holomorphic endomorphisms of W are scalars. (i.e., $H^0(V, \underline{\text{End}}\, W) = \mathbb{C}$).

Theorem 4.5. Let V be a compact complex manifold and let W be a simple holomorphic bundle on V with $H^2(V, \underline{\text{End}}\, W) = 0$. Then there exists a complex manifold T and a holomorphic vector bundle $\mathcal{W}$ on $T \times V$ such that

i) For every $t \in T$, W_t is a simple bundle.

ii) W is isomorphic W_{t_0} for some $t_0 \in T$.

iii) The infinitesimal deformation map $\eta_t : T_t \to H^1(U, \underline{\text{End}}\, W_t)$ is an isomorphism for all t.

iv) If t_1 and t_2 are distinct points of T, then W_{t_1} and W_{t_2} are not isomorphic.

For example if V is a compact Riemann surface, the condition on H^2 is automatically satisfied. Suppose we consider the set $\mathcal{M}$ of isomorphism classes of simple bundles on V("the moduli space of simple bundles"). If W is a simple bundle on V, complex coordinates in a neighbourhood of t_0 in T can be used to give complex coordinates at the isomorphism class of W on $\mathcal{M}$. By the completeness property, these coordinates will patch up to give a complex structure on $\mathcal{M}$ but the resulting manifold is not in general Hausdorff.

A deeper study of the global moduli problem is possible in the frame-work of Algebraic Geometry.

5. Differentiable families of complex structures

In this section we shall define the notion of complex structures depending differentiably on a parameter. We shall later outline proofs of the analogues of Theorems 4.1 and 4.4 in this case.

Definition 5.1. Let $p: V \to T$ be a smooth proper surjective map of maximal rank, where V and T are smooth manifolds. Assume that for each $t \in T$ $p^{-1}(t)$ has a structure of a complex manifold, denoted by V_t, compatible with the differentiable structure on $p^{-1}(t)$ satisfying the following condition: There is an open covering $\{U\}$ of V and diffeomorphisms $\phi_\alpha : U_\alpha \to p(U_\alpha) \times W_\alpha$, where W_α is an open subset in $\mathbb{C}^n$, with commutativity in the diagram

$$\begin{array}{ccc} U_\alpha & \xrightarrow{\ \phi_\alpha\ } & p(U_\alpha) \times W_\alpha \\ \downarrow p & & \downarrow \text{projection onto the first factor} \\ p(U_\alpha) & \longrightarrow & p(U_\alpha) \end{array}$$

and such that the map $p_W \circ \phi_\alpha | V_t \cap U_\alpha \to W$ is a complex analytic isomorphism for $t \in p(U_\alpha)$ (Here p_W denotes the projection $p(U_\alpha) \times W_\alpha \to W_\alpha$ and on $V_t \cap U_\alpha$ we take the complex structure induced from V_t). We then say that $p: V \to T$ is a differentiable family of complex structures parametrised by the (smooth) manifold T.

We can also define the notion of a differentiable family of holomorphic bundles W_t on V_t parametrised by T. Then Theorem 4.4 makes sense in this context. An example of such a family is W_t, where W_t is the holomorphic tangent bundle of V_t.

In the proofs we shall make use of the Dolbeault complex and the theory of elliptic operators operating on sections of smooth vector bundles on compact smooth manifolds.

6. Laplacian and harmonic forms. A finite dimensional complex for Dolbeault cohomology

Let W be a holomorphic vector bundle on a compact complex manifold V. Let E^q denote the vector space of smooth differential forms of type (o,q) with coefficients in W. We then have the Dolbeault complex

$$\to E^q \xrightarrow{\bar\partial} E^{q+1} \to$$

and the cohomology of this complex at the qth stage, $H^q(E)$ is canonically isomorphic to $H^q(V,\underline{W})$, the qth cohomology group of V with coefficients in the sheaf $\underline{W}$ of holomorphic sections of W.

If we introduce hermitian metrics on V and on W, we can consider the operator

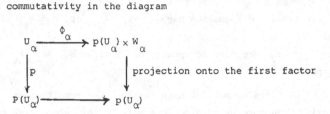

$\vartheta : E^q \to E^{q-1}$ adjoint to $\bar\partial$ and the Laplacian $\Delta = \bar\partial\vartheta + \vartheta\bar\partial : E^q \to E^q$. Let $H^q \subset E^q$ be the subspace of harmonic forms i.e. the space of elements $\omega \in E^q$ satisfying the condition $\Delta\omega = 0$ (This condition is equivalent to the conditions $\bar\partial\omega = \vartheta\omega = 0$). Then, by Hodge theory, the natural map $H^q \to H^q(E)$ is an isomorphism. Moreover we have the Hodge decomposition: for $\omega \in E^q$,

$$\omega = H\omega + \bar\partial\vartheta G\omega + \vartheta\bar\partial G\omega$$

where $H: E^q \to H^q$ is the orthogonal projection onto H^q and $G: E^q \to E^q$ is a continuous linear operator.

We shall define a certain complex of finite dimensional vector spaces which has the same cohomology as that of the Dolbeault complex. We say that a complex number λ is an eigenvalue of Δ on E^q if the space $H^q_\lambda = \{\omega \in E^q \mid \Delta\omega = \lambda\omega\}$ is different from zero. The spaces H^q_λ are finite dimensional vector spaces and the set of eigenvalues form a discrete set in $\mathbb{R}$. Let r be a positive real number and let B^q denote the finite dimensional subspace of E^q spanned by the eigenspaces $H^q_{\lambda_i}$ with $|\lambda_i| < r$. Then $B^q = \bigoplus_{|\lambda_i|<r} H^q_{\lambda_i}$ the sum being direct. Note that $H^q \subset B^q$.

<u>Proposition 6.1</u>. The operator $\bar\partial$ maps B^q into B^{q+1} and the cohomology of the complex $\{B^q, \bar\partial\}$ is isomorphic to the Dolbeault cohomology.

<u>Proof</u>. Since Δ commutes with $\bar\partial$ it is immediate that $\bar\partial$ maps B^q into B^{q+1}. Let E^q denote the cohomology space of $\{B^q, \bar\partial\}$ at the qth stage. If $\varphi \in H^q$, then $\varphi \in B^q$ and $\bar\partial\varphi = \vartheta\varphi = 0$. We thus have a natural map $H^q \to E^q$ and we shall prove that it is an isomorphism. The map is injective: For if, $\varphi \in H^q$, $\varphi = \bar\partial\psi$, $\psi \in B^{q-1}$ then $(\varphi, \varphi) = (\varphi, \bar\partial\psi) = (\vartheta\varphi, \psi) = 0$ so that $\varphi = 0$. [Here (φ, φ) denotes the scalar product on forms defined by using the metrics]. To prove that the map is surjective, let $\omega \in B^q$ with $\bar\partial\omega = 0$. Let H be the projection of B^q onto H^q with respect to the decomposition $B^q = \bigoplus_{|\lambda_i|<r} H^q_{\lambda_i}$. Then $\omega - H\omega = \Sigma\,\omega_i$, where $\omega_i \in H^q_{\lambda_i}$ with $\lambda_i \neq 0$. Since $\bar\partial(\omega - H\omega) = 0$, we have $\Sigma\,\bar\partial\omega_i = 0$. Since $\bar\partial\omega_i \in H^{q+1}_{\lambda_i}$ it follows that $\bar\partial\omega_i = 0$ for every i, as the sum is direct. Now $\Delta\omega_i = \lambda_i\omega_i$ so that $\bar\partial\vartheta\omega_i = \lambda_i\omega_i$, using $\bar\partial\omega_i = 0$. Hence $\omega_i = \bar\partial(\lambda_i^{-1}\vartheta\omega_i)$. Since $\Delta\vartheta\omega_i = \vartheta\Delta\omega_i = \lambda_i\vartheta\omega_i$, we have $\lambda_i^{-1}\vartheta\omega_i \in B^{q-1}$. This proves the surjectivity.

<u>Corollary to Proposition 6.1.</u>

$$\Sigma\,(-1)^q \dim H^q(V,\underline{W}) = \Sigma\,(-1)^q \dim B^q .$$

<u>Remark 6.1</u>. If L_2^q denotes the space of square integrable (0,q) forms with coefficients in W, Δ defines (in a unique way) a self adjoint operator on L_2^q. The

eigenvalues are precisely points of the spectrum of this operator. If λ is not eigenvalue let $G(\lambda)$ denote the resolvent at λ. Suppose, with the above notation, the circle $|z| = r$ does not contain any eigenvalue. Then the operator $P^q = P^q_r = -\frac{1}{2\pi i}\int_{|z|=r} G(\lambda)\, d\lambda$ is the orthogonal projection onto B^q, the subspace spanned by the eigenspaces with eigenvalues λ_i with $|\lambda_i| < r$. This follows from the remark that if P_{λ_i} denotes the orthogonal projection onto H_{λ_i}, the eigenspace corresponding to λ_i, then $G(\lambda) = \sum_i \frac{P_{\lambda_i}}{\lambda_i - \lambda}$.

7. Sobolev spaces and elliptic operators

Let E be a smooth vector bundle of rank N on a compact smooth manifold V of (real) dimension n. Let $k \geq 0$ be an integer. We shall denote by $S^{(k)}(E)$ (or simply by S^k) the topological vector space of sections of E which are of Sobolev class k. In more detail, let $\{U_i\}$ be a finite open covering by relatively compact coordinate neighbourhoods over which we have chosen trivialisations of E. Let ω be a section of E. Then $\omega \in S^k(E)$ if the following condition is satisfied: Let ω be represented in U_i by functions $\omega^i_1, \ldots, \omega^i_N$; for each i, the functions ω^i_j $(j=1,\ldots N)$ and their derivatives upto order k, taken in the sense of distributions in $\mathbb{R}^n$, are square integrable, with respect to the Lebesgue measure on $\mathbb{R}^n$, on compact subsets of U_i. Let now $\{\varphi_i\}$ be a smooth partition of unity subordinate to the covering U_i. Define

$$\|\omega\|_k = \sum_{i,j,\alpha} \|D^\alpha \varphi_i \psi^i_j\|_{L^2(\mathbb{R}^n)}$$

where $D^\alpha = (\frac{\partial}{\partial x_1})^{\alpha_1} \ldots (\frac{\partial}{\partial x_n})^{\alpha_n}$, the summation extending over $1 \leq j \leq N$, the indices of the open cover and over $\alpha = (\alpha_1, \ldots \alpha_n)$ with $|\alpha| = \alpha_1 + \ldots + \alpha_n \leq k$. This norm makes $S^k(E)$ into Banach space. Different choices of $\{U_i\}$ etc give equivalent norms.

In a similar way we can introduce a Banach space structure on the the space of $C^k(E)$ of k-times differentiable sections of E.

Theorem 7.1. (Sobolev Lemma). There exists an integer k_o, depending only n, such that $S^k(E)$ is contained in $C^{k-k_o}(E)$ and the inclusion map $S^k(E) \to C^{k-k_o}(E)$ is continuous.

Let A be a linear differential operator with smooth coefficients acting on smooth section of E. Then A defines a continuous linear operator $A: S^k(E) \to S^{k-m}(E)$ for $k \geq m$, where m is the order of A.

Theorem 7.2. Let A be a linear elliptic operator of order m with smooth coefficients acting on sections of E. Then

1) $A:S^k(E) \to S^{k-m}(E)$ has finite dimensional kernel and the image $A(S^k(E))$ is closed in $S^{k-m}(E)$ for $k \geq m$. Moreover if $\omega \in S^k(E)$ and $A\omega = 0$, then ω is a smooth section of E.

2) If $\omega \in S^o(E)$ and $A\omega \in S^k(E)$, then $\omega \in S^{k+m}(E)$.

Remark 7.1. With the notation of § 6, let $E = \Lambda T^{o,1} \otimes W$, where $T^{o,1}$ is the anti-holomorphic tangent bundle of V. If $\lambda \in \mathbb{C}$, then λ is not in the spectrum of $\Delta^{(q)}$ (the Laplacian on (o,q) forms with coefficients in W) if and only if $\Delta^{(q)}-\lambda : S^k(E) \to S^{k-2}(E)$ is an isomorphism for $k \geq 2$.

8. Some results from functional analysis

Let B_1 and B_2 be two Banach spaces. Let U be an open set in $\mathbb{R}^p$. If $T(t): B_1 \to B_2$ is a family of continuous linear operators, $t \in U$, we say that $T(t)$ depends differentiably on t, if for every $f \in B_1$, the function $t \mapsto T_t f$ is a smooth function from U into B_2.

Proposition 8.1. Let $T(t) : B_1 \to B_2$ depend differentiably on t. Suppose that for $t_o \in U$, the kernel of $T(t_o)$ is finite dimensional and the image of $T(t_o)$ is closed. Then there exists a neighbourhood U' of t_o such that for $t \in U'$ we have

$$\dim \ker T(t) \leq \dim \ker T(t_o)$$

As an easy consequence of Proposition 8.1 we have

Proposition 8.2. Let B be a Banach space and let $T(t)$ be a family of (continuous) projections in B onto finite dimensional subspaces depending differentiably on t. If $t_o \in U$, we have a neighbourhood U' of t_o such that for $t \in U'$, the dimension of the image of $T(t)$ is constant.

Proposition 8.3. Let $T(t) : B_1 \to B_2$ be a family of continuous linear operators depending differentiably on $t \in U$.

i) If for all $t \in U$, $T(t)$ is an isomorphism, then the family $T(t)^{-1}: B_2 \to B_1$ depends differentiably on $t \in U$.

2) If for $t_o \in U$, $T(t_o)$ is an isomorphism, there exists a neighbourhood U' of t_o such that $T(t)$ is an isomorphism for $t \in U'$.

9. Families of elliptic operators

Let $\mathbb{E}$ be a vector bundle on $V \times U$, where V is a compact smooth manifold and U a ball in $\mathbb{R}^p$. Let $E_t = \mathbb{E}\,|\,V \times t$; E_t may be considered as a vector bundle on V. We can choose isomorphism $\varphi_t : E_t \to E_{t_o}$ depending differentiably on t. That is, there is a commutative diagram

$$\begin{array}{ccc} \mathbb{E} & \xrightarrow{\Phi} & E_{t_o} \\ \downarrow & & \downarrow \\ V \times U & \xrightarrow[p_V]{} & U \end{array}$$

where Φ is smooth and $\varphi_t = \Phi | E_t \rightarrow E_{t_o}$ is a vector bundle isomorphism for $t \in U$ (Thus $\mathbb{E}$ is isomorphic to the pull back of E_{t_o} by the map $p_V: V \times U \to V$.) We may also assume that φ_{t_o} is the identity map.

Let A_t be a family of linear elliptic operators A_t of order m, operating on sections of E_t depending differentiably on t. (i.e., when we express the differential operators in local coordinates, the coefficients are smooth functions in (x,t), $x \in V, t \in U$.) Now using the isomorphisms φ_t we can consider A_t as a differential operator $\tilde{A}_t$ operating on sections of E_{t_o}. Let S_t^k denote the Sobolev space of sections of order k of E_t. The operators $\tilde{A}_t : S_{t_o}^k \to S_{t_o}^{k-m}$, $k \geq m$, is a family of continuous linear operators depending differentiably on t, as is easy to verify. Moreover $\dim \ker A_t$ = dimension of the space of smooth solutions of the equation $A_t \omega = 0$ (Th. 7.2,1). Using Theorem 7.1(1) and Proposition 8.1, we obtain

Theorem 9.1. Let H_t denote the space of smooth sections ω of E_t satisfying the equation $A_t \omega = 0$. If $t_o \in U$, there exists a neighbourhood U' of t_o such that, for $t \in U'$,

$$\dim \, H_t \leq \dim \, H_{t_o} .$$

10. Proof of the semi-continuity theorem.

Let $V \to T$ be a differentiable family of complex structures and let $W = \{ W_t \}$ be a differentiable family of holomorphic bundles W_t on V_t. Let $E_t^q : W_t \otimes \overset{q}{\Lambda} (T_t^{o,1})^*$ where $T_t^{o,1}$ is the anti-holomorphic tangent bundle of V_t. Choose hermitian metrics on V_t hnd W_t depending differentiably on t and form the Laplacian Δ_t^q acting on sections of E_t^q. We have (§ 6) $\dim \ker \Delta_t^q = \dim H^q(V_t, \underline{W}_t)$. We obtain from Theorem 9.1,

Theorem 10.1. If $t_o \in T$, there exists a neighbourhood U of t_o such that $\dim H^q(V_t, \underline{W}_t) \leq \dim H^q(V_{t_o}, W_{t_o})$ for $t \in U$.

11. Invariance of the Euler characteristic

We use the notations of § 6 and § 10. Let C be a circle $|z| = r$ in the complex plane which does not contain any eigenvalue of $\Delta_{t_o}^q$ for any q. For

$\lambda \in \mathbb{C}, \Delta^q_{t_o} - \lambda: S^2(E^q_{t_o}) \to S^o(E^q_{t_o})$ is an isomorphism (Remark 7.1). Using Proposition 8.3 we see that there exists a neighbourhood U' of t_o such that for (t,λ) $U'\times C$, $\widetilde{\Delta}^q_t - \lambda: S^2(E^q_{t_o}) \to S^o(E^q_{t_o})$ is an isomorphism. Consider the operators

$$\widetilde{P}^q_t = -\frac{1}{2\pi i}\int_{|\lambda|=r} (\widetilde{\Delta}^q_t - \lambda)^{-1} d\lambda .$$

These operators depend differentiably on t by Proposition 8.3. If φ^q_t denotes the isomorphism $S^k(E^q_t) \to S^k(E^q_{t_o})$ induced by the isomorphism $\varphi^q_t : E^q_t \to E^q_{t_o}$, we see using Remarks 6.1 and 7.1 that $\widetilde{P}_t = \varphi_t \circ P_t \circ \varphi_t^{-1}$, where P_t is the orthogonal projection onto the space B^q_t. By Proposition 8.2, the dimension of B^q_t is constant in a neighbourhood of t_o. By Corollary to Proposition 6.1, we get

Theorem 11.1. The function

$$t \mapsto \Sigma\, (-1)^q \dim H^q(V_t, \underline{W}_t)$$

is constant.

12. Proof of the rigidity theorem

We shall prove

Theorem 12.1. Let $p: \mathcal{V} \to T$ be a differentiable family of complex structures. Assume that for $t_o \in T, H^1(V_{t_o}, \Theta) = 0$. Then there exists a neighbourhood U of t_o and a smooth map $\Phi : p^{-1}(U) \to V_{t_o} \times U$ with commutativity in the diagram

$$\begin{array}{ccc} p^{-1}(U) & \xrightarrow{\ \Phi\ } & V_{t_o} \times U \\ \big\downarrow p & & \big\downarrow p_U \\ U & \xrightarrow{\ \mathrm{Id}\ } & U \end{array}$$

such that $\Phi | V_t \to V_{t_o}$ is a complex analytic isomorphism for $t \in U$.

First we make

Definition 12.1. Let $\widetilde{X}$ be a projectable vector field over an open set of $\mathcal{V}$. We say that $\widetilde{X}$ is an A-vector field if the local 1-parameter group of automorphisms. $\exp(s\widetilde{X})$, generated by $\widetilde{X}$ maps fibres holomorphically into fibres.

Remark 12.1. If $\widetilde{X}$ is a projectable vector field, the operation of Lie derivation, $\theta_{\widetilde{X}}$, by $\widetilde{X}$ of tensors along the fibres is well defined. If $\mathcal{J} = \{J_t\}$ denotes the almost complex structure tensor along the fibres, the above condition means $\theta_{\widetilde{X}} \mathcal{J} = 0$. If X_1 and X_2 are A-vector fields, so are $X_1 \pm X_2$.

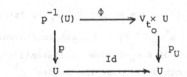

Remark 12.2. If V is a complex manifold with almost complex structure tensor J and X is a (real) vector field on V such that $\theta_X J = 0$, then $\frac{X-iJX}{2}$ is holomorphic vector field.

Referring to the proof of Proposition 1.1, Theorem 12.1 will be proved once we have

Proposition 12.1. There exists a neighbourhood N of t_o such that every smooth v vector field on N can be lifted to a projectable A-vector field on $p^{-1}(N)$.

For the proof we need

Proposition 12.2. Let $\mathbb{E}^q$ be the bundle $\{E_t = T_t^{1,0} \otimes \overset{q}{\Lambda} T_t^{*0,1}\}$ where $T_t^{1,0}$ (resp. $T_t^{*0,1}$) denotes the holomorphic tangen bundle (resp. bundle of anti-holomorphic 1-forms) of V_t. Assume that $H^1(V_{t_o}, \Theta) = 0$ for some $t_o \in T$. Then there exists a neighbourhood U of t_o with the following property: Let ψ be a smooth section of $\mathbb{E}^1$ over $p^{-1}(N)$ such that, if ψ_t denotes the restriction of ψ to V_t, we have $\bar\partial(t)\psi_t = 0$. Then there exists a smooth section φ of $\mathbb{E}^o$ such that $\bar\partial(t)\varphi_t = \psi_t$, where $\varphi_t = \varphi|V_t$.

Proof. By hypothesis, $\Delta^1_{t_o} : S^2(E^1_{t_o}) \to S^o(E^1_{t_o})$ is an isomorphism (see Remark 7.1). By Proposition 8.3, $\tilde\Delta^1_t : S^2(E^1_{t_o}) \to S^o(E^1_{t_o})$ is an isomorphism for t in a neighbourhood U of t_o. Using Remark 7.1 and Proposition 8.3, we see that for all $k \geq 2$ $\tilde\Delta^1_t : S^k(E^1_{t_o}) \to S^{k-2}(E^1_{t_o})$ is an isomorphism and the inverse $\tilde G_t : S^{k-2}(E^1_{t_o}) \to S^k(E^1_{t_o})$ depends differentiably on $t \in U$. If $\tilde\psi_t$ denotes the section of $E^1_{t_o}$ corresponding to ψ_t under the isomorphism φ_t, the function $t \mapsto \psi_t$ is a smooth function with values in $S^{k-2}(E^1_{t_o})$, for $k \geq 2$. Using the Sobolev Lemma (Theorem 7.1) we see that $t \mapsto \tilde G_t\psi_t$ is a smooth function with values in $C^{k-k_o-2}(E^1_{t_o})$, (since $t \mapsto \tilde G_t\tilde\psi_t$ is a smooth function with values in $S^{k-2}(E^1_{t_o})$)], for all sufficiently large k [$C^{k-k_o-2}(E^1_{t_o})$ denotes the space of $k-k_o-2$ times differentiable sections of $E^1_{t_o}$].This shows that $\{\tilde G_t\tilde\psi_t\}$ define a smooth section of the bundle $E^1_{t_o} \times U$. Now $\Delta_t : S^k_t(E^1_t) \to S^{k-2}(E^1_t)$ is an isomorphism for $t \in U$, let G_t be the inverse. Since $(G_t\psi_t) = \tilde G_t\tilde\psi_t$, we see that $\{G_t\psi_t\}$ define a smooth section of $\mathbb{E}^1$. If $\varphi_t = \vartheta_t G_t\psi_t$, the $\{\varphi_t\}$ define a smooth section of $\mathbb{E}^o$. We have using $\bar\partial(t)\psi_t = 0$, $\bar\partial(t)\varphi_t = \bar\partial(t)\vartheta_t G_t\psi_t = \Delta_t G_t\psi_t = \psi_t$.

Proof of Proposition 12.1. Let $\{U_\alpha\}$ be the open cover in the Definition 5.1. Let $\tilde X_\alpha$ be the projectable vector field on U_α corresponding to the "horizontal lift' $(X,0)$ in $p(U_\alpha)\times W_\alpha$. Choose a neighbourhood U satisfying the condition of Proposition 12.2. Consider the vector field $\theta_{\alpha\beta} = \tilde X_\beta - \tilde X_\alpha$ tangential to the fibres. Let $L_t^{\alpha\beta} = \tilde X_\beta - \tilde X_\alpha|V_t \cap U_\alpha \cap U_\beta$. Since $\tilde X_\beta, \tilde X_\alpha$ are A-vector fields so is $\theta_{\alpha\beta}$. This means that $L_t^{\alpha\beta}$ generate a local 1-parameter group of automorphisms of V_t. Consider $\theta'_{\alpha\beta}(t) = \frac{1}{2}\{J_tL_t^{\alpha\beta} - iJ_tL_t^{\alpha\beta}\}$, where J_t is the almost complex structure tensor on V_t.

The vector field $\theta'_{\alpha\beta}(t)$ is a holomorphic vector field on V_t and $\{\theta'_{\alpha\beta}(t)\}$ define a smooth section $\theta'_{\alpha\beta}$ of $\mathbb{E}^0$ over $U_\alpha \cap U_\beta$. Find, using a partition of unity, smooth section $\{Y_\alpha\}$ of $\mathbb{E}^0$ in $\{U_\alpha\}$ such that $Y_\beta - Y_\alpha = \theta'_{\alpha\beta}$. Let $Y_\alpha(t) = Y_\alpha | V_t \times U_\alpha$. If $\psi^\alpha_t = \bar\partial(t)\, Y_\alpha(t)$ we have $\psi^\alpha_t = \psi^\beta_t$ in $U_\alpha \cap U_\beta$, as $\bar\partial(t)\, (Y_\beta(t) - Y_\alpha(t) = 0$ in $U_\alpha \cap U_\beta$; Hence $\{\psi^\alpha_t\}$ define a global section ψ of $\mathbb{E}^1$ on $p^{-1}(U)$. By Prop. 12.2 we can find a smooth section φ of $\mathbb{E}^0$ such that $\bar\partial(t)\, \varphi_t = \psi_t$. Then $Y_\alpha - \varphi$ is a smooth section of $\mathbb{E}$ in U_α. If $\eta_\alpha(t) = Y_\alpha - \varphi | V_t \times U_\alpha$ we have $\bar\partial(t)\, \eta_\alpha(t) = 0$, so that $\eta_\alpha(t)$ is a holomorphic vector field on $V_t \times U_\alpha$. Let $Z_\alpha(t) = \dfrac{\eta_\alpha(t) + \overline{\eta_\alpha(t)}}{2}$ and Z_α the vector field in U_α defined by $\{Z_\alpha(t)\}$. Then $\tilde X_\alpha - Z_\alpha = \tilde X_\beta - Z_\beta$ in $U_\beta \cap U_\beta$ and hence there is a globally defined vector field $\tilde X$ with $\tilde X | U_\alpha = \tilde X_\alpha - Z_\alpha$ and projecting into X. Since $\tilde X_\alpha$ and Z_α are A-vector fields, so is $\tilde X$.

References

1. K. Kodaira and D.C. Spencer: On deformations of complex analytic structures I,II. Annals of Mathematics, 67(1958), 328-466.

2. K. Kodaira and D.C Spencer: A theorem of completeness for analytic fibre spaces Acta Math. 100(1958), 281-294.

3. K. Kodaira, L. Nirenberg and D.C. Spencer: On the existence of deformations of complex analytic structures, Ann. of Math. 68(1958), 450-459.

4. K. Kodaira and D.C. Spencer: On deformations of complex analytic structures III, Ann. of Math. 71(1960), 43-76.

5. M.S. Narasimhan and C.S. Seshadri: Stable and unitary vector bundles on a Compact Riemann surface, Ann. of Math. 82(1965), 540-567.

6. M.S. Narasimhan and R.R. Simha: Manifolds with ample canonical class, Inventiones Math. 5(1968), 120-128.

7. Y.T. Siu : Dimensions of sheaf cohomology groups under holomorphic deformation, Math. Ann. 192(1971), 203-215.

Introduction to Value Distribution Theory of Meromorphic Maps

Dedicated to the memory of Aldo Andreotti*

Wilhelm Stoll

Preface

Value distribution of functions of one complex variable has grown into a huge theory. In several complex variables the theory has not been developed to such an extent, but considerable progress has been made. In these notes an introduction to some aspects of the several variable theory is provided. The two Main Theorems are proved for meromorphic maps $f : W \to \mathbb{P}(V)$ from a complex vector space W into complex projective space, where the intersection of f(W) with hyperplanes is studied. The First Main Theorem is established along standard lines. Vitter's Lemma of the Logarithmic Derivative is proved by a method of Biancofiore and Stoll. The Second Main Theorem and the Defect Relation is obtained by the original method of H. Cartan following ideas of Vitter. The value distribution theory of Carlson, Griffiths and King for holomorphic maps from parabolic manifolds into projective algebraic varieties is outlined without proof. These results are specialized to holomorphic maps $f : W \to \mathbb{P}(V)$ where the intersection of f(W) with hypersurfaces of degree p are studied. Here the Carlson, Griffiths, King theory has to require that rank f = dim $\mathbb{P}(V)$. It is a difficult open question if this restriction can be removed. We formulate a recent result of A. Biancofiore which gives a partial answer and which is the best answer available at present.

Thus the notes lead from the beginning to the frontier of today's research. Although many topics had to be omitted, they will familiarize the careful reader with basic topics and ideas in several complex variables value distribution theory.

These notes constitute the background for the lectures which will emphasize more the ideas and results and provide less proofs. It is hoped that lectures and notes will inform and create new interest in the subject matter.

Wilhelm Stoll
University of Notre Dame
Notre Dame, IN 46556 USA
Spring 1980

*) This work was partially supported by an NSF grant MCS 80-03257.

Letters

Latin Capital	Latin Small	German Capital	German Small	Greek Capital	Greek Small
A	a	$\mathfrak{A}$	$\mathfrak{a}$	Α	α
B	b	$\mathfrak{B}$	$\mathfrak{b}$	Β	β
C	c	$\mathfrak{C}$	$\mathfrak{c}$	Γ	γ
D	d	$\mathfrak{D}$	$\mathfrak{d}$	Δ	δ
E	e	$\mathfrak{E}$	$\mathfrak{e}$	Ε	ε
F	f	$\mathfrak{F}$	$\mathfrak{f}$	Ζ	ζ
G	g	$\mathfrak{G}$	$\mathfrak{g}$	Η	η
H	h	$\mathfrak{H}$	$\mathfrak{h}$	Θ	θ
I	i	$\mathfrak{I}$	$\mathfrak{i}$	Ι	ι
J	j	$\mathfrak{J}$	$\mathfrak{j}$	Κ	κ
K	k	$\mathfrak{K}$	$\mathfrak{k}$	Λ	λ
L	l	$\mathfrak{L}$	$\mathfrak{l}$	Μ	μ
M	m	$\mathfrak{M}$	$\mathfrak{m}$	Ν	ν
N	n	$\mathfrak{N}$	$\mathfrak{n}$	Ξ	ξ
O	o	$\mathfrak{O}$	$\mathfrak{o}$	Ο	ο
P	p	$\mathfrak{P}$	$\mathfrak{p}$	Π	π
Q	q	$\mathfrak{Q}$	$\mathfrak{q}$	Ρ	ρ
R	r	$\mathfrak{R}$	$\mathfrak{r}$	Σ	σ, ς
S	s	$\mathfrak{S}$	$\mathfrak{s}$	Τ	τ
T	t	$\mathfrak{T}$	$\mathfrak{t}$	Υ	υ
U	u	$\mathfrak{U}$	$\mathfrak{u}$	Φ	φ
V	v	$\mathfrak{V}$	$\mathfrak{v}$	Χ	χ
W	w	$\mathfrak{W}$	$\mathfrak{w}$	Ψ	ψ
X	x	$\mathfrak{X}$	$\mathfrak{x}$	Ω	ω
Y	y	$\mathfrak{Y}$	$\mathfrak{y}$		
Z	z	$\mathfrak{Z}$	$\mathfrak{z}$		

CONTENTS

1. Review of value distribution in one variable.

We shall study the behavior of holomorphic maps

$$f : \mathbb{C}^m \to \mathbb{P}_n \tag{1.1}$$

where $\mathbb{C}^m$ is the space of m complex variables and where $\mathbb{P}_n$ is the complex projective space of dimension n.

First let us review the classical case $m = n = 1$. Then $\mathbb{P}_1$ is the Riemann sphere which can be visualized as the euclidean sphere of diameter 1 in $\mathbb{R}^3$ or as the closed plane $\mathbb{C} \cup \{\infty\} = \overline{\overline{\mathbb{C}}}$. If a and w are on the sphere, let $\|a,w\|$ be the chordal distance from a to b measured through space. Obviously

$$0 \le \|a,w\| \le 1. \tag{1.2}$$

If we identify $\mathbb{P}_1 = \overline{\overline{\mathbb{C}}}$, this distance is computed for $a \in \mathbb{C}$ and $w \in \mathbb{C}$ by

$$\|a,w\| = \frac{|a - w|}{\sqrt{1 + |a|^2}\,\sqrt{1 + |w|^2}} \tag{1.3}$$

$$\|a,\infty\| = \frac{1}{\sqrt{1 + |a|^2}}\ . \tag{1.4}$$

Of course $\|\infty,\infty\| = 0$.

There exists one and only one rotation invariant volume element Ω on $\mathbb{P}_1$ with total measure

$$\int_{\mathbb{P}_1} \Omega = 1. \tag{1.5}$$

Take $a \in \mathbb{P}_1$. As an exterior differential form Ω can be computed at every $w \in \mathbb{P}_1 - \{a\}$ by

$$(1.6) \qquad \Omega(w) = dd^c \log \frac{1}{\|a,w\|^2} .$$

As on every complex manifold the exterior derivative splits into $d = \partial + \bar{\partial}$ and twists to

$$(1.7) \qquad d^c = \frac{i}{4\pi} (\bar{\partial} - \partial).$$

On $\mathbb{C}$, the form Ω is given by

$$(1.8) \qquad \Omega(w) = \frac{i}{2\pi} \frac{dw \wedge d\bar{w}}{(1 + |w|^2)^2} .$$

For $r > 0$ and any subset S of $\mathbb{C}$ denote

$$(1.9) \qquad \mathbb{C}(r) = \{z \in \mathbb{C} \mid |z| < r\} \qquad \mathbb{C}[r] = \{z \in \mathbb{C} \mid |z| \le r\}$$

$$(1.10) \qquad \mathbb{C}\langle r\rangle = \{z \in \mathbb{C} \mid |z| = r\} \qquad \mathbb{C}_* = \mathbb{C} - \{0\}.$$

Let f be a meromorphic function on $\mathbb{C}$. Then f can be regarded as a holomorphic map

$$(1.11) \qquad f : \mathbb{C} \to \mathbb{P}_1.$$

The form Ω is pulled back to

$$(1.12) \qquad f^*(\Omega) = \frac{i}{2\pi} \frac{df \wedge d\bar{f}}{(1 + |f|^2)^2} = \frac{|f'|^2}{(1 + |f|^2)^2} \frac{i}{2\pi} dz \wedge d\bar{z} \ge 0.$$

The covering volume of the disc defines the <u>spherical image function</u>

$$(1.13) \qquad A_f(r) = \int_{\mathbb{C}(r)} f^*(\Omega) \ge 0 \qquad \forall\ r > 0.$$

Obviously $A_f(r) > 0$ if and only if f is not constant. As a growth measure of f the <u>characteristic function of</u> f is introduced for $0 < s < r$ by

$$(1.14) \qquad T_f(r,s) = \int_s^r A_f(t)\,\frac{dt}{t}\,.$$

Easily, we see

$$(1.15) \qquad 0 \le \lim_{r\to\infty} \frac{T_f(r,s)}{\log r} = \lim_{t\to\infty} A_f(t) = A_f(\infty) \le +\infty$$

where $A_f(\infty) > 0$ if and only if f is not constant, and $A_f(\infty) = \infty$ if and only if f is transcendental.

If f is not constant, the fibers of f consist of isolated points with multiplicities. We formalize the concept. Let M be a Riemann surface. A function $\nu : M \to \mathbb{Z}$ is said to be a <u>divisor on</u> M iff its support

$$(1.16) \qquad \operatorname{supp} \nu = \{z \in M \mid \nu(z) \neq 0\}$$

is a closed set of isolated points in M. The set $\mathfrak{D}(M)$ of all divisors on M is a module under addition. The divisor is said to be non-negative iff $\nu(z) \ge 0$ for all $z \in M$. Let $\nu : \mathbb{C} \to \mathbb{Z}$ be a divisor, then the <u>counting function</u> of ν for $r \ge 0$ is defined by

$$(1.17) \qquad n_\nu(r) = \sum_{z\in\mathbb{C}(r)} \nu(z)$$

and the <u>valence function</u> of ν for $0 < s < r$ is defined by

$$(1.18) \qquad N_\nu(r,s) = \int_s^r n_\nu(t)\,\frac{dt}{t}\,.$$

Here $n_\nu(r)$ and $N_\nu(r,s)$ are additive in ν. If $\nu \ge 0$, they are non-negative and increase with r.

If $a \in \mathbb{P}_1$ and $z_0 \in \mathbb{C}$, let $\mu_f^a(z_0) \ge 0$ be the a-multiplicity of f at z_0. Then $\mu_f^a : \mathbb{C} \to \mathbb{Z}$ is a non-negative divisor. Define

$$(1.19) \qquad n_f(r,a) = n_{\mu_f^a}(r) \qquad\qquad N_f(r,s;a) = N_{\mu_f^a}(r,s).$$

Also $\mu_f = \mu_f^0 - \mu_f^\infty$ is called the <u>divisor</u> <u>of</u> f. Define

$$n_f(r) = n_{\mu_f}(r) \qquad N_f(r,s) = N_{\mu_f}(r,s). \tag{1.20}$$

For $r > 0$, let σ_r be the rotation invariant measure of the circle $\mathbb{C}\langle r\rangle$ such that

$$\int_{\mathbb{C}\langle r\rangle} \sigma_r = 1. \tag{1.21}$$

If the circle is parametrized by $z = re^{i\phi}$ then $\sigma_r = (1/2\pi)\, d\phi$.

If $0 < s < r$, the <u>Jensen</u> <u>formula</u> asserts

$$\boxed{N_f(r,s) = \int_{\mathbb{C}\langle r\rangle} \log|f|\sigma_r - \int_{\mathbb{C}\langle s\rangle} \log|f|\sigma_s} \;. \tag{1.22}$$

The <u>compensation</u> <u>function</u> of f for $a \in \mathbb{P}_1$ is defined for $r > 0$ by

$$m_f(r,a) = \int_{\mathbb{C}\langle r\rangle} \log \frac{1}{\|f,a\|}\, \sigma_r \geq 0. \tag{1.23}$$

The <u>First</u> <u>Main</u> <u>Theorem</u> states

$$\boxed{T_f(r,s) = N_f(r,s;a) + m_f(r,a) - m_f(s,a)} \tag{1.24}$$

if $0 < s < r$ and $a \in \mathbb{P}_1$ and $f^{-1}(a) \neq \mathbb{C}$. Since $m_f(r,a) \geq 0$ the growth measure $N_f(r,s;a)$ of the fiber a is bounded by the growth measure $T_f(r,s)$ of f plus a constant

$$N_f(r,s;a) \leq T_f(r,s) + m_f(s,a) \qquad \forall\, r > s. \tag{1.25}$$

In fact $\int_{a\in\mathbb{P}_1} \log \frac{1}{\|a,w\|}\,\Omega(a)$ is invariant under the rotations of $\mathbb{P}_1$. Hence

$$\int_{a\in\mathbb{P}_1} \log \frac{1}{\|a,w\|}\,\Omega(a) = \int_{a\in\mathbb{P}_1} \log \frac{1}{\|a,\infty\|}\,\Omega(a)$$

$$= \int_{a\in\mathbb{C}} \log \sqrt{1+|a|^2}\,\frac{i}{2\pi}\,\frac{da\wedge d\bar{a}}{(1+|a|^2)^2} = \int_0^{2\pi}\int_0^{\infty} \frac{1}{2}\log(1+t^2)\,\frac{1}{\pi}\,\frac{t\,dt\,d\phi}{(1+t^2)^2}$$

$$= \frac{1}{2}\int_0^{\infty} \log(1+r)\,\frac{dr}{(1+r)^2} = \frac{1}{2}\int_0^{\infty} \frac{1}{1+r}\,\frac{dr}{1+r} = \frac{1}{2}\,.$$

Therefore

$$\int_{a\in\mathbb{P}_1} m_f(r,a)\Omega(a) = \int_{a\in\mathbb{P}_1}\int_{\mathbb{C}\langle r\rangle} \log \frac{1}{\|f,a\|}\,\sigma_r\Omega(a)$$

$$= \int_{\mathbb{C}\langle r\rangle}\int_{a\in\mathbb{P}_1} \log \frac{1}{\|f,a\|}\,\Omega(a)\sigma_r = \frac{1}{2}\,.$$

Hence

(1.26) $$\int_{a\in\mathbb{P}_1} m_f(r,a)\Omega(a) = \frac{1}{2} \qquad \forall\, r > 0.$$

Therefore the First Main Theorem implies

(1.27) $$\boxed{T_f(r,s) = \int_{a\in\mathbb{P}_1} N_f(r,s;a)\Omega(a)} \qquad \forall\, 0 < s < r.$$

Assume that f is not constant, then $A_f(\infty) > 0$. By (1.15) $T_f(r,s) \to \infty$ for $r \to \infty$. The First Main Theorem implies

(1.28) $$0 \le \delta_f(a) = \liminf_{r\to\infty} \frac{m_f(r,a)}{T_f(r,s)} = 1 - \limsup_{r\to\infty} \frac{N_f(r,s;a)}{T_f(r,s)} \le 1$$

where $\delta_f(a)$ is called the <u>defect</u> <u>of</u> f <u>for</u> a. If $f^{-1}(a) = \emptyset$, then $N_f(r,s;a) = 0$ and $\delta_f(a) = 1$. Moreover f is rational if and only if $T_f(r,s) = O(\log r)$ and $N_f(r,s;a) = O(\log r)$ if and only if $f^{-1}(a)$ is finite. Hence $\delta_f(a) = 1$, if f is transcendental and $f^{-1}(a)$ is finite.

If $S \neq \emptyset$ is a finite subset of $\mathbb{P}_1$, the <u>Defect</u> <u>Relation</u> asserts

$$\boxed{\sum_{a\in S} \delta_f(a) \leq 2} \;. \tag{1.29}$$

Hence f omits at most two values, which is Picard's Theorem. Moreover, $\delta_f(a) > 0$ for at most countable many values $a \in \mathbb{P}_1$. The Defect Relation is an immediate consequence of the <u>Second</u> <u>Main</u> <u>Theorem</u>.

$$\boxed{N_\rho(r,s) + \sum_{a\in S} m_f(r,a) \stackrel{\leq}{\cdot} 2T_f(r,s) + c\log(rT_f(r,s))} \tag{1.30}$$

where $\stackrel{\leq}{\cdot}$ means the inequality holds outside a set of finite measure. Here $c > 0$ is a constant. There are holomorphic functions g and h on $\mathbb{C}$ with $(g(z),h(z)) \neq (0,0)$ for all $z \in \mathbb{C}$ such that $hf = g$. Then ρ is the divisor of $g'h - h'g$.

We shall extend these results to meromorphic maps $f : \mathbb{C}^m \to \mathbb{P}_n$ where a represents a hyperplane in $\mathbb{P}_n$. First some preparations have to be made.

2. Hermitian geometry.

First some notations. Let S be an ordered set and let M be a subset. For $a \in S$ and $b \in S$ with $a \le b$ define

$$M[a,b] = \{x \in M \mid a \le x \le b\} \tag{2.1}$$

$$M(a,b) = \{x \in M \mid a < x < b\} \tag{2.2}$$

$$M[a,b) = \{x \in M \mid a \le x < b\} \tag{2.3}$$

$$M(a,b] = \{x \in M \mid a < x \le b\}. \tag{2.4}$$

Let V be a complex vector space of dimension n + 1. Define

$$V_* = V - \{0\}. \tag{2.5}$$

The set V^* of linear functions $\alpha : V \to \mathbb{C}$ is again a vector space of dimension n + 1 called the <u>dual vector space</u>. For the exterior product we have the identification

$$\bigwedge_{p+1} V^* = (\bigwedge_{p+1} V)^*. \tag{2.6}$$

If $\mathfrak{x} \in V_*$, define $\mathbb{P}(\mathfrak{x}) = \mathbb{C}\mathfrak{x}$ as the line spanned by $\mathfrak{x}$. For $A \subseteq W$, define

$$\mathbb{P}(A) = \{\mathbb{P}(\mathfrak{x}) \mid 0 \neq \mathfrak{x} \in A\}. \tag{2.7}$$

Then $\mathbb{P}(V)$ is a compact, connected, complex manifold of dimension n and

$$\mathbb{P} : V_* \to \mathbb{P}(V) \tag{2.8}$$

is a surjective, holomorphic map. To each $\alpha \in V_*^*$ we can assign a coordinate patch. Define

$$V_\alpha = \{\mathfrak{x} \in V \mid \alpha(\mathfrak{x}) \neq 0\} \qquad \dot{V}_\alpha = \{\mathfrak{x} \in V \mid \alpha(\mathfrak{x}) = 1\}. \tag{2.9}$$

Then $\dot{V}_\alpha$ is an affine plane of dimension n biholomorphically equivalent to $\mathbb{C}^n$. Also $\mathbb{P}(V_\alpha) = \mathbb{P}(\dot{V}_\alpha)$ is open in $\mathbb{P}(V)$. The restriction

$$\mathbb{P}_\alpha = \mathbb{P}|\dot{V}_\alpha : \dot{V}_\alpha \to \mathbb{P}(V_\alpha) \tag{2.10}$$

is biholomorphic and provides "local coordinates". If $\alpha \in V_*^*$ and $\beta \in V_*^*$, then

$$\mathbb{P}_\alpha^{-1}(\mathbb{P}(V_\alpha) \cap \mathbb{P}(V_\beta)) = \dot{V}_\alpha \cap V_\beta. \tag{2.11}$$

The transition map $\mathbb{P}_\beta^{-1} \circ \mathbb{P}_\alpha : \dot{V}_\alpha \cap V_\beta \to \dot{V}_\beta \cap V_\alpha$ is given by

$$\mathbb{P}_\beta^{-1}(\mathbb{P}_\alpha(\mathfrak{x})) = \frac{\mathfrak{x}}{\beta(\mathfrak{x})}. \tag{2.12}$$

If L is a linear subspace of dimension $p + 1$ of V, then $\mathbb{P}(L)$ is smoothly embedded into $\mathbb{P}(V)$ and is called a p-<u>dimensional</u> <u>projective</u> <u>plane</u>. We will parameterize these planes.

The <u>Grassmann</u> <u>cone</u> of order p is defined by

$$\tilde{G}_p(V) = \{\mathfrak{x}_0 \wedge \ldots \wedge \mathfrak{x}_p \in \textstyle\bigwedge_{p+1} V \mid \mathfrak{x}_j \in V \ \forall\ j \in \mathbb{Z}[0,p]\}. \tag{2.13}$$

Then $G_p(V) = \mathbb{P}(\tilde{G}_p(V))$ is a connected, compact, complex submanifold of dimension

$$\dim G_p(V) = (n - p)(p + 1) \tag{2.14}$$

of $\mathbb{P}(\bigwedge_{p+1} V)$. Here $G_p(V)$ is called the <u>Grassmann</u> <u>manifold</u> <u>of</u> <u>order</u> p.

Take $x \in G_p(V)$. Then $x = \mathbb{P}(\mathfrak{x})$ where $0 \neq \mathfrak{x} = \mathfrak{x}_0 \wedge \ldots \wedge \mathfrak{x}_p \in \tilde{G}_p(V)$ Then

$$E(x) = E(\mathfrak{x}) = \{\mathfrak{z} \in V \mid \mathfrak{z} \wedge \mathfrak{x} = 0\} = \mathbb{C}\mathfrak{x}_0 + \ldots + \mathbb{C}\mathfrak{x}_p \tag{2.15}$$

is a linear subspace of dimension $p + 1$ of V with base $\mathfrak{x}_0, \ldots, \mathfrak{x}_p$ and

(2.16) $$\ddot{E}(x) = \ddot{E}(\mathfrak{x}) = \mathbb{P}(E(x))$$

is a p-dimensional, projective plane. The map $x \mapsto \ddot{E}(x)$ is a bijective parameterization of the set of p-dimensional projective planes in $\mathbb{P}(V)$.

Take $q = n - p - 1 \geq 0$. Pick $a \in G_q(V^*)$. Then $\alpha = \alpha_0 \wedge \ldots \wedge \alpha_q \neq 0$ exists with $\mathbb{P}(\alpha) = a$. Then

(2.17) $$E[a] = E[\alpha] = \{\mathfrak{z} \in V \mid \alpha \llcorner \mathfrak{z} = 0\} = \bigcap_{j=0}^{q} \alpha_j^{-1}(0)$$

is a (p + 1)-dimensional, linear subspace of V and

(2.18) $$\ddot{E}[a] = \ddot{E}[\alpha] = \mathbb{P}(E[a])$$

is a p-dimensional, projective plane in $\mathbb{P}(V)$. The map $a \mapsto \ddot{E}[a]$ is a bijective parameterization of the set of p-dimensional projective planes in $\mathbb{P}(V)$.

A biholomorphic map $\delta : G_p(V) \to G_q(V^*)$, called the <u>duality</u>, is defined by

(2.19) $$E(x) = E[\delta(x)] \qquad \forall\, x \in G_p(V).$$

Let V be a complex vector space of dimension n + 1. A <u>hermitian inner product</u> on V is a function $(\mid) : V \times V \to \mathbb{C}$ such that

(2.20) $$(\lambda\mathfrak{x} + \mu\mathfrak{y} \mid \mathfrak{z}) = \lambda(\mathfrak{x}, \mathfrak{z}) + \mu(\mathfrak{y} \mid \mathfrak{z})$$

(2.21) $$(\mathfrak{x} \mid \mathfrak{y}) = \overline{(\mathfrak{y} \mid \mathfrak{x})}$$

(2.22) $$(\mathfrak{x} \mid \mathfrak{x}) > 0 \qquad \text{if } \mathfrak{x} \neq 0$$

for $\mathfrak{x}$, $\mathfrak{y}$ and $\mathfrak{z}$ in V and λ, μ in $\mathbb{C}$. The <u>norm</u> is defined by

(2.23) $$\|\mathfrak{x}\| = \sqrt{(\mathfrak{x} \mid \mathfrak{x})} \qquad \forall\, \mathfrak{x} \in V.$$

We have the standard properties

(2.24) $$\|\mathfrak{x} + \mathfrak{y}\| \leq \|\mathfrak{x}\| + \|\mathfrak{y}\| \qquad \text{(Triangle inequality)}$$

(2.25) $$\|\lambda \mathfrak{x}\| = |\lambda| \, \|\mathfrak{x}\|$$

(2.26) $$\|(\mathfrak{x}|\mathfrak{y})\| \leq \|\mathfrak{x}\| \, \|\mathfrak{y}\| \qquad \text{(Schwarz inequality)}$$

for all $\mathfrak{x}$, $\mathfrak{y}$, $\mathfrak{z}$ in V and $\lambda \in \mathbb{R}$. If $A \subseteq V$ and $r > 0$, define

(2.27) $$A(r) = \{\mathfrak{x} \in A \mid \|\mathfrak{x}\| < r\}$$

(2.28) $$A[r] = \{\mathfrak{x} \in A \mid \|\mathfrak{x}\| \leq r\}$$

(2.29) $$A\langle r\rangle = \{\mathfrak{x} \in A \mid \|\mathfrak{x}\| = r\}.$$

The vector space V together with an assigned hermitian, inner product is called a <u>hermitian vector space</u>. It induces natural hermitian products on V* and $\bigwedge_{p+1} V$.

Take α and β in V*. Unique vectors $\mathfrak{a}$ and $\mathfrak{b}$ in V exist such that $\alpha(\mathfrak{x}) = (\mathfrak{x}|\mathfrak{a})$ and $\beta(\mathfrak{x}) = (\mathfrak{x}|\mathfrak{b})$ for all $\mathfrak{x} \in V$. Then $(\ |\ ) : V^* \times V^* \to \mathbb{C}$ is a hermitian inner product on V* defined by

(2.30) $$(\alpha|\beta) = (\mathfrak{b}|\mathfrak{a}).$$

A hermitian inner product $(\ |\ ) : (\bigwedge_{p+1} V) \times (\bigwedge_{p+1} V) \to \mathbb{C}$ is defined by

(2.31) $$(\mathfrak{x}_0 \wedge \ldots \wedge \mathfrak{x}_p | \mathfrak{y}_0 \wedge \ldots \wedge \mathfrak{y}_p) = \det(\mathfrak{x}_j | \mathfrak{y}_k).$$

(Observe, that the extension from $\tilde{G}_p(V) \times \tilde{G}_p(V)$ to $(\bigwedge_{p+1} V) \times (\bigwedge_{p+1} V)$ by bi-additivity is unique.) The hermitian inner product on $(\bigwedge_{p+1} V)^*$ induced from $\bigwedge_{p+1} V$ and the hermitian inner product on $\bigwedge_{p+1} V^*$ induced from V* agree under the identification (3.6).

Take $a \in \mathbb{P}(V^*)$ and $x \in \mathbb{P}(V)$. Then $\alpha \in V_*^*$ and $\mathfrak{x} \in V_*$ exist with $\mathbb{P}(\alpha) = a$ and $\mathbb{P}(\mathfrak{x}) = x$. The <u>projective distance</u> from $\ddot{E}[a]$ to x in $\mathbb{P}(V)$ is defined by

$$(2.32) \qquad 0 \le \|x,a\| = \|a,x\| = \frac{|\alpha(\mathfrak{x})|}{\|\alpha\| \ \|\mathfrak{x}\|} \le 1.$$

Then

$$(2.33) \qquad \ddot{E}[a] = \{x \in \mathbb{P}(V) \mid \|a,x\| = 0\}.$$

If $V = \mathbb{C}^{n+1}$, we always use the hermitian inner product

$$(2.34) \qquad (\mathfrak{x}|\mathfrak{y}) = x_0\overline{y}_0 + \ldots + x_n\overline{y}_n.$$

We shall have to consider differential forms. The calculus of differential forms is assumed to be known. However some remarks on positive and non-negative forms may be called for.

Let M be a complex manifold of pure dimension m. To each $x \in M$ there are attached the real tangent space $T_x(M)$, the complex tangent space $\mathfrak{T}_x(M)$, the conjugate complex tangent space $\overline{\mathfrak{T}}_x(M)$ and the complexified tangent space $T_x^c(M) = \mathfrak{T}_x(M) \oplus \overline{\mathfrak{T}}_x(M) \supseteq T_x(M)$. A form ψ of bidegree (p,q) attaches to each $x \in M$ a linear map

$$(2.35) \qquad \psi(x) : (\bigwedge_p \mathfrak{T}_x(M)) \otimes (\bigwedge_q \overline{\mathfrak{T}}_x(M)) \to \mathbb{C}.$$

If p = q, then ψ is said to be <u>non-negative</u> (<u>respectively positive</u>) iff

$$(2.36) \qquad (x, \mathfrak{y}_1 \wedge \ldots \wedge \mathfrak{y}_p \otimes \overline{i\mathfrak{y}_1 \wedge \ldots \wedge i\mathfrak{y}_p}) \ge 0 \qquad (\text{resp.} > 0)$$

for any set of linearly independent vectors $\mathfrak{y}_1, \ldots, \mathfrak{y}_p$ in $\mathfrak{T}_x(M)$. We have the following properties.

(1) If $\psi \ge 0$ (resp. $\psi > 0$), if $\chi \ge 0$ and $0 < a \in \mathbb{R}$, then $a\psi + \chi \ge 0$ (resp. > 0).

(2) If $p < m$, if $\psi \ge 0$ (resp. > 0) has bidegree (p,p) and if $\chi \ge 0$ (resp. > 0) has bidegree (1,1), then $\psi \wedge \chi \ge 0$ (resp. > 0).

(3) If $\psi \ge 0$, if $f : N \to M$ is holomorphic, then $f^*(\psi) \ge 0$.

(4) If $\psi > 0$, if $f : N \to M$ is holomorphic and smooth, then $f^*(\psi) > 0$.

A form is <u>closed</u>, if $d\psi = 0$. A form ψ is <u>exact</u>, if $\psi = d\chi$. Observe $d \circ d = 0$.

Let V be a hermitian vector space of dimension $n + 1$. Define

$$(2.37) \qquad \tau = \tau_V : V \to \mathbb{R}_+ \quad \text{by} \quad \tau(\mathfrak{x}) = \|\mathfrak{x}\|^2.$$

Take $\mathfrak{x} \in V$. Observe $\mathfrak{T}_{\mathfrak{x}}(V) = V$. For $\mathfrak{y} \in V$ and $\mathfrak{z} \in V$, we compute

$$(2.38) \qquad \partial\tau(\mathfrak{x},\mathfrak{y}) = (\mathfrak{y}|\mathfrak{x}) \qquad \bar{\partial}\tau(\mathfrak{x},\bar{\mathfrak{y}}) = (\mathfrak{x}|\mathfrak{y})$$

$$(2.39) \qquad dd^c\tau(\mathfrak{x},\mathfrak{y},\overline{i\mathfrak{z}}) = \frac{i}{2\pi}\,\partial\bar{\partial}\tau(\mathfrak{x},\mathfrak{y},\overline{i\mathfrak{z}}) = \frac{1}{2\pi}\,(\mathfrak{y}|\mathfrak{z})$$

$$(2.40) \qquad d\tau \wedge d^c\tau(\mathfrak{x},\mathfrak{y},\overline{i\mathfrak{z}}) = \frac{i}{2\pi}\,\partial\tau \wedge \bar{\partial}\tau(\mathfrak{x},\mathfrak{y},\overline{i\mathfrak{z}}) = \frac{1}{2\pi}\,(\mathfrak{x}|\mathfrak{z})(\mathfrak{y}|\mathfrak{x})$$

$$(2.41) \qquad (dd^c \log \tau)(\mathfrak{x},\mathfrak{y},\overline{i\mathfrak{z}}) = \frac{1}{2\pi}\,\|\mathfrak{x}\|^{-4}(\mathfrak{x} \wedge \mathfrak{y}|\mathfrak{x} \wedge \mathfrak{z}).$$

In particular

$$(2.42) \qquad dd^c\tau(\mathfrak{x},\mathfrak{y},\overline{i\mathfrak{y}}) = \frac{1}{2\pi}\,\|\mathfrak{y}\|^2 > 0 \qquad \text{if } \mathfrak{y} \in V_*$$

$$(2.43) \qquad (dd^c \log \tau)(\mathfrak{x},\mathfrak{y},\overline{i\mathfrak{y}}) = \frac{1}{2\pi}\,\frac{\|\mathfrak{x} \wedge \mathfrak{y}\|^2}{\|\mathfrak{x}\|^4} \geq 0 \qquad \forall\, \mathfrak{y} \in V.$$

Therefore

$$(2.44) \qquad \upsilon = \upsilon_V = dd^c\tau > 0 \qquad \omega = \omega_V = dd^c \log \tau \geq 0.$$

A trivial calculation shows

(2.45) $$\boxed{\tau^2\omega = \tau\upsilon - p\, d\tau \wedge d^c\tau}$$

(2.46) $$\boxed{\tau^{p+1}\omega^p = \tau\upsilon^p - p\, d\tau \wedge d^c\tau \wedge \upsilon^{p-1}} \qquad \forall\, p \in \mathbb{N}.$$

Let M be any smooth, complex submanifold of V_*. Let $j : M \to V_*$ be the inclusion map. Take $\mathfrak{x} \in M$. Then $\mathfrak{T}_{\mathfrak{x}}(M) \subseteq \mathfrak{T}_{\mathfrak{x}}(V) = V$. Hence if $\mathfrak{x} \notin \mathfrak{T}_{\mathfrak{x}}(M)$, then $\mathfrak{x} \wedge \mathfrak{y} \neq 0$ for all $0 \neq \mathfrak{y} \in \mathfrak{T}_{\mathfrak{x}}(M)$ and (2.43) shows that $j^*(\omega)$ is positive at $\mathfrak{x}$.

Take $\alpha \in V_*^*$. Let $j_\alpha : \dot{V}_\alpha \to V_*$ be the inclusion. Then $E[\alpha]$ is the tangent space to $\dot{V}_\alpha$ at every $\mathfrak{x} \in \dot{V}_\alpha$. Hence $j_\alpha^*(\omega) > 0$. Since $\mathbb{P}_\alpha : \dot{V}_\alpha \to \mathbb{P}(V_\alpha)$ is biholomorphic, a closed, positive form of bidegree (1,1) is defined by

(2.47) $$\Omega_\alpha = (\mathbb{P}_\alpha^{-1})^*(j_\alpha^*(\omega)).$$

We have $\mathbb{P}_\alpha = \mathbb{P} \circ j_\alpha$ on $\dot{V}_\alpha$. Take $\beta \in V_*^*$. Define $g_\beta : V_\beta \to V$ by $g_\beta(\mathfrak{x}) = \mathfrak{x}/\beta(\mathfrak{x})$. Then $\tau \circ g_\beta = |\beta|^{-2}\tau$. Therefore

(2.48) $$\begin{aligned} g_\beta^*(\omega) &= g_\beta^*\, dd^c \log\tau = dd^c \log \tau \circ g_\beta \\ &= dd^c \log\tau - dd^c \log|\beta|^2 = \omega. \end{aligned}$$

Obviously $j_\beta \circ \mathbb{P}_\beta^{-1} \circ \mathbb{P} = g_\beta$ on V_β. On $\mathbb{P}(V_\alpha) \cap \mathbb{P}(V_\beta)$ we obtain

(2.49) $$\begin{aligned} \Omega_\beta &= (j_\beta \circ \mathbb{P}_\beta^{-1})^*(\omega) = (j_\beta \circ \mathbb{P}_\beta^{-1} \circ \mathbb{P} \circ j_\alpha \circ \mathbb{P}_\alpha^{-1})^*(\omega) \\ &= (j_\alpha \circ \mathbb{P}_\alpha^{-1})^*(g_\beta^*(\omega)) = (j_\alpha \circ \mathbb{P}_\alpha^{-1})^*(\omega) = \Omega_\alpha. \end{aligned}$$

Therefore one and only one closed, positive form of bidegree (1,1) on $\mathbb{P}(V)$ exists such that

(2.50) $$\Omega | \mathbb{P}(V_\alpha) = \Omega_\alpha.$$

The form Ω defines a Kaehler metric on $\mathbb{P}(V)$ which is called the Fubini-Study metric associated to τ. On $\mathbb{P}(V_\alpha)$ we have

$$(2.51)\quad \mathbb{P}^*(\Omega) = \mathbb{P}^* \circ (j_\alpha \circ \mathbb{P}_\alpha^{-1})^*(\omega) = (j_\alpha \circ \mathbb{P}_\alpha^{-1} \circ \mathbb{P})^*(\omega) = g_\alpha^*(\omega) = \omega.$$

Since $\mathbb{P}$ is surjective, $\mathbb{P}^*$ is injective. Hence Ω can also be determined by

$$(2.52)\quad \boxed{\mathbb{P}^*(\Omega) = \omega}\ .$$

Because $\mathbb{P}(V)$ has dimension n, we have $\Omega^{n+1} = 0$ and therefore $\omega^{n+1} = \mathbb{P}^*(\Omega^{n+1}) = 0$. Hence

$$(2.53)\quad \boxed{\omega^{n+1} - 0}\ .$$

Now (2.46) implies immediately

$$(2.54)\quad \tau\upsilon^{n+1} = (n + 1)\, d\tau \wedge d^c\tau \wedge \upsilon^n.$$

Define

$$(2.55)\quad \sigma = \sigma_V = d^c \log \tau \wedge \omega^n.$$

Then (2.54) and (2.55) yield

$$(2.56)\quad \boxed{\upsilon^{n+1} = (n + 1)\tau^n\, d\tau \wedge \sigma} \qquad \boxed{d\sigma = 0}\ .$$

Lemma 2.1.
Let $F : V_* \to \mathbb{C}$ be a continuous function such that $F(r\mathfrak{z}) = F(\mathfrak{z})$ for all $\mathfrak{z} \in V_*$ and all $r > 0$. Then

$$(2.57)\quad \int_{V\langle r\rangle} F\sigma = \frac{1}{(n + 1)!} \int_V Fe^{-\tau}\upsilon^{n+1}.$$

Moreover

$$(2.58)\qquad \boxed{\int_{V\langle r\rangle} \sigma = 1} \qquad \boxed{\int_{V[r]} \upsilon^{n+1} = \int_{V(r)} \upsilon^{n+1} = r^{2n+2}} \qquad \forall\, r > 0.$$

<u>Proof</u>. If $0 < s < r$, Stokes Theorem implies

$$\int_{V\langle r\rangle} \sigma - \int_{V\langle s\rangle} \sigma = \int_{V(r)-V[s]} d\sigma = 0.$$

A constant c exists such that $\int_{V\langle r\rangle} \sigma = c$ for all $r > 0$. Now, (2.56) and fiber integration implies

$$(2.59)\qquad \int_V Fe^{-\tau}\upsilon^{n+1} = (n+1)\int_V Fe^{-\tau}\tau^n \, dt \wedge \sigma$$

$$= (n+1)\int_0^\infty \Big(\int_{V\langle 1\rangle} F\sigma\Big) e^{-t^2} t^{2n} \, dt^2$$

$$= (n+1)\int_0^\infty e^{-x} x^n \, dx \int_{V\langle 1\rangle} F\sigma = (n+1)! \int_{V\langle r\rangle} F\sigma.$$

Take $F = 1$. Select an orthonormal base of V. Then

$$(2.60)\qquad \upsilon = \frac{i}{2\pi}\sum_{\mu=0}^{n} dz_\mu \wedge d\bar{z}_\mu \qquad \upsilon^{n+1} = (n+1)!\Big(\frac{i}{2\pi}\Big)^{n+1} \prod_{\mu=0}^{n} dz_\mu \wedge d\bar{z}_\mu$$

$$(2.61)\qquad c = \int_{V\langle r\rangle} \sigma = \frac{1}{(n+1)!}\int_V e^{-\tau}\upsilon^{n+1}$$

$$= \Big(\frac{i}{2\pi}\Big)^{n+1} \int_{\mathbb{C}^{n+1}} \prod_{\mu=0}^{n} e^{-|z_\mu|^2} dz_\mu \wedge d\bar{z}_\mu$$

$$= \Big(\int_{\mathbb{C}} \frac{i}{2\pi} e^{-|z|^2} dz \wedge d\bar{z}\Big)^{n+1} = \Big(\frac{1}{\pi}\int_0^\infty e^{-t^2} t \, dt \int_0^{2\pi} d\phi\Big)^{n+1} = 1.$$

If $r > 0$, then

$$(2.62)\qquad \int_{V[r]} \upsilon^{n+1} = \int_{V[r]} (n+1)\tau^n\, d\tau \wedge \sigma$$

$$= \int_0^r (n+1)t^{2n}\, dt^2 = \int_0^{r^2} (n+1)x^n\, dx = r^{2n+2}$$

q.e.d.

Lemma 2.2.
Let $F : \mathbb{P}(V) \to \mathbb{C}$ be continuous. Take $r > 0$, then

$$(2.63)\qquad \boxed{\int_{V\langle r\rangle} (F \circ \mathbb{P})\sigma = \int_{\mathbb{P}(V)} F\Omega^n}\,.$$

Proof. Fiber integration implies

$$(2.64)\qquad \int_{V\langle r\rangle} (F \circ \mathbb{P})\sigma = \int_{V\langle r\rangle} (F \circ \mathbb{P})\, d^c \log \tau \wedge \mathbb{P}^*(\Omega^n)$$

$$= \int_{x\in\mathbb{P}(V)} \left(\int_{E(x)\langle r\rangle} d^c \log \tau \right) F\Omega^n$$

$$= \int_{\mathbb{P}(V)} \frac{1}{2\pi}\int_0^{2\pi} d\phi\, F\Omega^n = \int_{\mathbb{P}(V)} F\Omega^n$$

q.e.d.

In particular we obtain

$$(2.65)\qquad \boxed{\int_{\mathbb{P}(V)} \Omega^n = 1}\,.$$

Lemma 2.3.
Take $r > 0$. Let $F : V\langle r\rangle \to \mathbb{C}$ be a function such that $F\sigma$ is integrable over $V\langle r\rangle$. Then

$$(2.66)\qquad \int_{V\langle r\rangle} F\sigma = \int_{V\langle 1\rangle} \frac{1}{2\pi}\int_0^{2\pi} F(re^{i\phi}\mathfrak{z})\, d\phi\, \sigma(\mathfrak{z}).$$

Proof. For each $\phi \in \mathbb{R}$, we have

$$\int_{V\langle r\rangle} F\sigma = \int_{V\langle 1\rangle} F(re^{i\phi}\mathfrak{z})\sigma(re^{i\phi}\mathfrak{z}) = \int_{V\langle 1\rangle} F(re^{i\phi}\mathfrak{z})\sigma(\mathfrak{z}).$$

Integrate!

$$\int_{V\langle r\rangle} F\sigma = \frac{1}{2\pi}\int_0^{2\pi}\int_{V\langle 1\rangle} F(re^{i\phi}\mathfrak{z})\sigma(\mathfrak{z})\, d\phi = \int_{V\langle 1\rangle} \frac{1}{2\pi}\int_0^{2\pi} F(re^{i\phi}\mathfrak{z})\, d\phi\ \sigma(\mathfrak{z})$$

q.e.d.

Until now we have considered hermitian vector spaces without designating a particular base. Sometimes it is convenient to fix an orthonormal base and to identify V with $\mathbb{C}^{n+1}$. On $\mathbb{C}^{n+1}$ we always use the inner hermitian product defined by

(2.67) $$(\mathfrak{z}|\mathfrak{w}) = z_0\overline{w}_0 + \ldots + z_n\overline{w}_n.$$

We also denote $\mathbb{P}_n = \mathbb{P}(\mathbb{C}^{n+1})$.

For the following lemma we will let $m = n + 1$ be the dimension. We abbreviate

(2.68) $$\tau = \tau_{\mathbb{C}^m} \qquad \tilde{\tau} = \tau_{\mathbb{C}^{m-1}} \qquad \tilde{\upsilon} = dd^c\tilde{\tau} \qquad \sigma_m = \sigma_{\mathbb{C}^m}.$$

Lemma 2.4. ([3])
Take $r > 0$. Define $p = \sqrt{r^2 - \tilde{\tau}}$ on $\mathbb{C}^{m-1}[r]$. Let $h : \mathbb{C}^m\langle r\rangle \to \mathbb{C}$ be a function such that $h\sigma_m$ is integrable over $\mathbb{C}^m\langle r\rangle$. Define $p = \sqrt{r^2 - \tilde{\tau}}$

(2.69) $$\int_{\mathbb{C}^m\langle r\rangle} h\sigma_m = r^{2-2m}\int_{\mathbb{C}^{m-1}[r]}\left(\int_{\mathbb{C}\langle p(w)\rangle} h(w,\zeta)\sigma_1(\zeta)\right)\tilde{\upsilon}^{m-1}(w).$$

Proof. The set $E = \{(z_1,\ldots,z_n) \in \mathbb{C}^m\langle r\rangle \mid 0 \le z_m \in \mathbb{R}\}$ has measure zero on $\mathbb{C}^m\langle r\rangle$, and $S = \mathbb{C}^m\langle r\rangle - E$ is open in $\mathbb{C}^m\langle r\rangle$. A bijective map of class C^∞

(2.70) $$g : \mathbb{R}(0,2\pi) \times \mathbb{C}^{m-1}(r) \to S$$

is given by

(2.71) $$g(\phi,w) = (w,p(w)e^{i\phi}).$$

If $w = (z_1,\ldots,z_{m-1}) \in \mathbb{C}^{m-1}(r)$ and $z_j = x_j + iy_j$ where x_j, y_j are real variables, then ϕ, x_1, y_1, ..., x_{m-1}, y_{m-1} are real parameters of S. Observe that g is perpendicular to g_ϕ, g_{x_1}, g_{y_1}, ..., $g_{x_{m-1}}$, $g_{y_{m-1}}$ with

(2.72) $$\det(g,g_\phi,g_{x_1},g_{y_1},\ldots,g_{x_{m-1}},g_{y_{m-1}}) = r^2 > 0.$$

Hence g is an orientation preserving diffeomorphism, provided $\mathbb{C}^m\langle r\rangle$ is oriented to the exterior. Now we have

$$d^c\tau = \frac{i}{4\pi}\sum_{\mu=1}^{m}(z_\mu\, d\bar{z}_\mu - \bar{z}_\mu\, dz_\mu)$$

$$\begin{aligned} g^*(d^c\tau) &= d^c\tilde{\tau} + \frac{i}{4\pi}(pe^{i\phi}\, d(pe^{-i\phi}) - pe^{-i\phi}\, d(pe^{i\phi})) \\ &= d^c\tilde{\tau} + \frac{1}{2\pi}p^2\, d\phi = d^c\tilde{\tau} + \frac{1}{2\pi}(r^2 - \tilde{\tau})\, d\phi \end{aligned}$$

$$g^*(\upsilon) = g^*(dd^c\tau) = dg^*(d^c\tau) = dd^c\tilde{\tau} - \frac{1}{2\pi}d\tilde{\tau}\wedge d\phi$$

$$g^*(\upsilon^{m-1}) = \tilde{\upsilon}^{m-1} - \frac{m-1}{2\pi}d\tilde{\tau}\wedge d\phi\wedge\tilde{\upsilon}^{m-2}$$

$$\begin{aligned} g^*(d^c\tau\wedge\upsilon^{m-1}) &= \frac{1}{2\pi}p^2\, d\phi\wedge\tilde{\upsilon}^{m-1} + \frac{m-1}{2\pi}d\phi\wedge d\tilde{\tau}\wedge d^c\tilde{\tau}\wedge\tilde{\upsilon}^{m-2} \\ &= \frac{1}{2\pi}p^2\, d\phi\wedge\tilde{\upsilon}^{m-1} + \frac{\tilde{\tau}}{2\pi}d\phi\wedge\tilde{\upsilon}^{m-1} \\ &= \frac{r^2}{2\pi}d\phi\wedge\tilde{\upsilon}^{m-1}. \end{aligned}$$

Since $\tau\circ g = r^2$, we obtain

$$g^*(\sigma_m) = \frac{1}{r^{2m}} g^*(d^c\tau \wedge \upsilon^{m-1}) = \frac{1}{2\pi} \frac{1}{r^{2m-2}} d\phi \wedge \tilde{\upsilon}^{m-1}.$$

Fubini's Theorem implies

$$\int_{\mathbb{C}^m\langle r\rangle} h\sigma_m = r^{2-2m} \int_{\mathbb{C}^{m-1}[r]} \left(\frac{1}{2\pi}\int_0^{2\pi} h(w,p(w)e^{i\phi})\, d\phi\right)\tilde{\upsilon}^{m-1}$$

q.e.d.

We shall return to the Fubini-Study form Ω. Again let V be a hermitian vector space of dimension n + 1. Take $a \in \mathbb{P}(V^*)$. Then $\alpha \in V_*^*$ exists such that $\mathbb{P}(\alpha) = a$. Then

$$\|w,a\| = \frac{|\alpha(w)|}{\|w\|\ \|\alpha\|}.$$

Hence

$$\mathbb{P}^*(dd^c \log\|\square,a\|^2) = dd^c \log|\alpha|^2 - dd^c \log\tau = -\omega = -\mathbb{P}^*(\Omega).$$

Therefore

(2.73) $$\boxed{\Omega = dd^c \log \frac{1}{\|\square,a\|^2}} \qquad \text{on } \mathbb{P}(V) - \ddot{E}[a].$$

A smooth complex submanifold of dimension n + 1 of $\mathbb{P}(V) \times V$ is defined by

(2.74) $$S_0(V) = \{(x,\mathfrak{z}) \in \mathbb{P}(V) \times V \mid \mathfrak{z} \in E(x)\}.$$

Let $\sigma : S_0(V) \to V$ be the projection defined by $\sigma(x,\mathfrak{z}) = \mathfrak{z}$. Then $\sigma^{-1}(0) = \mathbb{P}(V) \times \{0\}$ and

(2.75) $$\sigma : S_0(V) - \mathbb{P}(V) \times \{0\} \to V_*$$

is biholomorphic. Hence σ blows up 0 to the projective space $\mathbb{P}(V)$ and leaves V_* unchanged. This process is called <u>the</u> σ-<u>process</u>.

Let $\pi : S_0(V) \to \mathbb{P}(V)$ be the projection defined by $\pi(x,\mathfrak{x}) = x$. Then $\pi : S_0(V) \to \mathbb{P}(V)$ is a holomorphic line bundle, called the <u>tautological bundle</u>, since over $x \in \mathbb{P}(V)$ the fiber is given by

$$(2.76) \qquad S_0(V)_x = \pi^{-1}(x) = \{x\} \times E(x) \approx E(x).$$

Usually we shall identify $S_0(V)_x = E(x)$ as fiber. Here $S_0(V)$ is a subbundle of the trivial bundle $\mathbb{P}(V) \times V$. Let $Q_0(V)$ be the <u>quotient bundle</u>. We have the exact sequence of holomorphic vector bundles

$$(2.77) \qquad 0 \to S_0(V) \underset{\iota}{\to} \mathbb{P}(V) \times V \underset{\rho}{\to} Q_0(V) \to 0.$$

The dual holomorphic line bundle $H = H(V) = S_0(V)^*$ is called the <u>hyperplane section bundle</u>. We have the dual exact sequence

$$(2.78) \qquad 0 \longrightarrow Q_0(V)^* \xrightarrow[\rho^*]{} \mathbb{P}(V) \times V^* \xrightarrow[\eta=\iota^*]{} H \longrightarrow 0.$$

Take $(x,\alpha) \in \mathbb{P}(V) \times V^*$. Then $\eta(x,\alpha) = \eta_x(\alpha) \in H_x = S_0(V)_*^* = E(x)^*$ is given by

$$(2.79) \qquad \eta_x(\alpha) = \alpha | E(x).$$

For each $\alpha \in V^*$ a holomorphic section $\hat{\alpha} = \eta(\alpha)$ of H is given whose zero set is $\ddot{E}[\alpha]$.

We claim that all holomorphic sections of H are given that way. Let s be a holomorphic section of H. Then for each $x \in \mathbb{P}(V)$ a linear map $s_x : E(x) \to \mathbb{C}$ is given. A holomorphic function $\alpha : V \to \mathbb{C}$ is defined by $\alpha(\mathfrak{x}) = s_{\mathbb{P}(\mathfrak{x})}(\mathfrak{x})$. Take $\lambda \in \mathbb{C}_*$. Then $\alpha(\lambda\mathfrak{x}) = s_{\mathbb{P}(\lambda\mathfrak{x})}(\lambda\mathfrak{x}) = \lambda s_{\mathbb{P}(\mathfrak{x})}(\mathfrak{x}) = \lambda\alpha(\mathfrak{x})$. A holomorphic function which is homogeneous of degree 1 is linear. Hence $\alpha \in V^*$. If $x \in \mathbb{P}(V)$, then $\hat{\alpha}_x = \eta_x(\alpha) = \alpha|E(x) = s_x$. Therefore $\hat{\alpha} = s$. The claim is proved.

The hermitian inner product on V defines a hermitian metric along the fibers of trivial bundle $\mathbb{P}(V) \times V$ which restricts to a hermitian metric θ along the fibers of $S_0(V)$ and the dual metric κ on H. Hence a first Chern form $c(H,\kappa)$ is defined. We shall compute this form.

If $(x,\mathfrak{x}) \in S_0(V)$, then $\mathfrak{x} \in E(x)$ and $\|\mathfrak{x}\|_\theta = \|\mathfrak{x}\|$ is the θ-length of $\mathfrak{x}$. Take $\alpha \in V_*^*$, then a section $s_\alpha : \mathbb{P}(V_\alpha) \to S_0(V)$ is defined by

$$s_\alpha(x) = \frac{\mathfrak{x}\|\alpha\|}{\alpha(\mathfrak{x})} \qquad \text{if } 0 \neq \mathfrak{x} \in E(x). \tag{2.80}$$

Then

$$\|s_\alpha(x)\|_\theta = \frac{\|\mathfrak{x}\| \ \|\alpha\|}{|\alpha(\mathfrak{x})|} = \frac{1}{\|a,x\|}\,. \tag{2.81}$$

Hence

$$c(S_0(V),\theta) = -dd^c \log\|s_\alpha\|_\theta^2 = dd^c \log\|a,\square\|^2 = -\Omega.$$

Since $c(H,\kappa) = -c(S_0(V),\theta)$, we obtain

$$\boxed{c(H,\kappa) = \Omega}\,. \tag{2.82}$$

3. Divisors and the Jensen formula.

First we shall review some known properties of analytic sets. Let M be a complex manifold. A local coordinate system or chart is a biholomorphic map $\alpha : U \to U'$ from an open subset U of M onto an open subset U' of $\mathbb{C}^m$. Then $\dim_x M = m$ for all $x \in U$.

A subset A of M is said to be analytic if for every point $p \in M$ there exists an open neighborhood U of p and holomorphic functions $f_1, \ldots, f_s$ on U such that

$$(3.1) \qquad A \cap U = \bigcap_{j=1}^{s} f_j^{-1}(0).$$

An analytic set is closed. A point p of an analytic set A is said to be regular or simple if there exists an open neighborhood U of p such that $U \cap A$ is a smooth complex submanifold of U. A non-simple point is called singular. The set $\mathfrak{R}(A)$ of all regular points of A is open and dense in the topology of A. The set $\Sigma(A) = A - \mathfrak{R}(A)$ of all singular points is analytic and nowhere dense in A. If $p \in A$, then $\dim_p A = \dim_p \mathfrak{R}(A)$ is trivially defined. If $p \in \Sigma(A)$, the dimension of A at p is defined by

$$(3.2) \qquad \dim_p A = \limsup_{\substack{x \to p \\ x \in \mathfrak{R}(A)}} \dim_x A.$$

Define $\dim A = \sup_{x \in A} \dim_x A$ as the dimension of A. Also A is said to be pure dimensional if $\dim_x A = \dim A$ for all $x \in A$. We have $\dim_x A = \dim_x M - 1$ for all $x \in A$, iff for each $p \in A$ there exists an open neighborhood U and a holomorphic function $f : U \to \mathbb{C}$ such that $U \cap A = f^{-1}(0)$ and such that $f^{-1}(0)$ does not contain an interior point.

An analytic set A is said to be reducible if there are analytic sets A_1 and A_2 such that $A = A_1 \cup A_2$ and $A_1 \neq A \neq A_2$. If A is not reducible, then A is called irreducible. An irreducible analytic set is pure dimensional. An irreducible analytic set B is said to be a branch of the analytic set A, if $B \subseteq A$ and if for every irreducible analytic set C with $B \subseteq C \subseteq A$ we have $B = C$. A set B is a branch of A, if and only if there is a connectivity component N of $\mathfrak{R}(A)$ with $B = \overline{N}$. The family $\mathfrak{B}(A)$ of branches of A is locally finite with

$$(3.3) \qquad A = \bigcup_{B \in \mathfrak{B}(A)} B.$$

If $\mathfrak{C} \subseteq \mathfrak{B}(A)$, then $D = \bigcup_{B \in \mathfrak{C}} B$ is analytic with $\mathfrak{B}(D) = \mathfrak{C}$.

An analytic set A is <u>thin</u> in the analytic set B if $B \supseteq A$ and if $B - A$ is dense in B. An analytic set is <u>thin</u> if it is thin in M.

Let S be a thin analytic set in M. Let N be a complex manifold. Let $f : M - S \to N$ be a holomorphic map. Then

$$(3.4) \qquad \Gamma_f = \overline{\{(z, f(z)) \mid z \in M - S\}}$$

is called the <u>closed graph</u> of f in $M \times N$. Let $\pi : \Gamma_f \to M$ be the projection. Then f is said to be <u>meromorphic</u> iff Γ_f is an analytic set in $M \times N$ and if the map $\pi : \Gamma_f \to M$ is <u>proper</u> (i.e. the inverse image of compact sets is compact). A largest open subset H_f of M exists such that f continues to a holomorphic map $f : H_f \to N$. Then H_f is called the domain of holomorphy of f and the complement I_f is called <u>the indeterminacy of</u> f. The set I_f is analytic with

$$(3.5) \qquad \dim_x I_f \leq \dim_x M - 2 \qquad \forall\, x \in I_f.$$

Let V be a complex vector space of dimension $n + 1$. Then there is an alternative description of meromorphic maps into $\mathbb{P}(V)$. Let M be a complex manifold. Let S be a thin analytic set in M. Let $f : M - S \to \mathbb{P}(V)$ be a holomorphic map. Let U be an open subset of M. Let $\mathfrak{v} : U \to V$ be a holomorphic vector function such that $\mathfrak{v}^{-1}(0)$ is thin in U. Then $\mathfrak{v}$ is said to be a <u>representation</u> of f if $\mathbb{P} \circ \mathfrak{v} = f$ on $U - \mathfrak{v}^{-1}(0) \cup S$. The representation is <u>reduced</u> if $\dim_x \mathfrak{v}^{-1}(0) \leq \dim_x M - 2$ for all $x \in U$. A representation is said to be <u>global</u> if $U = M$. Then we have the following facts.

(1) The holomorphic map $f : M - S \to \mathbb{P}(V)$ is meromorphic, if and only if for every $p \in S$ there exists an open neighborhood U of p and a representation $\mathfrak{v} : U \to V$ of f.

(2) If $f : M - S \to \mathbb{P}(V)$ is meromorphic on M, then for every $p \in M$ there exists an open neighborhood U of p and a reduced representation $\mathfrak{v} : U \to V$ of f.

(3) If $f : M - S \to \mathbb{P}(V)$ is meromorphic on M, and if M is a vector space or an open ball in a vector space, then there exists a global reduced representation of f.

(4) If $\mathfrak{v} : U \to V$ is a reduced representation of f, then $I_f \cap U = \mathfrak{v}^{-1}(0)$.

(5) If $\mathfrak{v} : U \to V$ is a reduced representation of f, if $\tilde{\mathfrak{v}} : \tilde{U} \to V$ is a representation of f with $\tilde{U} \cap U \neq \emptyset$, then one and only one holomorphic function $h : U \cap \tilde{U} \to \mathbb{C}$ exists with $\tilde{\mathfrak{v}} = h\mathfrak{v}$. If $\tilde{\mathfrak{v}}$ is reduced $h(x) \neq 0$ for all $x \in \tilde{U} \cap U$.

A meromorphic function is a meromorphic map into $\mathbb{P}_1$ which is not identically ∞. To fix this concept precisely, we have to identify $\mathbb{P}_1 = \mathbb{P}(\mathbb{C}^2) = \mathbb{C} \cup \{\infty\}$. We identify

$$\text{(3.6)} \qquad \mathbb{P}(1,z) = z \quad \forall\, z \in \mathbb{C} \qquad \mathbb{P}(0,1) = \infty.$$

We identify $\mathbb{P}((\mathbb{C}^2)^*) = \mathbb{C} \cup \{\infty\}$ by $\mathbb{C}^2 = (\mathbb{C}^2)^*$ and

$$\text{(3.7)} \qquad \mathbb{P}(z,-1) = z \quad \forall\, z \in \mathbb{C} \qquad \mathbb{P}(1,0) = \infty.$$

If $a = \mathbb{P}(\alpha_1,\alpha_2) \in \mathbb{P}((\mathbb{C}^2)^*)$ and $z = \mathbb{P}(z_1,z_2) \in \mathbb{P}(\mathbb{C}^2)$, and if $a \neq \infty \neq z$ then $a = -\dfrac{\alpha_1}{\alpha_2}$ and $z = \dfrac{z_2}{z_1}$ we have

$$\text{(3.8)} \qquad \|a,z\| = \frac{|\alpha(z)|}{\|z\|\ \|\alpha\|}$$

$$= \frac{|\alpha_1 z_1 + \alpha_2 z_2|}{\sqrt{|z_1|^2 + |z_2|^2}\,\sqrt{|\alpha_1|^2 + |\alpha_2|^2}} = \frac{|a - z|}{\sqrt{1 + |a|^2}\,\sqrt{1 + |z|^2}}\,.$$

Hence (1.3) and (2.32) agree.

Let $f : M - S \to \mathbb{P}_1$ be a holomorphic map. Let $\mathfrak{w} = (h,g) : U \to \mathbb{C}^2$ be a representation. Then $\mathbb{P} \circ \mathfrak{w} = f$ on $U - (S \cup \mathfrak{w}^{-1}(0))$. Assume that $h^{-1}(0)$ is thin on U, then $f(z) = g(z)/h(z)$ on $U - S \cup h^{-1}(0)$. If $f = g/h$ where g and h are holomorphic on U and if $h^{-1}(0)$ is thin, then $\mathfrak{w} = (h,g)$ is a representation of f. The map $f : M - S \to \mathbb{P}_1$ is said to be a meromorphic function if $f^{-1}(\infty) \cap (M - S)$ is thin analytic and iff f is meromorphic on M. This is the case if and only if for every point $p \in M$ there exists an open neighborhood U of p and holomorphic functions g and h on U such that $h^{-1}(0)$ is thin analytic in U and such that $f = g/h$ on $U - (S \cup h^{-1}(0))$. Then g and h and U can be taken such that

$$\dim_x g^{-1}(0) \cap h^{-1}(0) \leq \dim_x M - 2 \quad \forall\, x \in g^{-1}(0) \cap h^{-1}(0). \tag{3.9}$$

Then $I_f \cap U = g^{-1}(0) \cap h^{-1}(0)$. Also there exists a largest open set H_f^0 of M such that $f : U_f^0 \to \mathbb{C}$ is holomorphic. The set $P_f = M - H_f^0$ is thin analytic and is called the pole set of f. We have $I_f \subseteq P_f$. Under the assumption (3.9) we have $P_f \cap U = h^{-1}(0)$.

Let $f : M \to \mathbb{C}$ be a holomorphic function. Take $p \in M$. Let $\alpha = (\alpha_1, \ldots, \alpha_m) : U \to U'$ be a local coordinate system where U is an open, connected neighborhood of p. Then there exists a homogeneous polynomial P_j of degree j in $\alpha_1, \ldots, \alpha_m$ for each $j \geq 0$ such that

$$f = \sum_{j=q}^{\infty} P_j \tag{3.10}$$

converges uniformly to f on some open neighborhood U_0 of p in U. Assume that $f|U \not\equiv 0$, then q can be taken such that $P_q \not\equiv 0$. The number $q \geq 0$ is independent of the choice of α and is called the <u>zero</u>

<u>multiplicity</u> of f at p and denoted by $\mu_f^0(p) = q$. If $a \in \mathbb{C}$ and if $f|U \not\equiv a$, then the a-<u>multiplicity of f at</u> p is defined by

$$\mu_f^a(p) = \mu_{f-a}(p). \tag{3.11}$$

The following properties of zero multiplicities are easily verified.

(1) If f and g are holomorphic and not identically zero at p, then

$$\mu_{f\cdot g}^0(p) = \mu_f^0(a) + \mu_g^0(a)$$

$$\mu_{f+g}^0(p) \geq \mu_f^0(a) + \mu_g^0(a) \qquad \text{if } f + g \not\equiv 0 \text{ at a.}$$

(2) If f and g are holomorphic on M and nowhere identically zero, then $\mu_f^0(p) = \mu_g^0(p)$ for all $p \in M$ if and only if there exists a holomorphic function $h : M \to \mathbb{C}_*$ such that $f = gh$.

(3) If f is holomorphic and nowhere identically zero on M, if $N = f^{-1}(0)$ and if $p \in \mathfrak{R}(N)$, then there exists an open neighborhood U of p with $N \cap U = \mathfrak{R}(N) \cap U$ such that $\mu_f^0|N \cap U$ is constant. Hence μ_f^0 is locally constant on $\mathfrak{R}(N)$.

There are several ways to define divisors. On manifolds, this can be done by the divisor multiplicity function.

An integer valued function $\nu : M \to \mathbb{Z}$ is called a <u>divisor</u> if and only if for every point $p \in M$ there exists an open, connected neighborhood U of p and holomorphic functions $f \not\equiv 0$ and $g \not\equiv 0$ such that

$$\nu|U = \mu_f^0 - \mu_g^0. \tag{3.12}$$

The divisor is called <u>non-negative</u> iff $\nu \geq 0$ in which case $g \equiv 1$ can be taken such that $\nu|U = \mu_f^0$. The set $\mathfrak{D}(M)$ of divisors is a module under addition. Let $\mathfrak{D}_+(M)$ be the subset of non-negative divisors on M. If

$\nu \in \mathfrak{D}(M)$ is a divisor, the <u>support</u> of ν is defined by

$$\text{supp}\,\nu = \overline{\{z \in M \mid \nu(z) \neq 0\}}. \tag{3.13}$$

If $\nu \equiv 0$, then supp $\nu = \emptyset$. If $\nu \not\equiv 0$, then supp ν is an analytic set of pure codimension 1, i.e. $\dim_x \text{supp}\,\nu = \dim_x M - 1$ for all $x \in \text{supp}\,\nu$. Moreover ν is locally constant on $\mathfrak{R}(\text{supp}\,\nu)$. If $\nu \geq 0$, then

$$\text{supp}\,\nu = \{z \in M \mid \nu(z) > 0\}. \tag{3.14}$$

Let N be a connected complex manifold. Let $\phi : N \to M$ be a holomorphic map. Let ν be a divisor on M such that $\phi(N) \not\subseteq \text{supp}\,\nu$. Then there exists a <u>pullback divisor</u> $\phi^*(\nu)$ on N which is uniquely defined by the following property: Take $p \in N$. Then there exists an open connected neighborhood U of $\phi(p)$ in M and holomorphic functions $f \not\equiv 0$ and $g \not\equiv 0$ such that $\nu|U = \mu_f^0 - \mu_g^0$. Then $\tilde{U} = \phi^{-1}(U)$ is open in N and $f \circ \phi$ and $g \circ \phi$ are holomorphic and nowhere identically zero on $\tilde{U}$. Then

$$\phi^*(\nu)|\tilde{U} = \mu_{f\circ\phi}^0 - \mu_{g\circ\phi}^0. \tag{3.15}$$

In general $\phi^*(\nu) \neq \nu \circ \phi$, but if $\phi : N \to M$ is biholomorphic, then $\phi^*(\nu) = \nu \circ \phi$.

Now, consider some examples of divisors. Let M be an m-dimensional, connected complex manifold. Let f be a meromorphic function. Take $a \in \mathbb{P}_1$. If $f \not\equiv a$, then a non-negative divisor μ_f^a is defined: Take $p \in M$. Then there exists an open connected neighborhood U of p and holomorphic functions $g \not\equiv 0$ and $h \not\equiv 0$ on U such that $\dim g^{-1}(0) \cap h^{-1}(0) \leq m - 2$ and such that $hf = g$ on U. Then

$$\mu_f^a|U = \mu_{g-ah}^0 \qquad \text{if } a \neq \infty \tag{3.16}$$

$$\mu_f^\infty|U = \mu_h^0 \qquad \text{if } a = \infty. \tag{3.17}$$

The divisor μ_f^a is called the <u>a-divisor of</u> f. If $a = \infty$, the divisor μ_f^∞ is called the <u>pole divisor of</u> f. If $f \not\equiv 0$, then

(3.18) $$\mu_f = \mu_f^0 - \mu_f^\infty$$

is called the <u>divisor of</u> f. Moreover f is holomorphic, if and only if $\mu_f \geq 0$.

Let V be a complex vector space of dimension n + 1. Let $0 \not\equiv \mathfrak{v} : M \to V$ be a holomorphic vector function. Then a non-negative divisor $\nu_{\mathfrak{v}}$ called the <u>greatest common divisor</u> of $\mathfrak{v}$ is assigned. Take any $p \in M$. Then there exists an open, connected neighborhood U of p and a holomorphic vector function $\mathfrak{w} : U \to V$ and a holomorphic function $g : U \to \mathbb{C}$ such that $\mathfrak{v}|U = g\mathfrak{w}$ and such that $\dim \mathfrak{w}^{-1}(0) \leq m - 2$. Then $\nu_{\mathfrak{v}}|U = \mu_g^0$.

Let $f : M \to \mathbb{P}(V)$ be a meromorphic map. Take $a \in \mathbb{P}(V^*)$. Assume that $f(M) \not\subseteq \ddot{E}[a]$. Then there is assigned a non-negative divisor μ_f^a. Take $\alpha \in V_*^*$ with $\mathbb{P}(\alpha) = a$. Take any $p \in M$. Then there exists an open, connected neighborhood U of p and a reduced representation $\mathfrak{v} : U \to V$. Then $\alpha \circ \mathfrak{v} \not\equiv 0$ is a holomorphic function on U. We have

(3.19) $$\mu_f^a|U = \mu_{\alpha \circ \mathfrak{v}}^0.$$

We want to introduce certain integrals over divisors. First we explain integration over analytic sets. Let A be an analytic subset of pure dimension p of M. Let $j : \mathfrak{R}(A) \to M$ be the inclusion. Let S be a subset of A such that $S \cap \mathfrak{R}(A)$ is measurable on $\mathfrak{R}(A)$. Let ψ be a form of bidegree (p,p) on M such that $j^*(\psi)$ is integrable over $S \cap \mathfrak{R}(A)$. Define

(3.20) $$\int_S \psi = \int_{\mathfrak{R}(A) \cap S} j^*(\psi).$$

If ψ is continuous at every point of S, if S is compact, then $\int_S \psi$ exists although $\mathfrak{R}(A) \cap S$ may not be compact.

In respect to later applications, we make a slight change of notation. Let W be a hermitian vector space of dimension $m > 0$. Let ν be a divisor on W. Define $A = \operatorname{supp} \nu$. If $r > 0$, then

$$\sigma_\nu(r) = \int_{A[r]} \nu\upsilon^{m-1} \qquad \text{if } m > 1 \tag{3.21}$$

$$\sigma_\nu(r) = \sum_{z \in A[r]} \nu(z) \qquad \text{if } m = 1 \tag{3.22}$$

is the volume of A[r] with multiplicities. The <u>counting function</u> n_ν of ν is defined by

$$n_\nu(r) = \frac{\sigma_\nu(r)}{r^{2m-2}}\,. \tag{3.23}$$

Obviously, $n_\nu(r) \geq 0$ and $\sigma_\nu(r) \geq 0$ if $\nu \geq 0$. Also if ν and μ are divisors, then

$$\sigma_{\mu+\nu}(r) = \sigma_\mu(r) + \sigma_\nu(r) \qquad n_{\mu+\nu}(r) = n_\mu(r) + n_\nu(r). \tag{3.24}$$

If $m > 1$, then A<r> has measure zero on A. Hence $n_\nu(r)$ is continuous in r and A[r] in (3.21) can be replaced by A(r). However if $m = 1$ this does not remain true and $\sigma_\nu = n_\nu$ is only one sided continuous from the right.

<u>Proposition 3.1.</u> ([63] proof Satz 10)
Let $f \not\equiv 0$ be a meromorphic function on the m-dimensional, hermitian vector space W. For $\mathfrak{z} \in W$, define

$$f'(\mathfrak{z}) = \frac{d}{d\lambda} f(\lambda\mathfrak{z})\big|_{\lambda=1}. \tag{3.25}$$

Then $\frac{f'}{f}\sigma$ is integrable over W<r> for almost all r with

$$\boxed{\int_{W\langle r\rangle} \frac{f'}{f}\sigma = n_{\mu_f}(r)}\,. \tag{3.26}$$

Proof. We take an orthonormal base of W and identify $W = \mathbb{C}^m$. Then

$$(3.27)\qquad f'(\mathfrak{z}) = \sum_{j=1}^{m} f_{z_j}(\mathfrak{z})z_j.$$

The Weierstrass preparation theorem can be used to show that $z_j\left(\frac{f_{z_j}}{f}\right)\upsilon^m$ is locally integrable. Take any $R > 0$. Now (2.56) with $n = m - 1$ implies that

$$(3.28)\qquad \int_{W[R]} z_j\left(\frac{f_{z_j}}{f}\right)\upsilon^m = \int_0^R\left(\int_{W\langle t\rangle} z_j \frac{f_{z_j}}{f}\,\sigma\right)mt^{2m-1}\,dt$$

exists. Hence for almost all $r > 0$ the integral

$$(3.29)\qquad L_j(r) = \int_{W\langle r\rangle} z_j \frac{f_{z_j}}{f}\,\sigma$$

exists, where

$$(3.30)\qquad \sum_{j=1}^{m} L_j(r) = \int_{W\langle r\rangle} \frac{f'}{f}\,\sigma.$$

We adopt the notations of Lemma 2.4, which implies

$$(3.31)\qquad L_m(r) = r^{2-2m}\int_{\mathbb{C}^{m-1}[r]}\int_{\mathbb{C}\langle p(w)\rangle} z_m \frac{f_{z_m}(w,z_m)}{f(w,z_m)}\,\sigma_1(z_m)\tilde{\upsilon}^{m-1}(w)$$

$$= r^{2-2m}\int_{\mathbb{C}^{m-1}[r]}\left(\frac{1}{2\pi i}\int_{\mathbb{C}\langle p(w)\rangle} \frac{f_{z_m}(w,z_m)}{f(w,z_m)}\,dz_m\right)\tilde{\upsilon}^{m-1}(w)$$

$$= r^{2-2m}\int_{\mathbb{C}^{m-1}[r]} n_{\nu_m(w)}(p(w))\tilde{\upsilon}^{m-1}(w)$$

where $\nu_m(w)$ is the divisor of $f(w,\square)$.

Let $\pi : A \to \mathbb{C}^{m-1}$ be the projection $\pi(z_1,\ldots,z_m) = (z_1,\ldots,z_{m-1})$. Define

$$(3.32) \qquad \rho_j = \left(\frac{1}{2\pi}\right)^{m-1}(m-1)! \bigwedge_{j\neq\lambda=1}^{m} dz_\lambda \wedge d\bar{z}_\lambda .$$

Then $\rho_m = \pi^*(\tilde{\upsilon}^{m-1})$. Fiber integration implies easily

$$(3.33) \quad L_m(r) = r^{2-2m} \int_{\mathbb{C}^{m-1}[r]} n_{\nu_m(w)}(p(w))\tilde{\upsilon}^{m-1}(w) = r^{2-2m} \int_{A[r]} \nu_f \rho_m$$

and by symmetry we may replace m by j. Since $\upsilon^{m-1} = \sum_{j=1}^{m} \rho_j$, we have

$$(3.34) \qquad \int_{W\langle r\rangle} \frac{f'}{f}\sigma = \sum_{j=1}^{m} L_j(r) = r^{2-2m} \sum_{j=1}^{m} \int_{A[r]} \nu_f \rho_j = n_{\nu_f}(r)$$

q.e.d.

The result is an extension of the argument principle to several variables. I do not know if (3.26) holds for all $r > 0$.

Let ν be a divisor on the m-dimensional hermitian vector space W with $m > 1$. Define $A = \operatorname{supp} \nu$. Take $\mathfrak{x} \in W_*$. Define $j_\mathfrak{x} : \mathbb{C} \to W$ by $j_\mathfrak{x}(z) = z\mathfrak{x}$ for all $z \in \mathbb{C}$. Then $j_\mathfrak{x}(\mathbb{C}) = E(\mathfrak{x})$ is the linear subspace spanned by $\mathfrak{x}$. Moreover ν is said to be restrictable to $E(\mathfrak{x})$ if $E(\mathfrak{x}) \nsubseteq A$. If so, the pullback divisor $\nu[\mathfrak{x}] = j_\mathfrak{x}^*(\nu)$ exists. If $z \in \mathbb{C}$ write also $\nu(z,\mathfrak{x}) = \nu[\mathfrak{x}](z)$. The counting function of $\nu[\mathfrak{x}]$ is also denoted by

$$(3.35) \qquad n_{\nu[\mathfrak{x}]}(r) = n_\nu[r,\mathfrak{x}] = \sum_{|z|\le r} \nu(z,\mathfrak{x}).$$

<u>Proposition</u> <u>3</u>.<u>2</u>. ([63], Kneser [34])
Let ν be a divisor on the m-dimensional, hermitian vector space W with $m > 1$. Take $r > 0$. Then

$$(3.36) \qquad n_\nu(r) = \int_{W\langle 1\rangle} n_\nu[r,\mathfrak{x}]\sigma(\mathfrak{x})$$

$$(3.37) \qquad n_\nu(r) \to \nu(0) \qquad \text{for } r \to 0.$$

There exists a thin analytic set K in W such that $\nu(0,\mathfrak{x}) = \nu(0)$ for all $\mathfrak{x} \in W_* - K$.

Proof. First assume that ν is non-negative. Then there exists a holomorphic function f on W with $\nu = \mu_f^0$. There exist homogeneous polynomials P_j of degree j such that

$$(3.38) \qquad f = \sum_{j=q}^{\infty} P_j$$

converges uniformly on every compact subset of W with $P_q \not\equiv 0$. Here $q = \nu(0)$. Then $K = P_q^{-1}(0)$ is thin analytic. Take $\mathfrak{x} \in W - K$. Then

$$f(z\mathfrak{x}) = \sum_{j=q}^{\infty} P_j(\mathfrak{x})z^j \qquad \text{with } P_q(\mathfrak{x}) \not\equiv 0.$$

Hence $\nu(0) = q = \nu(0,\mathfrak{x})$. By (3.25) we have $f'(z\mathfrak{x}) = z\,\frac{d}{dz}\,f(z\mathfrak{x})$. Therefore (3.26) implies

$$(3.39) \quad n_\nu(r) = \int_{W\langle r\rangle} \frac{f'}{f}\,\sigma = \int_{W\langle 1\rangle} \frac{1}{2\pi}\int_0^{2\pi} \frac{f'(re^{i\phi}\mathfrak{x})}{f(re^{i\phi}\mathfrak{x})}\,d\phi\;\sigma(\mathfrak{x})$$
$$= \int_{W\langle 1\rangle} \frac{1}{2\pi i}\int_{\mathbb{C}\langle 1\rangle} \frac{f'(z\mathfrak{x})}{f(z\mathfrak{x})}\,dz\;\sigma(\mathfrak{x}) = \int_{W\langle 1\rangle} n_\nu[r,\mathfrak{x}]\sigma(\mathfrak{x})$$

which proves (3.36) for almost all r. Since $n_\nu[r,\mathfrak{x}] \to n_\nu[r_0,\mathfrak{x}]$ for $r \to r_0$ with $r > r_0$ monotonically and since n_ν is continuous we obtain (3.36) for all $r > 0$. Since $n_\nu[r,\mathfrak{x}] \to \nu(0,\mathfrak{x})$ for $r \to 0$ with $r > 0$ monotonically and since $\nu(0,\mathfrak{x}) = \nu(0)$ for almost all $\mathfrak{x} \in W\langle 1\rangle$ we have

$$\lim_{0<r\to 0} n_\nu(r) = \int_{W\langle 1\rangle} \nu(0,\mathfrak{x})\sigma(\mathfrak{x}) = \nu(0).$$

If ν is any divisor, then there exist entire holomorphic functions g and h with $\dim g^{-1}(0) \cap h^{-1}(0) \le m - 2$ such that $\nu = \mu_g^0 - \mu_h^0$. We obtain (3.36) and (3.37) by subtraction and $\nu(0) = \nu(0,\mathfrak{x})$ whenever $\mu_g^0(0) = \mu_g^0[0,\mathfrak{x}]$ and $\mu_h^0(0) = \mu_h^0[0,\mathfrak{x}]$ which is true outside a thin analytic set. q.e.d.

Lemma 3.3. ([63] Satz 10, Kneser [34])
Let ν be a divisor on the m-dimensional hermitian vector space W with $m > 1$. Define $A = \operatorname{supp} \nu$. Take $r > 0$. Then

$$n_\nu(r) = \int_{A[r]} \nu\omega^{m-1} + \nu(0). \tag{3.40}$$

Moreover n_ν increases if $\nu \geq 0$.

Proof. By Tung [84], Stokes Theorem applies to the open subset $A(r)$ of A. On $A\langle r\rangle$ we have $d\tau = 0$. Hence $\omega = r^{-2}\upsilon$ and $d^c \log \tau = r^{-2}\, d^c\tau$ on $A\langle r\rangle$. Since ν is locally constant on $\Re(A)$ we obtain if $0 < s < r$

$$\begin{aligned}
\int_{A(r)-A[s]} \nu\omega^{m-1} &= \int_{A(r)-A[s]} d(\nu(d^c \log \tau) \wedge \omega^{m-2}) \\
&= \int_{A\langle r\rangle} \nu\, d^c \log \tau \wedge \omega^{m-2} - \int_{A\langle s\rangle} \nu\, d^c \log \tau \wedge \omega^{m-2} \\
&= \frac{1}{r^{2m-2}} \int_{A\langle r\rangle} \nu\, d^c\tau \wedge \upsilon^{m-1} - \frac{1}{s^{2m-2}} \int_{A\langle s\rangle} \nu\, d^c\tau \wedge \upsilon^{m-1} \\
&= \frac{1}{r^{2m-2}} \int_{A[r]} \nu\, dd^c\tau \wedge \upsilon^{m-1} - \frac{1}{s^{2m-2}} \int_{A[s]} \nu\, d^c d\tau \wedge \upsilon^{m-1} \\
&= n_\nu(r) - n_\nu(s).
\end{aligned}$$

Now $s \to 0$ implies (3.40). Trivially, $n_\nu(r)$ increases if $\nu \geq 0$ since $\omega^{m-1} \geq 0$; q.e.d.

Lemma 3.4. (Kneser [34])
Let ν be a divisor on an m-dimensional, hermitian vector space W. Define $A = \operatorname{supp} \nu$. Take $0 < s < r$. Let $f : \mathbb{R}[s,r] \to \mathbb{C}$ be a function of class C^1. Then

$$(3.41)\qquad \int_{A[r]-A[s]} \nu f(\sqrt{\tau})\upsilon^{m-1} = \sigma_\nu(r)f(r) - \sigma_\nu(s)f(s) - \int_s^r \sigma_\nu(t)f'(t)\,dt$$

$$(3.42)\qquad \int_{A[r]-A[s]} \nu f(\sqrt{\tau})\omega^{m-1} = n_\nu(r)f(r) - n_\nu(s)f(s) - \int_s^r n_\nu(t)f'(t)\,dt.$$

Proof. Define

$$(3.43)\qquad \chi(t,x) = \begin{cases} 1 & \text{if } s \le t \le x \le r \\ 0 & \text{if } s \le x < t \le r. \end{cases}$$

Then

$$\int_{A[r]-A[s]} \nu f(\sqrt{\tau})\upsilon^{m-1}$$

$$= \int_{A[r]-A[s]} \nu \int_s^{\sqrt{\tau}} f'(t)\,dt\,\upsilon^{m-1} + f(s)(\sigma_\nu(r) - \sigma_\nu(s))$$

$$= \int_{A[r]-A[s]} \nu \int_s^r \chi(t,\sqrt{\tau})f'(t)\,dt\,\upsilon^{m-1} + f(s)(\sigma_\nu(r) - \sigma_\nu(s))$$

$$= \int_s^r \Big(\int_{A[r]-A[s]} \nu\chi(t,\sqrt{\tau})\upsilon^{m-1}\Big) f'(t)\,dt + f(s)(\sigma_\nu(r) - \sigma_\nu(s))$$

$$= \int_s^r \int_{A[r]-A[t]} \nu\upsilon^{m-1} f'(t)\,dt + f(s)(\sigma_\nu(r) - \sigma_\nu(s))$$

$$= \int_s^r (\sigma_\nu(r) - \sigma_\nu(t))f'(t)\,dt + f(s)\sigma_\nu(r) - f(s)\sigma_\nu(s)$$

$$= \sigma_\nu(r)f(r) - f(s)\sigma_\nu(s) - \int_s^r \sigma_\nu(t)f'(t)\,dt.$$

Similarly we have

$$\int_{A[r]-A[s]} \nu f(\sqrt{\tau})\omega^{m-1}$$

$$= \int_{A[r]-A[s]} \nu \int_s^r \chi(t,\sqrt{\tau})f'(t)\,dt\,\omega^{m-1} + f(s)(n_\nu(r) - n_\nu(s))$$

$$= \int_s^r \int_{A[r]-A[s]} \nu\chi(t,\sqrt{\tau})\omega^{m-1}f'(t)\,dt + f(s)(n_\nu(r) - n_\nu(s))$$

$$= \int_s^r \Big(\int_{A[r]-A[t]} \nu\omega^{m-1}\Big)f'(t)\,dt + f(s)(n_\nu(r) - n_\nu(s))$$

$$= \int_s^r (n_\nu(r) - n_\nu(t))f'(t)\,dt + f(s)(n_\nu(r) - n_\nu(s))$$

$$= n_\nu(r)f(r) - n_\nu(s)f(s) - \int_s^r n_\nu(t)f'(t)\,dt$$

q.e.d.

Let ν be a divisor on the hermitian vector space of dimension m. The <u>valence function</u> N_ν of ν is defined for $0 < s < r$ by

$$N_\nu(r,s) = \int_s^r n_\nu(t)\,\frac{dt}{t} \, . \tag{3.44}$$

Obviously, if $\nu \geq 0$, then $N_\nu(r,s) \geq 0$ increases with r.

If $0 < s < r$ define the functions ϕ_{rs} and ψ_{rs} on $\mathbb{R}[0,r]$ by

$$0 \leq \phi_{rs}(x) = \begin{cases} \log \frac{r}{x} & \text{if } s \leq x \leq r \\ \log \frac{r}{s} & \text{if } 0 \leq x \leq s \end{cases} \tag{3.45}$$

$$0 \leq \psi_{rs}(x) = \begin{cases} \frac{1}{2m-2}\left(\frac{1}{x^{2m-2}} - \frac{1}{r^{2m-2}}\right) \\ \frac{1}{2m-2}\left(\frac{1}{s^{2m-2}} - \frac{1}{r^{2m-2}}\right). \end{cases} \tag{3.46}$$

<u>Lemma 3.5.</u> (Kneser [34])
Let ν be a divisor on the m-dimensional hermitian vector space W with

$m > 1$. Take $0 < s < r$. Then

$$(3.47)\qquad \boxed{N_\nu(r,s) = \int_{A[r]} \nu\psi_{rs}(\sqrt{\tau})\upsilon^{m-1} = \int_{A[r]} \nu\phi_{rs}(\sqrt{\tau})\omega^{m-1} + \phi_{rs}(0)\nu(0)}\,.$$

Proof. (3.41) implies

$$\int_{A[r]} \nu\psi_{rs}(\sqrt{\tau})\upsilon^{m-1} = \int_{A[r]-A[s]} \nu\psi_{rs}(\sqrt{\tau})\upsilon^{m-1} + \psi_{rs}(0)\sigma_\nu(s)$$

$$= \psi_{rs}(r)\sigma_\nu(r) - \psi_{rs}(s)\sigma_\nu(s) - \int_s^r \sigma_\nu(t)\psi_{rs}'(t)\,dt + \psi_{rs}(s)\sigma_\nu(s)$$

$$= \int_s^r n_\nu(t)t^{2m-2}\,\frac{dt}{t^{2m-1}} = \int_s^r n_\nu(t)\,\frac{dt}{t} = N_\nu(r,s).$$

Also

$$\int_{A[r]} \nu\phi_{rs}(\sqrt{\tau})\omega^{m-1} + \phi_{rs}(0)\nu(0)$$

$$= \int_{A[r]-A[s]} \nu\phi_{rs}(\sqrt{\tau})\omega^{m-1} + \phi_{rs}(s)n_\nu(s)$$

$$= \phi_{rs}(r)n_\nu(r) - \phi_{rs}(s)n_\nu(s) - \int_s^r n_\nu(t)\phi_{rs}'(t)\,dt + \phi_{rs}(s)n_\nu(s)$$

$$= \int_s^r n_\nu(t)\,\frac{dt}{t} = N_\nu(r,s)$$

q.e.d.

Theorem 3.6. Jensen formula. ([65] Satz 6.5 and p. 156 No. 23) Let $f \not\equiv 0$ be a meromorphic function on the m-dimensional, hermitian vector space W. Take $0 < s < r$. Then

$$(3.48)\qquad \boxed{N_{\mu_f}(r,s) = \int_{W\langle r\rangle} \log|f|\sigma - \int_{W\langle s\rangle} \log|f|\sigma}\,.$$

<u>Proof</u>. We have

$$N_{\mu_f}(r,s) = \int_s^r n_{\mu_f}(t)\,\frac{dt}{t} = \int_s^r \int_{W\langle 1\rangle} n_{\mu_f}[t,\mathfrak{x}]\sigma(\mathfrak{x})\,\frac{dt}{t}$$

$$= \int_{W\langle 1\rangle} \int_s^r n_{\mu_f}[t,\mathfrak{x}]\,\frac{dt}{t}\,\sigma(\mathfrak{x})$$

$$= \int_{W\langle 1\rangle} \frac{1}{2\pi}\int_0^{2\pi} \log|f(re^{i\phi}\mathfrak{x})|\,d\phi\,\sigma(\mathfrak{x})$$

$$- \int_{W\langle 1\rangle} \frac{1}{2\pi}\int_0^{2\pi} \log|f(se^{i\phi}\mathfrak{x})|\,d\phi\,\sigma(\mathfrak{x})$$

$$= \int_{W\langle r\rangle} \log|f|\sigma - \int_{W\langle s\rangle} \log|f|\sigma$$

q.e.d.

4. The First Main Theorem.

Let V and W be hermitian vector spaces with dim V = n + 1 > 1 and dim W = m > 0. Let

$$(4.1) \qquad f : W \to \mathbb{P}(V)$$

be a meromorphic map. Take $a \in \mathbb{P}(V^*)$ and assume that $f(W) \nsubseteq \ddot{E}[a]$. Then the divisor $\mu_f^a \geq 0$ is defined. The counting function of f for a and the valence function of f for a are defined by

$$(4.2) \qquad n_f(r,a) = n_{\mu_f^a}(r) \qquad\qquad N_f(r,s;a) = N_{\mu_f^a}(r,s).$$

The compensation function of f for a is defined by

$$(4.3) \qquad m_f(r,a) = \int_{W\langle r\rangle} \log \frac{1}{\|f,a\|}\,\sigma \geq 0.$$

Let $\mathfrak{v} : W \to V$ be a reduced representation of f. Take $\alpha \in W_*^*$ with $\mathbb{P}(\alpha) = a$ and $\|\alpha\| = 1$. Then $\mu_f^a = \mu_{\alpha\circ\mathfrak{v}}^0$. Hence Jensen's formula implies

$$(4.4) \qquad N_f(r,s;a) = \int_{W\langle r\rangle} \log|\alpha\circ\mathfrak{v}|\sigma - \int_{W\langle s\rangle} \log|\alpha\circ\mathfrak{v}|\sigma$$

while

$$(4.5) \qquad m_f(r,a) - m_f(s,a) = \int_{W\langle r\rangle} \log\frac{\|\mathfrak{v}\|}{|\alpha\circ\mathfrak{v}|}\,\sigma - \int_{W\langle s\rangle} \log\frac{\|\mathfrak{v}\|}{|\alpha\circ\mathfrak{v}|}\,\sigma.$$

Therefore

$$(4.6) \qquad N_f(r,a) + m_f(r,a) - m_f(s,a) = \int_{W\langle r\rangle} \log\|\mathfrak{v}\|\sigma - \int_{W\langle s\rangle} \log\|\mathfrak{v}\|\sigma.$$

The <u>characteristic function of</u> f is defined for $0 < s < r$ by

(4.7) $$\boxed{T_f(r,s) = \int_{W\langle r\rangle} \log\|\mathfrak{v}\|\sigma - \int_{W\langle s\rangle} \log\|\mathfrak{v}\|\sigma} .$$

By (4.6) the characteristic does not depend on the choice of the reduced representation $\mathfrak{v}$ of f. We obtain

<u>Theorem 4.1. The First Main Theorem.</u> ([65], [64], Kneser [34]) Let V and W be hermitian vector spaces with $\dim V = n + 1 > 1$ and $\dim W = m > 0$. Let $f : W \to \mathbb{P}(V)$ be a meromorphic map. Take $a \in \mathbb{P}(V)$ such that $f(W) \not\subseteq \ddot{E}[a]$. Take $0 < s < r$. Then

(4.8) $$\boxed{T_f(r,s) = N_f(r,s;a) + m_f(r,a) - m_f(s,a)} .$$

The First Main Theorem becomes valuable only if other expressions and properties of the characteristic are found. First we have to define an <u>indeterminacy multiplicity of f at</u> 0. Let $\mathfrak{v} : W \to V$ be a reduced representation of f. For each $j \geq 0$ there exists a holomorphic vector function $\mathfrak{v}_j : W \to V$ which is homogeneous of degree j (i.e. $\mathfrak{v}_j(\lambda\mathfrak{x}) = \lambda^j\mathfrak{v}(\mathfrak{x})$ for all $\lambda \in \mathbb{C}$ and $\mathfrak{x} \in W$) such that

(4.9) $$\mathfrak{v} = \sum_{j=\nu}^{\infty} \mathfrak{v}_j$$

converges uniformly to $\mathfrak{v}$ on every compact subset of W and such that $\mathfrak{v}_\nu \not\equiv 0$. We want to show that ν does not depend on $\mathfrak{v}$. Let $\tilde{\mathfrak{v}} : W \to V$ be another reduced representation of f. Then there exists a holomorphic function $h : W \to \mathbb{C}$ such that $\tilde{\mathfrak{v}} = h\mathfrak{v}$. There are homogeneous polynomials $h_k : W \to \mathbb{C}$ of degree k such that

$$h = \sum_{k=0}^{\infty} h_k$$

converges uniformly on every compact subset of W where $h(0) = h_0 \not\equiv 0$. Then $\tilde{\mathfrak{v}}_p = \sum_{k+j=p} h_k\mathfrak{v}_j$ is holomorphic and homogeneous of degree p with

$\tilde{\mathfrak{u}}_\nu = h_0 \mathfrak{u}_\nu \not\equiv 0$. Then

$$\tilde{\mathfrak{u}} = h\mathfrak{u} = \sum_{p=\nu}^{\infty} \tilde{\mathfrak{u}}_p$$

converges uniformly on compact subsets of W. The number ν does not depend on $\mathfrak{u}$ but on f only. This number $\nu_f = \nu$ is called the <u>indeterminacy multiplicity of</u> f at 0.

The <u>spherical image</u> of f is defined for $t \geq 0$ by

$$(4.10) \qquad \boxed{A_f(t) = \int_{W[t]} f^*(\Omega_V) \wedge \omega_W^{m-1} + \nu_f \geq 0} \; .$$

Clearly A_f increases.

<u>Theorem 4.2.</u> ([65], [64], Kneser [34])
Let V and W be hermitian vector spaces with $\dim V = n + 1 > 1$ and $\dim W = m > 0$. Let $f : W \to \mathbb{P}(V)$ be a meromorphic map. Take $0 < s < r$. Then

$$(4.11) \qquad \boxed{T_f(r,s) = \int_s^r A_f(t) \frac{dt}{t}} \; .$$

<u>Proof</u>. <u>First we shall consider the case</u> m = 1. Let $\mathfrak{u} : W \to V_*$ be a reduced representation of f. Then $u = \log\|\mathfrak{u}\|^2$ is a function of class C^∞ on W. Since $f = \mathbb{P} \circ \mathfrak{u}$ we have

$$f^*(\Omega_V) = (\mathbb{P} \circ \mathfrak{u})^*(\Omega_V) = \mathfrak{u}^*(\mathbb{P}^*(\Omega_V)) = \mathfrak{u}^*(dd^c \log \tau_V)$$
$$= dd^c \log\|\mathfrak{u}\|^2 = dd^c u.$$

Abbreviate $\tau = \tau_W$. Then $\partial u \wedge \partial \log \tau$ has bidegree (2,0) and $\overline{\partial} u \wedge \overline{\partial} \log \tau$ has bidegree (0,2). Both are zero, since $\dim W = 1$. Therefore

$$du \wedge d^c \log \tau = \frac{i}{4\pi} (\partial u \wedge \bar\partial \log \tau + \partial \log \tau \wedge \bar\partial u) = d \log \tau \wedge d^c u.$$

If $0 < s < r$, Stokes Theorem and (3.43) imply

$$\begin{aligned}
T_f(r,s) &= \frac{1}{2} \int_{W\langle r\rangle} u\, d^c \log \tau - \frac{1}{2} \int_{W\langle s\rangle} u\, d^c \log \tau \\
&= \frac{1}{2} \int_{W(r)-W[s]} du \wedge d^c \log \tau \qquad (dd^c \log \tau = 0) \\
&= \frac{1}{2} \int_{W(r)-W[s]} d \log \tau \wedge d^c u \\
&= \int_{W(r)-W[s]} d\left(\log \frac{\sqrt{\tau}}{r}\, d^c u\right) + \int_{W(r)-W[s]} \log \frac{r}{\sqrt{\tau}}\, dd^c u \\
&= \log \frac{r}{s} \int_{W\langle s\rangle} d^c u + \int_{W(r)-W[s]} \int_{\sqrt{\tau}}^{r} \frac{dt}{t}\, dd^c u \\
&= \log \frac{r}{s} \int_{W[s]} dd^c u + \int_{W(r)-W[s]} \int_{s}^{r} \chi(t,\sqrt{\tau}) \frac{dt}{t}\, dd^c u \\
&= \log \frac{r}{s} \int_{W[s]} dd^c u + \int_{s}^{r} \int_{W(r)-W[s]} \chi(t,\sqrt{\tau})\, dd^c u \frac{dt}{t} \\
&= \log \frac{r}{s} \int_{W[s]} dd^c u + \int_{s}^{r} \int_{W[t]-W[s]} dd^c u \frac{dt}{t} \\
&= \int_{s}^{r} \int_{W[t]} dd^c u \frac{dt}{t} = \int_{s}^{r} A_f(t) \frac{dt}{t}.
\end{aligned}$$

Now, consider the case $m > 1$. Let $\mathfrak{v} : W \to V$ be a reduced representation. Then $I_f = \mathfrak{v}^{-1}(0)$ has at most dimension $m - 2$. Therefore $I_f' = \mathbb{P}(I_f)$ is a set of measure zero in $\mathbb{P}(W)$. Take the development (4.9). Then $\mathfrak{v}_\nu^{-1}(0) \neq W$ is an analytic cone in W. Hence $N = \mathbb{P}(\mathfrak{v}_\nu^{-1}(0))$ is a thin analytic subset of $\mathbb{P}(W)$. Therefore $S = N \cup I_f'$ has measure zero in $\mathbb{P}(W)$. Take $x \in \mathbb{P}(W) - S$. Then $\mathfrak{v}(\mathfrak{x}) \neq 0$ if $0 \neq \mathfrak{x} \in E(x)$ since $\mathfrak{x} \notin I_f$. At 0 the restriction $\mathfrak{v}|E(x)$ vanishes of order ν. Take $\mathfrak{x} \in E(x)$ with $\|\mathfrak{x}\| = 1$. A reduced representation $\mathfrak{v}_x : E(x) \to W$ of $f_x = f|E(x)$ is defined by

$$\mathfrak{w}_x(z\mathfrak{x}) = z^{-\nu}\mathfrak{w}(z\mathfrak{x}) \qquad \forall\ z \in \mathbb{C}.$$

Then

$$T_{f_x}(r,s) = \int_{E(x)\langle r\rangle} \log\|\mathfrak{w}_x\|\, d^c \log \tau - \int_{E(x)\langle s\rangle} \log\|\mathfrak{w}_x\|\, d^c \log \tau$$

$$= \int_{E(x)\langle r\rangle} \log\|\mathfrak{w}\|\, d^c \log \tau - \int_{E(x)\langle s\rangle} \log\|\mathfrak{w}\|\, d^c \log \tau - \nu \log \frac{r}{s}.$$

Since $f_x^*(\Omega_V) = dd^c \log\|\mathfrak{w}_x\|^2 = dd^c \log\|\mathfrak{w}|E(x)\|^2$, the first part of the proof implies

$$T_{f_x}(r,s) = \int_s^r \int_{E(x)[t]} f_x^*(\Omega_V)\, \frac{dt}{t} = \int_s^r \int_{E(x)[t]} f^*(\Omega_V)\, \frac{dt}{t}.$$

Now, Lemmata 2.2 and 2.3 imply

$$T_f(r,s) - \nu \log \frac{r}{s} = \int_{W\langle r\rangle} \log\|\mathfrak{w}\|\sigma - \int_{W\langle s\rangle} \log\|\mathfrak{w}\|\sigma - \nu \log \frac{r}{s}$$

$$= \int_{W\langle 1\rangle} \Big(\int_{E(x)\langle r\rangle} \log\|\mathfrak{w}\|\, d^c \log \tau - \int_{E(x)\langle s\rangle} \log\|\mathfrak{w}\|\, d^c \log \tau - \nu \log \frac{r}{s}\Big)\sigma(\mathfrak{x})$$

$$= \int_{W\langle 1\rangle} T_{f_{\mathbb{P}(\mathfrak{x})}}(r,s)\sigma(\mathfrak{x}) = \int_{\mathbb{P}(W)} T_{f_x}(r,s)\Omega_W^{m-1}$$

$$= \int_{\mathbb{P}(W)} \int_s^r \int_{E(x)[t]} f^*(\Omega_V)\, \frac{dt}{t}\, \Omega_W^{m-1}$$

$$= \int_s^r \int_{W[t]} f^*(\Omega_V) \wedge \omega_W^{m-1}\, \frac{dt}{t}$$

or

$$T_f(r,s) = \int_s^r \Big(\int_{W[t]} f^*(\Omega_V) \wedge \omega_W^{m-1} + \nu \Big) \frac{dt}{t}$$

q.e.d.

Proposition 4.3. ([65], [64], Kneser [34])
Let V and W be hermitian vector spaces with $\dim V = n + 1 > 1$ and with $\dim W = m > 1$. Let $f : W \to \mathbb{P}(V)$ be a meromorphic map. Then

$$(4.12) \qquad \boxed{A_f(t) = \frac{1}{t^{2m-2}} \int_{W[t]} f^*(\Omega_V) \wedge \upsilon_W^{m-1}} \qquad \forall\ t > 0.$$

Proof. First, assume that $0 \notin I_f$. Take $0 < s < r$. Then

$$(4.13) \qquad A_f(r) - A_f(s) = \int_{W[r]-W[s]} d(f^*(\Omega_V) \wedge (d^c \log \tau_W \wedge \omega^{m-2}))$$

$$= \int_{W\langle r\rangle} f^*(\Omega_V) \wedge d^c \log \tau_W \wedge \omega_W^{m-2} - \int_{W\langle s\rangle} f^*(\Omega_V) \wedge d^c \log \tau_W \wedge \omega_W^{m-2}$$

$$= \frac{1}{r^{2m-2}} \int_{W\langle r\rangle} f^*(\Omega_V) \wedge d^c \tau_W \wedge \upsilon_W^{m-2} - \frac{1}{s^{2m-2}} \int_{W\langle s\rangle} f^*(\Omega_V) \wedge d^c \tau_W \wedge \upsilon_W^{m-2}$$

$$= \frac{1}{r^{2m-2}} \int_{W(r)} f^*(\Omega_V) \wedge \upsilon_W^{m-1} - \frac{1}{s^{2m-2}} \int_{W(s)} f^*(\Omega_V) \wedge \upsilon_W^{m-1}.$$

Clearly (4.10) implies $A_f(s) \to \nu_f = 0$ if $s \to 0$. A number $s_0 > 0$ exists such that $I_f \cap W[s_0] = \emptyset$. Then $f^*(\Omega_V)$ is of class C^∞ on $W[s_0]$. A constant $c > 0$ exists such that

$$0 \le f^*(\Omega_V) \le c\upsilon_W \qquad \text{on } W[s_0].$$

Hence

$$\frac{1}{s^{2m-2}} \int_{W(s)} f^*(\Omega_V) \wedge \upsilon_W^{m-1} \le \frac{c}{s^{2m-2}} \int_{W(s)} \upsilon_W^m = cs^2.$$

Hence

$$\frac{1}{s^{2m-2}} \int_{W(s)} f^*(\Omega_V) \wedge \upsilon_W^{m-1} \to 0 \qquad \text{for } s \to 0.$$

Hence (4.13) implies with $s \to 0$

$$A_f(r) = \frac{1}{r^{2m-2}} \int_{W(r)} f^*(\Omega_V) \wedge \upsilon_W^{m-1} \qquad \forall\, r > 0$$

if $0 \notin I_f$.

Now assume that $0 \in I_f$. Take $\mathfrak{a} \in W\langle 1\rangle$ such that $E(\mathfrak{a}) \nsubseteq I_f$. A number $\rho > 0$ exists such that $\lambda\mathfrak{a} \notin I_f$ if $0 < |\lambda| < \rho$.

Define $f_\lambda : W \to \mathbb{P}(V)$ by $f_\lambda(\mathfrak{z}) = f(\lambda\mathfrak{a} + \mathfrak{z})$ for all $\mathfrak{z} \in W$. Let $\mathfrak{v} : W \to V$ be a reduced representation of f. Then a reduced representation of f_λ is defined by $\mathfrak{v}_\lambda(\mathfrak{z}) = \mathfrak{v}_\lambda(\lambda\mathfrak{a} + \mathfrak{z})$ for all $\mathfrak{z} \in W$. Then $\mathfrak{v}_\lambda(0) \neq 0$. Hence

$$A_{f_\lambda}(t) = \frac{1}{t^{2m-2}} \int_{W(t)} f_\lambda^*(\Omega_V) \wedge \upsilon_W^{m-1}$$
$$= \frac{1}{t^{2m-2}} \int_{\lambda\mathfrak{a}+W(t)} f^*(\Omega_V) \wedge \upsilon_W^{m-1}.$$

If $0 < s < r$, then

$$T_{f_\lambda}(r,s) = \int_s^r \int_{\lambda\mathfrak{a}+W(t)} f^*(\Omega_V) \wedge \upsilon_W^{m-1} \frac{dt}{t^{2m-1}}$$
$$\to \int_s^r \int_{W(t)} f^*(\Omega_V) \wedge \upsilon_W^{m-1} \frac{dt}{t^{2m-1}} \qquad \text{for } \lambda \to 0.$$

Also

$$T_{f_\lambda}(r,s) = \int_{W\langle r\rangle} \log\|\mathfrak{v}_\lambda\|\sigma - \int_{W\langle s\rangle} \log\|\mathfrak{v}_\lambda\|\sigma$$
$$\to \int_{W\langle r\rangle} \log\|\mathfrak{v}\|\sigma - \int_{W\langle s\rangle} \log\|\mathfrak{v}\|\sigma \qquad \text{for } \lambda \to 0$$
$$= T_f(r,s) = \int_s^r A_f(t)\,\frac{dt}{t}\,.$$

Hence

$$\int_s^r A_f(t)\,\frac{dt}{t} = \int_s^r \int_{W[t]} f^*(\Omega_V) \wedge \upsilon_W^{m-1}\,\frac{dt}{t^{2m-1}} \qquad \forall\ r > s > 0.$$

Differentiation for r implies

$$A_f(t) = \frac{1}{t^{2m-2}} \int_{W[t]} f^*(\Omega_V) \wedge \upsilon_W^{m-1}$$

q.e.d.

Originally, the characteristic was defined by means of a reduced representation (4.7). However any representation can be used with proper adjustment. Recall the definition of a greatest common divisor of a holomorphic vector function in section 3.

Theorem 4.4. ([65], [64])
Let V and W be hermitian vector spaces with $\dim V = n + 1 > 1$ and with $\dim W = m > 0$. Let $f : W \to \mathbb{P}(V)$ be a meromorphic map. Let $\mathfrak{w} : W \to V$ be a representation of f. Take $0 < s < r$. Then

$$(4.14)\qquad T_f(r,s) = \int_{W\langle r\rangle} \log\|\mathfrak{w}\|\sigma - \int_{W\langle s\rangle} \log\|\mathfrak{w}\|\sigma - N_{\nu_{\mathfrak{w}}}(r,s)$$

$$(4.15)\qquad T_f(r,s) \le \int_{W\langle r\rangle} \log\|\mathfrak{w}\|\sigma - \int_{W\langle s\rangle} \log\|\mathfrak{w}\|\sigma.$$

Proof. Let $\mathfrak{v} : W \to V$ be a reduced representation of f. A holomorphic function $h : V \to \mathbb{C}$ exists such that $\mathfrak{w} = h\mathfrak{v}$. Then $\nu_{\mathfrak{w}} = \mu_h^0$. Now (4.7) and the Jensen formula imply

$$T_f(r,s) = \int_{W\langle r\rangle} \log\|\mathfrak{v}\|\sigma - \int_{W\langle s\rangle} \log\|\mathfrak{v}\|\sigma$$

$$N_{\nu_{\mathfrak{w}}}(r,s) = \int_{W\langle r\rangle} \log|h|\sigma - \int_{W\langle s\rangle} \log|h|\sigma.$$

Addition implies (4.14). Since $N_{\nu_{\mathfrak{w}}}(r,s) \ge 0$, we obtain (4.15); q.e.d.

Lemma 4.5. (Weyl [87])
Take $w \in \mathbb{P}(V)$. Then

(4.16)
$$\boxed{\int_{\mathbb{P}(V^*)} \log \frac{1}{\|w,a\|}\, \Omega^n(a) = \frac{1}{2} \sum_{p=1}^{n} \frac{1}{p}}\,.$$

Proof. Take $\mathfrak{w} \in V\langle 1\rangle$ with $\mathbb{P}(\mathfrak{w}) = w$. Lemmata 2.1 and 2.2 imply

$$I = \int_{\mathbb{P}(V^*)} \log \frac{1}{\|w,a\|}\, \Omega^n(a) = \int_{V\langle 1\rangle} \log \frac{\|\alpha\|}{|\alpha(\mathfrak{w})|}\, \sigma(\alpha)$$
$$= \frac{1}{(n+1)!} \int_V e^{-\|\alpha\|^2} \log \frac{\|\alpha\|}{|\alpha(\mathfrak{w})|}\, \upsilon^{n+1}(\alpha).$$

Let $\mathfrak{e}_0, \ldots, \mathfrak{e}_n$ be an orthonormal base of V such that $\mathfrak{e}_0 = \mathfrak{w}$. Let $\varepsilon_0, \ldots, \varepsilon_n$ be the dual base. Then $\alpha = z_0\varepsilon_0 + \ldots + z_n\varepsilon_n$ and $\alpha(\mathfrak{w}) = z_0$. Introduce polar coordinates $z_j = r_j e^{i\phi_j}$ where $r_j > 0$ and $0 \le \phi_j < 2\pi$. Then

$$2I = \int_{\mathbb{C}^{n+1}} e^{-|z_0|^2-\ldots-|z_n|^2} \log \frac{|z_0|^2 + \ldots + |z_n|^2}{|z_0|^2} \left(\frac{i}{2\pi}\right)^{n+1} \bigwedge_{\lambda=0}^{n} dz_\lambda \wedge d\bar{z}_\lambda$$
$$= \int_0^\infty \cdots \int_0^\infty e^{-r_0^2-\ldots-r_n^2} \log \frac{r_0^2 + \ldots + r_n^2}{r_0^2}\, 2^{n+1} r_0 \cdots r_n \, dr_0 \cdots dr_n$$
$$= \int_0^\infty \cdots \int_0^\infty e^{-t_0-\ldots-t_n} \log \frac{t_0 + \ldots + t_n}{t_0}\, dt_0 \cdots dt_n.$$

Substitute

$$t_0 = s(x_1 + \ldots + x_n) \qquad \text{with } 0 < s < 1$$
$$t_\nu = (1 - s)x_\nu \qquad \text{for } \nu = 1, \ldots, n \quad 0 < x_\nu < \infty.$$

Then

$$t_0 + \ldots + t_n = x_1 + \ldots + x_n$$

$$s = \frac{t_0}{t_0 + \ldots + t_n} \qquad x_\nu = t_\nu \frac{t_n + \ldots + t_n}{t_1 + \ldots + t_n} \quad \text{for } \nu = 1, \ldots, n$$

$$\frac{\partial(t_0,\ldots,t_n)}{\partial(s,x_1,\ldots,x_n)} = (x_1 + \ldots + x_n)(1 - s)^{n-1}.$$

Hence

$$2I = \int_0^\infty \ldots \int_0^\infty e^{-x_1-\ldots-x_n}(x_1 + \ldots + x_n)\, dx_1 \ldots dx_n \int_0^1 (1 - s)^{n-1} \log \frac{1}{s}\, ds$$

$$= n \int_0^1 (1 - s)^{n-1} \log \frac{1}{s}\, ds = n \int_0^1 s^{n-1} \log \frac{1}{1 - s}\, ds$$

$$= \lim_{x\to 1} n \int_0^x \sum_{p=1}^{\infty} \frac{s^{p+n-1}}{p}\, ds = \lim_{x\to 1} n \sum_{p=1}^{\infty} \frac{x^{p+n}}{p(p + n)}$$

$$= \sum_{p=1}^{\infty} \frac{n}{p(p + n)} = \lim_{q\to\infty} \sum_{p=1}^{q} \frac{n}{p(p + n)}$$

$$= \lim_{q\to\infty}\left(\sum_{p=1}^{q} \frac{1}{p} - \sum_{p=1}^{q} \frac{1}{p + n}\right) = \lim_{q\to\infty}\left(\sum_{p=1}^{q} \frac{1}{p} - \sum_{p=n+1}^{q+n} \frac{1}{p}\right)$$

$$= \sum_{p=1}^{n} \frac{1}{p} + \lim_{q\to\infty} \sum_{p=1}^{n} \frac{1}{p + q} = \sum_{p=1}^{n} \frac{1}{p}$$

q.e.d.

Theorem 4.6. ([65] (9.13))
Let V and W be hermitian vector spaces with dim V = n + 1 and dim W = m. Let $f : W \to \mathbb{P}(V)$ be a meromorphic map. Then

$$(4.17) \qquad \boxed{\int_{\mathbb{P}(V^*)} m_f(r,a)\Omega_V^n(a) = \frac{1}{2} \sum_{p=1}^{n} \frac{1}{p}} \qquad \forall\, r > 0.$$

Proof. (4.16) implies

$$\begin{aligned}\int_{\mathbb{P}(V^*)} m_f(r,a)\Omega_V^n(a) &= \int_{\mathbb{P}(V^*)} \left(\int_{W\langle 1\rangle} \log \frac{1}{\|f,a\|}\, \sigma\right)\Omega_V^n(a) \\ &= \int_{W\langle 1\rangle} \left(\int_{\mathbb{P}(V^*)} \log \frac{1}{\|f,a\|}\, \Omega_V^n(a)\right)\sigma \\ &= \int_{W\langle 1\rangle} \frac{1}{2} \sum_{p=1}^{n} \frac{1}{p}\, \sigma \\ &= \frac{1}{2} \sum_{p=1}^{n} \frac{1}{p}\, \sigma\end{aligned}$$

q.e.d.

Theorem 4.7. ([65] Satz 9.2, Weyl [87])
Let V and W be hermitian vector spaces with $\dim V = n + 1 > 1$ and with $\dim W = m > 0$. Let $f : W \to \mathbb{P}(V)$ be a meromorphic map. Take $0 < s < r$. Then

$$\boxed{T_f(r,s) = \int_{\mathbb{P}(V^*)} N_f(r,s;a)\Omega_V^n(a)} . \tag{4.18}$$

Proof. Integrate the First Main Theorem

$$\begin{aligned}T_f(r,s) &= \int_{\mathbb{P}(V^*)} (N_f(r,s;a) + m_f(r,a) - m_f(s,a))\Omega_V^n(a) \\ &= \int_{\mathbb{P}(V^*)} N_f(r,s;a)\Omega_V^n(a) + \frac{1}{2}\sum_{p=1}^{n}\frac{1}{p} - \frac{1}{2}\sum_{p=1}^{n}\frac{1}{p} \\ &= \int_{\mathbb{P}(V^*)} N_f(r,s;a)\Omega_V^n(a)\end{aligned}$$

q.e.d.

Now we shall investigate the behavior of $T_f(r,s)$ for $r \to \infty$.

Lemma 4.8.
Let $g : \mathbb{R}^+ \to \mathbb{R}_+$ be an increasing function. Define

$$(4.19)\qquad 0 \le g(\infty) = \lim_{r\to\infty} g(r) \le \infty.$$

For $0 < s < r$ define

$$(4.20)\qquad G(r,s) = \int_s^r g(t)\,\frac{dt}{t}.$$

Then

$$(4.21)\qquad \frac{G(r,s)}{\log r} \to g(\infty) \qquad \text{for } r \to \infty.$$

Proof. Define $\rho : \mathbb{R}_+ \to \mathbb{R}(0,1]$ by $\rho(r) = 1$ if $0 \le r \le e$ and by $\rho(r) = 1/\sqrt{\log r}$ for $e \le r < \infty$. Then $\rho(r) \to 0$ for $r \to \infty$. If $r > e$, then

$$r^{\rho(r)} = e^{\rho(r)\log r} = e^{\sqrt{\log r}} \to \infty \qquad \text{for } r \to \infty.$$

Take $r_0 > s + e$ such that $r^{\rho(r)} > s$ for all $r \ge r_0$. If $r > r_0$, then

$$g(r^{\rho(r)})(1-\rho(r)) \le \frac{1}{\log r}\int_{r^{\rho(r)}}^r g(t)\,\frac{dt}{t}$$
$$\le \frac{G(r,s)}{\log r} \le g(r)\left(1 + \frac{|\log s|}{\log r}\right).$$

Now $r \to \infty$ implies (4.21); q.e.d.

Of course if $g(\infty) > 0$ then $G(r,s) \to \infty$ for $r \to \infty$.

Proposition 4.9.
Let V and W be hermitian vector spaces with $\dim V = n + 1$ and with $\dim W = m > 0$. Let $f : W \to \mathbb{P}(V)$ be a meromorphic map. Then

$$(4.22)\qquad \infty \ge A_f(\infty) = \lim_{r\to\infty} A_f(r) = \lim_{r\to\infty} \frac{T_f(r,s)}{\log r} \ge 0.$$

Moreover, $A_f(\infty) > 0$ if and only if f is not constant. In fact, if f is not constant, then $A_f(t) > 0$ for all $t > 0$ and $T_f(r,s) > 0$ for all

$r > s > 0$ and $T_f(r,s) \to \infty$ for $r \to \infty$. If f is constant, then $A_f(t) = 0$ for all $t > 0$ and $T_f(r,s) = 0$ for all $r > s > 0$.

Proof. Lemma 4.8 implies (4.22). If f is constant, $f^*(\Omega_V) = 0$. Hence $A_f(t) \equiv 0 = A_f(\infty)$ and $T_f(r,s) \equiv 0$. Assume that f is not constant. Then $a \in \mathbb{P}(V^*)$ exists such that $f(W) \not\subseteq \ddot{E}[a]$. The First Main Theorem implies

$$T_f(r,s) \geq N_f(r,s;a) - m_f(s,a)$$

$$(4.23) \qquad A_f(\infty) = \lim_{r\to\infty} \frac{T_f(r,s)}{\log r} \geq \lim_{r\to\infty} \frac{N_f(r,s;a)}{\log r} = n_f(\infty,a)$$

for all $a \in \mathbb{P}(V^*)$ with $f(W) \not\subseteq \ddot{E}[a]$. Take $a \in \mathbb{P}(V^*)$ such that $f(W) \cap \ddot{E}[a] \neq \emptyset$ but $f(W) \not\subseteq \ddot{E}[a]$. Then $n_f(t,a) > 0$ for some $t > 0$. Therefore $n_f(\infty,a) > 0$ and $A_f(\infty) > 0$ and $T_f(r,s) \to \infty$ for $r \to \infty$. Assume that $t_0 > 0$ exists such that $A_f(t_0) = 0$. Then $f^*(\Omega_V) \wedge \upsilon_W^{m-1}$ is identically zero on $W(t_0)$. Also this form is real analytic. Hence $f^*(\Omega_V) \wedge \upsilon_W^{m-1} \equiv 0$ on W which implies $A_f(t) = 0$ for all $t > 0$. Then $A_f(\infty) = 0$, which is wrong. Therefore $A_f(t) > 0$ for all $t > 0$. By (4.11), $T_f(r,s) > 0$ if $0 < s < r$; q.e.d.

Shortly, we will determine all those meromorphic maps f for which $A_f(\infty) < \infty$. However first we shall establish a Casorati-Weierstrass Theorem.

Let $f : W \to \mathbb{P}(V)$ be a non-constant, meromorphic map. Take $a \in \mathbb{P}(V^*)$ such that $f(W) \not\subseteq \ddot{E}[a]$. The Nevanlinna defect of f for a is defined by

$$(4.24) \qquad \delta_f(a) = \liminf_{r\to\infty} \frac{m_f(r,a)}{T_f(r,s)}$$

and does not depend on s. The First Main Theorem implies

$$\delta_f(a) = 1 - \limsup_{r\to\infty} \frac{N_f(r,s;a)}{T_f(r,s)} . \tag{4.25}$$

Then

$$0 \le \delta_f(a) \le 1. \tag{4.26}$$

If $f(W) \cap \ddot{E}[a] = \emptyset$, then $N_f(r,s;a) = 0$ and $\delta_f(a) = 1$.

Theorem 4.10. Casorati-Weierstrass.
Let $f : W \to \mathbb{P}(V)$ be a non-constant meromorphic map. Then $\delta_f(a) = 0$ for almost all $a \in \mathbb{P}(V^*)$. In particular, $f(W) \cap \ddot{E}[a] \neq \emptyset$ for almost all $a \in \mathbb{P}(V^*)$.

Proof. Fatou's Lemma implies

$$1 \ge \int_{\mathbb{P}(V^*)} (1 - \delta_f(a))\Omega_V^n(a) = \int_{\mathbb{P}(V^*)} \limsup_{r\to\infty} \frac{N_f(r,s;a)}{T_f(r,s)} \Omega_V^n(a)$$

$$\ge \limsup_{r\to\infty} \int_{\mathbb{P}(V^*)} \frac{N_f(r,s;a)}{T_f(r,s)} \Omega_V^n(a) = 1.$$

Hence

$$\int_{\mathbb{P}(V^*)} \delta_f(a)\Omega_V^n(a) = 0.$$

Therefore $\delta_f(a) = 0$ for almost all $a \in \mathbb{P}(V^*)$; q.e.d.

Now we want to show that f is rational if and only if $A_f(\infty) < \infty$. First we need some definitions.

Let W and V be hermitian vector spaces with $\dim V = n + 1 > 0$ and $\dim W = m > 0$. A holomorphic vector function $\mathfrak{v} : W \to V$ is said to be a homogeneous vector polynomial of degree p if and only if

$$\mathfrak{v}(\lambda\mathfrak{z}) = \lambda^p \mathfrak{v}(\mathfrak{z}) \qquad \forall\ \lambda \in \mathbb{C} \quad \forall\ \mathfrak{z} \in W. \tag{4.27}$$

A holomorphic vector function $\mathfrak{v} : W \to V$ is said to be a vector polynomial if and only if there are homogeneous vector polynomials $\mathfrak{v}_j : W \to V$ of degree j for $j = 0, 1, \ldots, p$ such that

$$(4.28) \qquad \mathfrak{v} = \mathfrak{v}_0 + \ldots + \mathfrak{v}_p.$$

If $\mathfrak{v}_p \not\equiv 0$, then $\mathfrak{v}$ is said to have <u>degree</u> p. If we take bases in W and V then $\mathfrak{v}$ is a vector polynomial if and only if the coordinate functions of $\mathfrak{v}$ are polynomials in the coordinates of W over $\mathbb{C}$. If $V = \mathbb{C}$, we speak of homogeneous polynomials and polynomials.

A meromorphic map $f : W \to \mathbb{P}(V)$ is said to be <u>rational</u> if and only if there exists a reduced representation $\mathfrak{v} : W \to V$ which is a vector polynomial. A non-rational map is called <u>transcendental</u>.

A divisor ν on W is said to be <u>affine algebraic</u> if and only if $\nu = \mu_f$ is the divisor of a rational meromorphic function $f : W \to \mathbb{P}_1$. If $\nu \geq 0$ then ν is affine algebraic, if and only if $\nu = \mu_f^0$ is the zero divisor of a polynomial $f \not\equiv 0$ on W.

If $\mathfrak{v} : W \to V$ is a holomorphic vector function and if $r \geq 0$, define

$$(4.29) \qquad M_{\mathfrak{v}}(r) = \mathrm{Max}\{\|\mathfrak{v}(\mathfrak{x})\| \mid \mathfrak{x} \in W\langle r\rangle\}.$$

The function $M_{\mathfrak{v}}$ is called the <u>maximum modulus</u> of $\mathfrak{v}$.

<u>Lemma 4.11</u>.
Let $\mathfrak{v} : W \to V$ be a holomorphic vector function. Then $\mathfrak{v}$ is a vector polynomial if and only if there are constants $c \geq 0$, $r_0 > 0$ and $p \geq 0$ such that $M_{\mathfrak{v}}(r) \leq cr^p$ for all $r \geq r_0$. If $\mathfrak{v} \not\equiv 0$, then $\deg \mathfrak{v} \leq p$.

<u>Proof</u>. (a) <u>Let $\mathfrak{v} \not\equiv 0$ be a vector polynomial of degree</u> p. Then there are homogeneous vector polynomials $\mathfrak{v}_j : W \to V$ of degree j for $j = 0, 1, \ldots, p$ such that (4.28) holds. Take $c = M_{\mathfrak{v}_0}(1) + \ldots + M_{\mathfrak{v}_p}(1) \geq 0$. Then $\mathfrak{x} \in W\langle 1\rangle$ exists such that $M_{\mathfrak{v}}(r) = \|\mathfrak{v}(r\mathfrak{x})\|$. Take $r_0 = 1$ and $r \geq r_0$, then

$$M_{\mathfrak{v}}(r) = \|\mathfrak{v}(r\mathfrak{x})\| = \|\sum_{j=0}^{p} \mathfrak{v}_j(\mathfrak{x})r^j\| \le \sum_{j=0}^{p} \|\mathfrak{v}_j(\mathfrak{x})\| r^j \le cr^p.$$

(b) <u>Assume that</u> $c \ge 0$, $r_0 > 0$ <u>and</u> $p \ge 0$ <u>exist such that</u> $M_{\mathfrak{v}}(r) \le cr^p$ <u>for all</u> $r \ge r_0$. For each $j \ge 0$, there exists a homogeneous vector polynomial $\mathfrak{v}_j : W \to V$ of degree j such that $\mathfrak{v} = \sum_{j=0}^{\infty} \mathfrak{v}_j$ where the series converges uniformly on every compact subset of W. Take $r \ge r_0$ and $\mathfrak{x} \in W$. Then

$$\mathfrak{v}(w\mathfrak{x}) = \sum_{j=0}^{\infty} \mathfrak{v}_j(\mathfrak{x})w^j \qquad \forall\ w \in \mathbb{C}.$$

Hence the Cauchy Integral Theorem implies

$$\mathfrak{v}_j(\mathfrak{x}) = \frac{1}{2\pi i} \int_{\mathbb{C}\langle r\rangle} \frac{\mathfrak{v}(w\mathfrak{x})}{w^{j+1}}\, dw$$

or

$$\|\mathfrak{v}_j(\mathfrak{x})\| \le \frac{1}{2\pi} \int_{\mathbb{C}(r)} \frac{\|\mathfrak{v}(w\mathfrak{x})\|}{|w|^{j+1}}\, |dw| \le \frac{M_{\mathfrak{v}}(r)}{r^j} \le \frac{c}{r^{j-p}}\,.$$

If $j > p$, then $r^{j-p} \to \infty$ for $r \to \infty$. Therefore $\|\mathfrak{v}_j(\mathfrak{x})\| = 0$. We conclude $\mathfrak{v}_j \equiv 0$ for all $j > p$. Hence $\mathfrak{v}$ is a vector polynomial; q.e.d.

The characterization of affine algebraic divisors can be proved by various methods. However all require some facts of several complex variables which for time limitations cannot be derived here. We take a method which operates in value distribution only, namely the existence of a canonical function. (See Stoll [64], Lelong [36], Ronkin [51] and the outline in Stoll [75] Theorem 6.3.)

<u>Theorem 4.12</u>.
Let $\nu \ge 0$ be a divisor on the hermitian vector space W of dimension m. Assume that $\nu(0) = 0$. Assume that there exists an integer $q \ge 0$ such that

$$(4.30)\qquad \int_1^\infty n_\nu(t)\,\frac{dt}{t^{q+2}} < \infty.$$

Then there exists a holomorphic function $h : W \to \mathbb{C}$ with $h(0) = 1$ such that $\nu = \mu_h^0$ is the zero divisor of h and such that for $\theta \in \mathbb{R}(0,1)$ and $0 < r \in \mathbb{R}$ we have the estimate

$$(4.31)\qquad \boxed{\log M_h(\theta r) \le \frac{c_0 r^q}{(1-\theta)^{3m}}\left(\int_0^r n_\nu(t)\,\frac{dt}{t^{q+1}} + r\int_r^\infty n_\nu(t)\,\frac{dt}{t^{q+2}}\right)}$$

where $c_0 = 72m(q+1)^2(2 + \log(1+q))$.

Theorem 4.13. ([66], Rutishauser [53])
Let $\nu \ge 0$ be a non-negative on the m-dimensional hermitian vector space W. Then ν is affine algebraic if and only if $n_\nu(\infty) < \infty$.

Proof. (a) Assume that $n_\nu(\infty) < \infty$. First consider the case $\nu(0) = 0$. Then Theorem 4.12 applies with $q = 0$. Hence a holomorphic function h on W exists with $h(0) = 1$ such that $\nu = \mu_h^0$. Also (4.31) holds with $q = 0$. A number $s > 0$ exists such that $W[s] \cap \operatorname{supp}\nu = \emptyset$. Take $r > s$ and $\theta = \frac{1}{2}$. Then

$$\log M_h\left(\frac{r}{2}\right) \le c_0 2^{3m}\left(\int_s^r n_\nu(t)\,\frac{dt}{t} + r\int_r^\infty n_\nu(t)\,\frac{dt}{t^2}\right) \le c_0 2^{3m} n_\nu(\infty)\left(1 + \log\frac{r}{s}\right).$$

Take $r_0 > s$. Replace r by 2r. Then a constant $p > 0$ exists such that

$$\log M_h(r) \le p\log r \quad\text{or}\quad M_h(r) \le r^p$$

if $r \ge r_0$. By Lemma 4.11 h is a polynomial. Hence $\nu = \mu_h^0$ is affine algebraic.

Now consider the case $\nu(0) > 0$. Define $N = \operatorname{supp}\nu$. Then $0 \in N$. Take $\mathfrak{a} \in V - N$. Define $a = \|\mathfrak{a}\| > 0$. A biholomorphic map $\lambda : W \to W$ is defined by $\lambda(\mathfrak{z}) = \mathfrak{z} + \mathfrak{a}$. Then $\mu = \lambda^*(\nu) = \nu \circ \lambda$ is a divisor on W with

$\mu(0) = \nu(\mathfrak{a}) = 0$. Define $M = \operatorname{supp} \nu$. Then $M = \lambda^{-1}(N)$ and $\lambda(M) = N$. Take $t \geq a$. Then $\lambda(M[t]) = N \cap \{\mathfrak{a} + W[t]\} = N_{\mathfrak{a}}[t]$ with $N_{\mathfrak{a}}[t] \subseteq N[t + a]$. Since $\lambda^*(\upsilon^{m-1}) = \upsilon^{m-1}$, we have

$$n_\mu(t) = \frac{1}{t^{2m-2}} \int_{M[t]} \mu\upsilon^{m-1} = \frac{1}{t^{2m-2}} \int_{N_{\mathfrak{a}}[t]} \nu\upsilon^{m-1} \leq \frac{1}{t^{2m-2}} \int_{N[t+a]} \nu\upsilon^{m-1}$$

$$= \left(1 + \frac{a}{t}\right)^{2m-2} n_\nu(t + a) \leq 2^{2m-2} n_\nu(\infty)$$

for all $t \geq a$. Therefore $n_\mu(\infty) < \infty$. A polynomial h on W exists such that $\mu = \mu_h^0$. Then

$$\nu = \mu \circ \lambda^{-1} = (\lambda^{-1})^*(\mu) = (\lambda^{-1})^*(\mu_h^0) = \mu^0_{h\circ\lambda^{-1}}$$

where $h \circ \lambda^{-1}$ is a polynomial on W. Therefore ν is affine algebraic.

(b) <u>Assume that</u> ν <u>is affine algebraic</u>. Then there exists a polynomial $h \not\equiv 0$ on W such that $\nu = \mu_h^0$. There are constants $c > 0$, $r_0 > 0$ and $p > 0$ such that $M_h(r) \leq cr^p$ for $r \geq r_0$. Take $r > s = r_0$. The Jensen formula implies

$$N_\nu(r,s) = \int_{W\langle r\rangle} \log|h|\sigma = \int_{W\langle s\rangle} \log|h|\sigma \leq \log M_h(r) - \int_{W\langle s\rangle} \log|h|\sigma$$

$$\leq p \log r + \log c - \int_{W\langle s\rangle} \log|h|\sigma.$$

Hence

$$n_\nu(\infty) = \lim_{r\to\infty} \frac{N_\nu(r,s)}{\log r} \leq p < \infty$$

q.e.d.

Since $\dfrac{N_\nu(r,s)}{\log r} \to n_\nu(\infty)$ for $r \to \infty$, the divisor $\nu \geq 0$ is affine algebraic, if and only if there are constants $r_0 > s + 1$ and $c > 0$ such that

(4.32) $$N_\nu(r,s) \leq c \log r \qquad \text{for all } r \geq r_0.$$

Lemma 4.14.
Let h be a holomorphic function on W with $h(0) = 0$. Let $P \not\equiv 0$ and $Q \not\equiv 0$ polynomials on W. Assume that for every $a \in \mathbb{C}$ the zero divisor of $aP + Qe^h$ is affine algebraic. Then $h \equiv 0$.

Proof. First assume $\dim W = 1$. Hence we can identify $W = \mathbb{C}$. Then $f = -\frac{Q}{P} e^h$ is a meromorphic function. Assume that $h \not\equiv 0$. Since $h(0) = 0$, the function h is not constant. Hence f is singular at ∞. By Picard's theorem, for all a except 2, there are infinitely many a-points of f. Hence $aP + Qe^h$ has a divisor whose support consists of infinite points, hence is not affine algebraic. Contradiction! Consequently $h \equiv 0$.

Now, assume that $\dim W > 1$. For each $x \in \mathbb{P}(W)$, a linear subspace $E(x)$ of W of dimension 1 is given. Let $j_x : E(x) \to W$ be the inclusion. There exists an open subset U of $\mathbb{P}(W)$, such that $P \circ j_x \not\equiv 0 \not\equiv Q \circ j_x$ for all $x \in U$. Then the divisor $\mu^0_{aP\circ j_x + Q\circ j_x e^{h\circ j_x}} = j_x^*(\mu^0_{aP+Qe^h})$ is affine algebraic for all $a \in \mathbb{C}$. Hence $h \circ j_x \equiv 0$ for all $x \in U$. Therefore $h|\mathbb{P}^{-1}(U) \equiv 0$ where $\mathbb{P}^{-1}(U)$ is open in W. Hence $h \equiv 0$, q.e.d.

Theorem 4.15. ([65] Satz 24.1)
Let V and W be hermitian vector spaces with $\dim V = n + 1 > 1$ and $\dim W = m > 0$. Let $f : W \to \mathbb{P}(V)$ be a meromorphic map. Then the following statements are equivalent.

(1) The map f is rational.

(2) Take $s > 0$. Then there exist constants $c > 0$ and $r_0 > s + 1$ such that

$$T_f(r,s) \leq c \log r \qquad \forall\ r \geq r_0.$$

(3) We have $A_f(\infty) < \infty$.

(4) For each $a \in \mathbb{P}(V^*)$ with $f(W) \subseteq \ddot{E}[a]$ the divisor μ_f^a introduced in (3.19) is affine algebraic.

<u>Proof</u>. (1) $\mapsto$ (2): There exists a reduced representation $\mathfrak{v} : W \to V$ of f, such that $\mathfrak{v}$ is a vector polynomial. Take $s > 0$. Constants $r_0 > s + 1$, $p > 0$ and $c_0 > 0$ exist such that $M_{\mathfrak{v}}(r) \le c_0 r^p$ for all $r \ge r_0$. Then a constant $c > p$ exists such that

$$T_f(r,s) = \int_{W\langle r\rangle} \log\|\mathfrak{v}\|\sigma - \int_{W\langle s\rangle} \log\|\mathfrak{v}\|\sigma \le \log M_{\mathfrak{v}}(r) - \int_{W\langle s\rangle} \log\|\mathfrak{v}\|\sigma$$
$$\le p \log r + \log c_0 - \int_{W\langle s\rangle} \log\|\mathfrak{v}\|\sigma \le c \log r$$

for all $r \ge r_0$.

(2) $\mapsto$ (3): We have $A_f(\infty) \le \lim\limits_{r\to\infty} \dfrac{T_f(r,s)}{\log r} \le c < \infty$.

(3) $\mapsto$ (4): Take any $a \in \mathbb{P}(V^*)$ with $f(W) \not\subseteq \ddot{E}[a]$. The First Main Theorem implies

$$N_f(r,s;a) \le T_f(r,s) + m_f(s,a).$$

Then

$$n_f(\infty,a) = \lim_{r\to\infty} \frac{N_f(r,s;a)}{\log r} \le \lim_{r\to\infty} \frac{T_f(r,s)}{\log r} = A_f(\infty) < \infty.$$

Therefore μ_f^a is affine algebraic.

(4) $\mapsto$ (1): Let $\mathfrak{v} : W \to V$ be a reduced representation. Since $\mathfrak{v} \not\equiv 0$, a base $\varepsilon_0, \ldots, \varepsilon_n$ of V^* exists such that $v_j = \varepsilon_j \circ \mathfrak{v} \not\equiv 0$ for all $j = 0, \ldots, n$. Because the zero divisor of $v_j = \varepsilon_j \circ \mathfrak{v}$ is affine algebraic, there exists a polynomial $w_j \not\equiv 0$ on W such that $\mu_{v_j}^0 = \mu_{w_j}^0$. Hence $v_j = w_j e^{g_j}$ where g_j is a holomorphic function on W.

We can choose w_j and g_j such that $g_j(0) = 0$. Then $h_j = g_j - g_0$ is a holomorphic function on W with $h_j(0) = 0$ and $h_0 \equiv 0$. Let $\mathfrak{e}_0, \ldots, \mathfrak{e}_n$ be the dual base of $\varepsilon_0, \ldots, \varepsilon_n$. Then $\mathfrak{v} = v_0\mathfrak{e}_0 + \ldots + v_n\mathfrak{e}_n$. A reduced representation $\mathfrak{m}$ of f is defined by

$$\mathfrak{m} = e^{-g_0}\mathfrak{v} = \sum_{j=0}^{n} w_j e^{h_j}\mathfrak{e}_j.$$

Take any $j \in \mathbb{N}[1,n]$. Take any $a \in \mathbb{C}$. Define $\alpha = a\varepsilon_0 + \varepsilon_j$. Then $\alpha \circ \mathfrak{m} = aw_0 + w_j e^{h_j}$ is either identically zero or has an affine algebraic zero divisor. Assume that for some a, we have $aw_0 + w_j e^{h_j} \equiv 0$. Since $w_j \not\equiv 0$, we conclude that $a \neq 0$. Define $Q = -w_j/w_0$. Then $Qe^{h_j} = a$. If Q is not constant, the rational function Q has a zero or pole, which is impossible. Hence $e^{h_j} = a/Q \neq 0$ is constant. Hence $e^{h_j}dh_j \equiv 0$. Therefore $dh_j \equiv 0$. Hence h_j is constant. Since $h_j(0) = 0$, we have $h_j \equiv 0$.

Hence we can assume $aw_0 + w_j e^{h_j} \not\equiv 0$ for all $a \in \mathbb{C}$. Then the zero divisor of $aw_0 + w_j e^{h_j}$ is affine algebraic. By Lemma 4.14 $h_j \equiv 0$. Hence $\mathfrak{m} = w_0\mathfrak{e}_0 + \ldots + w_n\mathfrak{e}_n$ is a vector polynomial and a reduced representation of f. Consequently, f is rational; q.e.d.

This theorem was first proved in [65] Satz 24.1. Compare also Griffiths-King [25], Proposition 5.9 and Carlson-Griffiths [10] Proposition 6.20.

5. The Lemma of the Logarithmic Derivative.

Almost all defects are zero. We want to improve this result. This is achieved by the defect relation. There are several proofs of the defect relation. We shall use the Lemma of the Logarithmic Derivative which at present is restricted to meromorphic maps $f : W \to \mathbb{P}(V)$ where W and V are hermitian vector spaces.

Some preparations are needed. For $x \geq 0$ define

$$\log^+ x = \begin{cases} 0 & \text{if } 0 \leq x \leq 1 \\ \log x & \text{if } x > 1. \end{cases} \tag{5.1}$$

If $x_j \geq 0$ for $j = 1, \ldots, p$, then

$$\log^+(x_1 \ldots x_p) \leq \log^+ x_1 + \ldots + \log^+ x_p \tag{5.2}$$

$$\log^+(x_1 + \ldots + x_p) \leq \log^+ x_1 + \ldots + \log^+ x_p + \log p. \tag{5.3}$$

Let $f \not\equiv 0$ be a meromorphic function on $\mathbb{C}$. First assume that $0 \neq f(0) \neq \infty$. Take $R > 0$ and $z \in \mathbb{C}(R)$. The <u>Poisson-Jensen formula</u> states

$$\log|f(z)| = \int_{\mathbb{C}\langle R\rangle} \frac{R^2 - |z|^2}{|\zeta - z|^2} \log|f(\zeta)|\sigma_1 + \sum_{u\in\mathbb{C}[R]} \nu_f(u) \log\left|\frac{R(u - z)}{R^2 - \bar{u}z}\right|. \tag{5.4}$$

Here $\operatorname{Re} \frac{\zeta + z}{\zeta - z} = \frac{R^2 - |z|^2}{|\zeta - z|^2}$ and $\frac{\zeta + z}{\zeta - z} - 1 = \frac{2z}{\zeta - z}$. Hence if we substitute $z = 0$ in (5.4) and subtract we obtain

$$\log\left|\frac{f(z)}{f(0)}\right| = \int_{\mathbb{C}\langle R\rangle} \left(\operatorname{Re} \frac{2z}{\zeta - z}\right) \log|f(\zeta)|\sigma_1 + \sum_{u\in\mathbb{C}[R]} \nu_f(u)\left(\log\left|1 - \frac{z}{u}\right| - \log\left|1 - \frac{\bar{u}z}{R^2}\right|\right). \tag{5.5}$$

A number $s \in \mathbb{R}(0,R]$ exists such that $\mathbb{C}(s) \cap \operatorname{supp} \nu_f = \emptyset$. Put $\log 1 = 0$; we obtain

$$(5.6)\qquad \log \frac{f(z)}{f(0)} = \int_{\mathbb{C}\langle R\rangle} \frac{2z}{\zeta - z} \log|f(\zeta)|\sigma_1 + \sum_{u\in\mathbb{C}[R]} \nu_f(u)\left(\log\left(1 - \frac{z}{u}\right) - \log\left(1 - \frac{\bar{u}z}{R^2}\right)\right)$$

for $z \in \mathbb{C}(s)$. Differentiation implies

$$(5.7)\qquad \frac{f'(z)}{f(z)} = \int_{\mathbb{C}\langle R\rangle} \frac{2\zeta}{(\zeta - z)^2} \log|f(\zeta)|\sigma_1 + \sum_{u\in\mathbb{C}[R]} \nu_f(u)\left(\frac{1}{z - u} + \frac{\bar{u}}{R^2 - \bar{u}z}\right)$$

for $z \in \mathbb{C}(s)$. However analytic continuation shows that (5.7) holds on $\mathbb{C}(R)$ as an equality between meromorphic functions. If f has a zero or pole at $0 \in \mathbb{C}(R)$ replace f by $f(z)\left(\frac{R}{z}\right)^{\nu_f(0)}$. Then we see that (5.7) holds also if $f(0) = 0$ or $f(0) = \infty$.

Lemma 5.1.
Let $A \neq \emptyset$ be a finite subset of $\mathbb{C}$. For $z \in \mathbb{C} - A$ define

$$(5.8)\qquad \eta_A(z) = \operatorname{Max}\left\{\frac{1}{|z - u|} \mid u \in A\right\}.$$

Then

$$(5.9)\qquad \int_{\mathbb{C}\langle r\rangle} \log^+(r\eta_A)\sigma_1 \le 2\log(\#A) + \frac{1}{2}.$$

For the proof see Gross [28] Lemma 3.5 p. 51-52 or Hayman [31] Lemma 2.2 p. 35-36.

Lemma 5.2.
Let $f \not\equiv 0$ be a meromorphic function on $\mathbb{C}$. If $0 < r < R$, then

$$(5.10)\qquad \int_{\mathbb{C}\langle r\rangle} \log^+\left|\frac{f'}{f}\right|\sigma_1$$

$$\le \log^+ m_f(R,0) + \log^+ m_f(R,\infty) + 3\log^+ n_f(R,0) + 3\log^+ n_f(R,\infty)$$

$$+ \log^+ \frac{R}{(R-r)^2} + \log^+ \frac{1}{r} + \log^+ \frac{r}{R-r} + 8\log 2.$$

Proof. Define $A = \mathbb{C}[R] \cap \operatorname{supp} \nu$. Then $\#A \le n_f(R,0) + n_f(R,\infty)$ and η_A is defined by (5.8). If $z \in \mathbb{C}[r] - \operatorname{supp} \nu$, then (5.7) implies

$$(5.11)\qquad \frac{|f'(z)|}{|f(z)|} \le \frac{2R}{(R-r)^2} \int_{\mathbb{C}\langle R\rangle} |\log|f||\sigma_1$$

$$+ (n_f(R,\infty) + n_f(R,0))\left(\eta_A(z) + \frac{1}{R-r}\right).$$

Here

$$(5.12)\qquad \int_{\mathbb{C}\langle R\rangle} |\log|f||\sigma_1 \le \int_{\mathbb{C}\langle R\rangle} \log\sqrt{1+|f|^2}\,\sigma_1 + \int_{\mathbb{C}\langle R\rangle} \log\frac{\sqrt{1+|f|^2}}{|f|}$$

$$= m_f(R,\infty) + m_f(R,0).$$

Hence (5.11) and (5.12) imply

$$\log^+ \frac{|f'(z)|}{|f(z)|} \le \log^+ m_f(R,\infty) + \log^+ m_f(R,0) + \log^+ n_f(R,\infty) + \log^+ n_f(R,0)$$

$$+ \log^+ \frac{1}{r} + \log^+(r\eta_A(z)) + \log^+\left(\frac{r}{R-r}\right) + \log^+ \frac{R}{(R-r)^2}$$

$$+ 8\log 2.$$

Integration over $\mathbb{C}\langle r\rangle$, the estimate of $\#A$ and (5.9) imply (5.10), q.e.d.

Consider $\mathbb{C}^m$ and define τ, $\tilde{\tau}$, $\tilde{\upsilon}$ and σ_m by (2.68). Take $r > 0$ and define $\rho = \sqrt{r^2 - \tilde{\tau}}$. For $w \in \mathbb{C}^{m-1}$ define $j_w : \mathbb{C} \to \mathbb{C}^m$ by $j_w(z) = (w,z)$.

Let f be a meromorphic function on $\mathbb{C}^m$. If $j_w(\mathbb{C}) \nsubseteq \operatorname{supp} \mu_f^\infty$ then $f_{[w]} = j_w^*(f) = f \circ j_w$ is defined. If $(w,z) \in \mathbb{C}^m - \operatorname{supp} \mu_f^\infty$ then $f_{[w]}(z) = f(w,z)$. Take $a \in \mathbb{P}_1$. If $j_w(\mathbb{C}) \nsubseteq \operatorname{supp} \mu_f^a$ the pullback $j_w^*(\mu_f^a)$

is defined with $j_w^*(\mu_f^a) = \mu_{f_{[w]}}^a$. If $j \in \mathbb{N}[1,m]$ define

$$(5.13) \qquad \upsilon_j = \frac{i}{2\pi} \sum_{\substack{\lambda=1 \\ \lambda \neq j}}^{m} dz_\lambda \wedge d\bar{z}_\lambda .$$

Then $\upsilon_j \leq \upsilon$. Take $j = m$ and let $\pi : \mathbb{C}^m \to \mathbb{C}^{m-1}$ be the projection $\pi(z_1,\ldots,z_m) = (z_1,\ldots,z_{m-1})$. Then $\pi^*(\tilde{\upsilon}) = \upsilon_m$. Define $A = \operatorname{supp} \mu_f^a$. Then

$$(5.14) \qquad r^{2m-2} n_f(r,a) = \int_{A[r]} \mu_f^a \upsilon^{m-1} \geq \int_{A[r]} \mu_f^a \upsilon_m^{m-1}$$

$$(5.15) \qquad \int_{A[r]} \mu_f^a \upsilon_m^{m-1} = \int_{\mathbb{C}^{m-1}[r]} n_{f_{[w]}}(p(w),a)\tilde{\upsilon}^{m-1}(w).$$

By Lemma 2.4 we have

$$\begin{aligned} m_f(r,a) &= \int_{\mathbb{C}^m\langle r\rangle} \log \frac{1}{\|a,f\|} \sigma_m \\ &= r^{2-2m} \int_{\mathbb{C}^{m-1}[r]} \int_{\mathbb{C}\langle p(w)\rangle} \log \frac{1}{\|a,f(w,z)\|} \sigma_1(z)\tilde{\upsilon}^{m-1}(w) \end{aligned}$$

or

$$(5.16) \qquad m_f(r,a) = r^{2-2m} \int_{\mathbb{C}^{m-1}[r]} m_{f_{[w]}}(p(w),a)\tilde{\upsilon}^{m-1}(w).$$

Still we need some technical lemmata.

<u>Lemma 5.3</u>.
If $r \geq 1$, then

$$(5.17) \qquad \frac{1}{r^{2m}} \int_{\mathbb{C}^m[r]} \log^+ \frac{1}{\sqrt{r^2 - \tau}} \upsilon^m \leq \frac{1}{2} \sum_{p=1}^{m} \frac{1}{p} .$$

<u>Proof</u>. We have

$$\frac{1}{r^{2m}} \int_{\mathbb{C}^m[r]} \log^+ \frac{1}{\sqrt{r^2 - \tau}}\, \upsilon^m = \int_{\mathbb{C}^m[1]} \log^+ \frac{1}{r\sqrt{1-\tau}}\, \upsilon^m \le \int_{\mathbb{C}^m[1]} \log \frac{1}{\sqrt{1-\tau}}\, \upsilon^m$$

$$= \frac{m}{2} \int_{\mathbb{C}^m[1]} \left(\log \frac{1}{1-\tau}\right)\tau^{m-1}\, d\tau \wedge \sigma$$

$$= \frac{m}{2} \int_0^1 \left(\log \frac{1}{1-t}\right) t^{m-1}\, dt = \frac{1}{2} \sum_{p=1}^{m} \frac{1}{p}$$

q.e.d.

<u>Lemma</u> <u>5</u>.<u>4</u>.
Let $h \ge 0$ be a non-negative, measurable function on $\mathbb{C}^m(r)$. Then

$$(5.18) \qquad \frac{1}{r^{2m}} \int_{\mathbb{C}^m(r)} \log^+ h\upsilon^m \le \log^+\left(\frac{1}{r^{2m}} \int_{\mathbb{C}^m(r)} h\upsilon^m\right) + \log 2.$$

<u>Proof</u>. By convexity, we have

$$\frac{1}{r^{2m}} \int_{\mathbb{C}^m(r)} \log^+ h\upsilon^m \le \frac{1}{r^{2m}} \int_{\mathbb{C}^m(r)} \log(1+h)\upsilon^m \le \log\left(\frac{1}{r^{2m}} \int_{\mathbb{C}^m(r)} (1+h)\upsilon^m\right)$$

$$= \log\left(1 + \frac{1}{r^{2m}} \int_{\mathbb{C}^m(r)} h\upsilon^m\right) \le \log^+\left(\frac{1}{r^{2m}} \int_{\mathbb{C}^m(r)} h\upsilon^m\right) + \log 2;$$

q.e.d.

<u>Lemma</u> <u>5</u>.<u>5</u>. ([3])
Let $f \not\equiv 0$ be a meromorphic function on $\mathbb{C}^m$ with $m > 1$. Take $R > r \ge 1$. Define

$$S_m(r,R) = \log^+ \frac{1}{\sqrt{R^2 - r^2}} + \log^+ \frac{R^2}{(R-r)^2} + \log^+ \frac{r}{R-r} + 16(m-1) \log \frac{R}{r}$$
$$+ \frac{1}{2} \sum_{p=1}^{m-1} \frac{1}{p} + 16 \log 2.$$

Then

$$(5.19)\quad \int_{\mathbb{C}^m\langle r\rangle} \log^+ \frac{|f_{z_m}|}{|f|}\sigma_m \le \log^+ m_f(R,0) + \log^+ m_f(R,\infty) + 3\log^+ n_f(R,0) + 3\log^+ n_f(R,\infty) + S_m(r,R).$$

<u>Proof</u>. Define $p = \sqrt{r^2 - \tilde{\tau}}$ on $\mathbb{C}^{m-1}[r]$ and $P = \sqrt{R^2 - \tilde{\tau}}$ on $\mathbb{C}^{m-1}[R]$. Lemma 2.4 and Lemma 5.2 imply

$$\begin{aligned}
\int_{\mathbb{C}^m\langle r\rangle} \log^+ \frac{|f_{z_m}|}{|f|}\sigma_m &= r^{2-2m}\int_{\mathbb{C}^{m-1}[r]}\int_{\mathbb{C}\langle p(w)\rangle} \log^+ \frac{|f'_{[w]}(z)|}{|f_{[w]}(z)|}\sigma_1(z)\tilde{\upsilon}^{m-1}(w)\\
&\le r^{2-2m}\int_{\mathbb{C}^{m-1}[r]} \log^+ m_{f_{[w]}}(P(w),0)\tilde{\upsilon}^{m-1}(w)\\
&+ r^{2-2m}\int_{\mathbb{C}^{m-1}[r]} \log^+ m_{f_{[w]}}(P(w),\infty)\tilde{\upsilon}^{m-1}(w)\\
&+ 3r^{2-2m}\int_{\mathbb{C}^{m-1}[r]} \log^+ n_{f_{[w]}}(P(w),0)\tilde{\upsilon}^{m-1}(w)\\
&+ 3r^{2-2m}\int_{\mathbb{C}^{m-1}[r]} \log^+ n_{f_{[w]}}(P(w),\infty)\tilde{\upsilon}^{m-1}(w)\\
&+ r^{2-2m}\int_{\mathbb{C}^{m-1}[r]} \log^+ \frac{P(w)}{(P(w)-p(w))^2}\tilde{\upsilon}^{m-1}(w)\\
&+ r^{2-2m}\int_{\mathbb{C}^{m-1}[r]} \log^+ \frac{1}{p(w)}\tilde{\upsilon}^{m-1}(w)\\
&+ r^{2-2m}\int_{\mathbb{C}^{m-1}[r]} \log^+ \frac{p(w)}{P(w)-p(w)}\tilde{\upsilon}^{m-1}(w)\\
&+ 8\log 2.
\end{aligned}$$

If $a \in \mathbb{P}_1$, Lemma 5.4 and (5.16) imply

$$
\begin{aligned}
& r^{2-2m} \int_{\mathbb{C}^{m-1}[r]} \log^+ m_{f_{[w]}}(P(w),a)\tilde{\upsilon}^{m-1}(w) \\
& \leq \log^+\Bigl(r^{2-2m} \int_{\mathbb{C}^{m-1}[r]} m_{f_{[w]}}(P(w),a)\tilde{\upsilon}^{m-1}(w)\Bigr) + \log 2 \\
& \leq \log^+\Bigl(R^{2-2m} \int_{\mathbb{C}^{m-1}[R]} m_{f_{[w]}}(P(w),a)\tilde{\upsilon}^{m-1}(w)\Bigr) + (2m-2)\log\frac{R}{r} + \log 2 \\
& = \log^+ m_f(R,a) + (2m-2)\log\frac{R}{r} + \log 2.
\end{aligned}
$$

Also Lemma 5.4, (5.15) and (5.14) imply

$$
\begin{aligned}
& r^{2-2m} \int_{\mathbb{C}^{m-1}[r]} \log^+ n_{f_{[w]}}(P(w),a)\tilde{\upsilon}^{m-1}(w) \\
& \leq \log^+\Bigl(r^{2-2m} \int_{\mathbb{C}^{m-1}[r]} n_{f_{[w]}}(P(w),a)\tilde{\upsilon}^{m-1}(w)\Bigr) + \log 2 \\
& \leq \log^+\Bigl(R^{2-2m} \int_{\mathbb{C}^{m-1}[r]} n_{f_{[w]}}(P(w),a)\tilde{\upsilon}^{m-1}(w)\Bigr) + (2m-2)\log\frac{R}{r} + \log 2 \\
& \leq \log^+ n_f(R,a) + (2m-2)\log\frac{R}{r} + \log 2.
\end{aligned}
$$

If $w \in \mathbb{C}^{m-1}[r]$, then $p(w) \leq \frac{r}{R} P(w)$ and $P(w) \geq \sqrt{R^2 - r^2}$. Hence

$$
\frac{P(w)}{(P(w)-p(w))^2} = \frac{1}{P(w)} \frac{1}{\left(1 - \frac{p(w)}{P(w)}\right)^2} \leq \frac{1}{\sqrt{R^2 - r^2}} \frac{1}{\left(1 - \frac{r}{R}\right)^2} .
$$

Hence

$$
r^{2-2m} \int_{\mathbb{C}^{m-1}[r]} \log^+ \frac{P(w)}{(P(w)-p(w))^2}\, \tilde{\upsilon}^{m-1}(w) \leq \log^+ \frac{1}{\sqrt{R - r^2}} + \log^+ \frac{R^2}{(R-r)^2} .
$$

Lemma 5.3 implies

$$
r^{2-2m} \int_{\mathbb{C}^{m-1}[r]} \log^+ \frac{1}{p(w)}\, \tilde{\upsilon}^{m-1}(w) \leq \frac{1}{2} \sum_{p=1}^{m-1} \frac{1}{p} .
$$

If $w \in \mathbb{C}^{m-1}[r]$, then $\frac{p(w)}{P(w) - p(w)} \le \frac{r}{R - r}$. Consequently

$$r^{2-2m} \int_{\mathbb{C}^{m-1}[r]} \log^+ \frac{p(w)}{P(w) - p(w)} \tilde{\upsilon}^{m-1}(w) \le \log^+ \frac{r}{R - r} .$$

Therefore we obtain

$$\int_{\mathbb{C}^m\langle r\rangle} \log^+ \frac{|f_{z_m}|}{|f|} \sigma_m$$

$$\le \log^+ m_f(R,0) + \log^+ m_f(R,\infty) + 3 \log^+ n_f(R,0) + 3 \log^+ n_f(R,\infty)$$

$$+ 16(m - 1) \log \frac{R}{r} + \log^+ \frac{1}{\sqrt{R^2 - r^2}} + \log^+ \frac{R^2}{(R - r)^2} + \log^+ \frac{r}{R - r}$$

$$+ \frac{1}{2} \sum_{p=1}^{m-1} \frac{1}{p} + 16 \log 2$$

q.e.d.

Now, it is easy to obtain a lemma of the logarithmic derivative without exceptional intervals.

Proposition 5.6. ([3])

Let $f \not\equiv 0$ be a meromorphic function on $\mathbb{C}^m$ with $m > 1$. Take $r > s > 0$ with $r > 1$ and $\theta > 1$. Then

$$(5.20) \qquad \int_{\mathbb{C}^m\langle r\rangle} \log^+ \frac{|f_{z_m}|}{|f|} \sigma_m$$

$$\le 8 \log^+ T_f(\theta r,s) + 4 \log^+ m_f(s,0) + 4 \log^+ m_f(s,\infty)$$

$$+ 8(2m - 1) \log \theta + 10 \log^+ \frac{1}{\theta - 1} + 33 \log 2 + \frac{1}{2} \sum_{p=1}^{m-1} \frac{1}{p} .$$

Proof. Take $R = \frac{1 + \theta}{2} r$. Then $r < R < \theta r$. Take $a \in \mathbb{P}_1$. Then

$$N_f(\theta r,s;a) \ge \int_R^{\theta r} n_f(t,a) \frac{dt}{t} \ge n_f(R,a) \frac{\theta r - R}{\theta r} = n_f(R,a) \frac{\theta - 1}{2\theta} .$$

The First Main Theorem implies

$$(5.21) \qquad \log^+ n_f(R,a) \le \log^+ \frac{2\theta}{\theta - 1} N_f(\theta r, s; a)$$
$$\le \log^+ T_f(\theta r, s) + \log^+ m_f(s,a)$$
$$+ \log \theta + \log^+ \frac{1}{\theta - 1} + 2 \log 2$$

$$(5.22) \qquad \log^+ m_f(R,a) \le \log^+ T_f(\theta r, s) + \log^+ m_f(s,a) + \log 2.$$

Also we have

$$(5.23) \qquad \log^+ \frac{R^2}{(R - r)^2} = 2 \log \frac{\theta + 1}{\theta - 1} \le 2 \log \theta + 2 \log^+ \frac{1}{\theta - 1} + 2 \log 2$$

$$(5.24) \qquad \log^+ \frac{1}{\sqrt{R^2 - r^2}} = \log^+ \left(\frac{1}{r} \frac{2}{\sqrt{(\theta + 1)^2 - 4}} \right) \le \frac{1}{2} \log^+ \frac{1}{\theta - 1}$$

$$(5.25) \qquad \log^+ \frac{r}{R - r} = \log^+ \frac{2}{\theta - 1} \le \log^+ \frac{1}{\theta - 1} + \log 2$$

$$(5.26) \qquad \log^+ \frac{R}{r} = \log \frac{\theta + 1}{2} \le \log \theta.$$

Now (5.21)-(5.26) and (5.19) imply (5.20); q.e.d.

Now, we want to eliminate θ. This is only possible if we restricted the validity of our inequalities. If $s > 0$ and if g and h are real valued functions on $\mathbb{R}[s,+\infty)$, we write $g \lessapprox h$ or $g(r) \lessapprox h(r)$ if and only if there exists a subset E of finite measure in $\mathbb{R}[s,\infty)$ such that $g(r) \le h(r)$ for all $r \in \mathbb{R}[s,\infty) - E$. If this is true, we can still conclude

$$(5.27) \qquad \liminf_{r \to \infty} g(r) \le \limsup_{r \to \infty} h(r).$$

We need a lemma of E. Borel [7] (see Hayman [31] Lemma 2.4 p. 38).

<u>Lemma 5.7.</u>
Take $r_0 \ge 0$. Let $h : \mathbb{R}[r_0,\infty) \to \mathbb{R}[1,\infty)$ be an increasing continuous function. Then the closed set

$$E = \left\{ r \in \mathbb{R}[r_0,\infty) \mid h\left(r + \frac{1}{h(r)}\right) \geq 2h(r) \right\} \tag{5.28}$$

has at most measure 2.

<u>Proof</u>. If $E = \emptyset$, the result is trivial. Assume that $E \neq \emptyset$. Sequences r_n, t_n are defined inductively by

$$r_1 = \operatorname{Min} E \qquad\qquad t_1 = r_1 + \frac{1}{h(r_1)} > r_1$$

$$r_2 = \operatorname{Min}(E \cap \mathbb{R}[t_1,\infty)) \qquad\qquad t_2 = r_2 + \frac{1}{h(r_2)} > r_2 \geq t_1$$

$$\ldots \qquad\qquad \ldots$$

$$r_n = \operatorname{Min}(E \cap \mathbb{R}[t_{n-1},\infty)) \qquad\qquad t_n = r_n + \frac{1}{h(r_n)} > r_n \geq t_{n-1}.$$

Assume that the sequence terminates with r_p where $p \in \mathbb{N}$. Then $E \cap \mathbb{R}[t_p,\infty) = \emptyset$ and $E \subseteq \mathbb{R}[r_1,t_p]$. Take any $r \in E$. Then a maximal index $n \in \mathbb{N}[1,p]$ exists such that $r_n \leq r$. If $n = p$, then $r \in \mathbb{R}[r_p,t_p)$. If $n < p$, then $r_n \leq r < r_{n+1}$. If $r \geq t_n$, then $r \in E \cap \mathbb{R}[t_n,\infty)$, hence $r_{n+1} \leq r$ which is impossible. Hence $r \leq t_n$. We have

$$E \subseteq \bigcup_{n=1}^{p} \mathbb{R}[r_n,t_n].$$

We have

$$h(r_n) \geq h(t_{n-1}) = h\left(r_{n-1} + \frac{1}{h(r_{n-1})}\right) \geq 2h(r_{n-1}).$$

Hence

$$h(r_n) \geq 2^{n-1}h(r_1) \geq 2^{n-1} \qquad \forall\, n = 1, \ldots, p.$$

Therefore

$$\int_E dr \leq \sum_{n=1}^{p} (t_n - r_n) = \sum_{n=1}^{p} \frac{1}{h(r_n)} \leq \sum_{n=1}^{p} \frac{1}{2^{n-1}} \leq 2.$$

Now assume that the sequence does not terminate. Then $\{r_n\}_{n\in\mathbb{N}}$ and $\{t_n\}_{n\in\mathbb{N}}$ are increasing sequences with $r_n < t_n \leq r_{n+1}$. Hence $r_n \to q$ and $t_n \to q$ for $n \to \infty$, where $q \leq \infty$. If $q < \infty$, then

$$0 = \lim_{n\to\infty} t_n - r_n = \lim_{n\to\infty} \frac{1}{h(r_n)} = \frac{1}{h(q)} > 0$$

which is impossible. Therefore $q = \infty$.

Take any $r \in E$. Then $n \in \mathbb{N}$ exists uniquely such that $r_n \leq r < r_{n+1}$. If $r \geq t_n$, then $r \in E \cap \mathbb{R}[t_n,+\infty)$. Hence $r_{n+1} \leq r$, which is impossible. Therefore $r < t_n$. Consequently

$$E \subseteq \bigcup_{n=1}^{\infty} \mathbb{R}[r_n,t_n].$$

As before we have $h(r_n) \geq 2^{n-1}$ for all $n \in \mathbb{N}$. Therefore

$$\int_E dr \leq \sum_{n=1}^{\infty} (t_n - r_n) = \sum_{n=1}^{\infty} \frac{1}{h(r_n)} \leq \sum_{n=1}^{\infty} \frac{1}{2^{n-1}} = 2$$

q.e.d.

<u>Theorem 5.8</u>. <u>Lemma of the Logarithmic Derivative</u>.
Let $f \not\equiv 0$ be a meromorphic function on $\mathbb{C}^m$ with $m > 1$. Take $s > 0$ and $j \in \mathbb{N}[1,m]$. Then

$$\text{(5.29)} \qquad \boxed{\int_{\mathbb{C}^m\langle r\rangle} \log^+ \frac{|f_{z_j}|}{|f|}\, \sigma_m \mathrel{\substack{<\\ \bullet}} 17 \log^+(rT_f(r,s))} \qquad (r > s).$$

Remark 1. If f is constant, the statement is trivial. If f is not constant, then $\log^+(rT_f(r,s))$ may be replaced by $\log(rT_f(r,s))$ since $rT_f(r,s) \to \infty$ for $r \to \infty$.

Remark 2. The theorem was first proved by Al Vitter [85] in 1977. The proof given here is due to Biancofiore and Stoll [3] and is more elementary than Vitter's.

Proof. We can assume that $j = m$ and that f is not constant. Then $r_0 > s + 1$ exists such that $T_f(r,s) > e$ for all $r \geq r_0$. Then a continuous increasing function $h \geq 1$ is defined on $\mathbb{R}[r_0,\infty)$ by $h(r) = \log T_f(r,s)$. By Lemma 5.7 we have

$$h\left(r + \frac{1}{h(r)}\right) \stackrel{\leq}{\cdot} 2h(r)$$

or

$$T_f\left(r + \frac{1}{\log T_f(r,s)}\ , s\right) \stackrel{\leq}{\cdot} T_f(r,s)^2.$$

If $r > r_0$, put $\theta = 1 + \frac{1}{r \log T_f(r,s)}$. Then $1 < \theta < 2$. Now (5.20) implies

$$\begin{aligned}\int_{\mathbb{C}^m\langle r\rangle} \log^+ \frac{|f_{z_m}|}{|f|}\,\sigma_m &\stackrel{\leq}{\cdot} 16 \log T_f(r,s) + 4 \log^+ m_f(s,0) + 4 \log^+ m_f(s,\infty) \\ &\quad + (8(2m-1) + 33) \log 2 + \frac{1}{2} \sum_{p=1}^{m-1} \frac{1}{p} \\ &\quad + 10 \log(r \log T_f(r,s)) \\ &\stackrel{\leq}{\cdot} 17 \log(rT_f(r,s))\end{aligned}$$

q.e.d.

Theorem 5.9. (Vitter [85])

Let $f \not\equiv 0$ be a meromorphic function on $\mathbb{C}^m$ with $m > 1$. Take $s > 0$ and

$j \in \mathbb{N}[1,m]$. Then

(5.30) $$\boxed{T_{f_{z_j}}(r,s) \leqq 2T_f(r,s) + 18 \log^+(rT_f(r,s))} \, .$$

Proof. W.l.o.g. we can assume that f is not constant. Let $g \not\equiv 0$ and $h \not\equiv 0$ be holomorphic functions on $\mathbb{C}^m$ such that $hf = g$ and such that $\dim h^{-1}(0) \cap g^{-1}(0) \leq m - 2$. Then (h,g) is a reduced representation of f. Also $\mu_h^0 = \mu_f^\infty$ and

$$h^2 f_{z_j} = g_{z_j} h - g h_{z_j} \not\equiv 0$$

is holomorphic on $\mathbb{C}^m$. Hence $\mathfrak{w} = (h^2, h^2 f_{z_j}) : \mathbb{C}^m \to \mathbb{C}^2$ is a representation of f_{z_j}. Now (4.16) implies

$$\begin{aligned}
T_{f_{z_j}}(r,s) &\leq \int_{\mathbb{C}^m\langle r\rangle} \log\|\mathfrak{w}\|\sigma - \int_{\mathbb{C}^m\langle s\rangle} \log\|\mathfrak{w}\|\sigma \\
&= \int_{\mathbb{C}^m\langle r\rangle} \log|h^2|\sigma - \int_{\mathbb{C}^m\langle s\rangle} \log|h^2|\sigma \\
&\quad + \int_{\mathbb{C}^m\langle r\rangle} \log\sqrt{1 + |f_{z_j}|^2}\,\sigma - \int_{\mathbb{C}^m\langle s\rangle} \log\sqrt{1 + |f_{z_j}|^2}\,\sigma \\
&\leq 2N_f(r,s;\infty) + \int_{\mathbb{C}^m\langle r\rangle} \log^+\frac{|f_{z_j}|}{|f|}\sigma + \int_{\mathbb{C}^m\langle r\rangle} \log^+|f|\sigma + \log 2 \\
&\leqq 2N_f(r,s;\infty) + 17 \log(rT_f(r,s)) + \int_{\mathbb{C}^m\langle r\rangle} \log\sqrt{1 + |f|^2}\,\sigma + \log 2 \\
&= 2N_f(r,s;\infty) + m_f(r,\infty) + 17 \log(rT_f(r,s)) + \log 2 \\
&= N_f(r,s;\infty) + T_f(r,s) + m_f(s,\infty) + 17 \log(rT_f(r,s)) + \log 2
\end{aligned}$$

$$T_{f_{z_j}}(r,s) \leqq 2T_f(r,s) + 2m_f(s,\infty) + 17\log(rT_f(r,s)) + \log 2$$
$$\leqq 2T_f(r,s) + 18\log(rT_f(r,s))$$

q.e.d.

We shall extend the Lemma of the Logarithmic Derivative to differential operators with constant coefficients and to meromorphic maps $f : W \to \mathbb{P}(V)$.

Let N be a complex manifold. Let V be a complex vector space. Let $\mathfrak{M}(N,V)$ be the vector space over $\mathbb{C}$ of all meromorphic vector functions $\mathfrak{w} : N \to V$. Let $\mathfrak{O}(N,V)$ be the linear subspace of all holomorphic vector functions $\mathfrak{w} : N \to V$. Then $\mathfrak{M}(N) = \mathfrak{M}(N,\mathbb{C})$ is the ring of meromorphic functions on N and $\mathfrak{O}(N) = \mathfrak{O}(N,\mathbb{C})$ is the subring of holomorphic functions on N. If N is connected, then $\mathfrak{M}(N)$ is a field and $\mathfrak{O}(N)$ is an integral domain. Moreover $\mathfrak{M}(N,V)$ is a vector space over $\mathfrak{M}(N)$ with

(5.31) $$\dim_{\mathfrak{M}(N)} \mathfrak{M}(N,V) = \dim_{\mathbb{C}} V.$$

Let W be a complex vector space of dimension $m > 0$. Let $N \neq \emptyset$ be an open subset in W. For every $\mathfrak{a} \in W$ we assign a $\mathbb{C}$-linear map

(5.32) $$D_{\mathfrak{a}} : \mathfrak{M}(N,V) \to \mathfrak{M}(N,V)$$

in the following manner. Take $\mathfrak{w} \in \mathfrak{M}(N,V)$. Let H be the largest open subset of N such that $\mathfrak{w}$ is holomorphic on H. Take $\mathfrak{z} \in H$. Define the directional derivative

(5.33) $$(D_{\mathfrak{a}}\mathfrak{w})(\mathfrak{z}) = \frac{d}{dz}\,\mathfrak{w}(\mathfrak{z} + z\mathfrak{a})\big|_{z=0}.$$

Then $D_{\mathfrak{a}}\mathfrak{w}$ is holomorphic on H and extends to a meromorphic vector function on N. The $\mathbb{C}$-linear map $D_{\mathfrak{a}}$ restricts to a $\mathbb{C}$-linear map

(5.34) $$D_{\mathfrak{a}} : \mathfrak{O}(N,V) \to \mathfrak{O}(N,V).$$

If $Q \neq \emptyset$ is an open subset of N, then $D_{\mathfrak{a}}(\mathfrak{w}|Q) = (D_{\mathfrak{a}}\mathfrak{w})|Q$ for $\mathfrak{w} \in \mathfrak{M}(N,V)$.

Hence $D_{\mathfrak{a}}$ commutes with restrictions. Of course $D_{\mathfrak{a}}$ is defined also if $V = \mathbb{C}$. In respect to the module structure $D_{\mathfrak{a}}$ is a derivation:

$$(5.35) \qquad D_{\mathfrak{a}}(g\mathfrak{w}) = (D_{\mathfrak{a}}g)\mathfrak{w} + g(D_{\mathfrak{a}}\mathfrak{w})$$

for all $g \in \mathfrak{M}(N)$ and $\mathfrak{w} \in \mathfrak{M}(N,V)$.

For any vector space Y let End Y = $\{\alpha \mid \alpha : Y \to Y\ \mathbb{C}\text{-linear}\}$ be the vector space of endomorphisms. Then $\mathfrak{D}_1 = \{D_{\mathfrak{a}} \mid \mathfrak{a} \in W\}$ is a linear subspace of End $\mathfrak{M}(N,W)$ and the map $\mathfrak{a} \mapsto D_{\mathfrak{a}}$ is a linear isomorphism of W onto $\mathfrak{D}_1$. If $\mathfrak{e}_1, \ldots, \mathfrak{e}_m$ is a base of W, then $D_{\mathfrak{e}_1}, \ldots, D_{\mathfrak{e}_m}$ is a base of $\mathfrak{D}_1$. If $W = \mathbb{C}^m$ and $\mathfrak{e}_j = (\delta_{1j}, \ldots, \delta_{mj})$ where δ_{kj} is the Kronecker symbol, then $D_{\mathfrak{e}_j} = \frac{\partial}{\partial z_j}$ is the jth partial derivative.

If $\mathfrak{a} \in W$ and $\mathfrak{b} \in W$, then $D_{\mathfrak{a}} \circ D_{\mathfrak{b}} = D_{\mathfrak{b}} \circ D_{\mathfrak{a}}$. Take $p \in \mathbb{N}$. Let $\mathfrak{D}_p$ be the linear subspace of End $\mathfrak{M}(N,W)$ over $\mathbb{C}$ generated over $\mathbb{C}$ by

$$(5.36) \qquad D_{\mathfrak{a}_p} \circ D_{\mathfrak{a}_{p-1}} \circ \ldots \circ D_{\mathfrak{a}_1} : \mathfrak{M}(N,V) \to \mathfrak{M}(N,V).$$

An element $D \in \mathfrak{D}_p$ is called a <u>differential operator of degree</u> p <u>with constant coefficients</u>.

If $\mathfrak{e}_1, \ldots, \mathfrak{e}_m$ is a base of W over $\mathbb{C}$, then every $D \in \mathfrak{D}_p$ can be written uniquely as

$$(5.37) \qquad D = \sum_{j_1,\ldots,j_p=1}^{m} a_{j_1\ldots j_p} D_{\mathfrak{e}_{j_1}} \circ \ldots \circ D_{\mathfrak{e}_{j_p}}$$

where $a_{j_1\ldots j_p} \in \mathbb{C}$ and where

$$(5.38) \qquad a_{j_{\pi(1)}\ldots j_{\pi(p)}} = a_{j_1\ldots j_p}$$

for all permutations $\pi : \mathbb{N}[1,p] \to \mathbb{N}[1,p]$.

If $W = \mathbb{C}^m$ and $\mathfrak{e}_j = (\delta_{1j},\ldots,\delta_{mj})$, then

(5.39) $$D_{\mathfrak{e}_{j_1}} \circ \ldots \circ D_{\mathfrak{e}_{j_p}} = \frac{\partial}{\partial z_{j_1}} \circ \ldots \circ \frac{\partial}{\partial z_{j_p}} = \frac{\partial^p}{\partial z_{j_1} \ldots \partial z_{j_p}} .$$

If $\mathfrak{w} \in \mathfrak{M}(N,V)$ and if $D \in \mathfrak{D}_p$ is given by (5.37) then

(5.40) $$D\mathfrak{w} = \sum_{j_1,\ldots,j_p=1}^{m} a_{j_1 \ldots j_p} \mathfrak{w}_{z_{j_1} \ldots z_{j_p}} .$$

Obviously, $\mathfrak{D}_p$ is linearly isomorphic to the pth symmetric tensor power of W.

Let $\mathfrak{D}_0$ be the linear subspace over $\mathbb{C}$ generated in End $\mathfrak{M}(G,W)$ by the identity on $\mathfrak{M}(G,W)$. Hence $\mathfrak{D}_0 = \mathbb{C}$ Id.

Theorem 5.10. The Lemma of the higher order Logarithmic Derivative Let W be a hermitian vector space of dimension $m > 0$. Let $f : W \to \mathbb{P}_1$ be a non-constant meromorphic function. Take $p \in \mathbb{N}$ and $D \in \mathfrak{D}_p$. Take $s > 0$. Then

(5.41) $$\int_{W\langle r\rangle} \log^+ \left|\frac{Df}{f}\right| \sigma \;\overset{\leq}{\bullet}\; 18pm^{p-1} \log(rT_f(r,s))$$

(5.42) $$T_{Df}(r,s) \;\overset{\leq}{\bullet}\; (p+1)T_f(r,s) + 18pm^{p-1} \log(rT_f(r,s))$$

where $r > s$. (Vitter [85])

Proof. The proof is accomplished by induction for p. Consider $p = 1$ first. We can pick an orthonormal base of W and identify $W = \mathbb{C}^m$ such that $D = a \frac{\partial}{\partial z_m}$ where $a \in \mathbb{C}$. Hence

$$\int_{W\langle r\rangle} \log^+\left|\frac{Df}{f}\right|\sigma = \int_{\mathbb{C}^m\langle r\rangle} \log^+\left(|a|\,\frac{|f_{z_m}|}{|f|}\right)\sigma$$

$$\mathrel{\dot{\le}} \log^+|a| + 17\log(rT_f(r,s)) \mathrel{\dot{\le}} 18\log(rT_f(r,s))$$

which proves (5.41) for p = 1. If a = 0, then (5.42) is trivial. Hence $a \neq 0$ can be assumed. Then $N_{(af_{z_m})}(r,s;\infty) = N_{f_{z_m}}(r,s;\infty)$ and

$$m_{af_{z_m}}(r,\infty) = \int_{\mathbb{C}^m\langle r\rangle} \log\sqrt{1+|af_{z_m}|^2}\,\sigma$$

$$\le \int_{\mathbb{C}^m\langle r\rangle} \log\sqrt{1+|f_{z_m}|^2}\,\sigma + \log^+|a|$$

$$= m_{f_{z_m}}(r,\infty) + \log^+|a|.$$

The First Main Theorem implies

$$T_{Df}(r,s) = N_{af_{z_m}}(r,s;\infty) + m_{af_{z_m}}(r,\infty) - m_{af_{z_m}}(s,\infty)$$

$$\le N_{f_{z_m}}(r,s;\infty) + m_{f_{z_m}}(r,\infty) + \log^+|a| - m_{af_{z_m}}(s,\infty)$$

$$= T_{f_{z_m}}(r,s) + \log^+|a| + m_{f_{z_m}}(s,\infty) - m_{af_{z_m}}(s,\infty)$$

$$\mathrel{\dot{\le}} 2T_f(r,s) + 17\log(rT_f(r,s)) + m_{f_{z_m}}(s,\infty) - m_{af_{z_m}}(s,\infty)$$

$$+ 2m_f(s,\infty) + \log 2$$

$$\mathrel{\dot{\le}} 2T_f(r,s) + 18\log(rT_f(r,s))$$

where the slightly sharper estimate in the proof of Theorem 5.9 was used. Hence (5.42) is proved for p = 1.

Take p > 1. Assume that (5.41) and (5.42) hold for p - 1. Take $D \in \mathfrak{D}_p$. Let $\mathfrak{e}_1, \ldots, \mathfrak{e}_m$ be an orthonormal base of W. By (5.37) for each $j \in \mathbb{N}[1,m]$ there exists $D_j \in \mathfrak{D}_{p-1}$ such that

$$D = \sum_{j=1}^{m} D_{\mathfrak{e}_j} \circ D_j .$$

Define $\mathfrak{J} = \{j \in \mathbb{N}[1,m] \mid D_j f \not\equiv 0\}$. Then

$$\int_{W\langle r\rangle} \log^+ \left|\frac{Df}{f}\right| \sigma \leq \sum_{j\in\mathfrak{J}} \int_{W\langle r\rangle} \log^+ \frac{|D_{\mathfrak{e}_j} D_j f|}{|f|} \sigma + \log \#\mathfrak{J}$$

$$\leq \sum_{j\in\mathfrak{J}} \int_{W\langle r\rangle} \log^+ \left|\frac{D_{\mathfrak{e}_j}(D_j f)}{|D_j f|}\right| \sigma + \sum_{j\in\mathfrak{J}} \int_{W\langle r\rangle} \log^+ \left|\frac{D_j f}{f}\right| \sigma + \log m$$

$$\underset{\bullet}{\leq} 17 \sum_{j\in\mathfrak{J}} \log(rT_{D_j f}(r,s)) + 18(p-1)m^{p-1} \log(rT_f(r,s)) + \log m$$

$$\underset{\bullet}{\leq} 17m \log(prT_f(r,s) + 18(p-1)m^{p-2} r \log(rT_f(r,s)))$$

$$+ 18(p-1)m^{p-1} \log(rT_f(r,s)) + \log m$$

$$\underset{\bullet}{\leq} 17m \log((p+1)rT_f(r,s)) + 18(p-1)m^{p-1} \log(rT_f(r,s)) + \log m$$

$$\underset{\bullet}{\leq} 18pm^{p-1} \log(rT_f(r,s)) - m \log T_f(r,s) + 17m \log(p+1) + \log m$$

$$\underset{\bullet}{\leq} 18pm^{p-1} \log(rT_f(r,s)) - \frac{m}{2} \log T_f(r,s)$$

which implies (5.41) for p.

There exist holomorphic functions $g \not\equiv 0$ and $h \not\equiv 0$ on W such that $hf = g$ and such that $\dim h^{-1}(0) \cap g^{-1}(0) \leq m - 2$. Then (h,g) is a reduced representation of f. Also $\mu_h^0 = \mu_f^\infty$. We claim that $h^{p+1}Df$ is holomorphic on W. For $p = 1$, this was shown in the proof of Theorem 5.9. Hence we can assume it is already shown for $p - 1$; i.e. we can assume that $h^p D_j f$ is holomorphic on W. Then

$$h^{p+1}D_f = \sum_{j=1}^{m} h^{p+1} D_{\mathfrak{e}_j}\left(\frac{h^p D_j f}{h^p}\right)$$

$$= \sum_{j=1}^{m} h D_{\mathfrak{e}_j}(h^p D_j f) - \sum_{j=1}^{m} p(D_j f) D_{\mathfrak{e}_j} h$$

is holomorphic, which proves the claim.

Therefore $\mathfrak{w} = (h^{p+1}, h^{p+1}Df)$ is a representation of Df. Now (4.15) implies

$$T_{Df}(r,s) \le \int_{W\langle r\rangle} \log\|\mathfrak{w}\|\sigma - \int_{W\langle s\rangle} \log\|\mathfrak{w}\|\sigma$$
$$= \int_{W\langle r\rangle} \log|h^{p+1}|\sigma - \int_{W\langle s\rangle} \log|h^{p+1}|\sigma$$
$$+ \int_{W\langle r\rangle} \log\sqrt{1+|Df|^2}\,\sigma - \int_{W\langle s\rangle} \log\sqrt{1+|Df|^2}\,\sigma$$

$$T_{Df}(r,s) \le (p+1)N_f(r,s;\infty) + \int_{W\langle r\rangle} \log^+\left|\frac{Df}{f}\right|\sigma + \int_{W\langle r\rangle} \log^+|f|\sigma + \log 2$$
$$\overset{\le}{\bullet}\ (p+1)N_f(r,s;\infty) + \int_{W\langle r\rangle} \log\sqrt{1+|f|^2}\,\sigma + \log 2$$
$$+ 18pm^{p-1}\log(rT_f(r,s)) - \frac{m}{2}\log T_f(r,s)$$
$$= (p+1)N_f(r,s;\infty) + m_f(r,\infty) + \log 2$$
$$+ 18pm^{p-1}\log(rT_f(r,s)) - \frac{m}{2}\log T_f(r,s)$$
$$\le (p+1)T_f(r,s) + (p+1)m_f(s,\infty) + \log 2$$
$$+ 18pm^{p-1}\log(rT_f(r,s)) - \frac{m}{2}\log T_f(r,s)$$
$$\overset{\le}{\bullet}\ (p+1)T_f(r,s) + 18pm^{p-1}\log(rT_f(r,s))$$

q.e.d.

Now, we will extend the Lemma of the Logarithmic Derivative to meromorphic maps.

Theorem 5.11.
Let V and W be hermitian vector spaces with $\dim V = n+1 > 1$ and $\dim W = m > 0$. Let $f : W \to \mathbb{P}(V)$ be a meromorphic map. Take α and $\beta \in V_*^*$ and define $a = \mathbb{P}(\alpha)$ and $b = \mathbb{P}(\beta)$. Assume that $f(W) \not\subseteq \ddot{E}[a]$ and $f(W) \not\subseteq \ddot{E}[b]$. Let $\mathfrak{v} : W \to \mathbb{P}(V)$ be a representation of f. Then $\alpha \circ \mathfrak{v} \not\equiv 0$ and $\beta \circ \mathfrak{v} \not\equiv 0$. A meromorphic function $F : W \to \mathbb{P}_1$ is defined by

$$F = \frac{\beta \circ \mathfrak{v}}{\alpha \circ \mathfrak{v}}. \tag{5.43}$$

Then F does not depend on the choice of the representation $\mathfrak{v}$. Let $D \in \mathfrak{D}_p$ be a differential operator of order $p \geq 1$ with constant coefficients. Take $s > 0$. Then there exists a constant c such that

(5.44) $$T_F(r,s) \leq T_f(r,s) + c \qquad \forall\ r > s$$

(5.45) $$\int_{W\langle r\rangle} \log^+ \frac{|DF|}{|F|}\,\sigma \leqq 19pm^{p-1} \log^+(rT_f(r,s))$$

(5.46) $$T_{DF}(r,s) \leqq (p+1)T_f(r,s) + 19pm^{p-1} \log^+(rT_f(r,s)).$$

Proof. If F is constant, the statement is trivial. Assume that F is not constant. Then f is not constant. If $\mathfrak{w}$ is any reduced representation, a holomorphic function $h \not\equiv 0$ exists on W such that $\mathfrak{v} = h\mathfrak{w}$. Then

$$F = \frac{\beta \circ \mathfrak{w}}{\alpha \circ \mathfrak{w}}.$$

Hence F does not depend on the choice of the representation $\mathfrak{v}$. W.l.o.g. we can assume that $\mathfrak{v}$ is reduced. Since $(\alpha \circ \mathfrak{v}, \beta \circ \mathfrak{v})$ is a representation of F, the Schwarz inequality implies

$$\begin{aligned} T_F(r,s) &\leq \int_{W\langle r\rangle} \log\sqrt{|\alpha\circ\mathfrak{v}|^2 + |\beta\circ\mathfrak{v}|^2}\,\sigma - \int_{W\langle s\rangle} \log\sqrt{|\alpha\circ\mathfrak{v}|^2 + |\beta\circ\mathfrak{v}|^2}\,\sigma \\ &\leq \int_{W\langle r\rangle} \log\|\mathfrak{v}\|\sigma + \log\sqrt{\|\alpha\|^2 + \|\beta\|^2} - \int_{W\langle s\rangle} \log\sqrt{|\alpha\circ\mathfrak{v}|^2 + |\beta\circ\mathfrak{v}|^2}\,\sigma \\ &= T_f(r,s) + c \end{aligned}$$

where c is a constant. Now (5.41) implies

$$\begin{aligned} \int_{W\langle r\rangle} \log^+ \frac{|DF|}{|F|}\,\sigma &\leqq 18pm^{p-1} \log(rT_F(r,s)) \\ &\leqq 18pm^{p-1}(\log r + \log(T_f(r,s) + c)) \\ &\leqq 19pm^{p-1} \log(rT_f(r,s)). \end{aligned}$$

In addition (5.42) implies

$$T_{DF}(r,s) \lesssim (p+1)T_F(r,s) + 18pm^{p-1}\log(rT_F(r,s))$$

$$\lesssim (p+1)T_f(r,s) + c(p+1) + 18pm^{p-1}\log r$$

$$+ 18pm^{p-1}\log T_f(r,s) + 18pm^{p-1}(\log^+|c| + \log 2)$$

$$\lesssim (p+1)T_f(r,s) + 19pm^{p-1}\log(rT_f(r,s))$$

q.e.d.

Theorem 5.12.
Let V and W be hermitian vector spaces with dim V = n + 1 and dim W = m > 0. Let $f : W \to \mathbb{P}(V)$ be a meromorphic map. Let $D_1, D_2, \ldots, D_q$ be differential operators of order 1 with constant coefficients. Let $\mathfrak{v} : W \to V$ be a reduced representation of f. Assume that $\mathfrak{w} = \mathfrak{v} \wedge D_1\mathfrak{v} \wedge \ldots \wedge D_q\mathfrak{v} \not\equiv 0$. Then a meromorphic map $w : W \to G_q(V)$ is defined by $w = \mathbb{P} \circ \mathfrak{w}$. Then

(5.47) $T_w(r,s) + N_{\nu_{\mathfrak{w}}}(r,s) \lesssim (q+1)T_f(r,s) + 20qn\log(rT_f(r,s))$.

Proof. Since $\mathfrak{w} \not\equiv 0$, the vector function $\mathfrak{v}$ is not constant. Hence f is not constant. Take $\alpha_0 \in V^*\langle 1\rangle$ such that $\alpha_0 \circ \mathfrak{v} \not\equiv 0$. Let $\alpha_0, \alpha_1, \ldots, \alpha_n$ be an orthonormal base of V^*. Let $\mathfrak{a}_0, \ldots, \mathfrak{a}_n$ be the dual base. For $\mu \in \mathbb{N}[1,n]$ define a meromorphic function $F_\mu = (\alpha_\mu \circ \mathfrak{v})/(\alpha_0 \circ \mathfrak{v})$ on W. A meromorphic vector function

$$\mathfrak{g} = \mathfrak{a}_0 + \sum_{\mu=1}^{n} F_\mu \mathfrak{a}_\mu : W \to V$$

is defined. Also a meromorphic map

$$\mathfrak{y} = \mathfrak{g} \wedge D_1\mathfrak{g} \wedge \ldots \wedge D_q\mathfrak{g} : W \to \tilde{G}_q(V)$$

is defined. We have $(\alpha_0 \circ \mathfrak{v})\mathfrak{g} = \mathfrak{v}$ and $(\alpha_0 \circ \mathfrak{v})^{q+1}\mathfrak{y} = \mathfrak{w} \not\equiv 0$. Hence $\mathfrak{y} \not\equiv 0$. If r > s > 0, then (4.13) implies

$$T_W(r,s) + N_{\nu_{\mathfrak{w}}}(r,s) = \int_{W\langle r\rangle} \log\|\mathfrak{w}\|\sigma - \int_{W\langle s\rangle} \log\|\mathfrak{w}\|\sigma$$

$$= (q+1)T_f(r,s) + \int_{W\langle r\rangle} \log \frac{\|\mathfrak{w}\|}{\|\mathfrak{v}\|^{q+1}}\,\sigma - \int_{W\langle s\rangle} \log \frac{\|\mathfrak{w}\|}{\|\mathfrak{v}\|^{q+1}}\,\sigma$$

$$= (q+1)T_f(r,s) + \int_{W\langle r\rangle} \log \frac{\|\mathfrak{y}\|}{\|\mathfrak{g}\|^{q+1}}\,\sigma - \int_{W\langle s\rangle} \log \frac{\|\mathfrak{y}\|}{\|\mathfrak{g}\|^{q+1}}\,\sigma.$$

Define $M = \{\mu \in \mathbb{N}[1,n] \mid F_\mu \not\equiv 0\}$. Then we have the estimate

$$\log \frac{\|\mathfrak{y}\|}{\|\mathfrak{g}\|^{q+1}} \le \log^+ \frac{\|\mathfrak{g} \wedge D_1\mathfrak{g} \wedge \ldots \wedge D_q\mathfrak{g}\|}{\|\mathfrak{g}\|^{q+1}}$$

$$\le \log^+ \frac{\|\mathfrak{g}\|\ \|D_1\mathfrak{g}\|\ \ldots\ \|D_q\mathfrak{g}\|}{\|\mathfrak{g}\|^{q+1}}$$

$$\le \sum_{j=1}^{q} \log^+ \frac{\|D_j\mathfrak{g}\|}{\|\mathfrak{g}\|}$$

$$= \sum_{j=1}^{q} \frac{1}{2} \log^+\left(\sum_{\mu=1}^{n} \frac{|D_jF_\mu|^2}{\|\mathfrak{g}\|^2}\right)$$

$$\le \sum_{j=1}^{q} \sum_{\mu=1}^{n} \log^+ \frac{|D_jF_\mu|}{\|\mathfrak{g}\|} + q \log n$$

$$\le \sum_{j=1}^{q} \sum_{\mu\in M} \log^+\left(\frac{|D_jF_\mu|}{|F_\mu|}\right) + q \log n.$$

Hence

$$T_W(r,s) + N_{\nu_{\mathfrak{w}}}(r,s)$$

$$\mathrel{\dot{\le}} (q+1)T_f(r,s) + \sum_{j=1}^{q} \sum_{\mu\in M} \int_{W\langle r\rangle} \log^+ \frac{|D_jF_\mu|}{|F_\mu|}\,\sigma + q\log n - \int_{W\langle s\rangle} \log \frac{\|\mathfrak{y}\|}{\|\mathfrak{g}\|^{q+1}}\,\sigma$$

$$\mathrel{\dot{\le}} (q+1)T_f(r,s) + 19qn\log(rT_f(r,s)) + q\log n - \int_{W\langle s\rangle} \log \frac{\|\mathfrak{y}\|}{\|\mathfrak{g}\|^{q+1}}\,\sigma$$

$$\mathrel{\dot{\le}} (q+1)T_f(r,s) + 20qn\log(rT_f(r,s))$$

q.e.d.

6. Associated maps.

The Lemma of the Logarithmic Derivative will be extended to higher order differential operators in connection with meromorphic maps. This leads to the definition of the associated maps. If $m > 1$, the associated maps will be introduced in accordance with Vitter [85] which differs from the associated maps introduced earlier [65], [74], [88].

Lemma 6.1.
Let N be a connected complex manifold of dimension m. Let V be a complex vector space of dimension $n + 1 \geq 1$. Let $\mathfrak{w}_j : N \to V$ be meromorphic vector functions for $j = 0, 1, \ldots, p$. Then $\mathfrak{w}_0, \ldots, \mathfrak{w}_p$ are linearly independent in the vector space $\mathfrak{M}(N,V)$ over the field $\mathfrak{M}(N)$ if and only if $\mathfrak{w}_0 \wedge \ldots \wedge \mathfrak{w}_p \not\equiv 0$ on N, which is the case if and only if there exists a point $x_0 \in N$ such that $\mathfrak{w}_j$ is holomorphic at x_0 for each $j = 0, 1, \ldots, p$ and such that $\mathfrak{w}_0(x_0), \ldots, \mathfrak{w}_p(x_0)$ are linearly independent in V over $\mathbb{C}$.

Proof. Trivially, the last two conditions are equivalent.

(a) Assume that there exists $x_0 \in N$ such that $\mathfrak{w}_j$ is holomorphic at x_0 for $j = 0, 1, \ldots, p$ and such that $\mathfrak{w}_0(x_0), \ldots, \mathfrak{w}_p(x_0)$ are linearly independent over $\mathbb{C}$. Then $\mathfrak{w}_0(x_0) \wedge \ldots \wedge \mathfrak{w}_p(x_0) \neq 0$. An open, connected neighborhood U of x_0 exists such that $\mathfrak{w}_j$ is holomorphic on U for $j = 0, 1, \ldots, p$ and such that $\mathfrak{w}_0(x) \wedge \ldots \wedge \mathfrak{w}_p(x) \neq 0$ for all $x \in U$. Hence $\mathfrak{w}_0(x), \ldots, \mathfrak{w}_p(x)$ are linearly independent in V over $\mathbb{C}$ for all $x \in U$.

Assume that meromorphic functions $h_0, h_1, \ldots, h_p$ are given on N such that $h_0\mathfrak{w}_0 + \ldots + h_p\mathfrak{w}_p \equiv 0$ on N. Let H be the largest open subset of N such that $h_0, h_1, \ldots, h_p$ are holomorphic on H. Then H is dense in N. Therefore $H \cap U \neq \emptyset$. Take any $x \in H \cap U$. Then $h_0(x)\mathfrak{w}_0(x) + \ldots + h_p(x)\mathfrak{w}_p(x) = 0$. Therefore $h_j(x) = 0$ for all $x \in H \cap U$ and $j = 0, \ldots, p$. Hence $h_j \equiv 0$ for $j = 0, 1, \ldots, p$ on N. Consequently $\mathfrak{w}_0, \ldots, \mathfrak{w}_p$ are linearly independent over $\mathfrak{M}(N)$.

(b) Assume that $\mathfrak{w}_0, \mathfrak{w}_1, \ldots, \mathfrak{w}_p$ are linearly independent over $\mathfrak{M}(N)$. Take $\mathfrak{w}_{p+1}, \ldots, \mathfrak{w}_n$ in $\mathfrak{M}(N,V)$ such that $\mathfrak{w}_0, \ldots, \mathfrak{w}_n$ is a base of $\mathfrak{M}(N,V)$ over $\mathfrak{M}(N)$. Let $\mathfrak{e}_0, \ldots, \mathfrak{e}_n$ be a base of V over $\mathbb{C}$. Then $\mathfrak{e}_0, \ldots, \mathfrak{e}_n$ is a base of $\mathfrak{M}(N,V)$ over $\mathfrak{M}(N)$. Therefore $W_{jk} \in \mathfrak{M}(N)$ exist such that

$$\mathfrak{w}_j = \sum_{k=0}^{n} W_{jk}\mathfrak{e}_k$$

where $\det W_{jk} \not\equiv 0$. Hence $\mathfrak{w}_0 \wedge \ldots \wedge \mathfrak{w}_n = (\det W_{jk})\mathfrak{e}_0 \wedge \ldots \wedge \mathfrak{e}_n \not\equiv 0$, q.e.d.

Let N be a connected complex manifold of dimension m. Let V be a complex vector space of dimension n + 1. Let L be a (p + 1)-dimensional linear subspace of $\mathfrak{M}(N,V)$ over $\mathfrak{M}(N)$. Let $\mathfrak{w}_0, \ldots, \mathfrak{w}_p$ be a base of L over $\mathfrak{M}(N)$. Then $\mathfrak{w}_0 \wedge \ldots \wedge \mathfrak{w}_p \not\equiv 0$. Therefore a meromorphic map

(6.1) $$\Lambda_L : N \to G_p(V)$$

by

(6.2) $$\Lambda_L = \mathbb{P} \circ \mathfrak{w}_0 \wedge \ldots \wedge \mathfrak{w}_p.$$

If $\tilde{\mathfrak{w}}_0, \ldots, \tilde{\mathfrak{w}}_p$ is another base, a meromorphic function $h \not\equiv 0$ exists such that

$$\tilde{\mathfrak{w}}_0 \wedge \ldots \wedge \tilde{\mathfrak{w}}_p = h\mathfrak{w}_0 \wedge \ldots \wedge \mathfrak{w}_p.$$

Hence $\mathbb{P} \circ \tilde{\mathfrak{w}}_0 \wedge \ldots \wedge \tilde{\mathfrak{w}}_p = \mathbb{P} \circ \mathfrak{w}_0 \wedge \ldots \wedge \mathfrak{w}_p$. Therefore Λ_L does not depend on the choice of the base of L.

If $f : N \to \mathbb{P}(V)$ is a meromorphic map, then f is said to be <u>linearly non-degenerate</u> if and only if $f(N) \not\subseteq \ddot{E}[a]$ for all $a \in \mathbb{P}(V^*)$. Now we will make the following <u>general assumptions</u>.

(A1) Let V be a hermitian vector space of dimension n + 1 > 1.

(A2) Let W be a hermitian vector space of dimension $m > 0$.

(A3) Let $f : W \to \mathbb{P}(V)$ be a linearly non-degenerate meromorphic map.

Assume that (A1)-(A3) hold. Let $\mathfrak{v} : W \to V$ be a reduced representation of f. Take $0 \leq p \in \mathbb{Z}$. Let $L_p(\mathfrak{v})$ be the vector space in $\mathfrak{M}(W,V)$ generated by $\bigcup_{q=0}^{p} \{D\mathfrak{v} \mid D \in \mathfrak{D}_q\}$ over $\mathfrak{M}(W)$. If $\tilde{\mathfrak{v}} : W \to V$ is another reduced representation of f, then there is a holomorphic function $h : W \to \mathbb{C}_*$ such that $\tilde{\mathfrak{v}} = h\mathfrak{v}$. By the chain rule $D\tilde{\mathfrak{v}} \in L_p(\mathfrak{v})$ and $D\mathfrak{v} \in L_p(\tilde{\mathfrak{v}})$ for all $D \in \mathfrak{D}_q$ with $0 \leq q \leq p$. Hence $L_p(\mathfrak{v}) = L_p(\tilde{\mathfrak{v}})$ does not depend on the choice of the reduced representation and is denoted by $L_p(f)$ and is called the pth associated <u>oscillating space</u> of f. Obviously

$$(6.3) \qquad 0 \neq L_0(f) \subseteq L_1(f) \subseteq \dots \subseteq L_p(f) \subseteq \mathfrak{M}(W,V)$$

which we call the <u>associated flag</u>. Define

$$(6.4) \qquad \ell_p(f) + 1 = \dim_{\mathfrak{M}(W)} L_p(f).$$

The pth <u>associated map</u> of f is defined by

$$(6.5) \qquad \boxed{\Lambda_p = \Lambda_{L_p(f)} : W \to G_{\ell_p(f)}(V)} \, .$$

Observe that $L_0(f)$ is generated by a reduced representation $\mathfrak{v} : W \to V$ of f. Hence $\mathfrak{v}$ is a base of $L_0(f)$ over $\mathfrak{M}(W)$. Hence

$$(6.6) \qquad \boxed{\Lambda_0 = \mathbb{P} \circ \mathfrak{v} = f} \qquad \boxed{\ell_0 = 0} \, .$$

<u>Lemma</u> <u>6</u>.<u>2</u>.
Assume that (A1)-(A3) hold. Then $p \in \mathbb{N}$ exists such that $L_p(f) = \mathfrak{M}(W,V)$.

Proof. Define $\mathbb{Z}_+ = \mathbb{Z}[0,\infty)$. If $\alpha = (\alpha_1,\ldots,\alpha_m) \in \mathbb{Z}_+^m$ and $\mathfrak{z} = (z_1,\ldots,z_m)$ define

$$\text{(6.7)} \qquad \alpha! = \alpha_1! \ldots \alpha_m! \qquad\qquad \mathfrak{z}^\alpha = z_1^{\alpha_1} \ldots z_m^{\alpha_m}$$

$$\text{(6.8)} \qquad S(k,m) = \{\alpha = (\alpha_1,\ldots,\alpha_m) \in \mathbb{Z}_+^m \mid \alpha_1 + \ldots + \alpha_m = k\}.$$

If $\mathfrak{e}_1, \ldots, \mathfrak{e}_m$ is a base of W and if $\alpha = (\alpha_1,\ldots,\alpha_m) \in \mathbb{Z}_+^m$ define

$$\text{(6.9)} \qquad D_{\mathfrak{e}}^\alpha = D_{\mathfrak{e}_1}^{\alpha_1} \circ D_{\mathfrak{e}_2}^{\alpha_2} \circ \ldots \circ D_{\mathfrak{e}_m}^{\alpha_m}.$$

Let $\mathfrak{v} : W \to V$ be a reduced representation of f. The Taylor series at $0 \in W$ of $\mathfrak{v}$ is given by

$$\mathfrak{v}(\mathfrak{z}) = \sum_{k=0}^{\infty} \sum_{\alpha\in S(k,m)} \frac{1}{\alpha!} (D_{\mathfrak{e}}^\alpha \mathfrak{v})(0)\mathfrak{z}^\alpha.$$

Since $\mathfrak{v}(W)$ is not contained in any linear subspace of V, the linear span of $\{(D^\alpha \mathfrak{v})(0) \mid \alpha \in \mathbb{Z}_+^m\}$ is V. Hence there are vectors $\alpha^{(0)}, \ldots, \alpha^{(n)}$ in $\mathbb{Z}_+^m$ such that $(D^{\alpha^{(0)}}\mathfrak{v})(0), \ldots, (D^{\alpha^{(n)}}\mathfrak{v})(0)$ are linearly independent over $\mathbb{C}$. Define $p = \operatorname*{Max}_{\mu=0,\ldots,n} \text{order } D^{\alpha(n)}$.

By Lemma 6.1, $D^{\alpha^{(0)}}\mathfrak{v}, \ldots, D^{\alpha^{(n)}}\mathfrak{v}$ are linearly independent over $\mathfrak{M}(W)$ in $L_p(f)$. Hence $\dim_{\mathfrak{M}(W)} L_p(f) \geq n + 1$. By (5.31) we have $\dim_{\mathfrak{M}(W)} L_p(f) = n + 1$. Therefore $L_p(f) = \mathfrak{M}(W,V)$, q.e.d.

The number $k = \operatorname{Min}\{p \in \mathbb{Z}_+ \mid L_p(f) = \mathfrak{M}(W,V)\}$ exists and is called the <u>span of</u> f. Then

$$\text{(6.10)} \qquad 0 \neq L_0(f) \subseteq L_1(f) \subseteq \ldots \subseteq L_k(f) = \mathfrak{M}(W,V)$$

$$\text{(6.11)} \qquad 0 = \ell_0(f) \leq \ell_1(f) \leq \ldots \leq \ell_k(f) = n.$$

Abbreviate $\ell_j = \ell_j(f)$. Then $\mathfrak{w}_0, \mathfrak{w}_1, \ldots, \mathfrak{w}_{\ell_p}$ is called a <u>flag base of $L_p(f)$ in respect to</u> f if and only if $\mathfrak{w}_0, \mathfrak{w}_1, \ldots, \mathfrak{w}_{\ell_j}$ is a base of $L_j(f)$ over $\mathfrak{M}(W)$ for $j = 0, 1, \ldots, p$. <u>If $\mathfrak{v} : W \to V$ is a reduced representation of f, then a flag base $\mathfrak{v}, D_1\mathfrak{v}, \ldots, D_{\ell_p}\mathfrak{v}$ of $L_p(f)$ in respect to f exists.</u>

<u>Lemma 6.3.</u>
Assume that (A1)-(A3) hold. Let $\mathfrak{v} : W \to V$ be a representation of f. Take $0 \not\equiv h \in \mathfrak{M}(W)$ and define $\mathfrak{w} = h\mathfrak{v}$. Take $p \in \mathbb{N}$. Abbreviate $\ell_j = \ell_j(f)$. Then $\mathfrak{v}, D_1\mathfrak{v}, \ldots, D_{\ell_p}\mathfrak{v}$ is a flag base of $L_p(f)$ in respect to f if and only if $\mathfrak{w}, D_1\mathfrak{w}, \ldots, D_{\ell_p}\mathfrak{w}$ is a flag base of $L_p(f)$ in respect to f. If so, then

$$(6.12) \qquad \boxed{\mathfrak{w} \wedge D_1\mathfrak{w} \wedge \ldots \wedge D_{\ell_p}\mathfrak{w} = h^{\ell_p+1}\, \mathfrak{v} \wedge D_1\mathfrak{v} \wedge \ldots \wedge D_{\ell_p}\mathfrak{v} \not\equiv 0} \,.$$

<u>Proof.</u> If $p = 0$, the statement is trivial. Assume that the statement is true for $p - 1 \geq 0$. If $\ell_{p-1} = \ell_p$, the statement is trivially true for p. Hence consider the case $\ell_p > \ell_{p-1}$. Assume that $\mathfrak{v}, D_1\mathfrak{v}, \ldots, D_{\ell_p}\mathfrak{v}$ is a flag base of $L_p(f)$ in respect to f. Then $\mathfrak{v}, D_1\mathfrak{v}, \ldots, D_{\ell_{p-1}}\mathfrak{v}$ is a flag base of $L_{p-1}(f)$ in respect to f. By induction $\mathfrak{w}, D_1\mathfrak{w}, \ldots, D_{\ell_{p-1}}\mathfrak{w}$ is a flag base of $L_{p-1}(f)$ and we have only to prove that $\mathfrak{w}, D_1\mathfrak{w}, \ldots, D_{\ell_p}\mathfrak{w}$ is a base of $L_p(f)$.

Observe that $\mathfrak{v}, D_1\mathfrak{v}, \ldots, D_{\ell_{p-1}}\mathfrak{v}$ is a base of $L_{p-1}(f)$. Take $\lambda \in \mathbb{N}(\ell_{p-1}, \ell_p]$. Then $\mathfrak{v}, D_1\mathfrak{v}, \ldots, D_{\ell_{p-1}}\mathfrak{v}, D_\lambda\mathfrak{v}$ are linearly independent over $\mathfrak{M}(W)$. Therefore $D_\lambda\mathfrak{v} \in L_p(f) - L_{p-1}(f)$. In particular D_λ has order p. The chain rule implies

$$D_\lambda(h\mathfrak{v}) - hD_\lambda\mathfrak{v} \in L_{p-1}(f) \quad \text{and} \quad D_\lambda(h\mathfrak{v}) \in L_p(f).$$

Hence by induction

$$\mathfrak{w} \wedge D_1\mathfrak{w} \wedge \ldots \wedge D_{\ell_p}\mathfrak{w}$$

$$= h^{\ell_{p-1}+1}\mathfrak{v} \wedge D_1\mathfrak{v} \wedge \ldots \wedge D_{\ell_{p-1}}\mathfrak{v} \wedge D_{\ell_{p-1}+1}(h\mathfrak{v}) \wedge \ldots \wedge D_{\ell_p}(h\mathfrak{v})$$

$$= h^{\ell_{p-1}+1}\mathfrak{v} \wedge D_1\mathfrak{v} \wedge \ldots \wedge D_{\ell_{p-1}}\mathfrak{v} \wedge hD_{\ell_{p-1}+1}\mathfrak{v} \wedge \ldots \wedge hD_{\ell_p}\mathfrak{v}$$

$$= h^{\ell_p+1}\mathfrak{v} \wedge D_1\mathfrak{v} \wedge \ldots \wedge D_{\ell_p}\mathfrak{v} \not\equiv 0.$$

Hence $\mathfrak{w}, D_1\mathfrak{w}, \ldots, D_{\ell_p}\mathfrak{w}$ are linearly independent over $\mathfrak{M}(W)$ and belong to $L_p(f)$ with $\dim_{\mathfrak{M}(W)} L_p(f) = \ell_p + 1$. Therefore $\mathfrak{w}, D_1\mathfrak{w}, \ldots, D_{\ell_p}\mathfrak{w}$ is a base of $L_p(f)$. Consequently, $\mathfrak{w}, D_1\mathfrak{w}, \ldots, D_{\ell_p}\mathfrak{w}$ is a flag base of $L_p(f)$ in respect to f.

If $\mathfrak{w}, D_1\mathfrak{w}, \ldots, D_{\ell_p}\mathfrak{w}$ is a flag base of $L_p(f)$ in respect to f, then the proof that $\mathfrak{v}, D_1\mathfrak{v}, \ldots, D_{\ell_p}\mathfrak{v}$ is a flag base of $L_p(f)$ in respect to f proceeds symmetrically, q.e.d.

Theorem 6.4. (Vitter [85])
Assume that (A1)-(A3) hold. Take $p \in \mathbb{N}$. Let Λ_p be the pth associated map and abbreviate $\ell = \ell_p(f)$. Let $\mathfrak{v} : W \to V$ be a reduced representation of f. Let $\mathfrak{v}, D_1\mathfrak{v}, \ldots, D_\ell\mathfrak{v}$ be a flag base of $L_p(V)$. Let $\rho \geq 0$ be the divisor of $\mathfrak{v} \wedge D_1\mathfrak{v} \wedge \ldots \wedge D_\ell\mathfrak{v}$. Take $s > 0$. Then

$$N_\rho(r,s) + T_{\Lambda_p}(r,s) \mathrel{\dot{\leqq}} (\ell+1)T_f(r,s) + 20\ell npm^{p-1}\log(rT_f(r,s)) \tag{6.13}$$

Proof. Let $\mathfrak{e}_0, \ldots, \mathfrak{e}_n$ be an orthonormal base of V. Let $\varepsilon_0, \ldots, \varepsilon_n$ be the dual base. Then $\varepsilon_\mu \circ \mathfrak{v} \not\equiv 0$ for $\mu = 0, 1, \ldots, n$. Define $F_\mu = \dfrac{\varepsilon_\mu \circ \mathfrak{v}}{\varepsilon_0 \circ \mathfrak{v}}$ for $\mu = 1, \ldots, n$. Then

$$\mathfrak{w} = \frac{\mathfrak{v}}{\varepsilon_0 \circ \mathfrak{v}} = \mathfrak{e}_0 + \sum_{\mu=1}^{n} F_\mu \mathfrak{e}_\mu$$

is a meromorphic vector function. Define

$$\mathfrak{x} = \mathfrak{v} \wedge D_1\mathfrak{v} \wedge \ldots \wedge D_\ell \mathfrak{v} \qquad \mathfrak{y} = \mathfrak{w} \wedge D_1\mathfrak{w} \wedge \ldots \wedge D_\ell \mathfrak{w}.$$

Then $(\varepsilon_0 \circ \mathfrak{v})^{\ell+1}\mathfrak{y} = \mathfrak{x}$ by Lemma 6.3. Moreover ρ is the divisor of $\mathfrak{x}$ and $\mathfrak{x}$ is a representation of Λ_p. Now, (4.13) implies

$$(6.14) \qquad N_\rho(r,s) + T_{\Lambda_p}(r,s) = \int_{W\langle r\rangle} \log\|\mathfrak{x}\|\sigma - \int_{W\langle s\rangle} \log\|\mathfrak{x}\|\sigma$$

$$= (\ell + 1)T_f(r,s) + \int_{W\langle r\rangle} \log \frac{\|\mathfrak{x}\|}{\|\mathfrak{v}\|^{\ell+1}}\sigma - \int_{W\langle s\rangle} \log \frac{\|\mathfrak{x}\|}{\|\mathfrak{v}\|^{\ell+1}}\sigma$$

$$= (\ell + 1)T_f(r,s) + \int_{W\langle r\rangle} \log \frac{\|\mathfrak{y}\|}{\|\mathfrak{w}\|^{\ell+1}} - \int_{W\langle s\rangle} \log \frac{\|\mathfrak{y}\|}{\|\mathfrak{w}\|^{\ell+1}}\sigma.$$

The Schwarz inequality implies

$$\|\mathfrak{y}\| = \|\mathfrak{w} \wedge D_1\mathfrak{w} \wedge \ldots \wedge D_\ell\mathfrak{w}\| \le \|\mathfrak{w}\| \; \|D_1\mathfrak{w}\| \ldots \|D_\ell\mathfrak{w}\|.$$

Also

$$\|\mathfrak{w}\| = \sqrt{1 + |F_1|^2 + \ldots + |F_n|^2} \ge \|F_\mu\| \qquad \text{for } \mu = 1, \ldots, n.$$

Therefore

$$\log \frac{\|\mathfrak{y}\|}{\|\mathfrak{w}\|^{\ell}} \le \sum_{j=1}^{\ell} \log \frac{\|D_j\mathfrak{w}\|}{\|\mathfrak{w}\|} \le \sum_{j=1}^{\ell} \frac{1}{2} \log^+ \sum_{\mu=1}^{n} \frac{|D_jF_\mu|^2}{\|\mathfrak{w}\|^2}$$

$$\le \sum_{j=1}^{\ell} \sum_{\mu=1}^{n} \log^+ \frac{|D_jF_\mu|}{|F_\mu|} + \frac{\ell}{2} \log n.$$

Let p_j be the order of D_j. Then $1 \le p_j \le p$. Theorem 5.11 implies

$$\int_{W\langle r\rangle} \log^+ \frac{|D_j F_\mu|}{|F_\mu|}\,\sigma \leqq 19 p_j m^{p_j - 1} \log(rT_f(r,s))$$

$$\leqq 19 p m^{p-1} \log(rT_f(r,s)).$$

Therefore

$$(6.15)\qquad \int_{W\langle r\rangle} \log \frac{|\mathfrak{g}|}{\|\mathfrak{w}\|^{\ell+1}}\,\sigma \leqq 19 \ell n p m^{p-1} \log(rT_f(r,s)).$$

Now (6.14) and (6.15) imply (6.13); q.e.d.

Theorem 6.5. The Lemma of the Logarithmic Derivative for associated maps.
Assume that (A1)-(A3) hold. Let k be the span of f. Let $\mathfrak{v} : W \to V$ be a reduced representation of f. Let $\mathfrak{v}, D_1\mathfrak{v}, \ldots, D_n\mathfrak{v}$ be a flag base of $L_k(f) = \mathfrak{M}(W,V)$ in respect to f. Let $\mathfrak{e}_0, \ldots, \mathfrak{e}_n$ be an orthonormal base of V. Let $\alpha_0, \ldots, \alpha_n$ be a base of V^*. For each $j \in \mathbb{Z}[0,n]$ define the holomorphic function $g_j = \alpha_j \circ \mathfrak{v}$. Then $g_j \not\equiv 0$. Also define the holomorphic vector function $\mathfrak{g} = g_0\mathfrak{e}_0 + \ldots + g_n\mathfrak{e}_n \not\equiv 0$. A meromorphic map $g : W \to \mathbb{P}(V)$ is defined by $g = \mathbb{P} \circ \mathfrak{g}$. Then $L_k(g) = \mathfrak{M}(W,V)$ and $\mathfrak{g}, D_1\mathfrak{g}, \ldots, D_n\mathfrak{g}$ is a flag base of $L_k(g) = \mathfrak{M}(W,V)$ in respect to g. Moreover

$$(6.16)\qquad \boxed{\int_{W\langle r\rangle} \log^+ \frac{|\mathfrak{g} \wedge D_1\mathfrak{g} \wedge \ldots \wedge D_n\mathfrak{g}|}{|g_0 g_1 \cdots g_n|}\,\sigma \leqq 20 n! n^2 k m^{k-1} \log(rT_f(r,s))}\,.$$

Proof. A linear isomorphism $\alpha : V \to V$ is defined by

$$\alpha(\mathfrak{x}) = \sum_{j=0}^{n} \alpha_j(\mathfrak{x})\mathfrak{e}_j \qquad \forall\ \mathfrak{x} \in V.$$

This isomorphism extends to an $\mathfrak{M}(W)$-linear isomorphism

$$\alpha : \mathfrak{M}(W,V) \to \mathfrak{M}(W,V) \quad \text{by} \quad \alpha(\mathfrak{w}) = \alpha \circ \mathfrak{w} \qquad \forall\ \mathfrak{w} \in \mathfrak{M}(W,V).$$

Then

$$0 \not\equiv \alpha(\mathfrak{v}) = \sum_{j=0}^{n} (\alpha_j \circ \mathfrak{v})\mathfrak{e}_j = \sum_{j=0}^{n} g_j \mathfrak{e}_j = \mathfrak{g}.$$

Hence a meromorphic map $g : W \to \mathbb{P}(V)$ is defined by $g = \mathbb{P} \circ \mathfrak{g}$ and $\mathfrak{g}$ is a representation of g with

$$\dim \mathfrak{g}^{-1}(0) = \dim \mathfrak{v}^{-1}(\alpha^{-1}(0)) = \dim \mathfrak{v}^{-1}(0) \le m - 2.$$

Therefore $\mathfrak{g}$ is a reduced representation of g. If $D \in \mathfrak{D}_q$ is any differential operator of order q with constant coefficients, then

$$D\mathfrak{g} = D\alpha(\mathfrak{v}) = D\alpha \circ \mathfrak{v} = \alpha \circ D\mathfrak{v} = \alpha(D\mathfrak{v}).$$

Therefore $L_q(f) = L_q(g)$ for all $q \in \mathbb{Z}_+$ and $\mathfrak{g}, D_1\mathfrak{g}, \ldots, D_n\mathfrak{g}$ is a flag base of $L_k(g) = \mathfrak{M}(W,V)$ in respect to g.

For each $j \in \mathbb{N}[1,n]$ a meromorphic function $F_j = g_j/g_0 \not\equiv 0$ is defined. Then

$$\mathfrak{p} = \mathfrak{e}_0 + \sum_{j=1}^{n} F_j \mathfrak{e}_j \in \mathfrak{M}(W,V)$$

is a meromorphic vector function with $g_0\mathfrak{p} = \mathfrak{g}$. Now (6.12) implies

$$\mathfrak{g} \wedge D_1\mathfrak{g} \wedge \ldots \wedge D_n\mathfrak{g} = g_0^{n+1}\mathfrak{p} \wedge D_1\mathfrak{p} \wedge \ldots \wedge D_n\mathfrak{p}.$$

Define

$$\Delta = \frac{\|\mathfrak{g} \wedge D_1\mathfrak{g} \wedge D_2\mathfrak{g} \wedge \ldots \wedge D_n\mathfrak{g}\|}{|g_0||g_1| \ldots |g_n|}.$$

Then

$$\Delta = \frac{\|\mathfrak{p} \wedge D_1\mathfrak{p} \wedge \ldots \wedge D_n\mathfrak{p}\|}{|F_1| \ldots |F_n|}$$

$$= \left| \frac{\begin{vmatrix} 1 & F_1 & \ldots & F_n \\ 0 & D_1F_1 & \ldots & D_1F_n \\ \vdots & \vdots & & \vdots \\ 0 & D_nF_1 & \ldots & D_nF_n \end{vmatrix}}{F_1F_2 \ldots F_n} \right|$$

$$= \left| \det\left(\frac{D_jF_q}{F_q}\right) \right| .$$

Abbreviate $A_{jq} = \frac{D_jF_q}{F_q}$. Let $\mathfrak{S}_n$ be the group of permutations of $\mathbb{N}[1,n]$. Then

$$\log^+\Delta = \log^+|\det(A_{jq})| = \log^+\left|\sum_{\pi\in\mathfrak{S}_n} \operatorname{sign}\pi\, A_{1\pi(1)} \ldots A_{n\pi(n)}\right|$$

$$\leq \log^+ \sum_{\pi\in\mathfrak{S}_n} |A_{1\pi(1)}| \ldots |A_{n\pi(n)}| \leq \sum_{\pi\in\mathfrak{S}_n} \sum_{j=1}^{n} \log^+|A_{j\pi(j)}| + \log n!$$

$$\leq n! \sum_{j=1}^{n} \sum_{q=1}^{n} \log^+ \frac{|D_jF_q|}{|F_q|} + \log n!$$

Let p_j be the order of D_j. Then $1 \leq p_j \leq k$. Theorem 5.11 implies

$$\int_{W\langle r\rangle} \log\left|\frac{D_jF_q}{F_q}\right| \sigma \lesssim 19p_j m^{p_j-1} \log(rT_f(r,s))$$

$$\lesssim 19km^{k-1} \log(rT_f(r,s)).$$

Hence

$$\int_{W\langle r\rangle} \log^+\Delta\sigma \lesssim 19n!n^2km^{k-1} \log(rT_f(r,s)) + \log n!$$

$$\lesssim 20n^2n!km^{k-1} \log(rT_f(r,s)) \qquad \text{q.e.d.}$$

As in the classical theory Theorem 6.5 will be the key to the proof of the Second Main Theorem.

7. The Defect Relation.

Let V be a complex vector space of dimension $n + 1 > 1$. Take $a_1, \ldots, a_q$ in $\mathbb{P}(V^*)$. Take $\alpha_j \in V_*^*$ with $\mathbb{P}(\alpha_j) = a_j$ for $j = 1, \ldots, q$. Then $a_1, \ldots, a_q$ or also $E[a_1], \ldots, E[a_q]$ or $\ddot{E}[a_1], \ldots, \ddot{E}[a_q]$ are said to be in general position if and only if for any selection $1 \leq j_0 < j_1 < \ldots < j_p \leq q$ of integers with $p \leq n$, the vectors $\alpha_{j_0}, \alpha_{j_1}, \ldots, \alpha_{j_p}$ are linearly independent over $\mathbb{C}$. Clearly, the definition is independent of the choices of α_j with $\mathbb{P}(\alpha_j) = a_j$. Clearly, if $q \geq n + 1$, this condition is already satisfied if $\alpha_{j_0}, \alpha_{j_1}, \ldots, \alpha_{j_n}$ are linearly independent for any selection $1 \leq j_0 < j_1 < \ldots < j_n \leq q$ of integers, in which case $\alpha_{j_0}, \ldots, \alpha_{j_n}$ is a base of V^*.

We shall make the general assumptions (A1)-(A3), which we repeat here and some additional assumptions.

(A1) Let V be a hermitian vector space of dimension $n + 1 > 1$.

(A2) Let W be a hermitian vector space of dimension $m > 0$.

(A3) Let $f : W \to \mathbb{P}(V)$ be a linearly non-degenerate meromorphic map.

(A4) Take $a_1, \ldots, a_q$ in $\mathbb{P}(V^*)$ with $q \geq n + 1$ such that the hyperplanes $\ddot{E}[a_1], \ldots, \ddot{E}[a_q]$ are in general position.

(A5) Let k be the span of f.

(A6) Let $\mathfrak{v} : W \to V$ be a reduced representation of f.

(A7) Let $\mathfrak{v}, D_1\mathfrak{v}, \ldots, D_n\mathfrak{v}$ be a flag base $L_k(f) = \mathfrak{M}(W,V)$ in respect to f. Let $\rho \geq 0$ be the divisor of $\mathfrak{y} = \mathfrak{v} \wedge D_1\mathfrak{v} \wedge \ldots \wedge D_n\mathfrak{v} \not\equiv 0$.

Given (A1)-(A5), we can choose $\mathfrak{v}$ and $D_1, \ldots, D_n$ such that (A6) and (A7) hold. The divisor ρ does not depend on the choice of $\mathfrak{v}$ by

Lemma 6.3 but may depend on the choice of the differential operators $D_1, \ldots, D_n$.

<u>Theorem 7.1</u>. <u>The Second Main Theorem</u>. (Vitter [85]) Assume that (A1)-(A7) hold. Define

$$c = 21q!n!n^2km^{k-1}. \tag{7.1}$$

Then

$$\boxed{(q-n-1)T_f(r,s) + N_\rho(r,s) \leqq \sum_{j=1}^{q} N_f(r,s;a_j) + c\log(rT_f(r,s))} \,. \tag{7.2}$$

<u>Proof</u>. Let $\mathfrak{T}$ be the set of all injective maps $\mu : \mathbb{Z}[0,n] \to \mathbb{N}[1,q]$. Then $\#\mathfrak{T} = q(q-1)\ldots(q-n) > 0$. Let $\mathfrak{e}_0, \mathfrak{e}_1, \ldots, \mathfrak{e}_n$ be an orthonormal base of V. Let $\varepsilon_0, \varepsilon_1, \ldots, \varepsilon_n$ be the dual base. Then $v_j = \varepsilon_j \circ \mathfrak{v} \not\equiv 0$ is a holomorphic function on W with

$$\mathfrak{v} = \sum_{j=0}^{n} v_j \mathfrak{e}_j \qquad\qquad \|\mathfrak{v}\|^2 = \sum_{j=0}^{n} |v_j|^2.$$

Take $\alpha_p \in V_*^*$ such that $a_p = \mathbb{P}(\alpha_p)$ for $p = 1, \ldots, q$. Take $\mu \in \mathfrak{T}$. Since $a_1, \ldots, a_q$ are in general position, $(\alpha_{\mu(0)}, \ldots, \alpha_{\mu(n)})$ is a base of V^*. Therefore $b_{jp}^{(\mu)} \in \mathbb{C}$ exist such that

$$\varepsilon_j = \sum_{p=0}^{n} b_{jp}^{(\mu)} \alpha_{\mu(p)} \qquad\qquad \forall\ j \in \mathbb{Z}[0,n]$$

where

$$B_\mu = |\det(b_{jp}^{(\mu)})| > 0.$$

For each $p \in \mathbb{N}[1,q]$, define the holomorphic function $g_p = \alpha_p \circ \mathfrak{v} \not\equiv 0$. For each $\mu \in \mathfrak{T}$, a holomorphic vector function

$$\mathfrak{g}_\mu = \sum_{p=0}^{n} g_{\mu(p)}\mathfrak{e}_p \not\equiv 0$$

is defined. Then

$$v_j = \varepsilon_j \circ \mathfrak{v} = \sum_{p=0}^{n} b_{jp}^{(\mu)}\alpha_{\mu(p)} \circ \mathfrak{v} = \sum_{p=0}^{n} b_{jp}^{(\mu)} g_{\mu(p)}$$

for $j = 0, \ldots, n$ and for each $\mu \in \mathfrak{T}$. Also we have

$$D_h v_j = \sum_{p=0}^{n} b_{jp}^{(\mu)} D_h g_{\mu(p)} \qquad \text{for } h = 1, \ldots, n.$$

Therefore

$$\begin{aligned}\|\mathfrak{B}\| &= \|\mathfrak{v} \wedge D_1\mathfrak{v} \wedge \ldots \wedge D_n\mathfrak{v}\| = |\det(D_h v_j)|\\ &= |\det(b_{jp}^{(\mu)}) \cdot \det(D_h g_{\mu(p)})| = B_\mu \|\mathfrak{g}_\mu \wedge D_1\mathfrak{g}_\mu \wedge \ldots \wedge D_n\mathfrak{g}_\mu\|.\end{aligned}$$

Define

$$D_\mu = \frac{\|\mathfrak{g}_\mu \wedge D_1\mathfrak{g}_\mu \wedge \ldots \wedge D_n\mathfrak{g}_\mu\|}{|g_{\mu(0)}g_{\mu(1)} \cdots g_{\mu(n)}|}.$$

Then

$$\|\mathfrak{B}\| = B_\mu D_\mu |g_{\mu(0)}||g_{\mu(1)}| \cdots |g_{\mu(n)}|.$$

Define

$$b = \mathrm{Max}\{|b_{jp}^{(\mu)}| \mid \mu \in \mathfrak{T},\ j \in \mathbb{Z}[0,n],\ p \in \mathbb{Z}[0,n]\} > 0$$

$$B = \mathrm{Max}\{B_\mu \mid \mu \in \mathfrak{T}\} > 0$$

$$c_0 = (q - n - 1) \log^+(b(n + 1)^2)$$

$$\Delta = \sum_{\mu\in\mathfrak{T}} B_\mu D_\mu.$$

<u>Claim</u>:

$$\boxed{\log \Delta + \log|g_1 \ldots g_q| - \log\|\mathfrak{y}\| \geq (q - n - 1)\log\|\mathfrak{v}\| + c_0} \, .$$

<u>Proof of the Claim</u>. Let $\mathfrak{S}_q$ be the set of all permutations of $\mathbb{N}[1,q]$. For each $\pi \in \mathfrak{S}_q$ define

$$A_\pi = \{\mathfrak{z} \in W \mid |g_{\pi(1)}(\mathfrak{z})| \leq |g_{\pi(2)}(\mathfrak{z})| \leq \ldots \leq |g_{\pi(q)}(\mathfrak{z})|\}.$$

Then

$$W = \bigcup_{\pi \in \mathfrak{S}_q} A_\pi .$$

Take $\mathfrak{z} \in W$. Then $\pi \in \mathfrak{S}_q$ exists such that $\mathfrak{z} \in A_\pi$. Take any $j \in \mathbb{N}[n + 1, q]$. Define $\mu \in \mathfrak{T}$ by $\mu(x) = \pi(x + 1)$ for $x = 0, 1, \ldots, n-1$ and by $\mu(n) = \pi(j)$. Then $|g_{\mu(p)}(\mathfrak{z})| \leq |g_{\pi(j)}(\mathfrak{z})|$ for $p = 0, 1, \ldots, n$. Therefore

$$\begin{aligned}\|\mathfrak{v}(\mathfrak{z})\|^2 &= \sum_{h=0}^{n} |v_h(\mathfrak{z})|^2 = \sum_{h=0}^{n} \Big| \sum_{p=0}^{n} b_{hp}^{(\mu)} g_{\mu(p)}(\mathfrak{z})\Big|^2 \\ &\leq \sum_{h=0}^{n} \Big(\sum_{p=0}^{n} |b_{hp}^{(\mu)}|^2\Big)\Big(\sum_{p=0}^{n} |g_{\mu(p)}(\mathfrak{z})|^2\Big) \\ &\leq b^2(n + 1)^3 |g_{\pi(j)}(\mathfrak{z})|^2\end{aligned}$$

$$\log\|\mathfrak{v}(\mathfrak{z})\| \leq \log|g_{\pi(j)}(\mathfrak{z})| + \log^+(b(n + 1)^2) \qquad \text{for } j = n+1, \ldots, q.$$

Define $\nu \in \mathfrak{T}$ by $\nu(x) = \pi(x + 1)$ for $x = 0, 1, \ldots, n$. Then

$$\begin{aligned}\Delta(\mathfrak{z})|g_1(\mathfrak{z}) \ldots g_q(\mathfrak{z})| &\geq B_\nu D_\nu |g_1(\mathfrak{z}) \ldots g_q(\mathfrak{z})| \\ &= \|\mathfrak{y}\|\, |g_{\pi(n+2)}(\mathfrak{z})| \ldots |g_{\pi(q)}(\mathfrak{z})|\end{aligned}$$

and

$$\log \Delta(\mathfrak{x}) + \log|g_1(\mathfrak{x}) \ldots g_q(\mathfrak{x})| - \log\|\mathfrak{y}(\mathfrak{x})\|$$

$$\geq \sum_{j=n+2}^{q} \log|g_{\pi(j)}(\mathfrak{x})| \geq (q-n-1) \log\|\mathfrak{v}(\mathfrak{x})\| - (q-n-1) \log^+(b(n+1)^2)$$

which proves the claim.

Take $\mu \in \mathfrak{T}$. Then Theorem 6.5 applies to $\mathfrak{g}_\mu$. Therefore

$$\int_{W\langle r\rangle} \log^+ D_\mu \sigma \lessapprox 20n!n^2 km^{k-1} \log(rT_f(r,s)).$$

We obtain

$$\begin{aligned} \int_{W\langle r\rangle} \log \Delta\sigma &\leq \int_{W\langle r\rangle} \log^+\Big(\sum_{\mu\in\mathfrak{T}} B_\mu D_\mu\Big)\sigma \\ &\leq \int_{W\langle r\rangle} \log^+\Big(\sum_{\mu\in\mathfrak{T}} D_\mu\Big)\sigma + \log^+ B \\ &\leq \sum_{\mu\in\mathfrak{T}} \int_{W\langle r\rangle} \log^+ D_\mu \sigma + \log q! + \log^+ B \\ &\lessapprox 20q!n!n^2 km^{k-1} \log(rT_f(r,s)) + \log q! + \log^+ B. \end{aligned}$$

The Jensen formula yields

$$N_f(r,s;a_j) = \int_{W\langle r\rangle} \log|g_j|\sigma - \int_{W\langle s\rangle} \log|g_j|\sigma$$

$$N_\rho(r,s) = \int_{W\langle r\rangle} \log\|\mathfrak{y}\|\sigma - \int_{W\langle s\rangle} \log\|\mathfrak{y}\|\sigma$$

for $0 < s < r$.

If c_1 and c_2 denote appropriate constants, we obtain

$$\begin{aligned}
&N_\rho(r,s) + (q - n - 1)T_f(r,s) \\
&= \int_{W\langle r\rangle} \log\|\mathfrak{y}\|\sigma - \int_{W\langle s\rangle} \log\|\mathfrak{y}\|\sigma + (q-n-1)\Big(\int_{W\langle r\rangle} \log\|\mathfrak{u}\|\sigma - \int_{W\langle s\rangle} \log\|\mathfrak{u}\|\sigma\Big) \\
&\leq \int_{W\langle r\rangle} \log \Delta\sigma + \sum_{j=1}^{q} \int_{W\langle r\rangle} \log|g_j|\sigma + c_1 \\
&\underset{\cdot}{\leqq} \sum_{j=1}^{q} N_f(r,s;a_j) + 20q!n!n^2km^{k-1} \log(rT_f(r,s)) + c_2 \\
&\underset{\cdot}{\leqq} \sum_{j=1}^{q} N_f(r,s;a_j) + 21q!n!n^2km^{k-1} \log(rT_f(r,s)),
\end{aligned}$$

q.e.d.

If we only assume (A1)-(A5) we do not have to be concerned with ρ. Hence Theorem 7.1 implies

Theorem 7.2. The Second Main Theorem.
Assume that (A1)-(A5) hold. Define $c > 0$ by (7.1). Then

$$(7.3) \qquad \boxed{(q - n - 1)T_f(r,s) \underset{\cdot}{\leqq} \sum_{j=1}^{q} N_f(r,s;a_j) + c \log(rT_f(r,s))}\ .$$

Under the assumptions (A1)-(A7) define the ramification defect by

$$(7.4) \qquad \Theta = \liminf_{r\to\infty} \frac{N_\rho(r,s)}{T_f(r,s)} \geq 0.$$

Since f is not constant $\log T_f(r,s)/T_f(r,s) \to 0$ for $r \to \infty$. If f is not rational then $T_f(r,s)/\log r \to \infty$ for $r \to \infty$. The First and the Second Main Theorem imply

$$N_\rho(r,s) + \sum_{j=1}^{q} m_f(r,a_j)$$

$$\leqq (n+1)T_f(r,s) + \sum_{j=1}^{q} m_f(s,a_j) + c \log r + c \log T_f(r,s)$$

$$\leqq (n+1)T_f(r,s) + (c+1) \log r + c \log T_f(r,s).$$

Hence

$$\begin{aligned}
\Theta + \sum_{j=1}^{q} \delta_f(a_j) &= \liminf_{r\to\infty} \frac{N_\rho(r,s)}{T_f(r,s)} + \sum_{j=1}^{q} \liminf_{r\to\infty} \frac{m_f(r,a_j)}{T_f(r,s)} \\
&\le \liminf_{r\to\infty}\left(\frac{N_\rho(r,s)}{T_f(r,s)} + \sum_{j=1}^{q} \frac{m_f(r,a_j)}{T_f(r,s)}\right) \\
&\le \limsup_{r\to\infty}\left(n + 1 + (c+1)\frac{\log r}{T_f(r,s)} + c\,\frac{\log T_f(r,s)}{T_f(r,s)}\right) \\
&= n + 1
\end{aligned}$$

if f is transcendental.

Theorem 7.3. The Defect Relation.
Assume that (A1)-(A7) hold. Assume that f is transcendental. Then

(7.5) $$\boxed{\Theta + \sum_{j=1}^{q} \delta_f(a_j) \le n + 1}\,.$$

Theorem 7.4. The Defect Relation.
Assume that (A1)-(A5) hold. Assume that f is transcendental. Then

(7.6) $$\boxed{\sum_{j=1}^{q} \delta_f(a_j) \le n + 1}\,.$$

The proof of the Defect Relation given here follows essentially Cartan's proof in the case m = 1 [11], which uses the Lemma of the Logarithmic Derivative and made it possible to carry out Cartan's proof if m > 1 (Vitter [85]). Biancofiore and Stoll [3] gave an elementary proof of Vitter's Theorem.

Ahlfors [1] introduced associated curves in a different manner and derived the defect relation for holomorphic curves and also for the associated curves in a geometric fashion. Joachim and Hermann Weyl extended this method to Riemann surfaces [87]. In [65] this approach was used to prove the two Main Theorems and the defect relation for meromorphic maps where M is a connected, non-compact complex manifold of dimension $m > 1$. In this fashion Theorems 7.1 to 7.4 were proved the first time under minor additional assumptions ([65]).

Carlson and Griffiths [9] and Griffiths and King [25] proved a defect relation for meromorphic maps where N is a projective algebraic variety and where $\dim M \geq \dim N = \operatorname{rank} f$. They prove the defect relation for zero divisors of sections of ample holomorphic line bundles. If $M = W$ and $N = \mathbb{P}(V)$ another proof of the defect relation (7.5) is obtained if $\dim M \geq \dim N = \operatorname{rank} f$. Cowen and Griffiths [15] used a curvature method to obtain the defect relation for holomorphic curves $f : \mathbb{C} \to \mathbb{P}(V)$. Wang [88] extended this proof to meromorphic maps $f : M \to \mathbb{P}(V)$ where M is a parabolic manifold of dimension $m \geq 1$. Shiffman [58] provides yet another proof of the defect relation for holomorphic curves.

Some consequences of Theorems 7.1-7.4 follow easily.

<u>Application 1</u>. (Borel [7])
Assume that (A1)-(A4) hold with $q = n + 2$. Assume that f is not rational. Then $j \in \mathbb{N}[1, n + 2]$ exists such that $f(W) \cap \ddot{E}[a_j] \neq \emptyset$.

<u>Proof</u>. If $f(W) \cap \ddot{E}[a_j] = \emptyset$ for all $j \in \mathbb{N}[1, n + 2]$, then $\delta(a_j) = 1$ for $j = 1, \ldots, n+2$. Hence (7.6) implies $n + 2 \leq n + 1$. Contradiction! q.e.d.

<u>Application 2</u>.
Assume that (A1)-(A4) hold with $q = n + 2$. Assume that $\mu_f^{a_j}$ is affine algebraic for $j = 1, \ldots, n+2$. Then f is rational.

<u>Proof</u>. Since f is not constant, $T_f(r,s) \to \infty$ for $r \to \infty$. Hence $r_0 > s$ exists such that $\log T_f(r,s) \leq \frac{1}{2} T_f(r,s)$ for all $r \geq r_0$. Now (7.3) implies

$$T_f(r,s) \leqq 2 \sum_{j=1}^{n+2} N_f(r,s;a_j) + 2c \log r$$

or

$$A_f(\infty) = \lim_{r\to\infty} \frac{T_f(r,s)}{\log r} \le 2 \sum_{j=1}^{n+1} \limsup_{r\to\infty} \frac{N_f(r,s;a_j)}{\log r} + 2c$$
$$= 2 \sum_{j=1}^{n+1} n_f(\infty;a_j) + 2c < \infty.$$

Therefore f is rational; q.e.d.

The <u>order</u> λ and the <u>lower order</u> μ of a meromorphic map $f : W \to \mathbb{P}(V)$ are defined by

(7.7) $$\lambda = \limsup_{r\to\infty} \frac{\log T_f(r,s)}{\log r} \qquad \mu = \liminf_{r\to\infty} \frac{\log T_f(r,s)}{\log r} .$$

In fact, there exists an extensive theory of maps and functions of finite order. (See Stoll [75] and Ronkin [52].)

<u>Application 3</u>.
Assume that (A1)-(A7) hold with $q = n + 1$. Assume that f has finite order and that $\mu_j^{a_j}$ is affine algebraic for $j = 1, \dots, n+1$. Then ρ is affine algebraic.

<u>Proof</u>. (7.2) implies

$$n_\rho(\infty) = \lim_{r\to\infty} \frac{N_\rho(r,s)}{\log r} \le \sum_{j=1}^{n+1} \limsup_{r\to\infty} \frac{N_f(r,s;a_j)}{\log r} + c + c\lambda$$
$$= \sum_{j=1}^{n+1} n_f(\infty,a_j) + c + c\lambda < \infty.$$

Therefore ρ is affine algebraic; q.e.d.

Clearly, the defect bound in (7.6) is the best possible. For instance let P_j be a polynomial of degree p_j for $j = 1, \dots, n+1$ with $0 = p_1 < \dots < p_{n+1}$. Let $\mathfrak{a}_1, \dots, \mathfrak{a}_{n+1}$ be a base of V and let

$\alpha_1, \ldots, \alpha_{n+1}$ be the dual base. Define $a_j = \mathbb{P}(\mathfrak{a}_j)$ for $j = n+2, \ldots, q$ such that $a_1, \ldots, a_q$ are in general position. Define $\mathfrak{v} = \sum_{j=1}^{n+1} e^{P_j} \mathfrak{a}_j$ and $f = \mathbb{P} \circ \mathfrak{v}$. Then $\delta_f(a_j) = 1$ for $j = 1, \ldots, n+1$ while (A1)-(A5) are satisfied.

Some of the few results on defects sums in several variables shall be mentioned here.

Theorem 7.5. (Noguchi [44])
Assume that (A1)-(A5) hold with $q = n + 1$. Assume that f has finite order λ, which is not an integer. Then there exists a positive constant $k(\lambda)$ with $0 < k(\lambda) \leq 1$ such that

$$\sum_{j=0}^{n} \delta_f(a_j) \leq 1 - k(\lambda) \tag{7.8}$$

where

$$k(\lambda) \geq \frac{2\Gamma\left(\frac{3}{4}\right)^4 |\sin(\pi\lambda)|}{\pi^2\lambda + \Gamma\left(\frac{3}{4}\right)^4 |\sin \pi\lambda|} \qquad \text{if } \lambda > 0 \tag{7.9}$$

$$k(\lambda) \geq 1 - \lambda \qquad \text{if } 0 \leq \lambda \leq 1. \tag{7.10}$$

In particular if f has order zero, then $\delta_f(a_j) = 1$ for $j = 1, \ldots, q+1$. Mori [39], [41] improves these results and extends them to meromorphic maps.

Theorem 7.6. (Fujimoto [16])
Assume that (A1)-(A4) hold with $q \geq 3n + 2$. Let $g : W \to \mathbb{P}(V)$ be a non constant meromorphic map. Assume that $\mu_f^{a_j} = \mu_g^{a_j}$ for all $j = 1, \ldots, q$. Then $f = g$.

Smiley [62] proves a number of similar results concerning the linear and algebraic dependence of meromorphic maps.

Theorem 7.7. (Sung [81])
Let $f : \mathbb{C} \to \mathbb{P}(V)$ be a linearly non-degenerate holomorphic map of finite lower order. Let $\{a_j\}_{j \in \mathbb{N}}$ be a sequence of points in $\mathbb{P}(V^*)$ such that $\ddot{E}[a_1], \ldots, \ddot{E}[a_q]$ are hyperplanes in general position for each $q \in \mathbb{N}$. Then

$$(7.11) \qquad \boxed{\sum_{j=1}^{\infty} \delta_f(a_j)^{\frac{1}{3}} < \infty} \; .$$

This extends a famous result of Weitsman [86] to holomorphic curves.

8. Nevanlinna Theory on parabolic manifolds.

The previous theory has been extended considerably. Levine [37] and Chern [12] were able to establish a First Main Theorem for projective planes of fixed dimension p. Later First Main Theorems were proved for holomorphic maps $f : M \to N$ between complex manifolds M and N, where N is compact and M is not compact. On N, a family $\mathfrak{A}$ of analytic sets is given. For a survey see [76]. Special types of families $\mathfrak{A}$ were considered with success. For instance Bott and Chern [8] studied the equidistribution of zeroes of sections of holomorphic vector bundles. Cowen [14] and Stoll [78] extended this theory to families of Schubert varieties defined by vector bundles. Here only a First Main Theorem was obtained and a Casorati-Weierstrass type theorem proved. However in the case of line bundles, Carlson and Griffiths [10] and Griffiths and King [25] achieved a major breakthrough and proved a powerful defect relation. In [77], this theory was formalized and extended to parabolic manifolds M. At present the theory is restricted to $\dim M \geq \operatorname{rank} f = \dim N$. Since this theory gives an excellent example of the methods by which value distribution has been extended to general holomorphic maps $f : M \to N$, an outline shall be given here.

Let N and E be complex manifolds. Let $\pi : E \to N$ be a surjective holomorphic map. For $x \in N$, let $E_x = \pi^{-1}(x)$ be the fiber of π over x. If $U \subseteq N$, denote $\pi^{-1}(U) = E|U$. A holomorphic map $s : U \to E$ is said to be a <u>holomorphic section of</u> E <u>over</u> U iff $s(x) \in E_x$ for all $x \in U$. Let $\Gamma(U,E)$ be the set of all holomorphic sections of E over U. Now $\pi : E \to N$ (or E alone) is said to be a <u>holomorphic vector bundle of</u> rank r <u>over</u> N, iff E_x is a complex vector space of dimension r for all $x \in N$ and if for each $a \in N$ there exists an open neighborhood U of a and holomorphic sections $s_j \in \Gamma(U,E)$ for $j = 1, \ldots, r$ such that $s_1(x), \ldots, s_r(x)$ is a base of E_x over $\mathbb{C}$ for each $x \in U$. Then $s_1, \ldots, s_r$ is called a <u>holomorphic frame</u> of E over U. If $U \neq \emptyset$ is open in N, then $\Gamma(U,E)$ is a complex vector space; if $s \in \Gamma(U,E)$, then the <u>zero set</u> $Z(s) = \{x \in N \mid s(x) = 0_x \in E\}$ is analytic in U. If N is compact, then $\Gamma(N,E)$ has finite dimension.

If $E_1, \ldots, E_p$ are holomorphic vector bundles on N, then there exist — uniquely up to natural isomorphisms — holomorphic vector bundles

$$(8.1) \qquad \bigoplus_{j=1}^{p} E_j \qquad \bigotimes_{j=1}^{p} E_j \qquad \bigwedge_p E \qquad E^*$$

such that fibers over each $x \in N$ are respectively given by

$$(8.2) \qquad \bigoplus_{j=1}^{p} E_{jx} \qquad \bigotimes_{j=1}^{p} E_{jx} \qquad \bigwedge_p E_x \qquad (E_x)^*.$$

The <u>holomorphic tangent bundle</u> $\mathfrak{T}(N)$ of N is a holomorphic vector bundle of rank n = dim N over N such that $\Gamma(U,\mathfrak{T}(N))$ is the vector space of <u>holomorphic vector fields</u> on N. The dual vector bundle $\mathfrak{T}(N)^*$ is called the <u>holomorphic cotangent bundle</u> and $\Gamma(U,\mathfrak{T}(N)^*)$ is the vector space of <u>holomorphic forms of bidegree (1,0) on</u> N. If n = dim N, then $K(N) = \bigwedge_n \mathfrak{T}(N)^*$ is called the <u>canonical bundle</u> of N.

If V is a complex vector space of dimension r, then N × V is a holomorphic vector bundle of rank r over N, where $\pi : N \times V \to V$ is the projection and where we identify $(N \times V)_x = \{x\} \times V = V$. The vector bundle N × V is called <u>trivial</u>.

A map $\kappa : E \oplus E \to \mathbb{C}$ of class C^∞ is said to be a <u>hermitian metric</u> along the fibers of E, if for each $x \in N$ the restriction $\kappa_x = \kappa : E_x \oplus E_x \to \mathbb{C}$ defines a hermitian inner product on E_x. The norm $\| \ \|_\kappa$ is defined by

$$(8.3) \qquad \|u\|_\kappa = \sqrt{\kappa(u,u)} \qquad u \in E_x \quad x \in N$$

and defines a norm on each E_x. Here κ induces hermitian metrics κ^* along the fibers of E* and $\bigwedge_p \kappa$ along the fibers of $\bigwedge_p E$ which are often again denoted by κ. If κ_j is a hermitian metric along the fibers of E_j for j = 1, ..., p, they induce hermitian metrics $\bigoplus_{j=1}^{p} \kappa_j$ along the fibers of $\bigoplus_{j=1}^{p} E_j$ and $\bigotimes_{j=1}^{p} \kappa_j$ along the fibers of $\bigotimes_{j=1}^{p} E_j$. A holomorphic vector bundle E together with a given hermitian metric κ along the fibers of E is also called a <u>hermitian vector bundle</u>.

A holomorphic vector bundle of rank 1 is called a <u>holomorphic line bundle</u>. Let L be a holomorphic line bundle over N. If $U \neq \emptyset$ is open in N, then $s \in \Gamma(U,L)$ is a holomorphic frame of L over U iff $Z(s) = \emptyset$. If $U \neq \emptyset$ is open and connected in N, if $0 \not\equiv s \in \Gamma(U,L)$, then s defines a divisor $\mu_s \geq 0$ with supp $\mu_s = Z(s)$ in the following manner. Take $a \in U$, then there is an open, connected neighborhood V of a and a holomorphic frame e of L over V and a holomorphic function $s_e \not\equiv 0$ on V such that $s|V = s_e e$. Then $\mu_s|V = \mu^0_{s_e}$.

Let κ be a hermitian metric along the fibers of the holomorphic line bundle L on N. Then there exists one and only one form $c(L,\kappa)$ of bidegree (1,1) on N, such that for each open subset $U \neq \emptyset$ of N and each holomorphic frame s of L on U we have

(8.4) $$c(L,\kappa) = -dd^c \log\|s\|^2_\kappa.$$

The form $c(L,\kappa)$ is called the <u>Chern form of L for</u> κ.

The line bundle is said to be <u>non-negative</u> (respectively <u>positive</u>) if κ exists such that $c(L,\kappa) \geq 0$ (respectively $c(L,\kappa) > 0$).

Let (L,κ) be a hermitian line bundle on the compact, connected complex manifold N of dimension n. Then $\Gamma(N,L)$ is a complex vector space of finite dimension $k + 1$. <u>Assume that</u> $k > 0$. Let V be the dual vector space of $\Gamma(N,L)$. Then $V^* = \Gamma(N,L)$. Take $p \in \mathbb{Z}[0,k]$. Take $a \in G_p(V^*)$. Then $a = \mathbb{P}(\mathfrak{a}_0 \wedge \ldots \wedge \mathfrak{a}_p)$ where $\mathfrak{a}_j \in \Gamma(N,L)$. Then

$$E_L[a] = Z(\mathfrak{a}_0) \cap \ldots \cap Z(\mathfrak{a}_p)$$

depends only on a and not on the choices of $\mathfrak{a}_0, \ldots, \mathfrak{a}_p$. For $j \in \mathbb{Z}[0,p]$ define

(8.5) $$\mathfrak{a}^j = \mathfrak{a}_0 \wedge \ldots \wedge \mathfrak{a}_{j-1} \wedge \mathfrak{a}_{j+1} \wedge \ldots \wedge \mathfrak{a}_p .$$

The evaluation map

(8.6) $$\varepsilon : N \times \bigwedge_{p+1} V^* \to L \otimes (N \times \bigwedge_p V^*)$$

is a bundle map defined by

(8.7) $$\varepsilon(x, \mathfrak{a}_0 \wedge \ldots \wedge \mathfrak{a}_p) = \sum_{j=0}^{p} (-1)^j \mathfrak{a}_j(x) \otimes (x, \mathfrak{a}^j).$$

Recall that $\mathfrak{a}_j \in \Gamma(N,L)$, hence $\mathfrak{a}_j(x) \in L_x$. If $p = 0$, then

(8.8) $$\varepsilon(x,\mathfrak{a}) = \mathfrak{a}(x) \qquad \forall\, \mathfrak{a} \in \Gamma(N,L).$$

If $\mathfrak{a} \in \bigwedge_{p+1} V$, a section $\varepsilon_{\mathfrak{a}} \in \Gamma(N, L \otimes \bigwedge_p V^*)$ is defined by $\varepsilon_{\mathfrak{a}}(x) = \varepsilon(x,\mathfrak{a})$. If $p = 0$, then $\varepsilon_{\mathfrak{a}} = \mathfrak{a} \in \Gamma(N,L)$.

Let ℓ be a hermitian inner product on V^*. It induces a hermitian metric along the fibers of $N \times \bigwedge_p V$ again denoted by ℓ. Hence $\kappa \otimes \ell$ is a hermitian metric along the fibers of $L \otimes (N \times \bigwedge_p V^*)$. Take $a \in G_p(V^*)$ and $\mathfrak{a} \in \tilde{G}_p(V^*)$ with $\mathbb{P}(\mathfrak{a}) = a$. Take $x \in N$. Define

(8.9) $$\boxed{0 \le \|a,x\|_\kappa = \frac{\|\varepsilon_{\mathfrak{a}}(x)\|_{\kappa\otimes\ell}}{\|\mathfrak{a}\|_\kappa}} .$$

We say that (κ,ℓ) is <u>distinguished</u> if $\|a,x\|_\kappa \le 1$ for all $x \in N$, all $a \in G_p(V^*)$ all $p \in \mathbb{Z}[0,k]$.

Observe that $G_k(V^*)$ consists of one and only one element which we shall denote by ∞. Then $E_L[\infty]$ is defined with

(8.10) $$E_L[\infty] = \bigcap_{s\in\Gamma(N,L)} Z(s).$$

Then $N_\infty = N - E_L[\infty]$ is open, connected and dense in N. The map $\varepsilon : N_\infty \times V^* \to L|N_\infty$ is surjective. Take $x \in N$. Then $\varepsilon_x : V^* \to L_x$ is a surjective linear map. Let S_x be the kernel and define

$$S_x^\perp = \{\mathfrak{a} \in V^* \mid \ell(\mathfrak{a},\mathfrak{b}) = 0 \ \forall\ \mathfrak{b} \in S_x\}.$$

Then $V^* = S_x \oplus S_x^\perp$ and $\varepsilon_x : S_x^\perp \to L_x$ is an isomorphism. Now the hermitian metric ℓ on $N \times V^*$ restricts to $S_x^\perp$ and carries over to L_x. This defines a hermitian metric ℓ along the fibers of $L|N_\infty$. Hence $\|a,x\|_\ell$ is defined with $0 \leq \|a,x\|_\ell \leq 1$. Hence (ℓ,ℓ) is distinguished on $L|N_\infty$. Moreover we have

$$\|a,x\|_\kappa = \|a,x\|_\ell \, \|\infty,x\|_\kappa. \tag{8.11}$$

Since N is compact, we can multiply ℓ (or κ) by a positive constant such that $\|\infty,x\|_\kappa \leq 1$. Then $\|a,x\|_\kappa \leq 1$ for all a and all x. Consequently, <u>we can assume without loss of generality that</u> (κ,ℓ) <u>is distinguished</u>.

Take $p \in \mathbb{Z}[0,k]$ and $a \in G_p(V^*)$. Then the form

$$\Phi_L[a] = c(L,\kappa) + dd^c \log\|a,\square\|_\kappa^2 \tag{8.12}$$

does not depend on κ and is non-negative with $\Phi_L[a]^{p+1} \equiv 0$. The <u>Chern-Levine</u> form of L for $a \in G_p(V^*)$ is defined by

$$\boxed{\Lambda_L[a]_\kappa = \log \frac{1}{\|a,\square\|_\kappa^2} \sum_{j=0}^{p} \Phi_L[a]^j \wedge c(L,\kappa)^{p-j}} \tag{8.13}$$

where $\Phi_L[a]^{p+1} \equiv 0$ implies

$$dd^c \Lambda_L[a] = c(L,\kappa)^{p+1} \qquad \text{on } N - E_L[a]. \tag{8.14}$$

If $p = 0$, observe that $\Phi_L[a] \equiv 0$ implies

$$(8.15)\qquad \boxed{c(L,\kappa) = dd^c \log \frac{1}{\|a,\square\|_\kappa^2}} \qquad \text{on } N - E_L[a]$$

for all $a \in \mathbb{P}(V^*)$ and that

$$(8.16)\qquad \boxed{\Lambda_L[a]_\kappa = \log \frac{1}{\|a,\square\|_\kappa^2}}\ .$$

Now the analogon to the balls in a vector space has to be introduced on the domain space.

Let M be a connected, non-compact, complex manifold of dimension m. Let τ be a non-negative function of class C^∞ on M. If $A \subseteq M$ and $r \geq 0$ define

$$(8.17)\qquad A[r] = \{x \in M \mid \tau(x) \leq r^2\} \qquad A(r) = \{x \in M \mid \tau(x) < r^2\}$$

$$(8.18)\qquad A\langle r\rangle = \{x \in M \mid \tau(x) = r^2\} \qquad M_* = M - M[0]$$

$$(8.19)\qquad \upsilon = dd^c\tau \qquad \omega = dd^c \log \tau$$

$$(8.20)\qquad \sigma_p = d^c \log \tau \wedge \omega^p \qquad \sigma = \sigma_{m-1}.$$

Then τ is said to be a <u>parabolic</u> <u>exhaustion</u> of M, and (M,τ) a <u>parabolic</u> <u>manifold</u> if M[r] is compact for all $r \geq 0$, if $\omega \geq 0$ and if $\upsilon^m \not\equiv 0 \equiv \omega^m$. Then $\upsilon \geq 0$ and $M[0] \neq \emptyset$. Moreover a positive constant $\varsigma > 0$ exists such that for $r > 0$

$$(8.21)\qquad \varsigma = \int_{M\langle r\rangle} \sigma \qquad \int_{M[r]} \upsilon^m = \varsigma r^{2m} = \int_{M(r)} \upsilon^m.$$

If W is a hermitian vector space with $\tau_W(\mathfrak{z}) = \|\mathfrak{z}\|^2$, then (W,τ_W) is a parabolic manifold with $\upsilon > 0$. Moreover, the following theorem holds.

Theorem 8.1. (Stoll [79])
If (M,τ) is a parabolic manifold of dimension m with $dd^c\tau > 0$ on M, then there exists a biholomorphic map $h : W \to M$ such that $\tau \circ h = \tau_W$.

If M is a connected complex manifold of dimension m, if $\pi : M \to W$ is a proper, surjective holomorphic map, then $(M, \tau_W \circ \pi)$ is a parabolic manifold, called a cover of (W,τ_W).

A connected affine algebraic manifold is a cover parabolic manifold of W. If (M_j,τ_j) are parabolic for j = 1, 2, if $\pi_j : M_1 \times M_2 \to M_j$ is the projection, then $(M_1 \times M_2, \tau_1 \circ \pi_1 + \tau_2 \circ \pi_2)$ is a parabolic manifold. A non-compact Riemann surface is parabolic, if and only if each subharmonic function bounded above is constant. Also $(\mathbb{C} - \mathbb{Z}) \times \mathbb{C}^{m-1}$ is parabolic but not affine algebraic.

Let (M,τ) be a parabolic manifold of dimension m. Let A be a pure q-dimensional analytic subset of M. Let $\nu : A \to \mathbb{Z}$ be an integral valued function which is locally constant on $\mathbb{R}(A)$. For $t > 0$ the counting function n_ν is defined by

$$(8.22) \qquad n_\nu(t) = \frac{1}{t^{2q}} \int_{A[t]} \nu\upsilon^q.$$

If $\nu \geq 0$, then n_ν increases. In general $n_\nu(t) \to n_\nu(0)$ for $t \to 0$ with

$$(8.23) \qquad n_\nu(t) = \int_{A[t]-A[0]} \nu\omega^q + n_\nu(0).$$

For $0 < s < r$, the valence function of ν is defined by

$$(8.24) \qquad N_\nu(r,s) = \int_s^r n_\nu(t)\,\frac{dt}{t}\,.$$

If ν is a divisor, then $q = m - 1$. If $\nu = \mu_f$ is the divisor of a meromorphic function $f \not\equiv 0$, the Jensen formula holds

(8.25) $$N_{\mu_f}(r,s) = \int_{M\langle r\rangle} \log|f|\sigma - \int_{M\langle s\rangle} \log|f|\sigma \qquad (0 < s < r).$$

If f is bounded, we see that f is constant. If f is not constant, $f(M)$ is dense in $\mathbb{P}_1$. If M is simply connected, and omits three points of $\mathbb{P}_1$, then f is constant.

Let $f : M \to N$ be a holomorphic map. Take $p \in \mathbb{Z}[0,m]$ and define $q = m - p \geq 0$. For $t > 0$ the <u>spherical image of f of order</u> p for (L,κ) is defined by

(8.26) $$\boxed{A_f^p(t) = \frac{1}{t^{2q}} \int_{M[t]} f^*(c(L,\kappa)^p) \wedge \upsilon^q}\ .$$

If $c(L,\kappa) \geq 0$, then $A_f^p(t) \geq 0$ increases and the limits

(8.27) $$A_f^p(0) = \lim_{0<t\to 0} A_f^p(t) \qquad 0 \leq A_f^p(\infty) = \lim_{0<t\to 0} A_f^p(t) \leq \infty$$

exist and we have

(8.28) $$\boxed{A_f^p(t) = \int_{M[t]-M[0]} f^*(c(L,\kappa)^p) \wedge \omega^q + A_f^p(0)}\ .$$

If $p = 0$, then $A_f^0(t) = c > 0$ is constant. If $p = 1$, abbreviate $A_f^1 = A_f$.

For $0 < s < r$, the <u>characteristic function of f of order</u> p for (L,κ) is defined by

(8.29) $$\boxed{T_f^p(r,s) = \int_s^r A_f^p(t)\,\frac{dt}{t}}\ .$$

If $c(L,\kappa) \geq 0$, then $T_f^p(r,s) \geq 0$ increases and $A_f^p(\infty) = \lim_{r\to\infty} \dfrac{T_f^p(r,s)}{\log r}$.

If $c(L,\kappa) > 0$ and if f is not constant, then $A_f^p(\infty) > 0$ and $T_f^p(r,s) \to \infty$

for $r \to \infty$. If $p = 0$, then $T_f^0(r,s) = c_0 \log \frac{r}{s}$; if $p = 1$, abbreviate $T_f^1 = T_f$.

Take $p \in \mathbb{N}[1,m]$. Take $a \in G_{p-1}(V^*)$. Then f is said to be L-*non-degenerate* for a iff $f^{-1}(E_L[a])$ is empty or pure q-dimensional. If $p = 1$, this is the case iff $f(M) \nsubseteq E_L[a]$. We say that f is L-*non-degenerate of order* $p - 1$, iff f is L-non-degenerate for all $a \in G_{p-1}(V^*)$.

Take $a \in G_{p-1}(V^*)$. Assume that f is L-non-degenerate for a. A multiplicity function $0 \le \theta_f^a[L] : f^{-1}(E_L[a]) \to \mathbb{N}$ can be assigned which is locally constant on $\mathfrak{R}(f^{-1}(E_L[a]))$. The *counting function* and the *valence function* of f for a in respect to L are defined for $t > 0$ and $0 < s < r$ by

(8.30) $$n_f(t;a) = n_f(t,a,L) = n_{\theta_f^a[L]}(t) \ge 0$$

(8.31) $$N_f(r,s;a) = N_f(r,s;a,L) = N_{\theta_f^a[L]}(r,s).$$

For $r > 0$, the *compensation function* of f for a in respect to L is defined by

(8.32) $$m_f(r,a) = m_f(r,a,L,\kappa) = \frac{1}{2} \int_{M\langle r\rangle} f^*(\Lambda_L[a]_\kappa) \wedge \sigma_q.$$

Actually, the integral exists only for almost all $r > 0$, but m_f can be extended to $\mathbb{R}^+$ as a function which is semi-continuous from the right. If $c(L,\kappa) \ge 0$ and if (κ,ℓ) is distinguished, then $m_f(r,a) \ge 0$.

For $0 < s < r$, the *deficit of* f *for* a *in respect to* L is defined by

(8.33) $$D_f(r,s;a) = D_f(r,s;a,L,\kappa) = \frac{1}{2} \int_{M[r]-M[s]} f^*(\Lambda_L[a]_\kappa) \wedge \omega^{q+1}.$$

If $c(L,\kappa) \geq 0$ and if (κ,ℓ) is distinguished, then $D_f(r,s;a) \geq 0$ increases in r.

The <u>First Main Theorem</u> states: If $0 < s < r$ and $a \in G_{p-1}(V^*)$, then

(8.34) $$\boxed{T_f^p(r,s) = N_f(r,s;a) + m_f(r,a) - m_f(s,a) - D_f(r,s;a)}\ .$$

Chern [12] first introduced the deficit term into the First Main Theorem. <u>If</u> $p = 1$, then $q + 1 = m$ and $\omega^m = 0$. Therefore

(8.35) $$D_f(r,s;a) \equiv 0 \qquad \forall\ a \in \mathbb{P}(V^*).$$

Also we have

(8.36) $$m_f(r,a) = \int_{M\langle r\rangle} \log \frac{1}{\|f,a\|_\kappa}\, \sigma \geq 0$$

if (κ,ℓ) is distinguished. Also $\alpha \in V_*^*$ exists with $\mathbb{P}(\alpha) = a$. Then $\theta_f^a[L] = f^*(\mu_\alpha)$ where μ_α is the zero divisor of the section α. Hence for $p = 1$, the <u>First Main Theorem</u> states

(8.37) $$\boxed{T_f(r,s) = N_f(r,s;a) + m_f(r,a) - m_f(s,a)}$$

if $0 < s < r$ and $a \in \mathbb{P}(V^*)$ such that $f(M) \nsubseteq E_L[a] = Z(\alpha)$.

Let us assume that $c(L,\kappa) \geq 0$, that (κ,ℓ) is distinguished and that $T_f^p(r,s) \to \infty$ for $r \to \infty$. Take $a \in G_{p-1}(V^*)$ and assume that f is L-non-degenerate for a. Then

(8.38) $$0 \leq \delta_f(a,L) = \liminf_{r\to\infty} \frac{m_f(r,a)}{T_f(r,s)}$$

is defined and called the <u>Nevanlinna defect of f for a in respect to L</u>. However the appearance of the deficit in (8.34) makes it impossible to conclude that $\delta_f(a,L) \le 1$. Only if $p = 1$, (8.37) implies

$$\delta_f(a,L) = 1 - \limsup_{r\to\infty} \frac{N_f(r,s;a)}{T_f(r,s)} \le 1 \qquad (p = 1). \tag{8.39}$$

Therefore, except for $p = 1$ our earlier proof of the Casorati-Weierstrass theorem cannot be repeated. Nevertheless such a theorem can be obtained.

We need T_f to be the integral average of N_f. However, N_f does not depend on κ while T_f does. Therefore we cannot expect T_f to be the integral average of N_f. However, here we are helped out by the following observation. Recall that the hermitian inner product ℓ on V induces a hermitian metric along the fibers of $L|N_\infty$ which defines $c(L|N_\infty,\ell) \ge 0$. Abbreviate $c(L,\ell) = c(L|N_\infty,\ell)$. It can be shown that if we replace $c(L,\kappa)$ by $c(L,\ell)$ the integrals A_f^p, T_f^p, A_f, T_f, m_f and D_f still exist and that the First Main Theorem remains valid although $c(L,\ell)$ is singular on $E_L[\infty]$. For simplicity we will denote the function formed with $c(L,\ell)$ by $A_f^p[t]$, $T_f^p[r,s]$, $A_f[t]$, $T_f[r,s]$, $m_f[r,a]$ and $D_f[r,s;a]$.

Let Ω_p be the Fubini-Study form induced by ℓ on $\mathbb{P}(\bigwedge_{p+1} V^*)$. Recall that $\dim V^* = k + 1$. Hence $G_p(V^*)$ has dimension

$$d(p,k) = \dim G_p(V^*) = (p + 1)(k - p) \tag{8.40}$$

and the degree of the projective algebraic variety $G_p(V)$ in $\mathbb{P}(\bigwedge_{p+1} V^*)$ is

$$D(p,k) = \deg G_p(V^*) = d(p,k)! \prod_{\mu=0}^{p} \frac{\mu!}{(k - \mu)!} \,. \tag{8.41}$$

Let $h : G_p(V^*) \to \mathbb{C}$ be a function such that $h\Omega_p^{d(p,k)}$ is integrable over $G_p(V^*)$. Then the <u>integral</u> <u>average</u> of h over $G_p(V^*)$ is defined by

(8.42)
$$I_p(h) = \frac{1}{D(p,k)} \int_{G_p(V^*)} h\Omega_p^{d(p,k)}.$$

Define

(8.43)
$$c(p,k) = \frac{1}{2} \sum_{\nu=1}^{p} \sum_{\mu=0}^{k-p} \frac{1}{\nu + \mu} .$$

Then we have

(8.44)
$$\boxed{I_{p-1}(N_f(r,s;a)) = T_f^p[r,s]}$$

(8.45)
$$\boxed{I_{p-1}(n_f(t;a)) = A_f^p[t]}$$

(8.46)
$$\boxed{I_{p-1}(m_f(r,a)) = c(p,k)A_f^{p-1}[r]}$$

(8.47)
$$\boxed{I_{p-1}(D_f(r,s;a)) = c(p,k)(A_f^{p-1}[r] - A_f^{p-1}[s])} .$$

Assume that f is L-non-degenerate of order p - 1. Then $T_f^p[r,s] \to \infty$ for $r \to \infty$. Define $\beta_p : G_{p-1}(V^*) \to \mathbb{Z}[0,1]$ by

$$\beta_p(a) = \begin{cases} 0 & \text{if } f(M) \cap E_L[a] = \emptyset \\ 1 & \text{if } f(M) \cap E_L[a] \neq \emptyset. \end{cases}$$

Then

$$0 \le b_p = b_p(f,L) = I_{p-1}(\beta_p) \le 1$$

is the probability for f(M) to intersect $E_L[a]$. The First Main Theorem implies

(8.48)
$$N_f(r,s;a) - \beta_p(a)T_f^p[r,s] \le m_a[s,f] + D_a[r,s,f]$$

$$(8.49)\qquad 1 \to b_p \le c(p,k)\,\frac{A_f^{p-1}[r]}{T_f^p[r,s]}$$

which is an estimate of the probability that $E_L[a] \cap f(M) = \emptyset$. Hence if

$$(8.50)\qquad \frac{A_f^{p-1}[r]}{T_f^p[r,s]} \to 0 \qquad \text{for } r \to \infty$$

then $b_p = 1$ and $f(M) \cap E_L[a] \neq \emptyset$ for almost all $a \in G_{p-1}(V^*)$. If $p = 1$, then $A_f^0[r] = c_0 > 0$ is a constant and (8.50) is satisfied. Therefore $f(M) \cap E_L[a] \neq \emptyset$ for almost all $a \in \mathbb{P}(V^*)$. _A Casorati-Weierstrass Theorem has been obtained._

If κ and $\tilde{\kappa}$ are two hermitian metrics along the fibers of L, then there is a constant $\gamma \ge 0$ such that

$$(8.51)\qquad |T_f(r,s,L,\kappa) - T_f(r,s,L,\tilde{\kappa})| \le \gamma \qquad \forall\, r > s.$$

Hence the growth of T_f does not depend on κ.

A defect relation exists only if $p = 1$. For its formulation some preparations are needed.

Let $\Omega > 0$ be a positive form of degree $2m$ on M. Then a form Ric Ω of bidegree (1,1) is associated. If U is an open subset of M and if $(z_1,\dots,z_m)$ are local holomorphic coordinates on U then

$$(8.52)\qquad \Omega = \Omega_{\mathfrak{z}}\left(\frac{1}{2\pi}\right)^m m!\; dz_1 \wedge d\bar{z}_1 \wedge \dots \wedge dz_m \wedge d\bar{z}_m$$

on U where $\Omega_{\mathfrak{z}} > 0$. Then

$$(8.53)\qquad \operatorname{Ric}\Omega|U = dd^c \log \Omega_{\mathfrak{z}}.$$

The form Ric Ω is called the _Ricci form_ of Ω. We associate the _Ricci function_

$$\text{(8.54)} \qquad \mathrm{Ric}(r,s,\Omega) = \int_s^r \int_{M[t]} \mathrm{Ric}\,\Omega \wedge \upsilon^{m-1} \frac{dt}{t^{2m-1}} .$$

Since $\upsilon^m \geq 0$ only, $\mathrm{Ric}(r,s,\upsilon^m)$ is not defined; however indirectly it can be done. A function $v \geq 0$ of class C^∞ exists such that $\upsilon^m = v\Omega$. Then

$$\text{(8.55)} \qquad \mathrm{Ric}_\tau(r,s) = \frac{1}{2} \int_{M\langle r\rangle} \log v\sigma - \frac{1}{2} \int_{M\langle s\rangle} \log v\sigma + \mathrm{Ric}(r,s,\Omega)$$

is defined for almost all $r > s$ and does not depend on the choice of Ω. The function $\mathrm{Ric}_\tau(r,s)$ is called the <u>Ricci function</u> of τ.

Let $f : M \to N$ be a holomorphic map. Let K_M and K_N be the canonical bundles of M and N. Let K_N^* be the dual bundle and let $f^*(K_N^*)$ be the pullback. Then $K_F = K_M \otimes f^*(K_N^*)$ is called the <u>Jacobian bundle</u> of f. A global holomorphic section $F \not\equiv 0$ of K_f is called a <u>Jacobian section</u>.

Let $\Omega_N^n(U)$ be the vector space of holomorphic forms of degree n on N. Define $\tilde{U} = f^{-1}(U)$. If $\tilde{U} \neq \emptyset$, then the Jacobian section F defines a linear map $F : \Omega_N^n(U) \to \Omega_M^m(\tilde{U})$. Let $(\cdot,\cdot)$ be the K_M-valued inner product between K_f and $f^*(K_N)$. Take $\psi \in \Omega^n(U)$. Then ψ defines a section ψ_f in $f^*(K_N)$ over $\tilde{U}$ and we have $F[\psi] = (F,\psi_f) \in \Omega_M^m(U)$. Let $A_N^q(U)$ be the vector space of forms of degree q and class C^∞ on U. Define

$$\text{(8.56)} \qquad i_p = (-1)^{\frac{p(p-1)}{2}} p!\left(\frac{i}{2\pi}\right)^p .$$

Then F extends to a linear map $F : A_N^{2n}(U) \to A_N^{2m}(\tilde{U})$ such that

$$\text{(8.57)} \qquad F[i_n\phi \wedge \overline{\chi}] = i_m F[\phi] \wedge \overline{F[\chi]} .$$

The Jacobian section F is said to <u>dominate</u> τ if for each $r > 0$ there exists a minimal constant $Y(r) \geq 1$ such that

$$n\left(\frac{F[\psi^n]}{\upsilon^m}\right)^{\frac{1}{n}} \upsilon^m \leq Y(r) f^*(\psi) \wedge \upsilon^{m-1} \tag{8.58}$$

on $M[r] \cap \{x \in \tilde{U} \mid \upsilon(x) > 0\}$ whenever U is open in N and $\tilde{U} = f^{-1}(U) \neq 0$ and whenever $0 \leq \psi$ is a non-negative form of bidegree (1,1). The function Y is called the dominator. Under reasonable assumptions F and Y exist. For instance if $m = n$ and rank $f = n$, then the differential of f determines a Jacobian section F which dominates τ with $Y \equiv m$. More generally, if $m \geq n =$ rank f, if (M,τ) is a covering parabolic manifold of the hermitian vector space W, that is, if there exists a proper, surjective holomorphic map $\beta : M \to W$ such that $\tau = \|\beta\|^2$, then there exists a Jacobian section $F \not\equiv 0$ which dominates τ with dominator $Y \equiv m$.

Let $\nu_1, \ldots, \nu_q$ be non-negative divisors. For each $x \in N$, take all divisors $\nu_{j_1}, \ldots, \nu_{j_p}$ with $x \in \text{supp}\, \nu_{j_\lambda}$. Let U be an open connected neighborhood of x and $f_\lambda \not\equiv 0$ be a holomorphic function on U with $\nu_{j_\mu} | U = \mu^0_{f_\lambda}$. Then $\nu_1, \ldots, \nu_q$ are said to be in general position iff $df_1 \wedge \ldots \wedge df_p \neq 0$ at x. If $a_j \in \mathbb{P}(V^*)$ for $j = 1, \ldots, q$, then $a_1, \ldots, a_q$ are said to be in general position iff the divisors $\theta_f^{a_1}[L], \ldots, \theta_f^{a_q}[L]$ are in general position.

We shall make the following general assumptions.

(B1) Let N be a connected, compact complex manifold of dimension $n > 0$.

(B2) Let L be a positive, holomorphic line bundle on N.

(B3) Let κ be a hermitian metric along the fibers of L such that $c(L,\kappa) > 0$.

(B4) Let V be the dual vector space to $\Gamma(N,L)$. Assume that $k + 1 = \dim V > 1$.

(B5) Assume that the evaluation map $\varepsilon : M \times V^* \to L$ is surjective.

(B6) Let ℓ be a hermitian inner product on V^* such that (κ, ℓ) are distinguished.

(B7) Let $a_1, \ldots, a_q$ in $\mathbb{P}(V^*)$ be in general position.

(B8) Let M be a connected, non-compact, complex manifold of dimension $m > 0$. Let τ be a parabolic exhaustion of M.

(B9) Let $f : M \to N$ be a holomorphic map.

(B10) Let $F \not\equiv 0$ be a Jacobian section which dominates τ with dominator Y.

Since L is positive, N is projective algebraic. (B10) implies

(8.59) $$m \geq n = \text{rank } f.$$

In particular $T_f(r,s) \to \infty$ for $r \to \infty$. Let μ_F be the zero-divisor of the section F. Then

(8.60) $$\Theta_F = \liminf_{r\to\infty} \frac{N_{\mu_f}(r,s)}{T_f(r,s)} \geq 0$$

is called the <u>ramification defect</u> of F. The <u>Ricci defect</u> of f is defined by

(8.61) $$R_f = \limsup_{r\to\infty} \frac{\mathrm{Ric}_\tau(r,s)}{T_f(r,s)} .$$

The <u>dominator defect</u> of F is defined by

(8.62) $$Y_F = \limsup_{r\to\infty} \frac{\log Y(r)}{T_f(r,s)} .$$

The <u>Nevanlinna defect</u> of $a \in \mathbb{P}(V^*)$ is defined by (8.38) or equivalently by (8.39).

If Q is any holomorphic line bundle on N, let [Q:L] be the infimum of all quotients v/w where $v \geq 0$ and $w \geq 0$ are integers such that $L^v \otimes (Q^*)^w$ is a non-negative line bundle. Then

Theorem 8.2. Defect relation.
Assume that (B1)-(B10) hold. Then

$$\boxed{\Theta_F + \sum_{j=1}^{q} \delta_f(a_j, L) \leq [K_N^*:L] + R_f + c_n Y_F} \quad . \tag{8.63}$$

9. Defect relation for hypersurfaces in complex projective space.

Let V and W be hermitian vector spaces with dim V = n + 1 > 1 and dim W = m > 0. Let $f : W \to \mathbb{P}(V)$ be a meromorphic map. Let H be the hyperplane section bundle on $\mathbb{P}(V)$. Take $p \in \mathbb{N}$. Then we will study the value distribution theory in respect to the line bundle H^p

$$(9.1) \qquad \begin{array}{c} H^p \\ \downarrow \\ f : W \to \mathbb{P}(V). \end{array}$$

We shall interpret (8.63) for this situation and explain some results of Biancofiore [4]. Let $V_{(p)} = \Gamma(\mathbb{P}(V), H^p)$ be the vector space of all holomorphic sections of H^p over $\mathbb{P}(V)$. We will identify $V_{(p)}$ with a number of vector spaces which enriches the theory but complicates matters.

Let $V_{[p]}$ be the vector space of all homogeneous polynomials of degree p on V, that is, $\alpha \in V_{[p]}$ if and only if $\alpha : V \to \mathbb{C}$ is a holomorphic function with $\alpha(\lambda \mathfrak{x}) = \lambda^p \alpha(\mathfrak{x})$ for all $\lambda \in \mathbb{C}$ and $\mathfrak{x} \in V$. We claim that we can identify $V_{(p)} = V_{[p]}$.

If $x \in \mathbb{P}(V)$, then $H_x = E(X)^*$ is the dual vector space of E(x). Take $0 \neq \beta \in E(x)$, then $H_x^p = \{z\beta^p \mid z \in \mathbb{C}\}$. Here $\beta : E(x) \to \mathbb{C}$ is linear hence $z\beta^p : E(x) \to \mathbb{C}$ is a homogeneous polynomial of degree p on E(x) and each homogeneous polynomial of degree p on E(x) can be written uniquely in the form $z\beta^p$. We see that $H_x^p = E(x)_{[p]}$ is the vector space of a homogeneous polynomial of degree p on E(x).

Take $\alpha \in V_{[p]}$. Then $\alpha|E(x)$ is a homogeneous polynomial of degree p on E(x). Hence $\alpha_x = \alpha|E(x) \in H_x^p$. A holomorphic section $\hat{\alpha} \in V_{(p)}$ is defined by $\hat{\alpha}(x) = \alpha_x$. Obviously, the map $\alpha \mapsto \hat{\alpha}$ is linear. Since $\hat{\alpha}(x)(\mathfrak{x}) = \alpha(\mathfrak{x})$ for all $\mathfrak{x} \in E(x)$ and $x \in \mathbb{P}(V)$ we have $\hat{\alpha} = 0$ if and only if $\alpha = 0$. The linear map $\alpha \mapsto \hat{\alpha}$ is injective. If $s \in V_{(p)}$ is a section, then $s_x : E(x) \to \mathbb{C}$ is a homogeneous polynomial of degree p and $\alpha : V_* \to \mathbb{C}$ defined by $\alpha(\mathfrak{x}) = s_{\mathbb{P}(\mathfrak{x})}(\mathfrak{x})$ is holomorphic. If $\mathfrak{x} \in V_*$ and $\lambda \in \mathbb{C}_*$, then

$$\alpha(\lambda\mathfrak{x}) = s_{\mathbb{P}(\lambda\mathfrak{x})}(\lambda\mathfrak{x}) = s_{\mathbb{P}(\mathfrak{x})}(\lambda\mathfrak{x}) = \lambda^p s_{\mathbb{P}(\mathfrak{x})}(\mathfrak{x}) = \lambda^p \alpha(\mathfrak{x}).$$

Hence α extends to a homogeneous polynomial of degree p on V. We have $\hat{\alpha}(x)(\mathfrak{x}) = \alpha(\mathfrak{x}) = s_x(\mathfrak{x})$ for all $\mathfrak{x} \in E(x)$ and $x \in \mathbb{P}(V)$. Hence $\hat{\alpha} = s$. The map $\alpha \mapsto \hat{\alpha}$ is a linear isomorphism by which we identify $V_{[p]} = V_{(p)}$ and $\alpha = \hat{\alpha}$. In particular $V_{(1)} = V_{[1]} = V^*$.

Take $a \in \mathbb{P}(V_{(p)})$. Take $0 \neq \alpha \in V_{(p)}$ with $a = \mathbb{P}(V_{(p)})$. Then

(9.2) $$Z(\alpha) = E_{H^p}[a] = \ddot{E}[a]$$

is called a <u>hypersurface of degree</u> p in $\mathbb{P}(V)$. The zero divisor $\mu_\alpha = \mu^a$ of the section α depends on a only with supp $\mu^a = \ddot{E}[a]$. If we interpret $\alpha \in V_{[p]}$ as a homogeneous polynomial, the zero divisor μ_α^0 of this polynomial is defined on V with $\mu_\alpha^0 = \mathbb{P}^*(\mu_\alpha)$. The analytic set $E[a] = \alpha^{-1}(0) = \text{supp}\,\mu_\alpha^0$ is called an <u>analytic cone</u> of degree p. Then $\ddot{E}[a] = \mathbb{P}(E[a])$.

The meromorphic map $f : W \to \mathbb{P}(V)$ is said to be <u>algebraically non-degenerate for</u> $a \in \mathbb{P}(V_{(p)})$ iff $f(W) \not\subseteq \ddot{E}[a]$ and f is said to be <u>algebraically non-degenerate of degree</u> p iff $f(W) \not\subseteq \ddot{E}[a]$ for all $a \in \mathbb{P}(V_{(p)})$. Also f is said to be <u>algebraically non-degenerate</u> iff f is algebraically non-degenerate of degree p for all $p \in \mathbb{N}$. Take $a \in \mathbb{P}(V_{(p)})$. Assume that f is algebraically non-degenerate for a. Let $\mathfrak{v} : W \to V$ be a reduced representation of f and take $0 \neq \alpha \in V_{[p]}$ with $a = \mathbb{P}(\alpha)$. Then $\alpha \circ \mathfrak{v} \not\equiv 0$ and the divisor $\mu_f^a = \mu_{\alpha\circ\mathfrak{v}}^0 \geq 0$ depends on a and f only. If f is holomorphic, then

$$\mu_f^a = \mu_{\alpha\circ\mathfrak{v}}^0 = \mathfrak{v}^*(\mu_\alpha^0) = \mathfrak{v}^*(\mathbb{P}^*(\mu_\alpha)) = (\mathbb{P} \circ \mathfrak{v})^*(\mu_\alpha) = f^*(\mu_\alpha)$$

is the pullback of the section divisor μ_α. In any case, the <u>counting function</u> and the <u>valence function</u> of f for a are defined by

$$(9.3)\qquad n_f(t,a) = n_{\mu_f^a}(t) = n_f(t,a,H^p) \geq 0 \qquad \forall\, t > 0$$

$$(9.4)\qquad N_f(r,s;a) = N_{\mu_f^a}(r,s) = N_f(r,s;a,H^p) \geq 0 \qquad \forall\, r > s > 0.$$

Take $x \in \mathbb{P}(V)$. Take $\mathfrak{x} \in V_*$ with $\mathbb{P}(\mathfrak{x}) = x$. Then $\alpha \in V_{[p]}$ exists such that $\alpha(\mathfrak{x}) \neq 0$. Then $x \notin \ddot{E}[a]$. Hence $x \notin Z(\alpha)$. Therefore the evaluation map

$$(9.5)\qquad \varepsilon : \mathbb{P}(V) \times V_{(p)} \to H^p$$

defined by $\varepsilon(x,\alpha) = \hat{\alpha}_x \in H_x$ is surjective. Hence any hermitian inner product on $V_{(p)}$ will induce a hermitian metric along the fibers of H^p. However we want to pick the hermitian inner product on $V_{(p)}$ canonically. Hence we consider another interpretation of $V_{(p)}$.

Let $\underset{p}{\circledast} V$ and $\underset{p}{\circledast} V^*$ be the pth symmetric tensor products. By multilinear algebra $\underset{p}{\circledast} V^* = V_{[p]}$ can be identified such that if $\alpha_j \in V^*$ for $j = 1, \ldots, p$ the symmetric tensor product $\alpha_1 \ldots \alpha_p$ is identified with the product $\alpha_1 \ldots \alpha_p$ of functions. Hence
$(\alpha_1 \ldots \alpha_p)(\mathfrak{x}) = \alpha_1(\mathfrak{x}) \ldots \alpha_p(\mathfrak{x})$. The hermitian inner product given on V^* induces a hermitian inner product on $\underset{p}{\circledast} V^*$. Let $\mathfrak{S}_p$ be the permutation group of $\mathbb{N}[1,p]$. Take $\alpha_j \in V^*$ and $\beta_j \in V^*$ for $j = 1, \ldots, p$, then the hermitian inner product on $\underset{p}{\circledast} V^*$ is uniquely defined by the condition

$$(9.6)\quad (\alpha_1 \ldots \alpha_p | \beta_1 \ldots \beta_p) = \frac{1}{p!} \sum_{\pi \in \mathfrak{S}_p} (\alpha_1 | \beta_{\pi(1)}) \ldots (\alpha_p | \beta_{\pi(p)}).$$

If $\alpha \in V^*$, then $\alpha^p \in V_{(p)}$ and $\|\alpha^p\| = \|\alpha\|^p$. Define

$$(9.7)\qquad S[p,n] = \{j = (j_0,\ldots,j_n) \in \mathbb{Z}_+^n \mid j_0 + \ldots + j_n = p\}.$$

If $j = (j_0,\ldots,j_n) \in \mathbb{Z}_+^n$ and if $\alpha = (\alpha_0,\ldots,\alpha_n)$ with $\alpha_\nu \in V^*$, define

$$(9.8)\qquad \alpha^j = \alpha_0^{j_0} \ldots \alpha_n^{j_n} \qquad\qquad j! = j_0! \ldots j_n!$$

$$(9.9)\qquad c_j = \sqrt{\frac{p!}{j!}} \qquad\qquad \text{if } j \in S[p,n].$$

If α is a base of V^*, then $\{c_j\alpha^j\}_{j\in S[p,n]}$ is a base of $\underset{p}{\otimes} V^*$. If α is an orthonormal base of V^*, then $\{c_j\alpha^j\}_{j\in S[p,n]}$ is an orthonormal base of $\underset{p}{\otimes} V^*$. In particular

$$(9.10)\qquad \dim V_{(p)} = \dim V_{[p]} = \dim \underset{p}{\otimes} V^* = \binom{n+p}{p}.$$

The identification $V_{(p)} = V_{[p]} = \underset{p}{\otimes} V^*$ provides hermitian inner products on $V_{(p)}$ and $V_{[p]}$. If $\alpha \in V_{[p]}$, then $\|\alpha\|$ is determined and can be shown to be the smallest non-negative constant such that

$$(9.11)\qquad |\alpha(\mathfrak{x})| \le \|\alpha\| \, \|\mathfrak{x}\|^p \qquad\qquad \forall\, \mathfrak{x} \in V.$$

The hermitian inner product defined on $V_{(p)}$ induces a hermitian metric κ_p along the fibers of H^p by (9.5) according to the procedure outlined between (8.10) and (8.11). Given $a \in \mathbb{P}(V_{(p)})$ and $x \in \mathbb{P}(V)$ we want to compute $\|a,x\|_\kappa$. Take $0 \neq \alpha \in V_{(p)}$ and $0 \neq \mathfrak{x} \in V$ with $\mathbb{P}(\alpha) = a$ and $\mathbb{P}(\mathfrak{x}) = x$. Then there exists an orthonormal base $\beta = (\beta_0,\ldots,\beta_n)$ of V such that $\beta_0(\mathfrak{x}) = \|\mathfrak{x}\|$ and $\beta_\nu(\mathfrak{x}) = 0$ for $\nu \in \mathbb{N}[1,n]$. Then $\{c_j\beta^j\}_{j\in S[p,n]}$ is an orthonormal base of $V_{(p)}$ and we have

$$(9.12)\qquad \alpha = \sum_{j\in S[p,n]} A_j c_j \beta^j .$$

Let $\varepsilon_x : V_{(p)} \to H_x^p$ be the restricted evaluation map and let S_x be the kernel of ε_x. Then S_x has codimension 1. Also $c_j\beta^j \in S_x$ for all $j \in S[p,n]$ except $j = (p,0\ldots0)$. Hence these $c_j\beta_j$ span S_x and $S_x^\perp$ is spanned by $c_{p0\ldots0}\beta_0^p$. Observe that $c_{p0\ldots0} = 1$. Hence

$$(9.13) \qquad \alpha = \tilde{\alpha} + \hat{\alpha} \qquad \tilde{\alpha} = A_{p0\ldots0}\beta_0^p \in S_x^\perp \qquad \hat{\alpha} = \alpha - \tilde{\alpha} \in S_x.$$

Therefore

$$(9.14) \qquad \|\alpha_x\|_{\kappa_p} = \|\tilde{\alpha}\| = |A_{p0\ldots0}|\,\|\beta_0^p\| = |A_{p0\ldots0}|\,\|\beta_0\|^p = |A_{p0\ldots0}|.$$

Also $\alpha(\mathfrak{x}) = A_{p0\ldots0}\beta_0(\mathfrak{x})^p = A_{p0\ldots0}\|\mathfrak{x}\|^p$. Therefore

$$(9.15) \qquad \|\alpha_x\|_{\kappa_p} = \frac{|\alpha(\mathfrak{x})|}{\|\mathfrak{x}\|^p}.$$

Now (8.9) implies

$$(9.16) \qquad \boxed{0 \le \|a,x\|_{\kappa_p} = \frac{|\alpha(\mathfrak{x})|}{\|\alpha\|\;\|\mathfrak{x}\|^p} \le 1}\,.$$

Abbreviate $\|a,x\| = \|a,x\|_{\kappa_p}$. If f is algebraically non-degenerate for $a \in \mathbb{P}(V_{(p)})$, the <u>compensation</u> <u>function</u> <u>of</u> f <u>for</u> a is defined for $r > 0$ by

$$(9.17) \qquad m_f(r,a) = \int_{W\langle r\rangle} \log \frac{1}{\|a,x\|}\,\sigma.$$

For $p = 1$, the hermitian metric κ_1 is exactly the hermitian metric κ in section 2. Hence (2.82) gives $c(H,\kappa_1) = \Omega$. Now, $\kappa_1 = \kappa$ defines a hermitian metric κ^p on H^p. If we take $x \in \mathbb{P}(V)$ and $\alpha \in V_{(p)}$ and proceed as above (9.12)-(9.16), then $\alpha_x = \varepsilon_x(\alpha) = A_{p0\ldots0}\beta_{0x}^p$ and

$$\|\alpha_x\|_{\kappa^p} = |A_{p0\ldots0}|\,\|\beta^p_{0x}\|_{\kappa^p} = |A_{p0\ldots0}|\,\|\beta_0\|^p_\kappa = |A_{p0\ldots0}|.$$

Hence $\|\alpha_x\|_{\kappa^p} = \|\alpha_x\|_{\kappa_p}$ for all $\alpha \in V_{(p)}$. Because the evaluation map is surjective and because the hermitian metrics are uniquely defined by their norms we have $\kappa^p = \kappa_p$. Therefore

$$(9.18) \qquad c(H^p,\kappa_p) = c(H^p,\kappa^p) = pc(H,\kappa) = p\Omega.$$

This can also be obtained directly from (9.16). On $\mathbb{P}(V) - \ddot{E}[a]$ we have

$$\mathbb{P}^*(c(H^p,\kappa_p)) = -dd^c \log\|a,x\|^2_{\kappa_p} = dd^c \log \frac{\|\alpha\|^2\ \|\mathfrak{x}\|^{2p}}{|\alpha(\mathfrak{x})|^2}$$
$$= p\ dd^c \log\|\mathfrak{x}\|^2 = p\mathbb{P}^*(\Omega) = \mathbb{P}^*(p\Omega).$$

Since $\mathbb{P}^*$ is injective, we have $c(H^p,\kappa_p) = p\Omega$. Therefore the <u>spherical image of f in respect to H^p is given by</u>

$$(9.19) \qquad A_f(t,H^p) = \frac{1}{t^{2m-2}} \int_{W[t]} f^*(c(H^p,\kappa_p))\upsilon^{m-1}$$
$$= \frac{p}{t^{2m-2}} \int_{W[t]} f^*(\Omega)\upsilon^{m-1} = pA_f(t).$$

The <u>characteristic</u> is given by

$$(9.20) \qquad \boxed{T_f(r,s;H^p) = pT_f(r,s)} \qquad \text{if } 0 < s < r.$$

If f is algebraically non-degenerate for $a \in \mathbb{P}(V_{(p)})$, then the <u>First Main Theorem</u> holds

$$(9.21) \qquad \boxed{pT_f(r,s) = N_f(r,s;a) + m_f(r,a) - m_f(s,a)}$$

for $r > s > 0$.

There are three possible ways to prove (9.21). First we can repeat the proof of Theorem 4.1 with the explicit definitions given here. No appeal to H^p would be necessary if we consider $V_{(p)} = V_{[p]}$ as the vector space of homogeneous polynomials of degree p. Second we need only to apply (8.37) to $L = H^p$. Since κ corresponds to ℓ, we obtain also the integral averaging formulas. Put $q = \binom{n+p}{p} - 1$, then

$$(9.22)\qquad pT_f(r,s) = \int_{\mathbb{P}(V_{(p)})} N_f(r,s;a)\Omega^q \qquad (0 < s < r)$$

$$(9.23)\qquad pA_f(t) = \int_{\mathbb{P}(V_{(p)})} n_f(t;a)\Omega^q \qquad (t > 0)$$

$$(9.24)\qquad \int_{\mathbb{P}(V_{(p)})} m_f(r,a)\Omega^q = \frac{1}{2}\sum_{\mu=1}^{q}\frac{1}{\mu} \qquad (r > 0).$$

A third proof is obtained by yet another interpretation of $V_{(p)}$. Recall that $V_{(p)} = \underset{p}{\otimes} V^*$. By multilinear algebra $\underset{p}{\otimes} V^* = (\underset{p}{\otimes} V)^*$. If $\langle\ ,\ \rangle$ denotes the inner product between dual spaces, and if $\mathfrak{S}_p$ is the permutation group on $\mathbb{N}[1,p]$, then the duality $\underset{p}{\otimes} V$ and $\underset{p}{\otimes} V^*$ is determined by

$$(9.25)\qquad \langle \alpha_1 \ldots \alpha_p, \mathfrak{x}_1 \ldots \mathfrak{x}_p\rangle = \frac{1}{p!}\sum_{\pi\in\mathfrak{S}_p} \alpha_1(\mathfrak{x}_{\pi(1)}) \ldots \alpha_p(\mathfrak{x}_{\pi(p)})$$

where $\alpha_j \in V^*$ and $\mathfrak{x}_j \in V$ for $j = 1, \ldots, p$. Observe also that the hermitian inner product already established on $\underset{p}{\otimes} V^*$ is dual to the hermitian inner product on $\underset{p}{\otimes} V$ induced by the hermitian inner product on V. Hence if $\mathfrak{a} = (\mathfrak{a}_0,\ldots,\mathfrak{a}_n)$ is an orthonormal base of V and if $(\alpha_0,\ldots,\alpha_n)$ is the dual base of V^*, then the orthonormal base $\{c_j\mathfrak{a}^j\}_{j\in S[p,n]}$ of $\underset{p}{\otimes} V$ is dual to the orthonormal base $\{c_j\alpha^j\}_{j\in S[p,n]}$.

If $\mathfrak{x} \in V$, then $\mathfrak{x}^p = \mathfrak{x} \ldots \mathfrak{x} \in \underset{p}{\otimes} V$. Hence a holomorphic map $\tilde{\phi} : V \to \underset{p}{\otimes} V$ is defined by $\tilde{\phi}(\mathfrak{x}) = \mathfrak{x}^p$. We have

$$(9.26) \qquad \|\tilde{\phi}(\mathfrak{x})\| = \|\mathfrak{x}\|^p \qquad \forall\, \mathfrak{x} \in V$$

$$(9.27) \qquad \langle \alpha, \tilde{\phi}(\mathfrak{x}) \rangle = \alpha(\mathfrak{x}) \qquad \forall\, \mathfrak{x} \in V \quad \forall\, \alpha \in \underset{p}{\otimes} V^*$$

$$(9.28) \qquad \tilde{\phi}(\lambda \mathfrak{x}) = \lambda^p \tilde{\phi}(\mathfrak{x}) \qquad \forall\, \lambda \in \mathbb{C} \quad \forall\, \mathfrak{x} \in V.$$

Also if $\mathfrak{x} \in V$ and $\mathfrak{y} \in V$, then $\tilde{\phi}(\mathfrak{x}) = \tilde{\phi}(\mathfrak{y})$ if and only if there is $\zeta \in \mathbb{C}\langle 1 \rangle$ with $\zeta^p = 1$ such that $\mathfrak{x} = \zeta \mathfrak{y}$. Therefore there exists one and only one injective, holomorphic map $\phi : \mathbb{P}(V) \to \mathbb{P}(\underset{p}{\otimes} V)$ called the <u>Veronese</u> map such that $\mathbb{P} \circ \tilde{\phi} = \phi \circ \mathbb{P}$.

Take $a \in \mathbb{P}(\underset{p}{\otimes} V^*)$ and $\alpha \in \underset{p}{\otimes} V^*$ with $\mathbb{P}(\alpha) = a$. Then a determines a linear subspace

$$F[a] = \{\mathfrak{z} \in \underset{p}{\otimes} V \mid \langle \alpha, \mathfrak{z} \rangle = 0\}$$

of codimension 1 in $\underset{p}{\otimes} V$ and a hyperplane $\ddot{F}[a] = \mathbb{P}(F[a])$ in $\mathbb{P}(\underset{p}{\otimes} V)$. Then

$$E[a] = \tilde{\phi}^{-1}(F[a]) \qquad \ddot{E}[a] = \phi^{-1}(\ddot{F}[a]).$$

Also $\phi \circ f : W \to \mathbb{P}(\underset{p}{\otimes} V)$ is a meromorphic map with $\mu^a_{\phi \circ f} = \mu^a_f$. Therefore

$$(9.29) \qquad n_{\phi \circ f}(t,a) = n_f(t,a) \qquad N_{\phi \circ f}(r,s;a) = N_f(r,s;a).$$

If $x \in \mathbb{P}(V)$, take $0 \neq \mathfrak{x} \in V$ such that $\mathbb{P}(\mathfrak{x}) = x$. Then

$$(9.30) \qquad \|a, \phi(x)\| = \frac{\langle \alpha, \tilde{\phi}(\mathfrak{x}) \rangle}{\|\alpha\| \ \|\tilde{\phi}(\mathfrak{x})\|} = \frac{|\alpha(\mathfrak{x})|}{\|\alpha\| \ \|\mathfrak{x}\|^p} = \|a, x\|.$$

Hence

$$m_{\phi\circ f}(r,a) = m_f(r,a) \qquad \forall\, r > 0. \tag{9.31}$$

Now, (9.26) and (4.7) imply immediately

$$T_{\phi\circ f}(r,s) = pT_f(r,s) = T_f(r,s,H^p). \tag{9.32}$$

By Theorem 4.1 we have

$$T_{\phi\circ f}(r,s) = N_{\phi\circ f}(r,s;a) + m_{\phi\circ f}(r,a) - m_{\phi\circ f}(s,a)$$

which yields

$$pT_f(r,s) = N_f(r,s;a) + m_f(r,a) - m_f(s,a).$$

Thus another proof of the First Main Theorem is obtained. Also (4.16) implies (9.24) and (4.17) implies (9.22) and (9.23) is obtained by differentiation.

Take $a \in \mathbb{P}(V_{(p)})$. Assume that f is not algebraically degenerate for a. The <u>Nevanlinna defect</u> is given by

$$0 \le \delta_f(a) = \liminf_{r\to\infty} \frac{m_f(r,s)}{pT_f(r,s)} = 1 - \limsup_{r\to\infty} \frac{N_f(r,s;a)}{pT_f(r,s)} \le 1. \tag{9.33}$$

Then

$$\delta_{\phi\circ f}(a) = \delta_f(a). \tag{9.34}$$

Therefore if f is not rational and if f is algebraically non-degenerate of order p and if $a_j \in \mathbb{P}(\underset{p}{\otimes} V^*)$ are given such that $F[a_1], \ldots, F[a_q]$ are in general position, then <u>the defect relation</u>

$$\boxed{\sum_{j=1}^{q} \delta_f(a_j) \le \binom{n+p}{p}}\,. \tag{9.35}$$

The problem is that $\binom{n+p}{p}$ is not the best possible number, and that we would like to require that $E[a_1]$, ..., $E[a_q]$ be in general position.

Assume that $m \geq n = \text{rank } f$. Then there exists a Jacobian section $F \not\equiv 0$ which dominates τ_W, such that $Y \equiv m$. Hence the defect $Y_F = 0$ vanishes. Also $\upsilon_W > 0$ implies $v = 1$ and $\Omega = \upsilon_W^m > 0$. Trivially Ric $\Omega = 0$ since υ_W^m has a constant Jacobian. Hence $R_f = 0$. Therefore (8.63) implies

$$\sum_{j=1}^{q} \delta_f(a_j) \leq [K^*_{\mathbb{P}(V)} : H^p].$$

Abbreviate $(H^*)^\nu = H^{-\nu}$ for $\nu \in \mathbb{Z}_+$. Then $K = H^{-n-1}$. Take integers $w > 0$ and $v \geq 0$ such that

$$H^{pv} \otimes K^w = H^{pv-(n+1)w}$$

is non-negative. Then $pv - n + 1w \geq 0$ or $\frac{v}{w} \geq \frac{n+1}{p}$. Therefore $[K^*_{\mathbb{P}(V)} : H^p] = \frac{n+1}{p}$. We have the defect relation:

Theorem 9.1. Defect relation.
Let V and W be hermitian vector spaces with $\dim V = n + 1 > 1$ and $\dim W = m \geq n$. Let $f : W \to \mathbb{P}(V)$ be a holomorphic map with $m \geq n = \text{rank } f$. Assume that f is not rational. Take $a_j \in \mathbb{P}(V_{[p]})$ such that $\ddot{E}[a_1]$, ..., $\ddot{E}[a_q]$ are in general position. Then

$$\boxed{\sum_{j=1}^{q} \delta_f(a_j) \leq \frac{n+1}{p}} \,. \tag{9.36}$$

For $p = 1$, we obtain (7.6) under the more restrictive assumption $m \geq n = \text{rank } f$. If $p > 1$, it is conjecture that (9.36) still holds if f is not algebraically degenerate. This conjecture is unproved, and is a difficult, outstanding problem in value distribution theory. Aldo Biancofiore [4] has constructed an example of a holomorphic map $f : W \to \mathbb{P}(V)$ which is not algebraically degenerate of order p, but algebraically degenerate of an order $p' > p$ where (9.36) is violated.

Shiffman [59] has shown that $\sum_{j=1}^{q} \delta_f(a_j) \leq 2n$ for a certain class of maps f where $\ddot{E}[a_1]$, ..., $\ddot{E}[a_q]$ do not need to be in general position, but where a_1, ..., a_q are distinct and where each point $x \in \mathbb{P}(V)$ is contained in at most n of the sets $\ddot{E}[a_1]$, ..., $\ddot{E}[a_q]$.

Biancofiore [4] has proved (9.36) for an essentially more general class of maps f if $\ddot{E}[a_1]$, ..., $\ddot{E}[a_q]$ are in general position. We shall state his result.

We will list the <u>assumptions</u> <u>for</u> <u>Biancofiore's</u> <u>Theorem</u>.

(1) Let V, W and X be hermitian vector spaces with $\dim V = n + 1 > 1$ and $\dim W = m > 0$ and $\dim X = N + 1$ where $N \geq n$. Take $2 \leq p \in \mathbb{N}$.

(2) Let $f : W \to \mathbb{P}(V)$ be a meromorphic map.

(3) Let $\mathfrak{v} : W \to V$ be a reduced representation of f.

(4) Let $g : W \to \mathbb{P}(X)$ be a meromorphic map algebraically, non-degenerate of order p.

(5) Let $\mathfrak{g} : W \to X$ be a reduced representation of g.

(6) Let $u : W \to \mathbb{C}$ be a holomorphic function with $u \not\equiv 0$.

(7) Let $\mathfrak{e}_0$, ..., $\mathfrak{e}_N$ be an orthonormal base of X and let ε_0, ..., ε_N be the dual base. Define $e_j = \mathbb{P}(\varepsilon_j)$ for j = 0, ..., N.

(8) Let $\phi : X \to V$ be a surjective linear map such that $\phi(\mathfrak{e}_j) \not\equiv 0$ for all j = 0, ..., N.

(9) Assume that $u\mathfrak{v} = \phi \circ \mathfrak{g}$. Hence $f = \mathbb{P}(\phi \circ \mathfrak{g})$.

(10) Assume that $\delta_g(e_j) = 1$ for j = 0, ..., N.

(11) Assume that

$$\frac{N_u(r,s,0)}{T_g(r,s)} \to 0 \qquad \text{for } r \to \infty. \tag{9.37}$$

(12) Take $a_j \in \mathbb{P}(V_{(p)})$ for $j = 1, \ldots, q$ with $q \geq n + 1$ such that $a_1, \ldots, a_q$ are in general position in $\mathbb{P}(V)$.

By a theorem of Mori [39] and Noguchi [44], g has either finite integral order or infinite order. Hence g is transcendental. Moreover Biancofiore shows that

$$\frac{T_f(r,s)}{T_g(r,s)} \to 1 \qquad \text{for } r \to \infty. \tag{9.38}$$

Hence f is transcendental.

<u>Theorem 9.2. Second Main Theorem and Defect relation.</u>
If (1)-(12) hold, then

$$\boxed{\sum_{j=1}^{q} m_f(r,a_j) \leqq (n+1)T_f(r,s) + o(T_f(r,s))} \tag{9.39}$$

$$\boxed{\sum_{j=1}^{q} \delta_f(a_j) \leq \frac{n+1}{p}}\,. \tag{9.40}$$

Biancofiore [4] has shown by example that this limit is sharp.

We will give an example of a holomorphic curve which is algebraically non-degenerate of degree 2 but algebraically degenerate of degree 3 for which (9.40) is violated. The example is due to Aldo Biancofiore. First we need a lemma.

<u>Lemma 9.3.</u>
Take $a_j \in \mathbb{R}$ for $j = 0, \ldots, n$ with $a_j \neq a_k$ if $j \neq k$. Define

$$A = \operatorname{Max}\{a_0,\ldots,a_n\} \qquad a = \operatorname{Min}\{a_0,\ldots,a_n\}.$$

Take $0 \neq p_j \in \mathbb{R}$ for $j = 0, \ldots, n$. Define

(9.41) $$S(r) = \frac{1}{2\pi} \int_{-\pi}^{+\pi} \log \left| \sum_{j=0}^{n} p_j e^{a_j r \cos \phi} \right| d\phi.$$

Then

$$\frac{S(r)}{r} \to \frac{(A - a)}{\pi} \qquad \text{for } r \to \infty.$$

Proof. W.l.o.g. we can assume that $a = a_0 < a_1 < \ldots < a_n = A$. Define

$$S_1(r) = \frac{1}{2\pi} \int_{-\frac{\pi}{2}}^{+\frac{\pi}{2}} \log \left| \sum_{j=0}^{n-1} p_j e^{(a_j - A) r \cos \phi} + p_n \right| d\phi$$

$$T_1(r) = \frac{1}{2\pi} \int_{-\frac{\pi}{2}}^{+\frac{\pi}{2}} \log e^{Ar \cos \phi} \, d\phi = \frac{Ar}{\pi}$$

$$S_2(r) = \frac{1}{2\pi} \int_{-\pi}^{-\frac{\pi}{2}} \log \left| \sum_{j=1}^{n} p_j e^{(a_j - a) r \cos \phi} + p_1 \right| d\phi$$

$$T_2(r) = \frac{1}{2\pi} \int_{-\pi}^{-\frac{\pi}{2}} \log e^{ar \cos \phi} \, d\phi = - \frac{ar}{2\pi}$$

$$S_3(r) = \frac{1}{2\pi} \int_{\frac{\pi}{2}}^{\pi} \log \left| \sum_{j=1}^{n} p_j e^{(a_j - a) r \cos \phi} + p_1 \right| d\phi$$

$$T_3(r) = \frac{1}{2\pi} \int_{\frac{\pi}{2}}^{\pi} \log e^{ar \cos \phi} \, d\phi = - \frac{ar}{2\pi} .$$

Then

$$S(r) = \sum_{j=1}^{3} (S_j(r) + T_j(r)).$$

If $\phi \in \mathbb{R}\left(-\frac{\pi}{2},\frac{\pi}{2}\right)$ and $j \in \mathbb{Z}[0, n - 1]$, then $(a_j - A)r \cos \phi < 0$. If $\phi \in \mathbb{R}\left(-\pi,-\frac{\pi}{2}\right) \cup \mathbb{R}\left(\frac{\pi}{2},\pi\right)$ and $j \in \mathbb{Z}[1,n]$, then $(a_j - a)r \cos \phi < 0$. Hence

$$S_1(r) \to \frac{1}{2} \log|p_n| \qquad S_j(r) \to \frac{1}{4} \log|p_1| \qquad \text{for } j = 2, 3 \quad \text{if } r \to \infty$$

and

$$\frac{S_j(r)}{r} \to 0 \qquad \text{for } r \to \infty \quad \text{and } j = 1, 2, 3$$

$$\frac{T_1(r)}{r} = \frac{A}{\pi} \qquad \frac{T_2(r)}{r} = \frac{T_3(r)}{r} = -\frac{a}{2\pi}.$$

Therefore

$$\frac{S(r)}{r} \to \frac{A - a}{\pi} \qquad \text{for } r \to \infty$$

q.e.d.

Define $\alpha_j \in \mathbb{C}^3_{[2]}$ by

$$\alpha_1(\mathfrak{z}) = z_0 z_2 - z_1^2 - z_1 z_2 + z_2^2$$

$$\alpha_2(\mathfrak{z}) = z_0 z_1 + z_0 z_2 + z_1^2$$

$$\alpha_3(\mathfrak{z}) = z_0^2 + z_1 z_2.$$

Define $a_j = \mathbb{P}(\alpha_j) \in \mathbb{P}(\mathbb{C}^{(3)}_{[2]})$. A computation shows that a_1, a_2, a_3 are in general position on $\mathbb{P}_2$.

A holomorphic map $f : \mathbb{C} \to \mathbb{P}_2$ and a reduced representation $\mathfrak{v} : \mathbb{C} \to \mathbb{C}^3$ are defined by

$$\mathfrak{v}(z) = (1, e^z, e^{2z} + e^{3z}) \qquad\qquad f = \mathbb{P} \circ \mathfrak{v}.$$

Then

$$\alpha_1(\mathfrak{v}(z)) = e^{5z}(2 + e^z)$$

$$\alpha_2(\mathfrak{v}(z)) = e^z(1 + e^z)^2$$

$$\alpha_3(\mathfrak{v}(z)) = 1 + e^{3z} + e^{4z}.$$

Assume that $\alpha \in \mathbb{C}^3_{(2)}$ exists such that $\alpha \circ \mathfrak{v} \equiv 0$. Then $\alpha = \sum_{0 \le i \le j \le 2} a_{ij} z_i z_j$ and

$$0 \equiv a_{00} + a_{01}e^z + a_{02}(e^{2z} + e^{3z}) + a_{11}e^{22}$$
$$+ a_{12}(e^{3z} + e^{4z}) + a_{22}(e^{4z} + 2e^{5z} + e^{6z})$$

$$0 \equiv a_{00} + a_{01}e^z + (a_{02} + a_{11})e^{2z} + (a_{12} + a_{02})e^{3z}$$
$$+ (a_{11} + a_{22})e^{4z} + 2a_{22}e^{5z} + a_{22}e^{6z}.$$

Hence $a_{00} = a_{01} = (a_{02} + a_{11}) = (a_{12} + a_{02}) = (a_{11} + a_{22}) = a_{22} = 0$ or $a_{ij} = 0$. Therefore $\alpha \equiv 0$. Consequently f is algebraically non-degenerate of degree 2. Define $\beta \in \mathbb{C}^3_{(3)}$ by

$$\beta(\mathfrak{z}) = z_0 z_1^2 + z_1^3 - z_0^2 z_2 \not\equiv 0.$$

Then

$$\beta \circ \mathfrak{v} \equiv 0.$$

Therefore f is algebraically degenerate of order 3. For $0 < s < r$ we have

$$T_f(r,s) = \int_{\mathbb{C}\langle r\rangle} \log\|\mathfrak{w}\|\sigma - \int_{\mathbb{C}\langle s\rangle} \log\|\mathfrak{w}\|\sigma$$

where

$$\int_{\mathbb{C}\langle r\rangle} \log\|\mathfrak{w}\|\sigma$$

$$= \frac{1}{2} \int_{\mathbb{C}\langle r\rangle} \log(1 + |e^z|^2 + |e^{4z}| + e^{2z+3\bar{z}} + e^{2\bar{z}+3z} + |e^{6z}|)\sigma$$

$$= \frac{1}{2}\,\frac{1}{2\pi} \int_0^{2\pi} \log|1 + e^{2r\cos\phi} + e^{4r\cos\phi} + 2e^{5r\cos\phi} + e^{6r\cos\phi}|\sigma.$$

Therefore

$$\boxed{\frac{T_f(r,s)}{r} \to \frac{3}{\pi}} \qquad \text{for } r \to \infty.$$

Also

$$N_f(r,s;a_1) = \int_{\mathbb{C}\langle r\rangle} \log|2 + e^z|\sigma - \int_{\mathbb{C}\langle s\rangle} \log|2 + e^z|\sigma$$

where

$$\int_{\mathbb{C}\langle r\rangle} \log|2 + e^z|\sigma = \frac{1}{2} \int_{\mathbb{C}\langle r\rangle} \log(4 + e^z + e^{\bar{z}} + |e^z|^2)\sigma$$

$$= \frac{1}{2}\,\frac{1}{2\pi} \int_0^{2\pi} \log(4 + 2e^{r\cos\phi} + e^{2r\cos\phi})\sigma.$$

Hence

$$\boxed{\frac{N_f(r,s;a_1)}{r} \to \frac{1}{\pi}} \qquad \text{for } r \to \infty.$$

Also

$$N_f(r,s;a_2) = 2 \int_{\mathbb{C}\langle r\rangle} \log|1 + e^z|\sigma - 2 \int_{\mathbb{C}\langle s\rangle} \log|1 + e^z|\sigma$$

where

$$2 \int_{\mathbb{C}\langle r\rangle} \log|2 + e^z|\sigma = \int_{\mathbb{C}\langle r\rangle} \log(1 + e^z + e^{\overline{z}} + |e^z|^2)\sigma$$

$$= \frac{1}{2\pi} \int_0^{2\pi} \log(1 + 2e^{r \cos \phi} + e^{2r \cos \phi})\, d\phi.$$

Hence

$$\frac{N_f(r,s;a_2)}{r} \to \frac{2}{\pi} \qquad \text{for } r \to \infty.$$

Also

$$N_f(r,s;a_3) = \int_{\mathbb{C}\langle r\rangle} \log|1 + e^{3z} + e^{4z}|\sigma - \int_{\mathbb{C}\langle s\rangle} \log|1 + e^{3z} + e^{4z}|\sigma$$

where

$$\int_{\mathbb{C}\langle r\rangle} \log|1 + e^{3z} + e^{4z}|\sigma$$

$$= \frac{1}{2} \int_{\mathbb{C}\langle r\rangle} \log|1 + |e^{3z}|^2 + |e^{4z}|^2 + 2\,\mathrm{Re}(e^{3z} + e^{4z} + e^{3z+4\overline{z}})|\sigma$$

$$= \frac{1}{2}\,\frac{1}{2\pi} \int_{\mathbb{C}\langle r\rangle}$$

$$\log|1 + 2e^{3r \cos \phi} + 2e^{4r \cos \phi} + 2e^{7r \cos \phi} + e^{6r \cos \phi} + e^{8r \cos \phi}|\sigma.$$

Hence

$$\boxed{\frac{N_f(r,s;a_3)}{r} \to \frac{4}{\pi}} \qquad \text{for } r \to \infty.$$

Consequently

$$\delta_f(a_1) = 1 - \lim_{r\to\infty} \frac{N_f(r,s;a_1)}{2T_f(r,s)} = 1 - \frac{1}{6} = \frac{5}{6}$$

$$\delta_f(a_2) = 1 - \lim_{r\to\infty} \frac{N_f(r,s;a_2)}{2T_f(r,s)} = 1 - \frac{2}{6} = \frac{2}{3}$$

$$\delta_f(a_3) = 1 - \lim_{r\to\infty} \frac{N_f(r,s;a_3)}{2T_f(r,s)} = 1 - \frac{4}{6} = \frac{1}{3}$$

$$\delta_f(a_1) + \delta_f(a_2) + \delta_f(a_3) = \frac{11}{6} > \frac{3}{2} = \frac{n+1}{p} .$$

References

[1] L. Ahlfors, The theory of meromorphic curves, Acta Soc. Sci. Fenn. Nova Ser. A 3 (4) (1941), 31 pp.

[2] S. Bergman, Über den Wertevorrat einer Function von zwei komplexen Veränderlichen, Math. Z. 36 (1932), 171-183.

[3] A. Biancofiore and W. Stoll, Another proof of the Lemma of the Logarithmic Derivative in several complex variables.

[4] A. Biancofiore, A hypersurface defect relation for a class of meromorphic maps.

[5] A. Biancofiore, A defect relation for holomorphic curves.

[6] A. Biancofiore, Second Main Theorem without exceptional intervals.

[7] E. Borel, Sur les zéros des fonctions entières, Acta Math. 20 (1897), 357-396.

[8] R. Bott and S. S. Chern, Hermitian vector bundles and the equidistribution of the zeroes of their holomorphic sections, Acta Math. 114 (1965), 71-112.

[9] J. Carlson and Ph. Griffiths, A defect relation for equidimensional holomorphic mappings between algebraic varieties, Ann. of Math. (2) 95 (1972), 567-584.

[10] J. Carlson and Ph. Griffiths, The order functions of entire holomorphic mappings, Value-Distribution Theory Part A (edited by R. O. Kujala and A. L. Vitter III) Pure and Appl. Math. 25, Marcel Dekker, New York, 1974, 225-248.

[11] H. Cartan, Sur les zéros des combinaisons linéaires de p fonctions holomorphes données, Mathematica (Cluj) 7 (1933), 5-31.

[12] S. S. Chern, The integrated form of the first main theorem for complex analytic mappings in several variables, Ann. of Math. (2) 71 (1960), 536-551.

[13] S. S. Chern, Holomorphic curves in the plane, Diff. Geom. in honor of K. Yano, Kinokunija, Tokyo, 1972, 72-94.

[14] M. Cowen, Hermitian vector bundles and value distribution for Schubert cycles, Trans. Amer. Math. Soc. 180 (1973), 189-228.

[15] M. Cowen and Ph. Griffiths, Holomorphic curves and metrics of negative curvature, J. Analyse Math. 29 (1976), 93-152.

[16] H. Fujimoto, The uniqueness problem of meromorphic maps into the complex projective space, Nagoya Math. J. 58 (1975), 1-23.

[17] H. Fujimoto, A uniqueness theorem of algebraically non-degenerate meromorphic maps into $\mathbb{P}^n(\mathbb{C})$, Nagoya Math. J. 64 (1976), 117-147.

[18] M. Green, Holomorphic maps into complex projective space omitting hyperplanes, Trans. Amer. Math. Soc. 169 (1972), 89-103.

[19] M. Green, Some Picard theorems for holomorphic maps to algebraic varieties, Amer. J. Math. 97 (1975), 43-75.

[20] M. Green, Some examples and counter-examples in value distribution theory for several variables, Compositio Math. 30 (1975), 317-322.

[21] W. H. Greub, Linear Algebra, Grundl. d. Math. Wiss. 97, Springer, 1967, 434 pp.

[22] W. H. Greub, Multilinear Algebra, Grundl. d. Math. Wiss. 136, Springer, 1967, 224 pp.

[23] Ph. Griffiths, Holomorphic mappings into canonical algebraic varieties, Ann. of Math. (2) 93 (1971), 439-458.

[24] Ph. Griffiths, Holomorphic mappings: Survey of some results and discussion of open problems, Bull. Amer. Math. Soc. 78 (1972), 374-382.

[25] Ph. Griffiths and J. King, Nevanlinna theory and holomorphic mappings between algebraic varieties, Acta Math. 130 (1973), 145-220.

[26] Ph. Griffiths, Some remarks on Nevanlinna theory, Value-Distribution Theory Part A (edited by R. O. Kujala and A. L. Vitter III) Pure and Appl. Math. 25, Marcel Dekker, New York, 1974, 1-11.

[27] Ph. Griffiths, Entire holomorphic mappings in one and several complex variables, Annals of Math. Studies 85, Princeton Univ. Press, Princeton, N.J., 1976, X + 99 pp.

[28] F. Gross, Factorization of meromorphic functions, Math. Res. Center Naval Research Lab., Washington D.C., 1972, 258 pp.

[29] L. Gruman, Value Distribution for holomorphic maps in $\mathbb{C}^n$, Math. Ann. 245 (1979), 199-218.

[30] Y. Hashimoto, On the deficiencies of algebroid functions, J. Math. Soc. Japan 31 (1979), 29-37.

[31] W. K. Hayman, Meromorphic functions, Oxford Math. Monographs, Clarendon Press, 1964, 191 pp.

[32] J. Hirschfelder, The first main theorem of value distribution in several variables, Invent. Math. 8 (1969), 1-33.

[33] Kataka Conference, Open problems in geometric function theory, Proceed. 5 Int. Symp. Division Math., The Taniguchi Foundation, (available from: Dept. of Math., Osaka Univ., Toyonaka, 560 Japan), 1978, 22 pp.

[34] H. Kneser, Zur Theorie der gebrochenen Funktionen mehrerer Veränderlichen, Jber. Deutsch Math.-Verein. 48 (1938), 1-28.

[35] P. Lelong, Sur l'extension aux fonctions entières de n variables, d'ordre fini, d'un dévelopment canonique de Weierstrass, C.R. Acad Sci. Paris 237 (1953), 865-867.

[36] P. Lelong, Fonctions entières (n-variables) et fonctions plurisousharmoniques d'ordre fini dans $\mathbb{C}^n$, J. Analyse Math. 12 (1964), 365-407.

[37] H. Levine, A theorem on holomorphic mappings into complex projective space, Ann. of Math. (2) 71 (1960), 529-535.

[38] R. Molzon and B. Shiffman, Average growth estimates for hyperplane sections of entire analytic sets.

[39] S. Mori, On the deficiencies of meromorphic mappings of $\mathbb{C}^m$ into $\mathbb{P}^N(\mathbb{C})$, Nagoya Math. J. 67 (1977), 165-176.

[40] S. Mori, The deficiencies and the order of holomorphic mappings of $\mathbb{C}^n$ into a compact complex manifold, Tôhoku Math. J. 31 (1979), 285-291.

[41] S. Mori, Holomorphic curves with maximal deficiency sum, Kodai Math. J. 2 (1979), 116-122.

[42] J. Murray, A second main theorem of value distribution theory on Stein manifolds with pseudoconvex exhaustion (1974 Notre Dame Thesis), 69 pp.

[43] R. Nevanlinna, Eindentige analytische Funktionen, Grundl. d. Math. Wiss. XLVC, 2 ed., Springer-Verlag, 1953, 379 pp.

[44] J. Noguchi, A relation between order and defects of meromorphic mappings of $\mathbb{C}^n$ into $\mathbb{P}^N(\mathbb{C})$, Nagoya Math. J. 59 (1975), 97-106.

[45] J. Noguchi, Meromorphic mappings of a covering space over $\mathbb{C}^m$ into a projective algebraic variety and defect relations, Hiroshima Math. J. 6 (1976), 265-280.

[46] J. Noguchi, Holomorphic curves in algebraic varieties, Hiroshima Math. J. 7 (1977), 833-853.

[47] J. Noguchi, Supplement to: Holomorphic curves in algebraic varieties, Hiroshima Math. J. 10 (1980), 229-231.

[48] T. Ochiai, On holomorphic curves in algebraic varieties with ample irregularity, Invent. Math. 43 (1977), 83-96.

[49] M. Ozawa, On the growth of algebroid functions with several deficiencies, Kodai Math. Sem. Rep. 22 (1970), 122-127.

[50] M. Ozawa, On the growth of algebroid functions with several deficiencies, II, Kodai Math. Sem. Rep. 22 (1970), 129-137.

[51] L. I. Ronkin, An analog of the canonical product for entire functions of several complex variables, Trudy Moskov. Mat. Obšč. 18 (1968), 105-146 translated as Trans. Moscow Math. Soc. 18 (1968), 117-160.

[52] L. I. Ronkin, Introduction to the theory of entire functions of several variables, Translations of Math. Monographs 44, Amer. Math. Soc., 1974, 272 pp.

[53] H. Rutishauser, Über die Folgen und Scharen von analytischen und meromorphen Functionen mehrerer Variabeln, sowie von analytischen Abbildungen, Acta Math. 83 (1950), 249-325.

[54] M.-H. Schwartz, Formules apparenté es a celles de Nevanlinna-Ahlfors pour certaines applications d'une variete a n-dimensions dans un autre, Bull. Soc. Math. France 82 (1954), 317-360.

[55] B. Shiffman, Holomorphic and meromorphic mappings and curvature, Math. Ann. 222 (1976), 171-194.

[56] B. Shiffman, Applications of geometric measure theory to value distribution theory for meromorphic maps, Value-Distribution Theory Part A (edited by R. O. Kujala and A. L. Vitter III) Pure and Appl. Math. 25, Marcel Dekker, New York, 1974, 63-95.

[57] B. Shiffman, Nevanlinna defect relations for singular divisors, Invent. Math. 31 (1975), 155-182.

[58] B. Shiffman, Holomorphic curves in algebraic manifolds, Bull. Amer. Math. Soc. 83 (1977), 553-568.

[59] B. Shiffman, On holomorphic curves and meromorphic maps in projective space, Indiana Univ. Math. J. 28 (1979), 627-641.

[60] H. Skoda, Solution a croissance du second problème de Cousin dans $\mathbb{C}^n$. Ann. Inst. Fourier (Grenoble) 21 (1971), 11-23.

[61] H. Skoda, Sous-ensembles analytiques d'ordre fini dans $\mathbb{C}^n$, Bull. Soc. Math. France 100 (1972), 353-408.

[62] L. Smiley, Dependence theorems for meromorphic maps (1979 Notre Dame Thesis), 57 pp.

[63] W. Stoll, Mehrfache Integrale auf komplexen Mannigfaltigkeiten, Math. Z. 57 (1952), 116-154.

[64] W. Stoll, Ganze Funktionen endlicher Ordnung mit gegebenen Nullstellenflächen, Math. Z. 57 (1953), 211-237.

[65] W. Stoll, Die beiden Hauptsätze der Wertverteilungstheorie bei Funktionen mehrerer komplexer Veränderlichen I, II, Acta Math. 90 (1953), 1-115, ibid. 92 (1954), 55-169.

[66] W. Stoll, Einige Bemerkungen zur Fortsetzbarkeit analytischer Mengen, Math. Z. 60 (1954), 287-304.

[67] W. Stoll, The growth of the area of a transcendental analytic set I, Math. Ann. 156 (1964), 47-78, II, Math. Ann. 156 (1964), 144-170.

[68] W. Stoll, The multiplicity of a holomorphic map, Invent. Math. 2 (1966), 154-218.

[69] W. Stoll, A general first main theorem of value distribution, Acta Math. 118 (1967), 111-191.

[70] W. Stoll, About value distribution of holomorphic maps into projective space, Acta Math. 123 (1969), 83-114.

[71] W. Stoll, Value distribution of holomorphic maps into compact complex manifolds, Lecture Notes in Mathematics 135, Springer-Verlag, 1970, 267 pp.

[72] W. Stoll, Value distribution of holomorphic maps, Several complex variables I, Maryland 1970 Lecture Notes in Mathematics 155, Springer-Verlag, 1970, 165-170.

[73] W. Stoll, Fiber integration and some of its applications, Symp. on Sev. Compl. Var., Park City, Utah 1970, Lecture Notes in Mathematics 184, Springer-Verlag, 1971, 109-120.

[74] W. Stoll, Deficit and Bezout estimates, Value-Distribution Theory Part B (edited by R. O. Kujala and A. L. Vitter III) Pure and Appl. Math. 25, Marcel Dekker, New York, 1973, 271 pp.

[75] W. Stoll, Holomorphic functions of finite order in several complex variables, Conf. Board Math. Scien. Reg. Conference Series in Math. 21, Amer. Math. Soc., 1974, 83 pp.

[76] W. Stoll, Aspects of value distribution theory in several complex variables, Bull. Amer. Math. Soc. 83 (1977), 166-183.

[77] W. Stoll, Value Distribution on parabolic spaces, Lecture Notes in Mathematics 600, Springer-Verlag, 1977, 216 pp.

[78] W. Stoll, A Casorati-Weierstrass Theorem for Schubert zeroes in semi-ample holomorphic vector bundles, Atti della Acc. Naz. d. Lincei Serie VIII 15 (1978), 63-90.

[79] W. Stoll, The characterization of strictly parabolic manifolds (to appear in Annali di Pisa).

[80] W. Stoll, The characterization of strictly parabolic spaces.

[81] C.-H. Sung, On the deficiencies of holomorphic curves in projective space.

[82] N. Toda, Sur les combinaisons exceptionelles de fonctions holomorphes; applications aux fonctions algébroides, Tôhoku Math. J. 22 (1970), 290-319.

[83] N. Toda, Sur le nombre de combinaisons exceptionelles; applications aux fonctions algébroides, Tôhoku Math. J. 22 (1970), 480-491.

[84] Ch. Tung, The first main theorem of value distribution on complex spaces, Atti della Acc. Naz. d. Lincei Serie VIII 15 (1979), 93-262.

[85] A. Vitter, The lemma of the logarithmic derivative in several complex variables, Duke Math. J. 44 (1977), 89-104.

[86] A. Weitsman, A theorem on Nevanlinna deficiencies, Acta Math. 128 (1972), 41-52.

[87] H. Weyl and J. Weyl, Meromorphic functions and analytic curves, Annals of Math. Studies 12, Princeton Univ. Press, Princeton, N.J., 1943, 269 pp.

[88] P.-M. Wong, Defect relations for meromorphic maps on parabolic manifolds (to appear in Duke Journal).

[89] H. Wu, Mappings of Riemann surfaces (Nevanlinna Theory), Proceed. of Symp. in Pure Math. 11 (1968), 480-552.

[90] H. Wu, Remarks on the first main theorem of equidistribution theory, I, J. Differential Geom. 2 (1968), 197-202, II, ibid. 2 (1968), 369-384, III, ibid. 3 (1969), 83-94, IV, ibid. (1969), 433-446.

[91] H. Wu, The equidistribution theory of holomorphic curves, Annals of Math. Studies 64, Princeton Univ. Press, Princeton, N.J., 1970, 219 pp.

Index

COMPACT HAUSDORFF TRANSVERSALLY HOLOMORPHIC FOLIATIONS

D. Sundararaman

This is a revised version of part of the lectures given by the author at Trieste Seminar on Complex Analysis and its Applications. The last part of this paper gives a report on the results obtained by Girbau-Haefliger-Sundararaman subsequent to the Seminar. The author would like to thank Professor A. Haefliger for his suggestions. The remaining part of the lectures of the author has appeared in [58]. The author thanks Centro de Investigación del I.P.N., México City, for hospitality during the writing of the paper.

§1. Holomorphic Foliations.

Let M denote a compact connected complex manifold of complex dimension n. A holomorphic foliation F of codimention q is a decomposition $F = \{L_\alpha\}_{\alpha \in \Lambda}$ of M into pairwise disjoint connected subsets L_α such that each point $x \in M$ has a neighborhood U and a holomorphic chart $\varphi : U \to P_r^n$ (:= Polydisc around the origin of $\mathbb{C}^n$ of radius r) with the property that the connected components of $\varphi(L_\alpha \cap U)$ are the form $\{z \in P_r^n | Z_{n-q+1} = \text{constant}, \ldots, Z_n = \text{constant}\}$. The L_α are called the leaves of the foliation. Each leaf L_α is a complex submanifold of M of complex dimension $n-q$; But the topology of the submanifold L_α may not be topology induced by M. This is because it is possible that a leaf may pass through a given chart infinitely often and may also accumulate on itself. It is possible for a foliation to have a dense leaf. We can introduce however another topology, called the leaf topology, on M to avoid this situation; the basis sets of the leaf topology are of the form $(t \in U \,|\, Z_1(t)$ constant$, \ldots, Z_q(t)=$constant$)$. With the leaf topology on M, the leaves become precisely the connected components of M. Consideration of leaf topology becomes necessary sometimes in order to have a better picture of the topology of the foliation. We do not consider leaf topology in this paper.

Holomorphic submersions, holomorphic fibrations and holomorphic vector fields provide simplest examples of holomorphic foliations. More complicated examples are given by integrable Pfaffian systems of partial differential equations on complex manifolds (see Gerard [14], Jouanolou [32]). Interesting examples are obtained by considering the complex Lie group actions on a complex manifold where all the orbits have the same dimension. Holomorphic actions of the additive group of complex numbers on complex manifolds where the orbits are complex tori (these are called periodic holomorphic flows) are especially important.

From the definition it is clear that the existence of holomorphic or differentiable foliation can not be taken for granted. The following lines of investigation are suggested:
Given a compact complex manifold M,

1. Does M admit holomorphic foliations (of a given codimension q)?

2. If M admits a holomorphic foliation $\mathfrak{F}$., What is the internal structure (the topology) of the foliation ? In what way the global behaviour of $\mathfrak{F}$ is related to the global properties of M ?

3. If M admits holomorphic folliations, how does a neighborhood of a given foliation look like in the space of all foliations? That is, Is there a deformation theory for holomorphic foliations analogous to the deformation theory of compact complex manifolds ? (Refer Kodaira-Spencer [34], Kuranishi [37], Kodaira-Morrow [35], Narasimhan [45], Sundararaman [58]).

4. If M admits holomorphic foliations can we construct, "a moduli space", which parametrises the holomorphic foliations of M (of fixed codimension)? Then study its properties (analogous to the Moduli Theory of compact complex manifolds: See especially Mumford [44]).

5. Discuss stability properties of a given holomorphic foliation of M (analogous to stability theory of smooth foliations

Reeb [51], Thurston [60], Langevin-Rosenberg [39], Epstein-Rosenberg [12], Edwards-Millet-Sullivan [9], Schweitzer [53]).

In regard to question 1 , Bott obtained topological obstructions for the existence of foliations. By taking the tangent spaces to the leaves, we get a holomorphic subbundle F of codimension q of the holomorphic tangent bundle $T_C M$ of M. This subbundle is integrable (involutive): for any two sections V_1, V_2 of F, their Lie bracket $[V_1, V_2]$ is also a section of F. Conversely by the Frobenins theorem an integrable holomorphic subbundle F of codimension q gives a foliation $\mathfrak{F}$ of codimension q of M. Thus a holomorphic foliation of codimension q can be defined as an integrable holomorphic subbundle of codimension q of $T_C M$. In order that M admits a holomorphic foliation of codimension q, $T_C M$ must admit first a holomorphic subbundle of codimension q, (holomorphic almost foliation) and this subbundle must be integrable (homotopic to an integrable subbundle). There are topological obstructions to the existence of subbundles (see eg. Sundararaman [58], chapter I). Assuming that $T_C M$ admits a holomorphic subbundle $\mathfrak{F}$, the first known obstructions to integrability are given by the

<u>Bott Vanishing Theorem:</u> Let M be a complex manifold of complex dimension n. Let F be an integrable holomorphic subbundle of complex codimension q of the holomorphic tangent bundle $T_C(M)$. Let $\nu(F) = T_C M/F$ be the quotient bundle (called the normal bundle to the foliation F). Then the polynomial Chern classes of $\nu(F)$ must be zero in dimensions greater than 2 q.

The proof of this theorem is not complicated it is based on the Chern Weil theory of characteristic classes. For the complete proof see Bott [3]. Using this theorem Bott has shown that the complex projective spaces $P^n(C)$ for n odd, do not admit holomorphic foliations. $P^n(C)$ however admit 'holomorphic foliations with singularities' For singular holomorphic foliations refer Bott-Baum [1].

Complete answers to the questions raised in 2 to 5 are not known as far as the author is aware of. A good program would be to consider these questions to start with for compact complex surfaces.

Deformation theory for foliations was initiated by Kodaira-Spencer [36]: the theory given by Kodaira-Spencer is in the general context of Multifoliate structures. Recently it has been proved by Hardrop [26] that every compact 3-manifold admits a smooth total foliation, which is a particular case of multifoliations. Following general deformation theory of Kodaira-Spencer, important results on deformations of transversally holomorphic foliations have been obtained recently: The last section of this paper gives a report on the work in progress of Girbau-Haefliger-Sundararaman on deformations of compact Hausdorff transversally holomorphic foliations.
Also refer Duchamp-Kalka [8], Gómez-Mont [18]. Holmann [30], [31] has studied stability properties of holomorphic foliations.

Apart from the importance of the theory of foliations as such, the methods of the theory ('foliation techniques') have been very useful and powerful (Refer for example Bedford-Kalka [2], Stoll [55])

§2. Compact Hausdorff foliations:

Smooth foliations with all leaves compact are important since they generalize periodic flows. Smooth foliations with a compact leaf, foliations with all leaves compact and foliations with no leaf compact have been extensively studied in recent years. Refer Epstein [10],[11], Edwards-Millet-Sullivan [9], Millet [41], Schweitzer [53], Vogt [64]. In the holomorphic category also, there are important examples of foliations with all leaves compact. There are also holomorphic foliations with no compact leaf. For example, if $w = 0$ is a completely integrable Pfaffian equation on the complex projective space $P^n(\mathbb{C})$ then the holomorphic foliation defined by $w = 0$ in the complement of the singular set of w in $P^n(\mathbb{C})$ has no compact leaf (for proof refer Gerard [14]). Call a foliation compact if all the leaves are compact.

The leaf space of a foliation $\mathfrak{F}$ of M is the set $M/\mathfrak{F}$ of all leaves provided with the quotient topology. In general the leaf space of a foliation has a complicated structure. An unpleasant situation is that the leaf space may not be Hausdorff even if the leaves are compact (Reeb [51] , Sullivan [55]). The leaf space of every smooth foliation of codimension 1 with compact leaves is Hausdorff (Edwards-Millet-Sullivan [9]). It can be shown that the leaf space is Hausdorff if each leaf is compact and stable (see Epstein [11]). A leaf L is said to be stable if each open neighborhood U of L contains an open saturated neighborhood U' of the leaf (that is U' is a union of leaves).

When the leaf space of a holomorphic foliation is Hausdorff, it has a complex structure: if in addition every leaf is compact and stable, the leaf space has a canonical complex structure:
Canonical in the sense that a function $f:V \to C$ of an open set of M/F is holomorphic iff $f \circ \pi$ is holomorphic on the open set $\pi^{-1}(V)$ of M where $\pi: M \to M/F$ is the projection map. Call a foliation Hausdorff if the leaf space is Hausdorff. If F is a holomorphic foliation of complex codimension one and if all the leaves are compact, then all the leaves must be stable and hence the foliation must be Hausdorff (see e. Holmann [30]). This is not true for higher codimensional foliations. Call a compact holomorphic foliation stable if all the leaves are stable.

Contrary to the smooth case, Holmann[31] has shown that any periodic holomorphic flow must be stable and any compact holomorphic foliation of a Kahler manifold (compact or not) must be stable. Recenty Th. Muller [43] has found an example of a compact holomorphic foliation of a noncompact complex manifold (necessarily non Kahler) which is unstable.

§ 3. Transversally holomorphic foliations.

Holomorphic foliations on a complex manifold M can be defined in an equivalent way by holomorphic Haefliger cocycles: Consider a covering $\{U_\alpha\}_{\alpha\in\Lambda}$ of M and holomorphic submersions $f_\alpha : U_\alpha \to C^q$, for each α, $\beta \in \Lambda$ with $U_\alpha \cap U_\beta \neq \emptyset$, for each $x \in U_\alpha \cap U_\beta$ there is $\gamma^x_{\alpha\beta} \in \Gamma_C$ (: =

the pseudo group of local biholomorphic maps of C^n) satisfying $f_\alpha = \gamma^x_{\alpha\beta} \circ f_\beta$ in a neighborhood of x and for $\alpha, \beta, \gamma \in \Lambda$, $x \in U_\alpha \cap U_\beta \cap U_\gamma$, $\gamma^x_{\alpha\gamma} = \gamma^x_{\alpha\beta}\gamma^x_{\beta\gamma}$ in a neighborhood of $f_\gamma(x)$.

A collection $\{U_\alpha, f_\alpha, (\gamma^x_{\alpha\beta})_{x \in U_\alpha \cap U_\beta}\}$ maximal with respect to the properties stated above is called a holomorphic Haefliger cocycle of codimension q. It is clear from the definition that a holomorphic foliation of codimension q determines a holomorphic Haefliger cocycle of codimension q and conversely a holomorphic Haefliger cocycle of co dimension q determines a holomorphic foliation of codimention q. More generally if f_α are not necessarily submersions, then we get a singular holomorphic foliation in the sence of Haefliger. Analogusly smooth foliations also can be defined in terms of differentiable Haefliger cocycles.

In between the categories of differentiable (smooth) and holomorphic foliations, lies the category of transversally holomorphic foliations. These were first introduced by Haefliger and are defined as follows: A codimension $2q$ foliation $\mathcal{F}$ of Y given by smooth Haefliger cocycle $\{U_\alpha, f_\alpha, (\gamma^x_{\alpha\beta})_{x \in U_\alpha \cap U_\beta}\}$. If the local diffeomorphisms $\gamma^x_{\alpha\beta}$ of R^{2q} are actually local biholomorphic maps of $C^q (:= R^{2q})$, we say $\mathcal{F}$ is a transversally holomorphic foliation of (complex) codimension q of the smooth manifold Y. Equivalently a transversally holomorphic foliation of Y is the pseudo group structure given by the pseudogroup $\Gamma_{t\cdot h}$ of local diffeomorphisms ψ of $R^n \times C^q$ such that if $\psi = (\phi_1, \ldots, \varphi_n, \psi_1, \ldots, \psi_q)$ then $(\partial \psi i / \partial \bar{z} j) = 0$ for $1 \leq 1$, $j \leq q$. Here $x^1, \ldots, x^n$ represent local cordinates in R^n and $z^1, \ldots, z^q$ represent local cordinates in C^q. In this case of transversally holomorphic foliations the leaves (which are smooth submanifolds) are attached together in a holomorphic manner.

Obviously any holomorphic foliation of a complex manifold can be viewed as transversally holomorphic foliation of the underlying differentiable manifold. The odd dimensional sphere S^{2r+1} has transversally holomorphic foliation given by the projection $S^{2r+1} \longrightarrow P^r(C)$. More interesting class of tranversally holomorphic foliations arise as follows: Let $\overline{Y}$ be the universal covering manifold of a differentiable manifold Y and regard the fundamental group $\pi_1(Y)$ as the group of covering transformations. Let M be any compact complex manifold of dimension q and let G = Aut M be the Lie group of holomorphic authomorphism group of M . Given a group homeomorphism $\gamma ; \pi_1(Y) \to G$. Then $\pi_1(Y)$ acts $\overline{Y} \times M$ as follows: $\lambda(\overline{y}, m) = (\lambda(\overline{y}), \gamma(\lambda) \circ m)$. The quotient of $\overline{Y} \times M$ by this action is a manifold X since $\pi_1(Y)$ acts as covering transformations on $\overline{Y}$. The natural transversally holomorphic foliation of the product $\overline{Y} \times M$ has leaves of the form $(Y \times m)_{m \in M}$ and X has the induced transversally holomorphic foliation. By considering the projection to the first factor X can be regarded as a fibre bundle over Y with fibres homeomorphic to M and the leaves of the foliation of X are transversal to the fibres. Conversly if $\mathfrak{F}$ is a transversally holomorphic foliation transverse to the fibres of a compact bundle $p : X \to Y$ with fibre M, then M inherits a complex structure and there is a group homomorphism (called the global holonomy homomorphism) $\gamma : \pi_1(Y) \to \text{Aut } M$ which is well defined up to conjugation and which caracterises the foliation.

Consider now a compact Hausdorff transversally holomorphic foliation $\mathfrak{F}$ of a compact smooth manifold Y of complex codimension q . Let X be the leaf space of $\mathfrak{F}$ then X is a compact normal complex space. The singularities of X are quotient singularities of the form $G \backslash \mathbb{C}^m$ where G is a finite subgroup of $GL(m, \mathbb{C})$. To see this we have to look at the local model of $\mathfrak{F}$ given by Epstein [11] [12]. The description is as follows: Given $\mathfrak{F}$ there exists a leaf L_0 with the following three properties:

(1) There exists a dense open set Y_0 of Y such that for each $y \in Y_0$, the leaf L_y passing through y is diffeomorphic to L_0 and has trivial holomony.

(2) For each leaf L , there exists a finite subgroup H , depending on L , of the unitary group U(q) which acts freely on the right on L_0 such that $L_0/H \approx L$.

Let D^q be the q-disk in $\mathbb{C}^k$. For each leaf L, the above group

H acts on $L_0 \times D^q$ by $h(x, y) = (x . h^1, h . y)$ for each $h \in H$, $x \in L_0$, $y \in D^q$. Let $L_0 \underset{H}{\times} D^q$ be the quotient. $L_0 \times D^q$ has the natural foliation given by the leaves $L_0 \times (y)$. This foliation is preserved by the action of H and hence $L_0 \underset{H}{\times} D^q$ inherits the foliation whose leaves are $p(L_0 \times (y))$ where $p : L_0 \times D^q \to L_0 \underset{H}{\times} D^q$ is the natural projection.

(3) There exists a diffeomorphism of $L_0 \underset{H}{\times} D^q$ onto a neighbourhhood of L, preserving the leaves of the foliation.

L_0 is called the generic leaf of $\mathfrak{F}$.

Thus we see that the leaf space X of a compact Hausdorff transversally holomorphic foliation $\mathfrak{F}$ of a compact smooth manifold Y is a compact complex V-manifold in the sense of Satake [52]. V-manifolds are referred to as Orbifolds by Thurston ([62], chapter 13). We don't get any additional information on X if $\mathfrak{F}$ is a compact Hausdorff holomorphic foliation of a compact complex manifold Y. The quotient map $q : Y \to X$ is a submersion in the category of V-manifolds.

§ 4. Deformations of compact complex V-manifolds and Deformations of compact Hausdorff transversally holomorphic foliations.

Let M be a complex manifold of dimension n. Let G be a properly discontinuos group of automorphisms of M. Then $G\backslash M$ is a normal complex space, with singularities at the fixed points of the action. Thus x is a singularity if and only if the isotropy G_x is nontrivial. The isotropy groups are finite groups. In a neighbourhood of $x \in M$, coordinates can be introduced so that the action of the finite group G_x is linear (Cartan [5]). Hence it sufficies to consider finite subgroups G of $G\ell(n, C)$ and study the singularity of the quotient $G\backslash C^n$ at the point 0 corresponding to the origin in C^n. A complex space X is a V-manifold if its local model is of the form $(G\backslash C^n, 0)$ where $G \subset G\ell(n, C)$ is a finite subgroup. The singularity determined by G is denoted by $V(G)$. Two finite subgroups G, H of $G\ell(n, C)$ are called V-equivalent if the singularities $V(G)$ and $V(H)$ are analytically the same. Linearly equivalent subgroups are V - equivalent. But two subgroups which are not linearly equivalent may be V-equivalent (consider subgroups of $G\ell(1, C)$). Call a finite subgroup G of $G\ell(n, C)$ small if no element of G has 1 as eigenvalue of multiplicity $(n-1)$.

Proposition 4.1 (Prill [50]). 1) Every n dimensional singularity of a quotient space is isomorphic to the singularity at 0 of $G\backslash C^n$ for some finite small subgroup of $G\ell(n, C)$.

2) Two finite small subroups of $G\ell(n,C)$ are V-equivalent if and only if they are linearly equivalent.

Prill has also given an estimate on the number of isomorphism classes of V-germs belonging to the same finite groups. The V-germ singularities are not rigid in dimension 2 but are rigid in higher dimensions. For details refer to Brieskorn [4], Schlessinger [54], Pinkham [48].

Globally V-manifolds (orbifolds) are defined in terms of uniformising coverings as follows:

Definition. A structure of Γ-V-manifold (or Γ- orbifold) on a topological space X is given by

i) an open covering $\{U_i\}_{i\ I}$ of X closed under finite intersection (we could ask that for any U_i and U_j there is $U_k \subset U_i \cap U_j$.)

ii) for each U_i, there is an open set $\tilde{U}_i$ of C^n, a finite group G_i of automorphisms of $\tilde{U}_i$ belonging to Γ, and a map $\pi_i : \tilde{U}_i \to U_i$ invariant by G_i (uniformizing map) inducing a homeomorphism of $G_i \backslash \tilde{U}_i$ on U_i,

iii) for $U_i \subset U_j$, there is an injective homomorphism $f_{ji} : G_i \to G_j$ and an element γ_{ji} of Γ mapping $\tilde{U}_i$ on an open set $\tilde{U}_j$ and which is f_{ji} equivariant

The notions of differentiable, analytic functions, riemannian metrics, differential forms, vector fields, bundles are defined for V-manifolds, by the corresponding notion in the uniformizing set $\tilde{U}_i$ in a compatible way with respect to the actions of G_i and the change of coordinates γ_{ji}.

When Γ is the pseudogroup of local complex analytic automorphisms of C^n, then a Γ-V-manifold will be called a complex V-manifold (or complex orbifold).

Any compact complex V-manifold X is the space of leaves of a transversally analytic compact Hausdorff foliation whose generic leaf is simply connected.

Indeed introduce on X a hermitian metric and let P be the principal bundle of unitary frames on X. As the isotropy subgroups of the points of X in the uniformizing coordinates neighbourhood U operates withoug fixed point on the bundle of frames, P is a non singular manifold. By considering an inclusion of U_n is SU_{n+1}, we can extend the structural group of P to SU_{n+1}, and so we obtain a fibre bundle over X with simply connected genericfibre.

<u>Holomorphic</u> <u>family</u> <u>of</u> <u>compact</u> <u>complex</u> V-<u>manifolds</u>: Let S be an analytic space not necessarily reduced.

Denote by Γ_S the pseudogroup of local complex analytic automorphisms of $S \times C^n$ over S, i.e. compatible with the projection p_S on S.

A holomorphic family of compact complex V-manifolds parametrized by a complex space S is a continuous proper mapp of a space X on S together with a structure of Γ_S-V-manifold on X such that the uniformizing maps π_i are compatible with the projections on S, namely $p\pi_i = p_S$.

In particular, for each $s \in S$, $X_s = p^{-1}(s)$ is naturally a compact complex V-manifold.

<u>Versal</u> <u>deformation</u> <u>space</u> <u>for</u> <u>complex</u> V-<u>manifolds</u>: Let $p:X \to S$ be a family of complex compact V-manifold and let $f:S' \to S$ be a morphism of analytic spaces. Then the induced family $f^*(X) \to S'$ is defined. The underlying space of $f^*(X)$ is the fiber product $S' \times_S X$ and the uniformizing charts are also the fiber products

$$f^{-1}(S_i) \times_S U_i \longrightarrow f^{-1}(S_i) \times_S U_i .$$

The family $p:X \to S$ is called <u>versal</u> at $s_o \in S$ if the following condition is satisfied. Let $p':X' \to S'$ be a holomorphic family of compact complex V-manifolds and let s'_o be a point of S' such that there is an isomorphism F_o of $p'^{-1}(s'_o)$ on $p^{-1}(s_o)$. Then there is a neighbourhood S'' of s'_o, a morphism $f:(S'',s'_o) \to (S,s_o)$ and an isomorphism F of $X'|_{S''}$ on $f(X)$ over S'' whose restriction to $p'^{-1}(s'_o)$ is F_o. Moreover the differential df of f at s'_o is unique.

The proof of the theorem of Kuranishi on the existence of a versal family goes through without essential change. So we state the following theorem, without giving the proof.

Theorem 4.2. Let X_o be a V-manifold, compact complex and let θ_{X_o} be the sheaf of germs of holomorphic vector fields on X_o. Then there is an analytic space S in a neighbourhood of 0 in $H^1(X_o,\theta_{X_o})$ and a holomorphic family $p:X \to S$ of compact compex V-manifolds versal at 0 and such that $p^{-1}(0)$ is isomorphic to X_o.

The existence of the analogue of Kuranishi family for any compact complex space is known. The proof is quite difficult see Douady [8], Grauert [19], Forster-Knorr [14].

Deformations of compact Hausdorff transversally holomorphic foliations.

A family of transversally holomorphic foliations $\mathfrak{F}_s$ on a compact smooth manifold Y parametrized by a complex analytic space S (not necessarily reduced) is given by

i) an open covering $\{S_i \times V_i\}$ of $S \times Y$, where S_i and V_i are open sets of S and Y,

ii) differentiable submersions $f_i:S_i \times V_i \to S \times C^n$ over S where the f_i are morphisms of ringed spaces and $f_i|s \times V_i$ is a differentiable submersion in $s \times C^n$.

iii) for each $(s,y) \in (S_i \times V_i) \cap (S_j \times V_j)$, there is a local biholo morphism γ_{ji} over S of $S \times C^n$ such that $f_j = \gamma_{ji} f_i$ on a neighbourhood of (s,y).

So for each $s \in S$ we have on Y a transversally holomorphic foliation F_s.

Theorem 4.3. There is a one to one correspondence between germs of deformations of compact Hausdorff transversally holomorphic foliations whose generic leaf has trivial one dimensional real cohomology and germs of deformations of the compact complex orbifolds of the leaves of these foliations.

As in the case of deformations of complex manifolds, the definitions of a versal (semiuniversal) family of deformations of a transversally holomorphic foliation (of a compact smooth manifold) can be given. In this case the existence of the analogue of the Kuranishi family has been proved by Duschamp and Kalka [8] under a very restrictive assumption. Recently Girbau has noticed that the Kodaira-Spencer resolution [36] works also in the case of transversally holomorphic foliation and this leads to a proof for the existence of the semiuniversal family of deformations for transversally holomorphic foliations.

REFERENCES

[1] Baum, P., Bott, R.: Singularities of holomorphic foliations, J. Diff. Geom. 7(1972), 279 - 342.

[2] Bedford, E., Kalka, M.: Foliations and complex Monge-Ampere equations. Comm.Pure. Appl.Math. (1980) 510 - 342.

[3] Bott, R.: Lectures on characteristic classes and foliations Springer Lecture Notes 279 (1972), 1 - 80.

[4] Brieskorn, E.: Rationale Singularitaten Komplexer Flachen, Invent. Math. 4 (1967 - 68) 336 - 358.

[5] Cartan, H.: Quotient d' un espace analytique par un groupe d'automorphisms, Algebraic Geometry and Topology (Lefchetz Volume), Princeton University Press (1957) 90 - 102.

[6] Cathelineau, J.L.: Deformations equivariantes d'espaces analytiques complexes compacts, Ann. Scient.Ec. Norm. Sup., 11 (1978) 391 - 406.

[7] Duchamp, T., Kalka, M: Deformation theory for holomorphic foliations, J.Diff.Geom. 14(1979) 317 - 337

[8] Douady, A.: Le probleme des modules locaux pour les espaces C- analytiques compacts, Ann.Sci. Ecole Norm.Sup. IV 4 (1974) 567 - 602.

[9] Edwards, R., Millet, K., Sullivan, D.: Foliations with all leaves compact, Topology 16 (1977) 13 - 32.

[10] Epstein, D.B.A.: Periodic flows on three manifolds Ann. of Math. 95 (1972) 68 - 82

[11] Epstein, D.B.A.: Foliations with all leaves compact. Ann. Inst Fourier, 26 (1976), 265 - 282.

[12] Epstein, D.B.A., Rosenberg, H.:Stability of compact foliations Springer Lecture NOtes 597 (1978)

[13] Epstein, D.B.A., Vogt, E.: A counter example to the periodic orbit conjecture in codimension 3, Ann. of Math 108 (1978), 539 - 552.

[14] Forster O, Knorr, K.: Konstruction Verseller Familien Kompakter Komplexer Raume. Springer Lecture Notes 705 (1979)

[15] Gerard, R.: Geometric theory of differential equations in the complex domain, Complex analysis and its applications, Vol. II, I.A.E.A. Vienna (1976) 47 -72.

[16] Gerard, R.: Feuilletages de Painleve, Bull.Soc.Mat., France 100 (1972) 47 - 72.

[17] Gerard, R., Jouanolou, J.P.: Etude de l'existence de feuilles compactes pour certain feuilletages analytiques complexes, C.R. Acad. de Paris 277 (1973).

[18] Gómez-Mont, X.: Transversal holomorphic structures in manifolds, Ph.D. Thesis, Princeton University, 1978.

[19] Grauert,H.: Der satz von Kuranishi fuer kompakte komplexe Raume Invent.Math. 25 (1974), 107 - 142.

[20] Haefliger, A. Structures feuilletees et cohomologic a valuer dans un faisceau de grupods, Comment. Math. Hevl. 32(1958), 249 - 329.

[21] Haefliger, A.: Varietes feuilletes, Ann. Ecole Norm. Sup. Pisa, 16(1962) 367 - 397.

[22] Haefliger, A.: Homotopy and Integrability, Springer Lecture Notes 197 (1971) 128 - 141.

[23] Haefliger, A.: Feuilletages sur les varietés ouvertes Topology 9 (1979) 183 - 194

[24] Hamilton, R.: Deformation theory of foliations, Cornell Univ. Preprint.

[25] Heitseh, J.: A cohomology for foliatod manifolds, Commont. Math. Helv. 50 (1975) 197 - 218.

[26] Hardrop, D.: All compact three dimensional manifolds admit total foliations, Memoirs A.M.S. 233 (1980)

[27] Hirsch, M.W.: Stability of compact leaves of foliations Dyna mical systems, edited by M.M. PEIXOTO, Academic Press (1973) 135 - 153.

[28] Holmann, H.: Holomorphic Blatterungen Komplexer Raume, Comment. Math. Hev. 47 (1972), 185 - 204.

[29] Holmann, H.: Anaytische periodische stromungen auf Kompakten Raumen, Comment. Math. Hev. 52 (1977), 251 - 257.

[30] Holmann, H.: On the stability of holomorphic foliations with all leaves compact, Springer Lecture Notes 683 (1977) 47 - 248

[31] Holmann, H.: On the stability of holomorphic foliations, Springer Lecture Notes 798 (1979) 192 - 202

[32] Jouanolou, J.P.: Equations de Pfaff Algebriques, Springer Lecture Notes 708 (1979).

[33] Jouanolou, J.P.: Fellilles compactes des feuilletages algebriques, Math. Ann. 241 (1979), 69 - 72.

[34] Kodaira, K., Spencer, D.C.: On deformations of complex analytic structures, I,II Ann. of Math. 67(1958) 328 - 466, III ibid 71 (1960) 43 - 76.

[35] Kodaira, K., Morrow, S.: Complex manifolds, Holt Reinhardt, 1971.

[36] Kodaira, K., Spencer, D.C.: Multifoliate structures, Ann. of Math. 74 (1961), 52 - 100.

[37] Kuranishi, M.: New Proof for the existence of locally complete families of complex structures, Proceedings of the Conference on Complex Analysis, Minneapolis (1965) 142 - 154.

[38] Kuranishi, M.: Deformations of Complex Structures, University of Montreal Press, Montreal, 1970.

[39] Langevin, R., Rosenberg, H.: On stability of compact leaves and fibrations, Topology 16 (1977) 107 - 111.

[40] Lawson, B.: Quantitive theory of foliations, Regional Conference Series No. 27 A.M.S. (1977).

[41] Millet, K.C.: Compact foliations,Springer-Verlag Lecture Notes 484 (1975) 277 - 287.

[42] Morrow, J., Kodaira,K .: Complex manifolds, Holt, Reinhart & Winston, Inc. 1971.

[43] Muller, Th.: Beispiel linear periodischen instabilen holomorphen stromung; L'Enseinm. Mathématique, 25 (1979) 309 - 312.

[44] Mumford, D.: Geometric Invariant Theory, Springer, Berlín-Heidelberg- New York (1965).

[45] Narasimhan, M.S.: Deformations of complex manifolds and vector Bundles, These Proceedings.

[46] Novikov, S.P.: Topology of foliations, Trans. Moscow Math. 14, A.M.S. Translations (1967) 268 - 304.

[47] Pasternack,J.: Foliations and compact Lie group actions, Comment. Math. Helv. 46 (1971), 467 - 477.

[48] Pinkham, H.C.: Deformations of quotient surface singularities, Proc. Symp. Pure.Maths. Vol. XXX A.M.S.(1977) 65 - 71.

[49] Plante, J.: Compact leaves in foliations of codimension one, Georgia Topology Conference.

[50] Prill, D.: Local classifications of quotients of complex manifolds by discrete groups. Duke Math. 34 (1967) 375 - 386.

[51] Reeb,G.: Sur certaines proprietes topologiques de varietes feuilletes, Actual Sci. Ind. 1183, Hermann, París (1952).

[52] Satake, I.: On a generalization of the notion of a manifold, Proc. Math. Acad.Sci., USA 42 (1956) 359 - 363.

[53] Schweitzer, P.A. Counter examples to Seifert conjecture and opening closed leaves of foliations, Ann. of Math., 100 (1976).

[54] Schlessinger, M.: Rigidity of quotient singularities, Invent. Math. 14 (1971) 17 - 26.

[55] Stoll, W.: The characterisation of strictly parabolic manifolds. Ann. Scuola. Norm. Sup. Pisa, VII (1980) 87 - 154.

[56] Sullivan, D.: A counter example to the periodic orbit conjecture, Publ. Math., I.H.E.S. 46 (1976).

[57] Sullivan, D.: A new flow, Bull. A.M.S., 82 (1976) 331- 332.

[58] Sundararaman, D.: Moduli,deformations and classifications of compact complex manifolds, Pitman Publishing Ltd., London, (1980).

[59] Suzuki, M.: Sur les integrales premieres de certains feuilletages analitiques complexes, Seminare Norguet, 1975-76 Springer Lecture Notes 670 (1978) 53 - 79.

[60] Thurston, W.: A generalization of the Reeb stability theorem, Topology (1974)

[61] Thurston, W.: Lectures on 3-manifolds, Princeton University Preprint.

[62] Thurston, W.: Foliations of 3-manifolds which are circle bundles, Ph.D. Thesis, U.C.L.A., California Berkeley, 1972.

[63] Vogt, E.: Foliations in codimension two with all leaves compact, Manuscripta Math. 18 (1976) 187 - 212.

[64] Vogt, E.: The first cohomology group of leaves and local stability of compact foliations, Manuscripta Math. 37 (1982) 229 - 267.

University of Hyderabad
Hyderabad 500 134
INDIA.

HOLOMORPHIC VECTORBUNDLES AND YANG MILLS FIELDS

Günther Trautmann

Dedicated to the memory of Aldo Andreotti

In terms of differential geometry a potential should be interpreted as a connection and its field as the curvature associated to the connection. In gauge theory one is lead to consider connections and curvatures in vectorbundles. The topic of these lectures is to describe the self-dual curvatures of SU(2)-connections of vectorbundles on S^4, which are called self-dual euclidean SU(2)-Yang Mills fields. In [1] it was shown that such fields are in a one to one correspondence with certain holomorphic vectorbundles on $\mathbb{P}_3(\mathbb{C})$, which are now called instantonbundles. By using the theory of moduli for algebraic vectorbundles on complex projective space explicit expressions for the euclidean SU(2)-Yang Mills fields can be derived from this correspondence. This procedure is described here only in the case of the instanton number $c_2 = 1$.

1. Vectorbundles and connections.

(1.1) The notion of a differentiable (holomorphic) $\mathbb{C}$-vectorbundle E on a differentiable (complex) manifold M is well known. If U is an open subset of M the restriction of E on U is denoted by $E|U$, and if $p \in M$ is a point the fibre of E at p is denoted by E_p. The script letter $\mathcal{E}$ is used to denote the sheaf of sections of E. Let r = rank E be the dimension of any fibre of E as $\mathbb{C}$-vectorspace. A local frame of E over the open set U is a sequence $e_1, \ldots, e_r \in \Gamma(U,E)$ of r sections such that $e_1(p), \ldots, e_r(p)$ is a basis of the vectorspace E_p for any $p \in U$. It is convenient to describe the local trivializations of E by such local frames. If for example T resp. T^* denotes the complexified tangent- resp. cotangent bundle on the differentiable manifold M and if $x_1, \ldots, x_n$ are local coordinates of M then $\frac{\partial}{\partial x_1}, \ldots, \frac{\partial}{\partial x_n}$ resp. $dx_1, \ldots, dx_n$ are local frames of T resp. T^*. If in addition E is a differentiable vectorbundle on M with a local frame $e_1, \ldots, e_r$ over the coordinate neighborhood U of $x_1, \ldots, x_n$, then the bundle $E \otimes \Lambda^k T^*$ has the local frame consisting of the sections $e_\alpha \otimes dx_{i_1} \wedge \ldots \wedge dx_{i_k}$ over U. In the following differentiable shall always mean C^∞.

(1.2) Let M be a differentiable manifold and E a differentiable $\mathbb{C}$-vectorbundle. A (linear) connection in E is a $\mathbb{C}$-linear homomorphism of the sheaves of sections

$$\nabla : E \to T^*\otimes E$$

satisfying

$$\nabla(fs) = f\nabla(s) + df\otimes s$$

whenever f is a (local) differentiable function and s is a (local) section of E. The connection ∇ can be uniquely extended to a $\mathbb{C}$-linear homomorphism

$$\nabla : \Lambda^p T^*\otimes E \to \Lambda^{p+1}T^*\otimes E$$

by the formula

$$\nabla(\omega\otimes s) = d\omega\otimes s + (-1)^p\omega\wedge\nabla(s) .$$

Here ω denotes a (local) differentiable p-form and s a (local) differentiable section, and $\omega\wedge\nabla(s)$ is defined as $\sum_k(\omega\wedge\omega_k)\otimes s_k$ if $\nabla(s)=\sum_k\omega_k\otimes s_k$.

By composition we obtain a homomorphism

$$\nabla^2 : E \to \Lambda^2 T^*\otimes E$$

which is called the curvature of the connection ∇. It is an easy exercise to show that ∇^2 actually is C^∞-linear, i.e. satisfies $\nabla^2(fs) = f\nabla^2(s)$ for any local differentiable function f and any local section of E.

(1.3) If $e_1,\dots,e_r$ is a local frame of the differentiable vectorbundle E, then to any connection ∇ there is assigned the local rxr-matrix $\omega = (\omega_{\alpha\beta})_{1\le\alpha,\beta\le r}$ of 1-forms $\omega_{\alpha\beta}$ defined by

$$\nabla e_\alpha = \sum_\beta \omega_{\alpha\beta}\otimes e_\beta \quad .$$

It is called the local connection matrix of ∇ with respect to the given frame. Similarly the local curvature matrix $\Omega = (\Omega_{\alpha\beta})$ of 2-forms $\Omega_{\alpha\beta}$ is defined by

$$\nabla^2 e_\alpha = \sum_\beta \Omega_{\alpha\beta}\otimes e_\beta .$$

It follows immediately that

$$\Omega_{\alpha\beta} = d\omega_{\alpha\beta} - \sum_\gamma \omega_{\alpha\gamma}\wedge\omega_{\gamma\beta} \quad ,$$

$$d\Omega_{\alpha\beta} = \sum_{\gamma} \omega_{\alpha\gamma} \wedge \Omega_{\gamma\beta} - \sum_{\gamma} \Omega_{\alpha\gamma} \wedge \omega_{\gamma\beta},$$

which in matrix notation reads as

$$\Omega = d\omega - \omega\wedge\omega$$

$$d\Omega = \omega\wedge\Omega - \Omega\wedge\omega \quad \text{(Bianchi-identity).}$$

Assume now that two local frames $e_1,\dots,e_r$ and $e_1',\dots,e_r'$ of $E|U$ are given and denote by ω,Ω resp. ω',Ω' the corresponding matrices of forms. Let $g = (g_{\alpha\beta})$ be the C^∞-transition matrix over U, which is defined by

$$e_\alpha' = \Sigma g_{\alpha\beta} e_\beta .$$

Then by the definition of the local forms we obtain the transformation law (gauge transformation)

$$\omega' = dg\, g^{-1} + g\omega g^{-1}$$

in obvious matrix notation.

Assume now that M is covered by open sets U_i and $E|U_i$ is trivialized by a frame $e_{i1},\dots,e_{ir}$ such that the transition matrices $g_{ij} = (g_{ij\alpha\beta})$ are defined over $U_i \cap U_j$ by $e_{i\alpha} = \sum_\beta g_{ij\alpha\beta} e_{j\beta}$. Then if for any i an $r\times r$-matrix ω_i of 1-forms on U_i is given such that on $U_i \cap U_j$ the transformation law

$$\omega_i = dg_{ij} g_{ij}^{-1} + g_{ij}\, \omega_j\, g_{ij}^{-1}$$

is satisfied, a unique connection ∇ on M is defined by $\nabla e_{i\alpha} = \sum_\beta \omega_{i\alpha\beta} \otimes e_{i\beta}$.

(1.4) It is well known that any differentiable $\mathbb{C}$-vectorbundle E on a differentiable manifold M admits a hermitian metric, i.e. for any $x \in M$ a hermitian metric $<\ ,\ >_x$ on E_x such that for any two sections s,t of E over an open set $U \subset M$ the function $<s,t>(x) := <s(x),t(x)>_x$ is differentiable on U. Then a sheaf map $<\ ,\ >: E\otimes E \to C^\infty$ is defined. Such hermitian metric can be extended to the exterior algebra as a sheaf map

$$(\wedge^p T^* \otimes E) \otimes (\wedge^q T^* \otimes E) \to \wedge^{p+q} T^*$$

by the formula (in obvious notation)

$$<\omega\otimes s, \omega'\otimes s'> = <s,s'>\omega\wedge\overline{\omega'}.$$

Moreover there exists always a connection ∇ on E which is compatible with the given metric, i.e. satisfies the product rule

$$d\langle s,t\rangle = \langle \nabla s,t\rangle + \langle s,\nabla t\rangle$$

for any local sections of E. This can be proved by partitian of unity, cf. [10]. Such a connection is called a metric connection. In this case if ω is the connection matrix of ∇ with respect to an orthonormal frame, then $\omega + \bar{\omega}^t = 0$.

(1.5) Assume now that $G \subset GL(r,\mathbb{C})$ is a Lie subgroup and E has a G-structure, i.e. there exists a covering U_i of M together with frames of E over U_i trivializing $E|U_i$ such that their transition matrices have values in G. (Then a frame of $E|U$ is called a frame of the G-structure, if the transition matrix with respect to any frame of the given ones has values in G). A connection ∇ on E is called a G-connection if for any G-frame its matrix ω has values in the Lie algebra $\mathfrak{g}$ of G. Again by partitian of unity one proves that there always exists a G-connection for any G-bundle.
If especially E has a SU(r)-structure together with a hermitian metric such that the SU(r)-frames are orthonormal, then any SU(r)-connection ∇ on E is also a metric connection, since for any local connection matrix ω with respect to an (orthonormal) frame of the SU(r)-structure we have $\omega + \bar{\omega}^t = 0$ and $\mathrm{trace}(\omega) = 0$, and the first of these conditions implies that ∇ is compatible with the metric.

(1.6) In the rest of this lecture we will only consider differentiable SU(2)-bundles on the manifold $M = S^4$. If ∇ is an SU(2)-connection in such a bundle and ω is the connection matrix of ∇ with respect to an orthonormal SU(2)-frame over $U \cong \mathbb{R}^4$ then ω can be written as

$$\omega = A_1 dt_1 + A_2 dt_2 + A_3 dt_3 + A_4 dt_4$$

in the coordinates $t_1,\ldots,t_4$, such that the 2x2-matrices A_μ are differentiable mappings

$$A \;:\mathbb{R}^4 \to \mathfrak{su}(2),$$

i.e. satisfy

$$A_\mu + \bar{A}_\mu^t = 0 \text{ and } \mathrm{trace}\, A_\mu = 0.$$

For its curvature form on U we obtain in the same way

$$\Omega = F_{12} dt_1 \wedge dt_2 + \ldots + F_{34} dt_3 \wedge dt_4$$

with

$$F_{\mu\nu} = \frac{\partial A_\nu}{\partial t_\mu} - \frac{\partial A_\mu}{\partial t_\nu} - (A_\mu A_\nu - A_\nu A_\mu) \ ,$$

such that also $F_{\mu\nu}$ is a differentiable mapping on $\mathbb{R}^4$ with values in $su(2)$. If we write $\Omega = (\Omega_{\alpha\beta})$ as a matrix of 2-forms, the second "Chern"-form $\sigma_2(\nabla)$ is defined by

$$\sigma_2(\nabla) = (2\pi i)^{-2}\det(\Omega) := -\frac{1}{4\pi^2}(\Omega_{11}\wedge\Omega_{22}-\Omega_{21}\wedge\Omega_{12}) \ ,$$

and this 4-form on $M = S^4$ is independent of the local curvature matrix Ω. With respect to the above chart $U \cong \mathbb{R}^4$ one obtains

$$\sigma_2(\nabla) = \frac{1}{4\pi^2}\,\mathrm{trace}(F_{12}F_{34}-F_{13}F_{24}+F_{23}F_{14})dt_1\wedge\ldots\wedge dt_4 \ .$$

Now it follows by the Chern-Weil-theory of characteristic classes, [5], that the second Chern class $c_2 = c_2(E)\in\mathbb{Z}$ of E is given by

$$c_2 = \int_{S^4}\sigma_2(\nabla) = (2\pi)^{-2}\int_{\mathbb{R}^4}\mathrm{trace}(F_{12}F_{34}-F_{13}F_{24}+F_{23}F_{14})dt_1\ldots dt_4$$

where we assume that $U = S^4\setminus\{\text{point}\}$.

(1.7) On the Riemannian manifold S^4 together with a fixed orientation the $*$-operator for differential forms is well defined. Since it is conformally invariant, it is not difficult to check that for a chart U of S^4 obtained by stereographic projection with coordinates $t_1,\ldots,t_4$ we obtain

$$\begin{aligned} *dt_1\wedge dt_2 &= dt_3\wedge dt_4 \\ *dt_1\wedge dt_3 &= -dt_2\wedge dt_4 \\ *dt_2\wedge dt_3 &= dt_1\wedge dt_4 \end{aligned}$$

with $** = \mathrm{id}$.

Conversely, using stereographic coordinates and the above formulas as local definitions of $*$, an operator $*$ for 2-forms on S^4 is invariantly defined. The SU(2) connection ∇ is called <u>self-dual</u>, if $*\Omega = \Omega$ for any local curvature matrix Ω. In terms of stereographic coordinates this reads as

$$F_{12} = F_{34} \ , \ F_{13} = -F_{24} \ , \ F_{23} = F_{14} \ .$$

<u>(1.8) Yang-Mills-equations:</u> Let E be a SU(2)-bundle on S^4 together with a SU(2)-connection ∇. The curvature ∇^2 is called a <u>Yang-Mills</u>

<u>field</u> or an <u>instanton</u>, if for one (and then any) local curvature matrix Ω of ∇ the identity

$$d(*\Omega) = \omega\wedge(*\Omega) - (*\Omega)\wedge\omega$$

is satisfied. If ∇^2 is self-dual, this equation is automatically satisfied because of the Bianchi identity. In terms of local stereographic coordinates the Yang-Mills equation is equivalent to the following system of nonlinear differential equations for $\mu = 1,\ldots,4$

$$\sum_{\lambda<\mu} \frac{\partial F_{\lambda\mu}}{\partial t_\lambda} - \sum_{\mu<\lambda} \frac{\partial F_{\mu\lambda}}{\partial t_\lambda} = \sum_{\lambda<\mu} [A_\lambda, F_{\lambda\mu}] - \sum_{\mu<\lambda} [A_\lambda, F_{\mu\lambda}] \ .$$

<u>(1.8.1) Remark:</u> The Yang-Mills equation for a potential ∇ as above can be derived by variation of the action integral (which is invariant) $|\Omega|^2 = -\int_{S^4} \text{trace } \Omega\wedge*\Omega = \int_{\mathbb{R}^4} \Sigma|f_{\mu\nu\alpha\beta}|^2 dt_1\ldots dt_4$ where $F_{\mu\nu} = (f_{\mu\nu\alpha\beta})$ and the sum is over all indices.

(1.8.2) The second Chern number $c_2 = c_2(E) = \int_{S^4} \sigma_2(\nabla)$ $= (2\pi)^{-2} \int_{\mathbb{R}^4} \text{trace } (F_{12}F_{34}-F_{13}F_{24}+F_{23}F_{14})dt_1\ldots dt_4$ is called the instanton number of the field Ω. It is determined by the bundle E. By decomposing Ω as $\Omega = \Omega^+ + \Omega^-$ with $\Omega^+ = \frac{1}{2}(\Omega+*\Omega)$, $\Omega^- = \frac{1}{2}(\Omega-*\Omega)$ one easily obtains $|\Omega|^2 \geq 8\pi^2|c_2|$. In this estimate equality holds iff $c_2>0$ and $\Omega = *\Omega$ or $c_2<0$ and $\Omega=-*\Omega$ or $c_2=0$ and $\Omega=0$, which is the trivial case. Therefore we consider the case $c_2>0$ and $\Omega = *\Omega$ only, since the second case is obtained by changing the orientation of S^4, [7],[9].

2. Fibering $\mathbb{P}_3(\mathbb{C}) \to S^4$.

(2.1) Let $x \neq y$ be points in $\mathbb{P}_3(\mathbb{C})$ with homogeneous coordinates $x_o,\dots,x_3$ resp. $y_o,\dots,y_3$. Then the point $p \in \mathbb{P}_5(\mathbb{C})$ with homogeneous coordinates

$$p_{ij} = x_i y_j - x_j y_i \quad \text{for } i<j$$

is uniquely determined by the line $\overline{x,y}$ through x and y and satisfies the equation

$$p_{o1}p_{23} - p_{o2}p_{13} + p_{o3}p_{12} = 0 \, .$$

By $\overline{x,y} \to p$ we thus obtain the Plücker-imbedding of the Graßmannian $Gr_1(\mathbb{P}_3(\mathbb{C}))$ into $\mathbb{P}_5(\mathbb{C})$ which so identifies with the nonsingular quadric $Q \subset \mathbb{P}_5(\mathbb{C})$ defined by the above equation. We write L_p for the line in $\mathbb{P}_3(\mathbb{C})$ corresponding to $p \in Q$.

Now if we consider the homogeneous coordinates $x_o,\dots,x_5$ of $\mathbb{P}_5(\mathbb{C})$ given by

$$x_o = p_{o1}+p_{23} \qquad x_2 = p_{13}+p_{o2} \qquad x_4 = p_{o3}-p_{12}$$

$$x_1 = p_{o1}-p_{23} \qquad x_3 = i(p_{13}-p_{o2}) \qquad x_5 = i(p_{o3}+p_{12}) \, ,$$

the equation for Q reads

$$x_o^2 = x_1^2+\dots+x_5^2 \, .$$

By this we easily obtain the identification

$$S^4 = Q \cap \{x \in \mathbb{P}_5(\mathbb{C}),\ x=\overline{x}\}$$

where $\overline{x}$ means complex conjugation. Moreover for $x \in S^4$ automatically $x_o \neq 0$.

(2.2) Let $\sigma: \mathbb{P}_3(\mathbb{C}) \to \mathbb{P}_3(\mathbb{C})$ be defined in homogeneous coordinates by $(z_o,\dots,z_3) \to (-\overline{z}_1,\overline{z}_o,\overline{z}_3,\overline{z}_2)$. Then σ is a diffeomorphism without fixed points and $\sigma^2 = id$. Now one can check that $p \in S^4$ if and only if L_p is a fixed line of σ, i.e. $\sigma(L_p) = L_p$, and that is the case if and only if for $z \in L_p$ also $\sigma z \in L_p$. Thus a mapping

$$P_3(C) \xrightarrow{\ \pi\ } S^4$$

can be defined by assigning $z \in \mathbb{P}_3(\mathbb{C})$ the point p corresponding to the line through z and σz. (This map turns out to be a differentiable fibre bundle with fibre $\mathbb{P}_1(\mathbb{C})$ and for $p \in S^4$ we have $L_p = \pi^{-1}(p)$.) The fibres of π are called the <u>real lines</u> of $\mathbb{P}_3(\mathbb{C})$.

(2.3) The open sets

$$V_1 = S^4 \cap \{p_{o1} \neq 0\} \quad \text{and} \quad V_2 = S^4 \cap \{p_{23} \neq 0\}$$

are coordinate neighborhoods of S^4 and cover S^4. If we choose the coordinate neighborhoods $U_i = \{z_i \neq 0\}$ in $\mathbb{P}_3(\mathbb{C})$, then

$$\pi^{-1} V_1 = U_o \cup U_1 \text{ and } \pi^{-1} V_2 = U_2 \cup U_3 .$$

Now on V_1 for example we obtain real coordinates by stereographic projection given as $t_1, \ldots, t_4$ with

$$t_\nu = \frac{x_{\nu+1}}{x_o + x_1} \quad \text{for} \quad \nu = 1, \ldots, 4.$$

Then $\pi | U_o$ for example is expressed in the coordinates t_μ and $z^*_\nu = z_\nu z_o^{-1}$ of U_o by

$$t_1 = \tfrac{1}{2}(p^*_{13} + p^*_{o2}) \qquad t_2 = \tfrac{i}{2}(p^*_{13} - p^*_{o2})$$

$$t_3 = \tfrac{1}{2}(p^*_{o3} - p^*_{12}) \qquad t_4 = \tfrac{i}{2}(p^*_{o3} + p^*_{12})$$

where $p^*_{ij} = p_{ij}(z) p_{o1}(z)^{-1}$ are functions in z^*_ν, $\bar{z}^*_\nu$, defined by

$$\begin{aligned}
p_{o1}(z) &= |z_o|^2 + |z_1|^2 \\
p_{o2}(z) &= -z_o \bar{z}_3 + z_2 \bar{z}_1 \\
p_{12}(z) &= -z_1 \bar{z}_3 - z_2 \bar{z}_o \\
p_{o3}(z) &= z_o \bar{z}_2 + z_3 \bar{z}_1 \\
p_{13}(z) &= z_1 \bar{z}_2 - z_3 \bar{z}_o \\
p_{23}(z) &= |z_2|^2 + |z_3|^2 ,
\end{aligned}$$

satisfying $p_{o1} = \bar{p}_{o1}$, $p_{o2} = \bar{p}_{13}$, $p_{12} = -\bar{p}_{o3}$, $p_{23} = \bar{p}_{23}$.

3. Instanton bundles.

It was the idea of Atiyah-Ward, [1], to use in addition to the holomorphic structure of vectorbundles on $\mathbb{P}_3(\mathbb{C})$ a certain real or symplectic structure on such bundles in order to describe the property of the fields to be $\mathfrak{su}(2)$-fields. Such a symplectic structure is by definition a holomorphic isomorphism $\tau: E \to \sigma^*\overline{E}$ with $\tau^2=-1$. Let us first describe the bundle $\sigma^*\overline{E}$ when E is a holomorphic vectorbundle on $\mathbb{P}_3(\mathbb{C})$, where σ denotes the involution of $\mathbb{P}_3(\mathbb{C})$ described above. We put $\sigma(o) = 1$, $\sigma(1) = o$, $\sigma(2) = 3$, $\sigma(3) = 2$ for the indices of the canonical covering $U_i=\{z_i \neq o\}$ of $\mathbb{P}_3(\mathbb{C})$ such that we have $\sigma^{-1}U_i = U_{\sigma(i)}$.

Now if $\{g_{ij}\}$ is a holomorphic cocycle of E with respect to the covering $\{U_i\}$, the bundle $\sigma^*\overline{E}$ has the cocycle $\{g^{\sigma}_{ij}\}$ with $g^{\sigma}_{ij}=\overline{g}_{\sigma(i),\sigma(j)}\circ\sigma$, which is again homomorphic. Moreover there is a canonical antilinear isomorphism $\Gamma(\mathbb{P}_3(\mathbb{C}),E) \to \Gamma(\mathbb{P}_3(\mathbb{C}),\sigma^*\overline{E})$ described by $\{s_i\} \leftrightarrow \{s^{\sigma}_i\}$ when s_i is a system of local expressions of a section s, i.e. a system of holomorphic r-tuples s_i on U_i such that $s_i = g_{ij}s_j$, $s^{\sigma}_i = \overline{s}_{\sigma(i)}\circ\sigma$.

A (holomorphic) <u>symplectic involution</u> on a holomorphic vectorbundle E on $\mathbb{P}_3(\mathbb{C})$ is now by definition a holomorphic isomorphism $\tau: E \to \sigma^*\overline{E}$ satisfying $\tau^2=-1$, i.e. $\tau^2 := \sigma^*(\overline{\tau})\circ\tau=-1$, where one should note that $\sigma^*\sigma^*\overline{\overline{E}} = E$. The theorem of Atiyah-Ward,[1], can now be stated as follows.

<u>Theorem (Atiyah-Ward):</u> There is a 1-1correspondence between

(I) the equivalence classes (under gauge transformation) of euclidean self-dual SU(2)-Yang Mills fields on S^4

(II) the isomorphism classes of holomorphic vectorbundles E of rank 2 on $\mathbb{P}_3(\mathbb{C})$ with the following properties:

  (i) For any "real" line $L_p \subset \mathbb{P}_3(\mathbb{C})$ the bundle $E|L_p$ is holomorphically trivial

  (ii) E has a (holomorphic) symplectic involution $\tau: E \to \sigma^*\overline{E}$ with $\tau^2=-1$.

(In this theorem ideas of Penrose have been realized in the special case described here.)

<u>Remark 1:</u> It is not difficult to show that bundles E of type (II) satisfy $c_1(E) = 0$ and $\Gamma(\mathbb{P}_3,E) = 0$, i.e. E has no nonzero holomorphic sections.

<u>Remark 2:</u> W. Barth has shown that bundles E of rank 2 with $c_1(E) = o$

and $\Gamma(\mathbb{P}_3,E) = o$ are "stable" and "simple", i.e. $Hom(\mathbb{P}_3,E,E) = \mathbb{C}$ in this context.

Remark 3: The involution $\tau: E \to \sigma^*\overline{E}$ is unique up to a constant $c\in\mathbb{C}$ with $|c| = 1$, if it exists. For if τ_1 and τ_2 are two such involutions, $\tau_1^{-1}\tau_2$ has to be a constant c as an endomorphism of E. By $\tau_i^2=-1$ it follows from $\tau_2= c\tau_1$ that $c\bar{c}= 1$.

Remark 4: (Vanishing theorem of Atiyah-Hitchin-Drinfeld-Manin)
If E is a holomorphic vectorbundle of type (II) then $H^1(\mathbb{P}_3,E(-2))= O$ and $H^1(\mathbb{P}_3,E(-2)\otimes E(-2))= O$. A proof has been explicated by Rawnsley [6]. Another proof is given by A. Douady in [4]. This theorem has as consequence that the space of coefficients of all possible self-dual SU(2)-Yang-Mills fields on S^4 with fixed instanton number c_2 is a real-analytic manifold of dimension $8c_2 - 3$.

The following terminology is now used: A holomorphic 2-bundle E on $\mathbb{P}_3(\mathbb{C})$ satisfying (i), (ii) is called a ("real") instanton bundle. A holomorphic 2-bundle E on $\mathbb{P}_3(\mathbb{C})$ with $c_1(E)= O$, $c_2(E)>O$, $\Gamma(\mathbb{P}_3,E)= O$, $H^1(\mathbb{P}_3,E(-2))= O$ is called a mathematical instanton bundle.

In this lecture I only want to indicate how a field of type (I) is obtained from a given bundle of type (II) and finally deduce (in the simplest case of instanton number $c_2= 1$) the general field by using moduli of holomorphic vectorbundles.

4. Yang-Mills field associated to an instanton bundle.

(4.1) Consider first the flag manifold Z consisting of all pairs (z,L) where $z \in \mathbb{P}_3(\mathbb{C})$ is a point and $L \subset \mathbb{P}_3(\mathbb{C})$ is a line with $z \in L$. Z is a compact complex submanifold of $\mathbb{P}_3 \times Q$ and from the definition of π it follows that π is the "restriction" of the natural projection q. If now E is a holomorphic vectorbundle on $\mathbb{P}_3(\mathbb{C})$, it follows from a comparison theorem in connection with Grauert's proper mapping theorem that the coherent sheaf $q_*p^*E=F$ is locally free on the open set of all $L \in Q$ such that $E|L=2\mathcal{O}_L=\mathcal{O}_L \oplus \mathcal{O}_L$ is holomorphically trivial, and the fibres of the underlying vectorbundle F are just given by

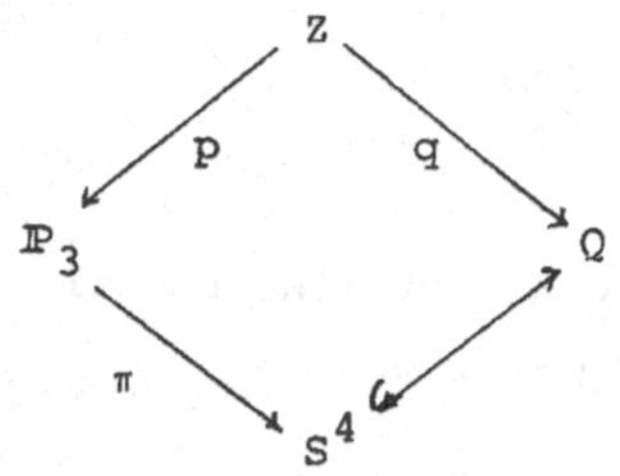

$$F_p = \Gamma(L_p, E|L_p) \cong \mathbb{C}^2 .$$

Since in our case S^4 is contained in this open set of lines we thus obtain a $\mathbb{C}$-vectorbundle F on S^4 which is real-analytic, such that

$$E \cong \pi^* F$$

with respect to the differentiable or real-analytic structure of E. It turns out that $c_2(E) = c_2(F)$. The wanted connection will now be a connection in F.

(4.2) Since the vectorbundle E has rank 2 and vanishing first Chern class it follows that $E \wedge E \cong \mathcal{O}$ such that by this isomorphism and the canonical map $E \otimes E \to E \wedge E$ we obtain a holomorphic nowhere degenerate symplectic form on $E \otimes E$ which shall be denoted by (,). The "symplectic" involution

$$\tau : E \to \sigma^* \overline{E}$$

now induces an involution $\tau \wedge \tau$ and $\tau_{\mathcal{O}}$ as indicated in the diagram

$$\begin{array}{ccccc} E \times E & \longrightarrow & E \wedge E & \longrightarrow & \mathcal{O} \\ \downarrow{\scriptstyle \tau \otimes \tau} & & \downarrow{\scriptstyle \tau \wedge \tau} & & \downarrow{\scriptstyle \tau_{\mathcal{O}}} \\ \sigma^*\overline{E} \otimes \sigma^*\overline{E} & \to & \sigma^*\overline{E} \wedge \sigma^*\overline{E} & \to & \sigma^*\overline{\mathcal{O}} \end{array}$$

satisfying $\tau_{\mathcal{O}}^2 = +1$.

Now $\tau_{\mathcal{O}}$ is up to a constant $c, |c| = 1$, the usual complex conjugation on $\mathcal{O}$ and thus we obtain the formula

$$(\tau s, \tau t) = c\overline{(s,t)}$$

for any two local differentiable or holomorphic sections. The involution τ also induces involutions

$$E|L_p \longrightarrow \sigma^*\overline{E}|L_p$$

for any real line and by this antilinear isomorphisms

$$\tau_p: F_p \longrightarrow F_p$$

with $\tau_p^2 = -1$, which moreover also fit together to an antilinear real-analytic bundle map

$$\tau: F \longrightarrow F \qquad \tau^2 = -1 .$$

Furthermore the symplectic form (,) is induced on F by $F_p = \Gamma(L_p, E|L_p)$

$$F \wedge F \cong C^\omega ,$$

and satisfying the formula

$$(\tau s, \tau t) = c\overline{(s,t)}$$

for local sections. Since τ is antilinear, by

$$<s,t>: = \frac{1}{\sqrt{c}}(s,\tau t)$$

a hermitian metric is defined on the bundle F which by definition is differentiable and real analytic.

(4.3) Since differentiably $E \cong \pi^*F$ the metric < , > can be pulled back to give a metric on E which is "constant" along the fibres $L_p = \pi^{-1}(p)$. Now it is well-known that on a holomorphic vectorbundle E with hermitian metric there is exactly one connection $\tilde{\nabla}$ which is compatible with the metric and such that any local connection matrix with respect to a holomorphic frame is of type (1,0) only, cf. [10]. Let $\tilde{\nabla}$ be this connection in our case. It will now be shown that $\tilde{\nabla} = \pi^*\nabla$ and ∇ is the Yang-Mills connection on F wanted.

If $e_1,\ldots,e_r$ is a holomorphic frame of $E|U$ and $h = (h_{\alpha\beta})$ is the metric matrix given by $h_{\alpha\beta} = <e_\alpha, e_\beta>$, then $\tilde{\omega} = \partial h \circ h^{-1}$ is the connection matrix of $\tilde{\nabla}$ on E.

(4.4) The bundle F on S^4 has a natural SU(2)-structure in the following way. Since it has a hermitian metric any frame can be orthonormalized by the usual process such that S^4 can be covered by open sets on which there are orthonormal frames. Then the transition matrices between two such frames automatically have values in U(2). Now since $\Lambda^2 F$ is trivial (in fact any line bundle on S^4 is trivial), the cocycle of F can be changed in such way, that $\det g_{ij} = 1$. Because of triviality $\det g_{ij} = a_i^{-1} a_j$ where a_i is a function on U_i without zeros. Then take $g'_{ij} = b_i g_{ij} b_j^{-1}$ with functions b_i satisfying $b_i^2 = a_i$.

(4.5) In order to obtain a connection ∇ on F with $\tilde{\nabla} = \pi^* \nabla$ we first construct local connection matrices $\tilde{\omega}$ of $\tilde{\nabla}$ which are with respect to orthonormal frames of E coming from F and which are no longer holomorphic. Let $V_1, V_2 \subset S^4$ be the coordinate neighborhoods defined in section 2. Since $V_\nu \cong \mathbb{R}^4$ it follows that $F|V_\nu$ is trivial. Then $F|V_\nu$ has an orthonormal frame giving a trivialization of $F|V_\nu$. By lifting this to E we obtain a differentiable orthonormal trivialization

$$E|U_0 \cup U_1 \xrightarrow[\psi_1]{} (U_0 \cup U_1) \times \mathbb{C}^2$$

$$E|U_2 \cup U_3 \xrightarrow[\psi_2]{} (U_2 \cup U_3) \times \mathbb{C}^2 \ .$$

On the other hand there are holomorphic frames of $E|U_i$ for any $U_i = \{z_i \neq 0\} \cong \mathbb{C}^3$, which induce holomorphic trivializations

$$E|U_i \xrightarrow[\varphi_i]{} U_i \times \mathbb{C}^2 .$$

Now let $h_i = \varphi_i \psi_\nu^{-1}$ where $i = 0,1$ in case $\nu = 1$ and $i = 2,3$ in case $\nu = 2$.

We want now to express $\tilde{\nabla}$ with respect to the frame of ψ_ν which gives us a connection matrix $\tilde{\omega}_1$ on V_1 and $\tilde{\omega}_2$ on V_2. Since the frame is orthonormal we have

$$\tilde{\omega}_\nu + \overline{\tilde{\omega}}_\nu^t = 0 \ .$$

On the other hand each 1-form η on $\mathbb{P}_3(\mathbb{C})$ can be uniquely written as $\eta' + \eta''$ where η' is the (1,0)-part and η'' the (0,1)-part. If we do this with $\tilde{\omega}_\nu$ and take into account the above identity, we get

$$\tilde{\omega}'_\nu = -\overline{\tilde{\omega}''^t_\nu} \ ,$$

since the conjugate of a (0,1)-form is a (1,0)-form. To know $\tilde{\omega}_\nu$ it is enough to know $\tilde{\omega}''_\nu$. Now considering h_i as a matrix function on U_i the frame of ψ_ν is obtained by the frame of φ_i by using h_i^t.
Hence the general transformation rule will finally give us

$$\tilde{\omega}''^t_\nu | U_i = h_i^{-1} \bar{\partial} h_i \ .$$

Now it is easily shown that $h_0^{-1} \bar{\partial} h_0 = h_1^{-1} \bar{\partial} h_1$ on $U_0 \cap U_1$ such that $\tilde{\omega}''_1$ by this is defined over $U_0 \cup U_1 = \pi^{-1} V_1$, and similarly $\tilde{\omega}''_2$ over $\pi^{-1} V_2$. (Actually $\tilde{\omega}''_\nu$ is indep. of the holomorphic trivialization φ_i used above)

(4.6) For any real line $L_p = \pi^{-1}(p)$ we have $\tilde{\omega}''_\nu | L_p = 0$, where here restriction means pulling back $\tilde{\omega}''_\nu$ via the imbedding $L_p \hookrightarrow \mathbb{P}_3$.
This follows from the fact that $h_i | L_p$ is holomorphic since $\psi_\nu | L_p$ is constant by definition, such that $\bar{\partial} h_i | L_p = \bar{\partial}(h_i | L_p) = 0$. Now by the above formulae we get $\tilde{\omega}_\nu | L_p = 0$. From this fact one can easily derive that $\tilde{\omega}_\nu$ is the pull-back of a 1-form ω_ν over $V_\nu \subset S^4$ i.e. $\tilde{\omega}_\nu = \pi^* \omega_\nu$.

Now if g_{12} is a transition matrix of the frames of ψ_1, ψ_2 resp. for the bundle F on S^4, $\tilde{g}_{12} = \pi^* g_{12}$ is the transition matrix for E of the frames of ψ_1, ψ_2. Since

$$\tilde{\omega}_1 = d\tilde{g}_{12}\, \tilde{g}_{12}^{-1} + \tilde{g}_{12}\, \tilde{\omega}_2\, \tilde{g}_{12} \qquad \text{on } \pi^{-1}(V_1 \cap V_2)$$

we obtain

$$\pi^*(\omega_1 - dg_{12}\, g_{12}^{-1} - g_{12}\, \omega_2\, g_{12}^{-1}) = 0 \qquad \text{on } V_1 \cap V_2 ,$$

from which it follows that (π is locally a product projection)

$$\omega_1 = dg_{12}\, g_{12}^{-1} + g_{12}\, \omega_2\, g_{12}^{-1} .$$

Hence by $\{\omega_1, \omega_2\}$ there is defined a connection ∇ on F such that

$$\tilde{\nabla} = \pi^* \nabla .$$

(4.7) It still has to be verified that ∇ is a SU(2)-connection. What is only known so far is that ∇ is compatible with the metric, i.e. $\omega_\nu + \bar{\omega}_\nu^t = 0$ by construction. To show that also trace$(\omega_\nu) = 0$ we have to use the symplectic involution $\tau : E \to \sigma^* \bar{E}$ again and the fact that τ is holomorphic. (Conversely if in the theorem of Atiyah-Ward F, ∇ are given, one has to use the $\mathfrak{su}(2)$-property of ∇ in order to prove that the morphism τ (defined in a certain way) is holomorphic.)

If s_1, s_2 is the orthonormal frame for $F|V_1$ for example, it follows that $s_2 = \pm \tau s_1$, since $\langle s, \tau s\rangle = o$ for any section, which comes from the definition of the metric. Hence we may choose $s_2 = \tau s_1$ (if $s_2 = -\tau s_1$ we have $s_1 = \tau s_2$ since $\tau^2 = -1$). Now it is easy to verify that the following diagrams are commutative

$$\begin{array}{ccc} E|\pi^{-1}V_\nu & \xrightarrow{\psi_\nu} & \pi^{-1}V_\nu \times \mathbb{C}^2 \\ \downarrow \tau & & \downarrow J_\nu = \begin{pmatrix} 0 & -1 \\ 1 & 0 \end{pmatrix} \\ \sigma^*\bar{E}|\pi^{-1}V_\nu & \xrightarrow[\psi_\nu]{} & \pi^{-1}V_\nu \times \mathbb{C}^2 \end{array} ,$$

where J as a mapping is considered antilinear. Defining $\varphi_i^\sigma = \overline{\varphi}_{\sigma(i)}\circ\sigma$ as the induced trivialization of $\sigma^*\overline{E}|U_i$ we also have the commutative diagrams

$$\begin{array}{ccc} U_i\times\mathbb{C}^2 & \xleftarrow{\ \varphi_i\ } & E|U_i \\ \Big\downarrow{\scriptstyle \tau_i} & & \Big\downarrow{\scriptstyle \tau} \\ U_i\times\mathbb{C}^2 & \xleftarrow[\ \varphi_i^\sigma\ ]{} & \sigma^*\overline{E}|U_i \end{array}$$

where τ_i is the holomorphic matrix defined by τ for the trivialization. Putting $h_i^\sigma = \varphi_i^\sigma\overline{\psi_\nu}^{-1} = \overline{h}_{\sigma(i)}\circ\sigma$ we obtain

$$\tau_i h_i = h_i^\sigma J_\nu \quad \text{on } U_i \subset V_\nu .$$

Now after applying the operator $\overline{\partial}$ to this equation and using $\overline{\partial}\tau_i = 0$ finally we obtain

$$J_\nu h_i^{-1}\overline{\partial}h_i = (h_i^\sigma)^{-1}\overline{\partial}h_i^\sigma J_\nu$$

as a matrix equation. If $\omega_\nu = \Sigma A_k dt_k$ in the local coordinates of the chart $V_\nu \subset S^4$ we obtain after checking the definition of $\tilde{\omega}_\nu''$

$$J_\nu A_k^t = \overline{A}_k^t J_\nu = -A_k J_\nu$$

for any k as a matrix identity. From this it follows that $\mathrm{trace}(A_k)=0$. Since $A_k + \overline{A}_k^t = 0$, we have checked that ∇ is a SU(2)-connection on F.

(4.8) Finally it follows that the curvature ∇^2 is also self-dual: First note that by the special form of the mapping $\pi\colon \mathbb{P}_3(\mathbb{C}) \to S^4$ for any local 2-form η on S^4 we have: $*\eta = \eta$ if and only if $\pi^*\eta$ is a form of type (1,1) with respect to complex structure of $\mathbb{P}_3$. Now by the definition of $\tilde{\omega}_\nu = \pi^*\omega_\nu$ we have $\tilde{\omega}_\nu''^{\,t} = h_i^{-1}\overline{\partial}h_i$ from which we got

$$\overline{\partial}\tilde{\omega}_\nu'' - \tilde{\omega}_\nu''\wedge\tilde{\omega}_\nu'' = 0$$

and similarly

$$\partial\tilde{\omega}_\nu' - \tilde{\omega}_\nu'\wedge\tilde{\omega}_\nu' = 0.$$

Since the local curvature matrix Ω_ν over V_ν is given by

$$\omega_\nu = d\omega_\nu - \omega_\nu\wedge\omega_\nu$$

it follows that

$$\begin{aligned} \pi^*\Omega_\nu &= d\tilde{\omega}_\nu - \tilde{\omega}_\mu\wedge\tilde{\omega}_\nu \\ &= \partial\tilde{\omega}_\nu' + \overline{\partial}\tilde{\omega}_\nu' + \partial\tilde{\omega}_\nu'' + \overline{\partial}\tilde{\omega}_\nu'' - \tilde{\omega}_\nu'\wedge\tilde{\omega}_\nu' - \tilde{\omega}_\nu'\wedge\tilde{\omega}_\nu'' - \tilde{\omega}_\nu''\wedge\tilde{\omega}_\nu' - \tilde{\omega}_\nu''\wedge\tilde{\omega}_\nu'' \\ &= \overline{\partial}\tilde{\omega}_\nu' + \partial\tilde{\omega}_\nu'' - \tilde{\omega}_\nu'\wedge\tilde{\omega}_\nu'' - \tilde{\omega}_\nu''\wedge\tilde{\omega}_\nu' \end{aligned}$$

is a form of type (1,1). Hence ∇^2 is self-dual.

5. Instanton bundles on $\mathbb{P}_3(\mathbb{C})$, cf.[9],[8],[3] .

(5.1) Let us recall some basic facts and notations for holomorphic vectorbundles on $\mathbb{P}_n(\mathbb{C})$. If E is such a bundle let $\mathcal{E}$ denote the locally free sheaf of holomorphic sections of E. Since $U_i = \{z_i \neq 0\} \cong \mathbb{C}^n$, it is well known that $\mathcal{E}|U_i \cong r\mathcal{O}|U_i$ where r=rank E and $r\mathcal{O}=\mathcal{O}\oplus\ldots\oplus\mathcal{O}$ (r times). Hence any holomorphic vectorbundle on $\mathbb{P}_n(\mathbb{C})$ is defined by a holomorphic cocycle $\{g_{ij}\}$ of holomorphic matrices $g_{ij}: U_i \cap U_j \to GL(r,\mathbb{C})$.

If $g_{ij} = (z_j z_i^{-1})^d$ then the corresponding bundle is denoted by $\mathcal{O}(d)$.

It is the "unique" line bundle with Chern class $c_1 = d \in \mathbb{Z}$.

If $\mathcal{E}$ is any holom. vectorbundle, $\mathcal{E}(d) = \mathcal{E}\otimes\mathcal{O}(d)$ is the "twisted" bundle by the number $d\in\mathbb{Z}$. If g_{ij} is a cocycle of $\mathcal{E}$, the bundle $\mathcal{E}(d)$ has the cocycle $g_{ij}(z_j z_i^{-1})^d$. We write $p\mathcal{E}(d) = \underbrace{\mathcal{E}(d)\oplus\ldots\oplus\mathcal{E}(d)}_{p} = (p\mathcal{E})\otimes\mathcal{O}(d)$. It follows from elementary calculations that

$$\dim_{\mathbb{C}}\Gamma(\mathbb{P}_n(\mathbb{C}),\mathcal{O}(d)) = \begin{cases} 0 & d<0 \\ \binom{n+d}{d} & 0\le d \end{cases} .$$

Moreover for $d\ge 0$ the space $\Gamma(\mathbb{P}_n(\mathbb{C}),\mathcal{O}(d))$ can be identified with the space of homogeneous polynomials of degree d in the homogeneous coordinates $z_0,\ldots,z_n$ of $\mathbb{P}_n(\mathbb{C})$.

(5.2) Using the description of $\Gamma(\mathbb{P}_n(\mathbb{C}),\mathcal{O}(d))$ one can easily derive that the space $\mathrm{Hom}(\mathbb{P}_n(\mathbb{C}),p\mathcal{O}(d),q\mathcal{O}(e))$ of global homomorphisms $p\mathcal{O}(d) \to q\mathcal{O}(e)$ is identified in a natural way with the space of pxq-matrices of homogeneous polynomials of degree e-d. If A is such a matrix the corresponding homomorphism is given by

$$\begin{array}{ccc} p\mathcal{O}(d)|U_i & \longrightarrow & q\mathcal{O}(e)|U_i \\ \downarrow\wr & & \downarrow\wr \\ q\mathcal{O}|U_i & \xrightarrow{A_i} & q\mathcal{O}|U_i \end{array}$$

with $A_i(f) = fz_i^{d-e}A$ in the local trivialization on U_i. Note that $z_i^{d-e}A$ is a well-defined holomorphic matrix function on U_i.

(5.3) Let us consider the sequence of homomorphisms over $\mathbb{P}_3(\mathbb{C})$ given by

$$0 \longrightarrow \mathcal{O}(-4) \xrightarrow[Z_3]{} 4\mathcal{O}(-3) \xrightarrow[Z_2]{} 6\mathcal{O}(-2) \xrightarrow[Z_1]{} 4\mathcal{O}(-1) \xrightarrow[Z_o]{} \mathcal{O} \longrightarrow 0$$

with

$$Z_3 = (-z_3 z_2 - z_1 z_o), \quad Z_2 = \begin{pmatrix} z_2 & -z_1 & z_o & 0 & 0 & 0 \\ z_3 & 0 & 0 & -z_1 & z_o & 0 \\ 0 & z_3 & 0 & -z_2 & 0 & z_o \\ 0 & 0 & z_3 & 0 & -z_2 & z_1 \end{pmatrix}, \quad Z_1 = \begin{pmatrix} -z_1 & z_o & 0 & 0 \\ -z_2 & 0 & z_o & 0 \\ 0 & -z_2 & z_1 & 0 \\ -z_3 & 0 & 0 & z_o \\ 0 & -z_3 & 0 & z_1 \\ 0 & 0 & -z_3 & z_2 \end{pmatrix}, \quad Z_o = \begin{pmatrix} z_o \\ z_1 \\ z_2 \\ z_3 \end{pmatrix}.$$

This sequence is the usual Hilbert-Koszul complex in terms of matrices of degree 1. It is exact (which is easily verified locally) and moreover we have

$$Im\ Z_k = \Omega^k ,$$

where here Ω^k denotes the locally free sheaf of holomorphic k-forms on $\mathbb{P}_3(\mathbb{C})$. Especially we have exact sequences

$$0 \longrightarrow \Omega^1 \longrightarrow 4\mathcal{O}(-1) \xrightarrow[Z_o]{} \mathcal{O} \longrightarrow 0$$

$$0 \longrightarrow \Omega^1(2) \rightarrow 4\mathcal{O}(1) \xrightarrow[Z_o]{} \mathcal{O}(2) \rightarrow 0 .$$

By the second sequence we can conclude that the space of sections $\Gamma(\mathbb{P}_3(\mathbb{C}), \Omega^1(2))$ has the basis

$$\omega_{o1} = (-z_1 z_o\ 0\ 0)$$
$$\vdots$$
$$\omega_{23} = (\ 0\ 0\ -z_3 z_2)$$

Here one has to use the fact that by the form of Z_o the map $4\Gamma(\mathbb{P}_3(\mathbb{C}), \mathcal{O}(1)) \xrightarrow[Z_o]{} \Gamma(\mathbb{P}_3(\mathbb{C}), \mathcal{O}(2))$ is surjective. One can also write $\omega_{ij} = z_i dz_j - z_j dz_i$ but one has to give a meaning first to the symbols on the right side.

(5.4) Important global invariants of holomorphic vectorbundles E on $\mathbb{P}_n(\mathbb{C})$ are the Chern classes $c_1(E), \ldots, c_r(E) \in \mathbb{Z}$ $(= H^{2i}(\mathbb{P}_n(\mathbb{C}), \mathbb{Z}),$ $i = 1, \ldots, n)$, which are only topological invariants of the underlying topological bundles, and the cohomology groups $H^i(\mathbb{P}_n(\mathbb{C}), \mathcal{E}(d))$ for $o \leq i \leq n$, $d \in \mathbb{Z}$, which are finite dimensional $\mathbb{C}$-vectorspaces. It turns out that the mathematical instanton bundles are already determined by the spaces $H^1(\mathbb{P}_3(\mathbb{C}), \mathcal{E}(-1))$, its "Serre-dual" $H^2(\mathbb{P}_3(\mathbb{C}), \mathcal{E}(-3))$ and certain operations on these spaces. To be more specific let us consider the case of $c_2 = 1$ in the following.

(5.5) Let E be a mathematical instanton bundle on $\mathbb{P}_3(\mathbb{C})$ with $c_1 = c_1(E) = 0$, $c_2 = c_2(E) = 1$, $\Gamma E = 0$, $H^1E(-2) = 0$.
Since $E \wedge E$ = line bundle of first Chern class $c_1(E)$ we get here $E\wedge E \cong \mathcal{O}$, such that under the natural pairing $E\otimes E \to E\wedge E \cong \mathcal{O}$ the bundle E identifies with its dual E^*.
It is not hard to show that in this case

$$\dim_{\mathbb{C}} H^1E(d) = 0 \quad \text{for } d \neq -1$$
$$\dim_{\mathbb{C}} H^2E(d) = 0 \quad \text{for } d \neq -3 ,$$

whereas the dimension of $H^1E(-1)$ and its Serre-dual $H^2E(-3)$ is one. (In general $\dim_{\mathbb{C}} H^1E(-1) = c_2$ for a mathematical instanton bundle.)

Now by looking at the exact sequence

$$0 \longrightarrow \Omega^1\otimes E(-1) \longrightarrow 4\mathcal{O}(-1)\otimes E(-1) \to E(-1) \longrightarrow 0 ,$$

which is obtained from (5.3) by tensoring with $E(-1)$, we obtain the exact sequence

$$\underset{\substack{\parallel \\ 0}}{H^1 4E(-2)} \to H^1E(-1) \xrightarrow{\ \delta\ } H^2\Omega^1 \times E(-1) \to \underset{\substack{\parallel \\ 0}}{H^2 4E(-2)}$$

such that δ is an isomorphism. (The same can be shown for any mathem. instanton bundle.)

(5.6) Remaining in the case of (5.5) we can now look at the homomorphism

$$\begin{array}{ccc} \Gamma\Omega^1(2)\otimes H^2E(-3) & \xrightarrow{\ \text{cup}\ } & H^2\Omega^1\otimes E(-1) \\ & {\scriptstyle\Phi}\searrow \quad \swarrow{\scriptstyle\delta} & \\ & H_1E(-1) & \end{array}$$

where cup is the usual canonical homomorphism, $\Phi = \delta^{-1}\circ \text{cup}$. Especially for any $\omega_{ij} \in \Gamma\Omega^1(2)$ we obtain a $\mathbb{C}$-linear map

$$\omega_{ij}: H^2E(-3) \to H^1E(-1)$$

defined by $\omega_{ij}(\xi) = \Phi(\omega_{ij}\otimes\xi)$.
Now let us fix a basis vector $\xi \in H^1E(-1)$ and $\eta \in H^2E(-3)$ of the one-dimensional vectorspaces. Then by

$$\omega_{ij}(\eta) = a_{ij}\xi$$

a matrix $(a_{01},\dots,a_{23})$ of 6 complex numbers is defined. It turns out that E itself is completely determined up to isomorphisms by the point

$[a] = [a_{01},\dots,a_{23}] \in \mathbb{P}_5(\mathbb{C})$, (5.7),(5.8). Moreover by looking at more canonical operations like the above, one finds that the condition

$$a_{01}a_{23} - a_{02}a_{13} + a_{03}a_{12} \neq 0$$

must be satisfied, i.e. $[a] \in \mathbb{P}_5(\mathbb{C})\setminus Q$, cf. section 2.

(5.7) Let $[a] \in \mathbb{P}_5(\mathbb{C})\setminus Q$ and let $E(a)$ be the kernel of the homomorphism

$$0 \to E(a) \longrightarrow 6\mathcal{O}(1) \xrightarrow{\begin{pmatrix} a_{01}-z_1 & z_0 & 0 & 0 \\ \vdots & \vdots & \vdots & \vdots \\ a_{23}\ 0 & 0 & -z_3 & z_2 \end{pmatrix}} \mathcal{O}(1)+4\mathcal{O}(2)$$

where we fix some representative a of [a]. It can be checked that $E(a)$ is locally free and that $E(a) \cong E(a')$ if and only if $a = \lambda a'$ for some $\lambda \neq 0$. Moreover if a is obtained from a given bundle E as in (5.6) then $E \cong E(a)$. Note that $E(a)$ is automatically a mathematical instantonbundle on $\mathbb{P}_3(\mathbb{C})$ with $c_2E(a) = 1$.

(5.8) Remark: By (5.6), (5.7) it is shown that the set $M_I(o,1)$ of isomorph classes of mathematical instantonbundles with $c_2 = 1$ can be identified with $\mathbb{P}_5(\mathbb{C})\setminus Q$. Moreover if we look at $S = \mathbb{C}^6\setminus\tilde{Q}$, where $\tilde{Q}$ is defined in the same way as Q, by allowing $(a_{01},\dots,a_{23})$ to be variables of S we obtain a bundle

$$0 \longrightarrow E \longrightarrow 6\mathcal{O}(1) \longrightarrow \mathcal{O}(1)\oplus 4\mathcal{O}(2)$$

over $S\times\mathbb{P}_3(\mathbb{C})$. This is a deformation family and is called a universal bundle for the family of all mathematical instanton bundles with $c_2=1$.

(5.9) Let $E = E(a)$ be as above for some $[a] \in \mathbb{P}_5(\mathbb{C})\setminus Q$. Let $p = [p_{01},\dots,p_{23}] \in Q$ and $L_p \subset \mathbb{P}_3(\mathbb{C})$ the line corresponding to p. Then using the definition of $E(a)$ one can show that

$$E|L_p \cong 2\mathcal{O}_{L_p}$$

(i.e. $E|L_p$ is analytically trivial) if and only if the expression

$$\delta_a(p) = a_{01}p_{23} - a_{02}p_{13} + a_{03}p_{12} + a_{12}p_{03} - a_{13}p_{02} + a_{23}p_{01} \neq 0.$$

Geometrically this means that p does not meet the polar hyperplane of [a] with respect to the quadric Q. See for example [9].

Remark: If $\delta_a(p) = 0$ then $E|L_p \cong \mathcal{O}_{L_p}(-1) \oplus \mathcal{O}_{L_p}(1)$.

6. Yang-Mills potentials for $c_2 = 1$.

(6.1) Let E now be an instanton bundle on $\mathbb{P}_3(\mathbb{C})$ such that $E \cong E(a)$ as in section 5. In addition there is a holomorphic "symplectic" involution

$$\tau : E \to \sigma^*\overline{E} .$$

This induces in a natural way antilinear involutions

$$H^1E(-1) \xrightarrow{\tau^1} H^1E(-1)$$
$$H^2E(-3) \xrightarrow[\tau^2]{} H^2E(-3)$$

such that $(\tau^\nu)^2 = \mathrm{id}$. The sign of $(\tau^\nu)^2$ depends on the twisting degree d of $E(d)$. Since these vectorspaces are $\cong \mathbb{C}$, we can choose the basis $\eta \in H^2E(-3)$ and $\xi \in H^1E(-1)$ to be "real", i.e. invariant under τ^2 resp. τ^1.

On $\Omega^1(2)$ and $\Gamma\Omega^1(2)$ there is also a natural antilinear involution Σ which can be induced by conjugating coordinates via the representation

$$0 \longrightarrow \Omega^1(2) \longrightarrow 4\mathcal{O}(1) \xrightarrow[Z_o]{} \mathcal{O}(2) \longrightarrow 0 ,$$

such that Σ induces an antilinear map

$$\Gamma\Omega^1(2) \xrightarrow{\Sigma} \Gamma\Omega^1(2)$$

given by

$$\begin{array}{lcl} \omega_{o1} & \longrightarrow & \omega_{o1} \\ \omega_{o2} & \longrightarrow & \omega_{13} \\ \omega_{12} & \longrightarrow & -\omega_{o3} \\ \omega_{o3} & \longrightarrow & -\omega_{12} \\ \omega_{13} & \longrightarrow & \omega_{o2} \\ \omega_{23} & \longrightarrow & \omega_{23} \end{array} \quad \text{, cf. [9].}$$

Moreover we do have the commutative diagrams

$$\begin{array}{ccc} \Gamma\Omega^1(2)\otimes H^2E(-3) & \xrightarrow{\Phi} & H^1E(-1) \\ \Big\downarrow \Sigma\otimes\tau^2 & & \Big\downarrow \tau^1 \\ \Gamma\Omega^1(2)\otimes H^2E(-3) & \xrightarrow[\Phi]{} & H^1E(-1) \end{array} ,$$

which can be shown by canonical arguments. Applying this to the operations ω_{ij} as defined in section 5 we get in obvious notations

$$\begin{array}{ccc} \omega_{ij}\otimes\eta & \longmapsto & a_{ij}\xi \\ \Big\updownarrow & & \Big\updownarrow \\ \pm\omega_{k\ell}\otimes\eta & \longmapsto & \pm a_{k\ell}\xi = \overline{a}_{ij}\xi \end{array}$$

hence
$$a_{o1} = \bar{a}_{o1} \qquad a_{12} = -\bar{a}_{o3}$$
$$a_{o2} = \bar{a}_{13} \qquad a_{23} = \bar{a}_{23} .$$
For more details see [9].

(6.2) So far we have shown that to an instanton bundle on $\mathbb{P}_3(\mathbb{C})$ with $c_2 = 1$ one can assign parameters $a_{o1},\dots,a_{23}\in\mathbb{C}$ satisfying

(i) $a_{o1}a_{23} - a_{o2}a_{13} + a_{o3}a_{12} \neq 0$

(ii) $\delta_a(p) = a_{o1}p_{23} - a_{o2}p_{13} +\dots \neq 0$ for $p\in S^4$

(iii) $a_{o1} = \bar{a}_{o1},\ a_{o2} = \bar{a}_{13},\ a_{12} = -\bar{a}_{o3},\ a_{23} = \bar{a}_{23}$.

Moreover it was shown in section 5 that $E \cong E(a)$. Now if we are given $a_{o1},\dots,a_{23}\in\mathbb{C}$ satisfying (i), (ii), (iii) we first get the mathematical instanton bundle $E(a)$ with $c_2E(a) = 1$. By (ii) and (5.9) it follows that $E(a)|L_p$ is trivial for any real line $p\in S^4$. Finally by looking at the representation of $E(a)$ in (5.7) one can define, by using (iii), a canonical holomorphic symplectic involution $\tau:E(a) \to \sigma^*\overline{E(a)}$, cf.[9]. This gives the result:

(6.2.1) The isomorphism classes of instanton bundles E on $\mathbb{P}_3(\mathbb{C})$ with $c_2 = 1$ are in a one to one correspondence with equivalence classes of systems $a_{o1},\dots,a_{23}\in\mathbb{C}$ satisfying (i), (ii), (iii), where $(a_{ij})\sim(b_{ij})$ iff $a_{ij}= \lambda b_{ij}$ for some $\lambda\in\mathbb{R}$, $\lambda\neq 0$.

<u>(6.2.2) Remark:</u> Using (iii) one can verify that
$d: = a_{o1}a_{23} - a_{o2}a_{13} + a_{o3}a_{12} > 0$ and that also $\delta_a(p) > 0$ for any $p\in S^4$.

(6.3) Now using the description of $E(a)$ one can calculate the corresponding field using the advice from section 4. It is not hard to find trivializations
$$U_o\times\mathbb{C}^2 \xleftarrow[\varphi_o]{} E|U_o,\ E|U_o\cup U_1 \xrightarrow[\psi_1]{} (U_o\cup U_1)\times\mathbb{C}^2$$
as in (4.5), such that
$$h: = \varphi_o\psi_1^{-1}$$
can be explicitly given by

$$h = \sqrt{\delta^*}\begin{pmatrix} h_{11} & h_{12} \\ h_{21} & h_{22} \end{pmatrix},$$

where δ^*, $h_{11},\ldots,h_{22}$ are the following functions:

$$\delta^* = \delta^*(p^*) = \frac{\delta_a(p)}{p_{o1}} = a_{o1}p^*_{23} + \ldots + a_{23} > 0$$

$$h_{11} = \frac{1}{\delta^*(p^*)}(a_{12} - a_{o2}z^*_1 + a_{o1}z^*_2)(p^*_{o3}a_{o1} - a_{o3}) - a_{o1}$$

$$h_{21} = \frac{1}{\delta^*(p^*)}(a_{12} - a_{o2}z^*_1 + a_{o1}z^*_2)(a_{o2} - p^*_{o2}a_{o1})$$

$$h_{12} = \frac{1}{\delta^*(p^*)}(-a_{13} + a_{o3}z^*_1 - a_{o1}z^*_3)(p^*_{o3}a_{o1} - a_{o3})$$

$$h_{22} = \frac{1}{\delta^*(p^*)}(-a_{13} + a_{o3}z^*_1 - a_{o1}z^*_3)(a_{o2} - p^*_{o2}\,a_{o1}) + a_{o1}.$$

Here we use the local coordinates $z^*_1 = z_1z_o^{-1},\ldots$ of $U_o \subset \mathbb{P}_3$ and $p^*_{o2} = p_{o2}p_{o1}^{-1},\ldots$ of $V_1 \subset S^4$, cf. (2.3). Note that p^*_{ij} can be considered as a function on $U_o(\cup U_1)$ by the projection π. (Here we have also assumed that f.e. $a_{o1} \neq 0$.)

(6.4) Now the connection matrix $\tilde{\omega}$ of $\tilde{\nabla}$ on $U_o \cup U_1$ with respect to ψ_1 is given by

$$\tilde{\omega}''^{t} = h^{-1}\bar{\partial}h$$

$$\tilde{\omega}' = -\overline{\tilde{\omega}''}^{t}$$

$$\tilde{\omega} = \tilde{\omega}' + \tilde{\omega}''$$

which after calculating has the form

$$\tilde{\omega} = \begin{pmatrix} \Theta_{11} & \Theta_{12} \\ \Theta_{21} & \Theta_{22} \end{pmatrix}$$

with

$$\Theta_{11} = \frac{1}{2\delta^*(p^*)}[(p^*_{13}a_{o1} - a_{13})dp^*_{o2} - (p^*_{o3}a_{o1} - a_{o3})dp^*_{12}$$
$$+ (p^*_{12}a_{o1} - a_{12})dp^*_{o3} - (p^*_{o2}a_{o1} - a_{o2})dp^*_{13}]$$

$$\Theta_{12} = \frac{1}{\delta^*(p^*)}[(p^*_{12}a_{o1} - a_{12})dp^*_{o2} - (p^*_{o2}a_{o1} - a_{o2})dp^*_{12}]$$

$$\Theta_{22} = -\Theta_{11},\ \Theta_{21} = -\Theta_{12}.$$

By this one immediately has the result that $\tilde{\omega} = \pi^* \omega$ for the same matrix ω of forms on $V_1 \subset S^4$, one has only to forget that p_{ij}^* are functions on $U_o \cup U_1$ and consider them as the coordinate functions on V_1, cf. (4.6). Thus we have obtained <u>an explicit expression of the connection matrix ω of any SU(2) Yang-Mills potential ∇ on S^4 with self-dual field and instanton number $c_2=1$</u>, when restricted to $V_1 \cong \mathbb{R}^4$.

(6.5) It is convenient to use other coordinates $t_1,\ldots,t_4$ on V_1 which are given by

$$t_\nu = \frac{x_{\nu+1}}{x_o+x_1} ,$$

where $x_o,\ldots,x_5$ are given by $p_{o1},\ldots,p_{23}$ as in (2.1). Then we have

$$t_1 - it_2 = p_{13}^* \qquad t_3 - it_4 = p_{o3}^*$$
$$t_1 + it_2 = p_{o2}^* \qquad t_3 + it_4 = -p_{12}^* .$$

If we use new parameters $a_1,\ldots,a_4$, d (where we assume $a_{o1}=1$) by the same transformation

$$a_1 - ia_2 = a_{13} \qquad a_3 - ia_4 = a_{o3}$$
$$a_1 + ia_2 = a_{o2} \qquad a_3 + ia_4 = -a_{12}$$
$$d = a_{23} - a_{o2}a_{13} + a_{o3}a_{12} > 0 ,$$

we obtain the following expressions for

$$\omega = A_1(t)dt_1 + \ldots + A_4(t)dt_4 ,$$

where

$$A_1(t) = \frac{1}{d+\|t-a\|^2}\begin{pmatrix} -i(t_2-a_2) & -(t_3-a_3)-i(t_4-a_4) \\ (t_3-a_3)-i(t_4-a_4) & i(t_2-a_2) \end{pmatrix}$$

$$A_2(t) = \frac{1}{d+\|t-a\|^2}\begin{pmatrix} i(t_1-a_1) & (t_4-a_4)-i(t_3-a_3) \\ -(t_4-a_4)-i(t_3-a_3) & -i(t_1-a_1) \end{pmatrix}$$

$$A_3(t) = \frac{1}{d+\|t-a\|^2}\begin{pmatrix} -i(t_4-a_4) & (t_1-a_1)+i(t_2-a_2) \\ -(t_1-a_1)+i(t_2-a_2) & i(t_4-a_4) \end{pmatrix}$$

$$A_4(t) = \frac{1}{d+\|t-a\|^2}\begin{pmatrix} i(t_3-a_3) & -(t_2-a_2)+i(t_1-a_1) \\ (t_2-a_2)+i(t_1-a_1) & -i(t_3-a_3) \end{pmatrix} .$$

This is exactly the general form of 1-instantons which had first been obtained by t'Hooft. Note that $d+\|t-a\|^2$ is nothing else then the function δ^* which by the above description has a geometrical meaning. Furthermore by the above transformation it follows that $a_1,\ldots,a_4$ are real parameters, which is equivalent to condition (6.2), (iii). The parameter space of all 1-instantons is thus the open upper halfspace $\mathbb{R}^5_+ = \{(d,a_1,\ldots,a_4)\in\mathbb{R}^5 | d>o\}$.

R e f e r e n c e s

[1] Atiyah-Ward, Instantons and Algebraic Geometry, Commun. Math. Phys. 55, 117-124 (1977)

[2] Atiyah-Hitchin-Drinfeld-Manin, Construction of Instantons, Phys. Letters 65 A, 185-187 (1978)

[3] Barth-Hulek, Monads and Moduli of vectorbundles, manuscr. math. 25, 323-347 (1978)

[4] Douady - Verdier, Les équations de Yang-Mills, Seminaire E.N.S. Paris 1977-78, astérisque 71-72 (1980

[5] Milnor-Stasheff, Characteristic classes, Princeton University Press 1974

[6] Rawnsley, On the Atiyah-Hitchin-Drinfeld-Manin vanishing theorem for cohomology groups of instanton bundles, Math. Ann. 241, 43-56 (1979)

[7] Rawnsley, Self-dual Yang-Mills fields, manuscript

[8] Okonek-Schneider-Spindler, Vectorbundles on complex projective space, Progr. in Math. 3, Birkhäuser Boston 1980

[9] Trautmann, Zur Berechnung von Yang-Mills-Potentialen durch holomorphe Vektorbündel, Proceedings of the Nice conference 1979, Progress in Math. 7, Birkhäuser Boston 1980

[9'] Trautmann, Moduli for vectorbundles on $\mathbb{P}_n(\mathbb{C})$, Math. Ann. 237, 167-186 (1978)

[10] Wells, Differential analysis on complex manifolds, Prentice Hall 1973.

LIST OF VISITORS

Name and Institute	Member State
D I R E C T O R S	
* A. ANDREOTTI Scuola Normale Superiore Piazza dei Cavalieri 56100 Pisa Italy	Italy
J. EELLS Mathematics Institute University of Warwick Coventry CV4 7AL U.K.	U.K./U.S.A.
I.M. SINGER (did not attend Seminar) Department of Mathematics University of California Berkeley, California 94720 U.S.A.	U.S.A.
G. VIDOSSICH Istituto di Matematica Università di Trieste Piazzale Europa 1 34127 Trieste Italy	Italy

* Professor Andreotti died on 21 February 1980.

Name and Institute	Member State

V I S I T O R S

E. ABBENA
Istituto di Geometria
Università di Torino
Via Principe Amedeo 8
Torino
Italy — Italy

E. ABDALLA
Departamento de Fisica Matematica
Instituto de Fisica
Universidade de Sao Paulo
Caixa Postal 20516
Sao Paulo
Brazil — Brazil

R.F.A. ABIODUN
Department of Mathematics
University of Ife
Ile-Ife
Nigeria — Nigeria

J. AHSAN (Associate)
Department of Mathematics Sciences
University of Petroleum and Minerals
U.P.M. No. 468
Dhahran
Saudi Arabia — Saudi Arabia/Pakistan

M.A. AL-BASSAM (Affiliate)
Department of Mathematics
Kuwait University
P.O.Box 5969
Kuwait — Kuwait/Iraq

Y.A.W. AL-HITI
Mathematics Department
College of Education
Basrah
Iraq — Iraq

A.H. AL-MOAJIL
University of Petroleum and Minerals
P.O.Box 294
Dhahran
Saudi Arabia — Saudi Arabia

S. ALPAY
Department of Mathematics
Middle East Technical University
Ankara
Turkey — Turkey

Name and Institute	Member State
G.D. AL-SABTI Department of Mathematics College of Science Basrah University Basrah Iraq	Iraq
A.S. ALVES (Affiliate) Departamento de Matematica Universidade de Coimbra 3000 Coimbra Portugal	Portugal
M. AMIN Department of Mathematics Government College Lahore Pakistan	Pakistan
S.A. ARAM Scientific Research Department University of Aden Aden People's Democratic Republic of Yemen	People's Democratic Republic of Yemen
M. ASADZADEH Department of Mathematics Chalmers University of Technology S-412 96 Göteborg Sweden	Sweden/Iran
M.B. ATTIA (Affiliate) Faculty of Engineering Ain Shams University Cairo Egypt	Egypt
A. AYTUNA Department of Mathematics Middle East Technical University Ankara Turkey	Turkey
B. BA Department of Mathematics Faculty of Science Université d'Abidjan 04 B.P. 322 Abidjan Ivory Coast	Ivory Coast/Niger

Name and Institute	Member State
O.P. BABELON L.P.T.H.E. Tour 16, 1er etage 4 place Jussieu 75230 Paris Cedex 05 France	France
A.G.A.G. BABIKER School of Mathematical Sciences University of Khartoum P.O.Box 321 Khartoum Sudan	Sudan
C. BACHELET (until November 1980) Section de Mathematiques Facult́é des Sciences 2 - 4 rue du Lievre C.P. 124 1211 Geneva 24 Switzerland	Switzerland/France
C. BADJI Departement de Mathematiques Facult́é des Sciences Universit́é de Dakar Dakar Senegal	Senegal
Y. BAHRAMPOUR Department of Mathematics Kerman University Kerman Iran	Iran
P. BAIRD Mathematics Institute University of Warwick Coventry CV4 7AL U.K.	U.K.
E.A. BANGUDU Department of Mathematics University of Ilorin P.M.B. 1515 Ilorin Nigeria	Nigeria
W.A. BASSALI (Affiliate) Department of Mathematics Faculty of Science University of Kuwait P.O.Box 5969 Kuwait	Kuwait/Egypt

Name and Institute	Member State
F. BATTELLI Istituto de Matematica "U. Dini" Viale Morgagni 67/A 50100 Firenze Italy	Italy
A. BELAL Faculty of Science Assiut University Assuan Egypt	Egypt
M. BEN SALEM Faculté des Sciences et Techniques Sfax Tunisia	Tunisia
B.G. BERNDTSSON Mathematics Department Chalmers University of Technology S-412 96 Göteborg Sweden	Sweden
A. BETTE Institute of Theoretical Physics University of Stockholm Vanadisv. 9 S-113 46 Stockholm Sweden	Sweden
A. BIANCOFIORE Istituto di Matematica Università di L'Aquila 67100 L'Aquila Italy	Italy
D.M. BOICHU Département de Mathematiques Institut des Sciences Exactes Université d'Oran Oran Algeria	Algeria/France
A. BOYOKAKSOY (Affiliate) Teorik Fizik Kürsüsü Istanbul Universitesi Fen Fakültesi Istanbul Turkey	Turkey

Name and Institute	Member State
F.F. BRACKX Rijksuniversiteit te Gent Seminarie voor Wiskundige Analyse St. Pietersnieuwstraat 39 B-9000 Gent Belgium	Belgium
A. BREZINI Department of Mathematics University of Oran Es Senia Oran Algeria	Algeria
N. BUCHDAHL Mathematical Institute University of Oxford 24-29 St. Giles Oxford OX1 3LS U.K.	U.K.
K. BUGAJSKA Instytut Fizyki Silesian University ul. Uniwersytecka 4 40-007 Katowice Poland	Poland
R. CADDEO Istituto Matematico Via Ospedale 72 09100 Cagliari Italy	Italy
E. CALABI Department of Mathematics Philadelphia, PA. U.S.A.	U.S.A.
R. CALVANI C S E L T Centro Studi e Laboratori Telecomunicazioni s.p.a. Via Guglielmo Reiss Romoli 274 10148 Torino Italy	Italy
A. CAVUS (Affiliate) Department of Mathematics Karadeniz Technical University Trabzon Turkey	Turkey

Name and Institute	Member State
R.P. COELHO (Affiliate) Instituto de Matematica Faculdade de Ciencias Universidade de Coimbra 3000 Coimbra Portugal	Portugal
K.T. DALE Telemark Distrikts Høgskole N-3800 Bø Norway	Norway
G. D'AMBRA Istituto Matematico Università Via Ospedale 72 09100 Cagliari Italy	Italy
H. DANIËLS Institute for Theoretical Physics University of Groningen P.O.Box 800 9700 AV Groningen The Netherlands	The Netherlands
M. D'APRILE Dipartimento di Matematica Università di Calabria C.P. 9 87030 - Roges (CS) Italy	Italy
G. DEDENE Departement Wiskunde Faculteit der Wetenschappen Katholieke Universiteit Leuven Celestijnenlaan 200 B B-3030 Leuven (Heverlee) Belgium	Belgium
E. DEEBA University of Petroleum and Minerals Dhahran Saudi Arabia	Saudi Arabia
C.L. DEPOLLIER Institut de Sciences Exactes Departement de Physique Université d'Oran Oran Es-Senia Algeria	Algeira/France

Name and Institute	Member State
L.H. DIEZ Department of Mathematics Universidad de Antioquia Apartado Aereo 1226 Medellin Colombia	Colombia
A.S. DUBSON (Private) (Chez Sibony) 11 rue de la Glaciere Paris 13 France	France/Argentina
M. EASTWOOD Mathematical Institute 24-29 St. Giles Oxford, OX1 3LS U.K.	U.K.
T.G. ELMROTH Mathematics Department Chalmers University of Technology S-412 96 Göteborg Sweden	Sweden
K.A. EL-SAMARRAI College of Science University of Baghdad Baghdad Iraq	Iraq
N. EL-SAMRA (Affiliate) Department of Mathematics Faculty of Women Ain Shams University Heliopolis Cairo Egypt	Egypt
S.A. ELSANOUSI School of Mathematical Sciences University of Khartoum P.O.Box 321 Khartoum Sudan	Sudan
D.B.A. EPSTEIN Mathematics Institute University of Warwick Coventry CV4 7AL U.K.	U.K.

Name and Institute	Member State
M. ESKIN Department of Mathematics Ben Gurion University of Negev Beer-Sheva Israel	Israel
A.A. FADLALLA Department of Mathematics Faculty of Science University of Cairo Giza, Cairo Egypt	Egypt
H.R. FARRAN (Affiliate) Department of Mathematics Faculty of Science University of Kuwait P.O.Box 5969 Kuwait	Kuwait/Lebanon
F. FAVILLI Istituto di Matematica Università di Pisa Via Buonarroti 2 56100 Pisa Italy	Italy
L.J. FERRO CASAS Universidad de los Andes A.A. 4976 Bogotá Colombia	Colombia
O. FORSTER Westfälische Wilhelms-Universität Münster Fachbereich 15 Mathematik Mathematisches Institut Roxeler Strasse 64 4400 Münster Fed. Rep. Germany	Fed. Rep. Germany
E. FOTSO Departement d'Informatique Generale Université Paris VII Tour 13, 1er etage 2 place Jussieu 75221 Paris Cedex 05 France	France/Cameroon

Name and Institute	Member State
B.A. FRIEDMAN Loomis Laboratory of Physics University of Illinois, U.C. Urbana, Illinois 61801 U.S.A.	U.S.A.
A. GAFFUR Department of Computer Science Al Fateh University P.O.Box 13531 Tripoli Libya	Libya
I. GARRO (Private) Nayal Amiri St. 39 Aleppo Syria	Syria
R. GATTAZZO Istituto di Matematica Applicata Via Belzoni 7 35100 Padova Italy	Italy
R. GIACHETTI I.N.F.N. Sezione de Firenze Large E. Fermi 2 51025 Firenze Italy	Italy
M.L. GINSBERG Mathematical Institute University of Oxford 24-29 St. Giles Oxford OX1 3LB U.K.	U.K./U.S.A.
E. GIUSTI Istituto di Matematica Università di Pisa 56100 Pisa Italy	Italy
J.F. GLAZEBROOK Mathematics Institute University of Warwick Coventry CV4 7AL U.K.	U.K.

Name and Institute	Member State
J. GLOBEVNIK Institute of Mathematics, Physics and Mechanics University of Ljubljana 19 Jadranska 61000 Ljubljana Yugoslavia	Yugoslavia
C. GORDON Department of Mathematics Lehigh University Bethlehem, Penn. 18015 U.S.A.	U.S.A.
A. GRAY Department of Mathematics University of Maryland College Park, Maryland 20742 U.S.A.	U.S.A.
G.G. GUSSI Institutul National Pentru Creatie Stiintofica si Tehnica Bd. Pacii 220 77538 Bucuresti Romania	Romania
M.S. HACHAICHI Université des Sciences et de la Technologie d'Alger B.P. No. 9 Dar el Beida (Alger) Algeria	Algeria
F. HAMZA Faculty of Science Sana's University Sana'a Yemen Arab Republic	Yemen Arab Republic/Egypt
T. HANGAN Université Sciences et Techniques de Lille I Villeneuve d'Ascq. France	France
W.N. HANNA (Affiliate) Faculty of Engineering Ain Shams University Cairo Egypt	Egypt

Name and Institute	Member State
V.L. HANSEN Mathematical Institute Technical University of Denmark Building 303 DK-2800 Lyngby Denmark	Denmark
Z. HAR'EL Department of Mathematics Israel Institute of Technology TECHNION Haifa 32000 Israel	Israel
G.A. HARRIS (until October 1981) Department of Mathematics Rice University Houston, Texas U.S.A. (Permanent) Department of Mathematics Texas Tech University Box 4319 Lubbock, Texas 79409 U.S.A.	U.S.A.
M. HUSSAIN Mathematics Department Government College Lahore Pakistan	Pakistan
E.M. IBRAHIM Department of Mathematics Faculty of Engineering Abassia Cairo Egypt	Egypt
M. IDA Istituto di Geometria Piazza di Porta S. Donato 5 40100 Bologna Italy	Italy
S.A. ILORI Department of Mathematics University of Ibadan Ibadan Nigeria	Nigeria

Name and Institute	Member State
C. IMORU Department of Mathematics University of Ife Ile-Ife Nigeria	Nigeria
A. JAKUBIEC Institute of Mathematical Methods in Physics University of Warsaw ul. Hoza 74 Warszawa 00-682 Poland	Poland
O.M. JUNEJA Departamento de Matematica Universidade Estadual de Londrina 86100 Londrina (PR) Brazil	Brazil/India
T.M. KARKAR Departement de Mathematiques Facultè des Sciences de Tunis Campus Universitaire du Belvèdere Tunis Tunisia	Tunisia
H.M. KHALIL (Affiliate) Physics Department University College for Girls Ain Shams University Cairo Egypt	Egypt
V.S. KLEIN (until September 1981) Department of Mathematics University of Kentucky Lexington, Kentucky 40506 U.S.A. (Permanent) Department of Mathematics Wellesley College Wellesley, Ma. 02181 U.S.A.	U.S.A.
N.H. KUIPER Director Institut des Hautes Etudes Scientifiques 91440 Bures-sur-Yvette France	France

Name and Institute	Member State
P.J.A. KUMLIN Department of Mathematics Chalmers University of Technology S-412 96 Göteborg Sweden	Sweden
A. KYRIAZIS (Affiliate) Mathematical Institute University of Athens 57 Solonos Street Athens 143 Greece	Greece
C. LACKSCHEWITZ Mathematisches Institut der Universität Bunsenstrasse 3/5 D-34 Göttingen Fed. Rep. Germany	Fed. Rep. Germany
S.P. LAM (until August 1982) Department of Pure Mathematics and Mathematical Statistics 16 Mill Lane Cambridge CB2 1SB U.K.	U.K./Hong Kong
A. LAMARI Université des Sciences et de la Technologie d'Alger B.P. No. 9 Dar el Beida (Alger) Algeria	Algeria
S. LARSSON Department of Mathematics Chalmers University of Technology S-412 96 Göteborg Sweden	Sweden
J.M. LAWRYNOWICZ Institute of Mathematics Polish Academy of Sciences Łódź Branch ul. Kilińskiego 86 PL-90-012 Łódź Poland	Poland
C.R. LEBRUN Department of Mathematics State University of New York Stoney Brook, N.Y. U.S.A.	U.S.A.

Name and Institute	U.S.A.
L. LEMAIRE Department of Mathematics Université Libre de Bruxelles C.P. 218 1050 Brussels Belgium	Belgium
L. LEMPERT Department of Analysis I Eötvös University Múzeum Krt. 6-8 1088 Budapest Hungary	Hungary
J.-L.C.F.G. LIEUTENANT Institut de Mathématiques Université de l'Etat a Liege 15 avenue des Tilleuls 4000 - Liege Belgium	Belgium
L.S.O. LIVERPOOL Department of Mathematics Fourah Bay College Freetown Sierra Leone	Sierra Leone
L. LUHAHI Faculté des Sciences B.P. 190 Kinshasa 11 Zaire	Zaire
D. LUMINET Mathematiques Université Libre de Bruxelles Bd. du Triomphe C.P. 214 Bruxelles Belgium	Belgium
C.H. LUTTERODT Department of Mathematics University of Cape Coast Cape Coast Ghana	Ghana
K. MAHARANA Physics Department Utkal University Bhubaneswar - 751004 India	India

Name and Institute	Member State
H. MAJIMA Department of Mathematics Faculty of Science University of Tokyo Tokyo Japan	Japan
P.M. MATZEU Istituto Matematico Via Ospedale 72 Cagliari Italy	Italy
Z. MILIGY (Affiliate) Department of Mathematics Ain Shams University Cairo Egypt	Egypt
V. MOLOTKOV (Affiliate) Institute of Nuclear Energy and Nuclear Research Sofia Bulgaria	Bulgaria
A.S. MSHIMBA Department of Mathematics University of Dar es Salaam P.O.Box 35062 Dar es Salaam Tanzania	Tanzania
M. NACINOVICH Istituto di Matematica Via Buonarroti 2 56100 Pisa Italy	Italy
S. NANDA Department of Mathematics Indian Institute of Technology Kharagpur India	India
M.S. NARASIMHAN Tata Institute for Fundamental Research Colaba Bombay India	India

Name and Institute	Member State
M.A. NASR Faculty of Science Mansoura University Mansoura Egypt	Egypt
A. NDUKA Physics Department University of Ife Ile-Ife Nigeria	Nigeria
B.-Y. NG Department of Mathematics University of Malaya Kuala Lumpur Malaysia	Malaysia
K.I. NOOR Mathematics Department Kerman University P.O.Box 333 Kerman Iran	Iran/Pakistan
M.A. NOOR Mathematics Department Kerman University P.O.Box 333 Kerman Iran	Iran/Pakistan
C.O. NWACHUKU Department of Mathematics University of Benin Benin City Nigeria	Nigeria
C.A. NYACK Physics Department University of the West Indies P.O.Box 64 Bridgetown Barbados	Barbados/Grenada
S.A. OBAID (Affiliate) Department of Mathematics University of Kuwait P.O.Box 5969 Kuwait	Kuwait/Jordan

Name and Institute	Member State
A. ODZIJEWICZ Division of Mathematical Methods in Physics University of Warsaw Hoza 74 00-682 Warsaw Poland	Poland
A. ORLANDI Istituto Matematico Via Buonarroti 2 56100 Pisa Italy	Italy
L.S. PANTA Instituto de Matematica Universidade Federal do Rio Grande do Sul Rua Sarmento Leite 425 90.000 Porto Alegre - RS Brazil	Brazil
L.C. PAPALOUCAS Institute of Mathematics University of Athens 57 Solonos Street Athens 143 Greece	Greece
R. PARVATHAM (until October 1980) Institut Fourier de Mathematiques Pures Boite Postale 116 38402 St. Martin d'Hères France	France/India
R. PERCACCI Istituto di Matematica Università di Trieste Piazzale Europa 1 34127 Trieste Italy	Italy
P. PFLUG Fachbereich 7 - Mathematik Gesamthochschule Wuppertal Gaussstrasse 20 Wuppertal-Elberfeld Fed. Rep. Germany	Fed. Rep. Germany
PHAM MAU QUAN Centre Scientifique et Polytechnique Université Paris-Nord Av. J.-B. Clément 93430 Villetaneuse France	France

Name and Institute	Member State
A.B. PHANUMPHAI Department of Physics Faculty of Science Ramkumbaeng University Bangkok 24 Thailand	Thailand
A.M. PIERZCHALSKI Institute of Mathematics Polish Academy of Sciences Lódź Branch Kilinskeigo 86 PL-90-012 Lódź Poland	Poland
R. PIGNONI Istituto di Matematica Via F. Buonarroti 2 56100 Pisa Italy	Italy
A. QADIR (Associate) Mathematics Department Quaid-i-Azam University Islamabad Pakistan	Pakistan
M. RADJABALIPOUR Institute of Mathematics University of Mazandaran P.O.Box 444 Babolsar Iran	Iran
R. RAMADAS Tata Institute of Fundamental Research Homi Bhahaba Road Bombay 5 India	India
E. RAMIREZ DE ARELLANO Departamento de Matematicas Centro de Investigacion del IPN Apartado 14740 Mexico 14, D.F. Mexico	Mexico
D.R.K.S. RAO Department of Mathematics Andhra University Andhra - Waltair India	India

Name and Institute	Member State
G.S. RAO Ramanujan Institute for Advanced Study in Mathematics University of Madras Madras 600 005 India	India
H.B. RIBEIRO-FILHO Departamento de Matematica Pontificia Universidade Católica do Rio de Janeiro Rua Marques de S. Vincent, 209 22453 Rio de Janeiro Brazil	Brazil
G. RITA Istituto di Matematica Applicata Via Belzoni 7 35100 Padova Italy	Italy
B.M. ROBLES E.S.F.M. - I.P.N. Edif # 6 Zacetenco Mexico 14, D.F. Mexico	Mexico
J.S. ROGULSKI Institute of Mathematics Warsaw Technical University Pl. Jednosci Robotniczej 1 00-661 Warszawa Poland	Poland
M.C. RONCONI Istituto di Matematica Applicata Via Belzoni 7 35100 Padova Italy	Italy
R. SAERENS (until end 1981) Department of Mathematics University of Washington C-138 Padelford Hall GN-50 Seattle, Wa. 98195 U.S.A. (Permanent) Dept. Wiskunde Vrije Universiteit Brussel Pleinlaan 2 B-1050 Brussels Belgium	U.S.A./Belgium

Name and Institute	Member State
Y.A. SAID Department of Physics and Mathematics College of Education University of Mosul Mosul Iraq	Iraq
G. SAITOTI Department of Mathematics University of Nairobi Box 30197 Nairobi Kenya	Kenya
R. SANCHEZ-PEREGRINO (until end 1981) Mathématiques 3e cycle Université Paris VII Tour 45-55 - 5e étage 2 place Jussieu 75005 Paris France (Permanent) Facultad de Ciencias UNAM Ciudad Universitaria Mexico, D.F. Mexico	France/Mexico
P.S. SCHNARE (Private) 11 Belmont Avenue Waterville, Maine U.S.A.	U.S.A.
H.C.J. SEALEY Mathematics Institute University of Warwick Coventry CV4 7AL U.K.	U.K.
C. SERIES Mathematics Institute University of Warwick Coventry CV4 7AL U.K.	U.K.
M.M. SHAHIN (Affiliate) Institute of Education for Girls Al-Shamia Kuwait	Kuwait/Egypt

Name and Institute	Member State
S.M. SHAHRTASH Mathematics Department University of Shiraz Shiraz Iran	Iran
M. SHALABY Department of Physics Ain Shams University Cairo Egypt	Egypt
S. SHERIF (Affiliate) Department of Mathematics University College for Women Ain Shams University Heliopolis Cairo Egypt	Egypt
O.P. SINGH Department of Mathematics Aligarh Muslim University Aligarh - 202110 India	India
M.L. SKWARCZYNSKI Mathematics Department Warsaw University P.K. i N. 9p Warsaw Poland	Poland
K.S. SOH (Affiliate) Department of Physics Education Seoul National University Gwanac Seoul Korea	Korea
I.H. SOLIMAN (Affiliate) Faculty of Engineering Ain Shams University Cairo Egypt	Egypt
V. SOUCEK Department of Mathematical Analysis Faculty of Mathematics and Physics Sokolovska 83 186 00 Praha Czechoslovakia	Czechoslovakia

Name and Institute	Member State
W. STOLL Department of Mathematics University of Notre Dame Post Office Box 398 Notre Dam, Indiana 46556 U.S.A.	U.S.A./Fed. Rep..Germany
R. STORA CERN CH-1211 Geneva 23 Switzerland	CERN/Switzerland
D. SUNDARARAMAN School of Mathematics, Computer and Information Sciences University of Hyderabad Hyderabad 500 001 India	India
A. SYED Department of Mathematics University of Karachi Karachi Pakistan	Pakistan
G.E. TANYI (Associate) Laboratory of Mathematical Modelling Faculty of Science B.P. 812 Yaounde United Republic of Cameroon	Cameroon
J. TARSKI Institut für Theoretische Physik Technische Universität Clausthal 3392 Clausthal-Zellerfeld Fed. Rep. Germany	Fed. Rep. Germany/U.S.A.
K. TILLEKERATNE Department of Mathematics Ruhunu University College Matara Sri Lanka	Sri Lanka
J.A. TORRES CHAZARO Departamento de Matematica Universidad Autonoma Metropolitana Ap. Postal 55-533 Mexico, D.F. Mexico	Mexico

Name and Institute	Member State
G.A. TOTH Mathematical Institute of the Hungarian Academy of Sciences H-1053 Budapest Reáltanoda u. 13-15 Hungary	Hungary
L.M. TOVAR SANCHEZ E.S.F.M. I.P.N. Edificio 6 Unidad D. Zacatenco Mexico, D.F. Mexico	Mexico
G. TRAUTMANN Mathematik Universität Kaiserslautern Postfach 3049 6750 Kaiserslautern Fed. Rep. Germany	Fed. Rep. Germany
G.F. TSAGAS Faculty of Technology University of Thessaloniki Thessaloniki Greece	Greece
Y.P. UPADHYAYA KOIRALA Mathematics Department Kirtipur Campus Kirtipur Kathmandu Nepal	Nepal
R. VALLEJO GARAMENDI E.S.F.M. I.P.N. Edificio 6 Unidad D. Zacatenco Mexico 14, D.F. Mexico	Mexico
F.J. VANHECKE Departamento de Fisica da UFRN Campus Universitario 59000 Natal - RN Brazil	Brazil/Belgium

Name and Institute	Member State
I.W. VAN SPIEGEL Twente University of Technology P.O.Box 217 7500 AE Enschede Netherlands	Netherlands
E. VASSILIOU Institute of Mathematics University of Athens 57 Solonos Street Athens 143 Greece	Greece
S. VERNIER PIRO Istituto Matematico Università di Cagliari Viale Merello 09100 Cagliari Italy	Italy
C.M. VIALLET L.P.T.H.E. Université Pierre et Marie Curie Tour 16, 1er étage 4 Place Jussieu 75230 Paris, Cedex 05 France	France
G. VIGNA SURIA (until October 1980) School of Mathematics University of Leeds Leeds LS2 9ST U.K.	U.K./Italy
G.D. VILLA SALVADOR E.S.F.M. I.P.N. Edificio 6 Unidad D. Zacatenco Mexico 14, D.F. Mexico	Mexico
M. VISINESCU Department of Fundamental Physics Institute for Physics and Nuclear Engineering P.O.Box 5206 Bucharest Romania	Romania

Name and Institute	Member State
A.A. WAWRZYNCZYK Department of Mathematical Methods in Physics Warsaw University Hoza 74 00-682 Warsaw Poland	Poland
G.G. WEILL Department of Mathematics Polytechnic Institute of New York 333 Jay Street Brooklyn, New York 11201 U.S.A.	U.S.A./France
C. WOOD Mathematics Institute University of Warwick Coventry CV4 7AL U.K.	U.K.
J.C. WOOD (until September 1981) Sonderforshuncsbereich Theoretische Mathematik Universität Bonn Berinnstrasse 4 5300 Bonn 1 Fed. Rep. Germany (Permanent) Department of Mathematics University of Leeds Leeds LS2 9JT U.K.	U.K.
M. YAHIA School of Mathematical Sciences University of Khartoum P.O.Box 321 Khartoum Sudan	Sudan
M. YOUSSOUF Department of Mathematics Al Fateh University Tripoli Libya	Libya/Egypt

Name and Institute	Member State
J. ZAFARANI Department of Mathematics University of Isfahan Isfahan Iran	Iran
S.I. ZAMAN Department of Mathematics University of Dacca Dacca - 2 Bangladesh	Bangladesh

. 787: Potential Theory, Copenhagen 1979. Proceedings, 1979. ited by C. Berg, G. Forst and B. Fuglede. VIII, 319 pages. 1980.

. 788: Topology Symposium, Siegen 1979. Proceedings, 1979. ted by U. Koschorke and W. D. Neumann. VIII, 495 pages. 1980.

. 789: J. E. Humphreys, Arithmetic Groups. VII, 158 pages. 1980.

. 790: W. Dicks, Groups, Trees and Projective Modules. IX, 127 jes. 1980.

. 791: K. W. Bauer and S. Ruscheweyh, Differential Operators Partial Differential Equations and Function Theoretic Applica- s. V, 258 pages. 1980.

. 792: Geometry and Differential Geometry. Proceedings, 1979. ted by R. Artzy and I. Vaisman. VI, 443 pages. 1980.

. 793: J. Renault, A Groupoid Approach to C*-Algebras. III, 160 jes. 1980.

. 794: Measure Theory, Oberwolfach 1979. Proceedings 1979. ted by D. Kölzow. XV, 573 pages. 1980.

. 795: Séminaire d'Algèbre Paul Dubreil et Marie-Paule Malliavin. ceedings 1979. Edited by M. P. Malliavin. V, 433 pages. 1980.

796: C. Constantinescu, Duality in Measure Theory. IV, 197 es. 1980.

797: S. Mäki, The Determination of Units in Real Cyclic Sextic ds. III, 198 pages. 1980.

. 798: Analytic Functions, Kozubnik 1979. Proceedings. Edited . Ławrynowicz. X, 476 pages. 1980.

799: Functional Differential Equations and Bifurcation. Proceed- s 1979. Edited by A. F. Izé. XXII, 409 pages. 1980.

800: M.-F. Vignéras, Arithmétique des Algèbres de Quaternions. 169 pages. 1980.

801: K. Floret, Weakly Compact Sets. VII, 123 pages. 1980.

802: J. Bair, R. Fourneau, Etude Géometrique des Espaces toriels II. VII, 283 pages. 1980.

803: F.-Y. Maeda, Dirichlet Integrals on Harmonic Spaces. X, pages. 1980.

804: M. Matsuda, First Order Algebraic Differential Equations. 11 pages. 1980.

805: O. Kowalski, Generalized Symmetric Spaces. XII, 187 es. 1980.

806: Burnside Groups. Proceedings, 1977. Edited by J. L. Men- e. V, 274 pages. 1980.

807: Fonctions de Plusieurs Variables Complexes IV. Pro- dings, 1979. Edited by F. Norguet. IX, 198 pages. 1980.

808: G. Maury et J. Raynaud, Ordres Maximaux au Sens de sano. VIII, 192 pages. 1980.

809: I. Gumowski and Ch. Mira, Recurences and Discrete amic Systems. VI, 272 pages. 1980.

810: Geometrical Approaches to Differential Equations. Pro- lings 1979. Edited by R. Martini. VII, 339 pages. 1980.

811: D. Normann, Recursion on the Countable Functionals. 191 pages. 1980.

812: Y. Namikawa, Toroidal Compactification of Siegel Spaces. 162 pages. 1980.

813: A. Campillo, Algebroid Curves in Positive Characteristic. 8 pages. 1980.

814: Séminaire de Théorie du Potentiel, Paris, No. 5. Proceed- Edited by F. Hirsch et G. Mokobodzki. IV, 239 pages. 1980.

315: P. J. Slodowy, Simple Singularities and Simple Algebraic ps. XI, 175 pages. 1980.

816: L. Stoica, Local Operators and Markov Processes. VIII, pages. 1980.

Vol. 817: L. Gerritzen, M. van der Put, Schottky Groups and Mumford Curves. VIII, 317 pages. 1980.

Vol. 818: S. Montgomery, Fixed Rings of Finite Automorphism Groups of Associative Rings. VII, 126 pages. 1980.

Vol. 819: Global Theory of Dynamical Systems. Proceedings, 1979. Edited by Z. Nitecki and C. Robinson. IX, 499 pages. 1980.

Vol. 820: W. Abikoff, The Real Analytic Theory of Teichmüller Space. VII, 144 pages. 1980.

Vol. 821: Statistique non Paramétrique Asymptotique. Proceedings, 1979. Edited by J.-P. Raoult. VII, 175 pages. 1980.

Vol. 822: Séminaire Pierre Lelong-Henri Skoda, (Analyse) Années 1978/79. Proceedings. Edited by P. Lelong et H. Skoda. VIII, 356 pages, 1980.

Vol. 823: J. Král, Integral Operators in Potential Theory. III, 171 pages. 1980.

Vol. 824: D. Frank Hsu, Cyclic Neofields and Combinatorial Designs. VI, 230 pages. 1980.

Vol. 825: Ring Theory, Antwerp 1980. Proceedings. Edited by F. van Oystaeyen. VII, 209 pages. 1980.

Vol. 826: Ph. G. Ciarlet et P. Rabier, Les Equations de von Kármán. VI, 181 pages. 1980.

Vol. 827: Ordinary and Partial Differential Equations. Proceedings, 1978. Edited by W. N. Everitt. XVI, 271 pages. 1980.

Vol. 828: Probability Theory on Vector Spaces II. Proceedings, 1979. Edited by A. Weron. XIII, 324 pages. 1980.

Vol. 829: Combinatorial Mathematics VII. Proceedings, 1979. Edited by R. W. Robinson et al.. X, 256 pages. 1980.

Vol. 830: J. A. Green, Polynomial Representations of GL_n. VI, 118 pages. 1980.

Vol. 831: Representation Theory I. Proceedings, 1979. Edited by V. Dlab and P. Gabriel. XIV, 373 pages. 1980.

Vol. 832: Representation Theory II. Proceedings, 1979. Edited by V. Dlab and P. Gabriel. XIV, 673 pages. 1980.

Vol. 833: Th. Jeulin, Semi-Martingales et Grossissement d'une Filtration. IX, 142 Seiten. 1980.

Vol. 834: Model Theory of Algebra and Arithmetic. Proceedings, 1979. Edited by L. Pacholski, J. Wierzejewski, and A. J. Wilkie. VI, 410 pages. 1980.

Vol. 835: H. Zieschang, E. Vogt and H.-D. Coldewey, Surfaces and Planar Discontinuous Groups. X, 334 pages. 1980.

Vol. 836: Differential Geometrical Methods in Mathematical Physics. Proceedings, 1979. Edited by P. L. García, A. Pérez-Rendón, and J. M. Souriau. XII, 538 pages. 1980.

Vol. 837: J. Meixner, F. W. Schäfke and G. Wolf, Mathieu Functions and Spheroidal Functions and their Mathematical Foundations Further Studies. VII, 126 pages. 1980.

Vol. 838: Global Differential Geometry and Global Analysis. Proceedings 1979. Edited by D. Ferus et al. XI, 299 pages. 1981.

Vol. 839: Cabal Seminar 77 – 79. Proceedings. Edited by A. S. Kechris, D. A. Martin and Y. N. Moschovakis. V, 274 pages. 1981.

Vol. 840: D. Henry, Geometric Theory of Semilinear Parabolic Equations. IV, 348 pages. 1981.

Vol. 841: A. Haraux, Nonlinear Evolution Equations- Global Behaviour of Solutions. XII, 313 pages. 1981.

Vol. 842: Séminaire Bourbaki vol. 1979/80. Exposés 543–560. IV, 317 pages. 1981.

Vol. 843: Functional Analysis, Holomorphy, and Approximation Theory. Proceedings. Edited by S. Machado. VI, 636 pages. 1981.

Vol. 844: Groupe de Brauer. Proceedings. Edited by M. Kervaire and M. Ojanguren. VII, 274 pages. 1981.

Vol. 845: A. Tannenbaum, Invariance and System Theory: Algebraic and Geometric Aspects. X, 161 pages. 1981.

Vol. 846: Ordinary and Partial Differential Equations, Proceedings. Edited by W. N. Everitt and B. D. Sleeman. XIV, 384 pages. 1981.

Vol. 847: U. Koschorke, Vector Fields and Other Vector Bundle Morphisms – A Singularity Approach. IV, 304 pages. 1981.

Vol. 848: Algebra, Carbondale 1980. Proceedings. Ed. by R. K. Amayo. VI, 298 pages. 1981.

Vol. 849: P. Major, Multiple Wiener-Itô Integrals. VII, 127 pages. 1981.

Vol. 850: Séminaire de Probabilités XV. 1979/80. Avec table générale des exposés de 1966/67 à 1978/79. Edited by J. Azéma and M. Yor. IV, 704 pages. 1981.

Vol. 851: Stochastic Integrals. Proceedings, 1980. Edited by D. Williams. IX, 540 pages. 1981.

Vol. 852: L. Schwartz, Geometry and Probability in Banach Spaces. X, 101 pages. 1981.

Vol. 853: N. Boboc, G. Bucur, A. Cornea, Order and Convexity in Potential Theory: H-Cones. IV, 286 pages. 1981.

Vol. 854: Algebraic K-Theory. Evanston 1980. Proceedings. Edited by E. M. Friedlander and M. R. Stein. V, 517 pages. 1981.

Vol. 855: Semigroups. Proceedings 1978. Edited by H. Jürgensen, M. Petrich and H. J. Weinert. V, 221 pages. 1981.

Vol. 856: R. Lascar, Propagation des Singularités des Solutions d'Equations Pseudo-Différentielles à Caractéristiques de Multiplicités Variables. VIII, 237 pages. 1981.

Vol. 857: M. Miyanishi. Non-complete Algebraic Surfaces. XVIII, 244 pages. 1981.

Vol. 858: E. A. Coddington, H. S. V. de Snoo: Regular Boundary Value Problems Associated with Pairs of Ordinary Differential Expressions. V, 225 pages. 1981.

Vol. 859: Logic Year 1979–80. Proceedings. Edited by M. Lerman, J. Schmerl and R. Soare. VIII, 326 pages. 1981.

Vol. 860: Probability in Banach Spaces III. Proceedings, 1980. Edited by A. Beck. VI, 329 pages. 1981.

Vol. 861: Analytical Methods in Probability Theory. Proceedings 1980. Edited by D. Dugué, E. Lukacs, V. K. Rohatgi. X, 183 pages. 1981.

Vol. 862: Algebraic Geometry. Proceedings 1980. Edited by A. Libgober and P. Wagreich. V, 281 pages. 1981.

Vol. 863: Processus Aléatoires à Deux Indices. Proceedings, 1980. Edited by H. Korezlioglu, G. Mazziotto and J. Szpirglas. V, 274 pages. 1981.

Vol. 864: Complex Analysis and Spectral Theory. Proceedings, 1979/80. Edited by V. P. Havin and N. K. Nikol'skii, VI, 480 pages. 1981.

Vol. 865: R. W. Bruggeman, Fourier Coefficients of Automorphic Forms. III, 201 pages. 1981.

Vol. 866: J.-M. Bismut, Mécanique Aléatoire. XVI, 563 pages. 1981.

Vol. 867: Séminaire d'Algèbre Paul Dubreil et Marie-Paule Malliavin. Proceedings, 1980. Edited by M.-P. Malliavin. V, 476 pages. 1981.

Vol. 868: Surfaces Algébriques. Proceedings 1976–78. Edited by J. Giraud, L. Illusie et M. Raynaud. V, 314 pages. 1981.

Vol. 869: A. V. Zelevinsky, Representations of Finite Classical Groups. IV, 184 pages. 1981.

Vol. 870: Shape Theory and Geometric Topology. Proceedings, 1981. Edited by S. Mardešić and J. Segal. V, 265 pages. 1981.

Vol. 871: Continuous Lattices. Proceedings, 1979. Edited by B. Banaschewski and R.-E. Hoffmann. X, 413 pages. 1981.

Vol. 872: Set Theory and Model Theory. Proceedings, 1979. Edited by R. B. Jensen and A. Prestel. V, 174 pages. 1981.

Vol. 873: Constructive Mathematics, Proceedings, 1980. Edited by F. Richman. VII, 347 pages. 1981.

Vol. 874: Abelian Group Theory. Proceedings, 1981. Edited by R. Göbel and E. Walker. XXI, 447 pages. 1981.

Vol. 875: H. Zieschang, Finite Groups of Mapping Classes of Surfaces. VIII, 340 pages. 1981.

Vol. 876: J. P. Bickel, N. El Karoui and M. Yor. Ecole d'Eté de Probabilités de Saint-Flour IX – 1979. Edited by P. L. Hennequin. XI, 280 pages. 1981.

Vol. 877: J. Erven, B.-J. Falkowski, Low Order Cohomology and Applications. VI, 126 pages. 1981.

Vol. 878: Numerical Solution of Nonlinear Equations. Proceedings, 1980. Edited by E. L. Allgower, K. Glashoff, and H.-O. Peitgen. XIV, 440 pages. 1981.

Vol. 879: V. V. Sazonov, Normal Approximation – Some Recent Advances. VII, 105 pages. 1981.

Vol. 880: Non Commutative Harmonic Analysis and Lie Groups. Proceedings, 1980. Edited by J. Carmona and M. Vergne. IV, 553 pages. 1981.

Vol. 881: R. Lutz, M. Goze, Nonstandard Analysis. XIV, 261 pages. 1981.

Vol. 882: Integral Representations and Applications. Proceedings, 1980. Edited by K. Roggenkamp. XII, 479 pages. 1981.

Vol. 883: Cylindric Set Algebras. By L. Henkin, J. D. Monk, A. Tarski, H. Andréka, and I. Németi. VII, 323 pages. 1981.

Vol. 884: Combinatorial Mathematics VIII. Proceedings, 1980. Edited by K. L. McAvaney. XIII, 359 pages. 1981.

Vol. 885: Combinatorics and Graph Theory. Edited by S. B. Rao. Proceedings, 1980. VII, 500 pages. 1981.

Vol. 886: Fixed Point Theory. Proceedings, 1980. Edited by E. Fadell and G. Fournier. XII, 511 pages. 1981.

Vol. 887: F. van Oystaeyen, A. Verschoren, Non-commutative Algebraic Geometry, VI, 404 pages. 1981.

Vol. 888: Padé Approximation and its Applications. Proceedings, 1980. Edited by M. G. de Bruin and H. van Rossum. VI, 383 pages. 1981.

Vol. 889: J. Bourgain, New Classes of $\mathcal{L}^p$-Spaces. V, 143 pages. 1981.

Vol. 890: Model Theory and Arithmetic. Proceedings, 1979/80. Edited by C. Berline, K. McAloon, and J.-P. Ressayre. VI, 306 pages. 1981.

Vol. 891: Logic Symposia, Hakone, 1979, 1980. Proceedings, 1979, 1980. Edited by G. H. Müller, G. Takeuti, and T. Tugué. XI, 394 pages. 1981.

Vol. 892: H. Cajar, Billingsley Dimension in Probability Spaces. III, 106 pages. 1981.

Vol. 893: Geometries and Groups. Proceedings. Edited by M. Aigner and D. Jungnickel. X, 250 pages. 1981.

Vol. 894: Geometry Symposium. Utrecht 1980, Proceedings. Edited by E. Looijenga, D. Siersma, and F. Takens. V, 153 pages. 1981.

Vol. 895: J.A. Hillman, Alexander Ideals of Links. V, 178 pages. 1981.

Vol. 896: B. Angéniol, Familles de Cycles Algébriques – Schéma de Chow. VI, 140 pages. 1981.

Vol. 897: W. Buchholz, S. Feferman, W. Pohlers, W. Sieg, Iterated Inductive Definitions and Subsystems of Analysis: Recent Proof-Theoretical Studies. V, 383 pages. 1981.

Vol. 898: Dynamical Systems and Turbulence, Warwick, 1980. Proceedings. Edited by D. Rand and L.-S. Young. VI, 390 pages. 1981.

Vol. 899: Analytic Number Theory. Proceedings, 1980. Edited by M.I. Knopp. X, 478 pages. 1981.